Springer Proceedings in Earth and Environmental Sciences

The series Springer Proceedings in Earth and Environmental Sciences publishes proceedings from scholarly meetings and workshops on all topics related to Environmental and Earth Sciences and related sciences. This series constitutes a comprehensive up-to-date source of reference on a field or subfield of relevance in Earth and Environmental Sciences. In addition to an overall evaluation of the interest, scientific quality, and timeliness of each proposal at the hands of the publisher, individual contributions are all refereed to the high quality standards of leading journals in the field. Thus, this series provides the research community with well-edited, authoritative reports on developments in the most exciting areas of environmental sciences, earth sciences and related fields.

Amalendu Sinha · Bhabesh Chandra Sarkar ·
Prabhat Kumar Mandal
Editors

Proceedings of the 10th Asian Mining Congress 2023

Roadmap for Best Mining Practices vis-à-vis Global Transformation

Springer

Editors
Amalendu Sinha
The Mining, Geological and Metallurgical Institute
Kolkata, India

Bhabesh Chandra Sarkar
The Mining, Geological and Metallurgical Institute
Kolkata, West Bengal, India

Prabhat Kumar Mandal
The Mining, Geological and Metallurgical Institute
Kolkata, West Bengal, India

ISSN 2524-342X ISSN 2524-3438 (electronic)
Springer Proceedings in Earth and Environmental Sciences
ISBN 978-3-031-46965-7 ISBN 978-3-031-46966-4 (eBook)
https://doi.org/10.1007/978-3-031-46966-4

This Springer imprint is published by the registered company Springer Nature Switzerland AG
The registered company address is: Gewerbestrasse 11, 6330 Cham, Switzerland

Paper in this product is recyclable.

Preface

The Mining, Geological and Metallurgical Institute of India (MGMI), established way back on 16 January 1906, is an internationally acclaimed professional body and is one of the oldest institutions of this kind in the world. Since inception, it has been working to promote and advance the mining and mineral industries. The Asian Mining Congress (AMC) and International Mining Exhibition (IME), held concurrently, are flagship international events organised by MGMI biennially. The 10th Asian Mining Congress and International Mining Exhibition, in this sequel, is held during 06–09 November 2023 in Kolkata, India. Organized by The Mining Geological and Metallurgical Institute of India.

The mining and mineral sectors currently require a serious introspection because their long-term sustainability largely depends on an inclusive growth of human civilisation. Increasing depth of mining and complexities of mineral deposits, efficient processing and utilisation of coal, minerals and metals and value addition pose major technical challenges. Added with these, the menace of combating climate change, energy transition and strong societal expectations are of serious concerns. At this juncture, it is very pertinent to brainstorm on best mining practices taking stock of the recent developments, availability of state-of-the-art technologies and equipment and relevant cutting-edge innovations. The increased processing power of computers and myriad interpretative software packages available in mining industry have revolutionised in processing and analysing vast amount of data that are of relevance, both in the present and future contexts. Keeping the need of the hour into the context, the 10th Asian Mining Congress has been organised with the theme of "Roadmap for Best Mining Practices vis-à-vis Global Transformation".

We are honoured to be associated with Springer Nature, a publishing house of global repute, for the publication of this congress proceedings volume. A total of 45 papers were received for the Springer Nature volume, out of which 39 papers were qualified for submission post plagiarism check and evaluation by reputed reviewers, who are learned and highly experienced in their respective fields of expertise. A wide spectrum of papers with contemporary relevance such as recent advancements in geostatistics and reserve estimation, challenges in underground and opencast mining, smart mining through adoption of AI/ML and IOT, rock excavation techniques, critical minerals and speciality materials, application of numerical modelling, waste utilisation, miners' health and safety, coal gasification, energy transition and many more vital topics are included in the congress proceedings volume. It is expected that the deliberations and discussions during the congress provide very valued inputs to redefine the roadmap for best mining practices in tune with the global transformation.

We thankfully acknowledge the valuable contributions made by the authors. We sincerely thank all the reviewers who have supported us by timely reviewing the papers to maintain the quality of the papers. Our sincere acknowledgements are due to one and all who have directly or indirectly supported us in our endeavour. Finally, we thank Springer Nature for value addition to this congress proceedings volume.

Amalendu Sinha
Bhabesh Chandra Sarkar
Prabhat Kumar Mandal

Organisation

Patron

Shri Amrit Lal Meena — IAS, Secretary, Ministry of Coal, GOI, India
Samir V. Kamat — Secretary DD (R&D) and Chairman DRDO, India

National Advisory Committee

Shri P. M. Prasad — Chairman, CIL & Chairman, National Advisory Committee, India

Principal Coordinators

B. Veera Reddy — President, MGMI, Director (Technical), Coal India Ltd., India

Shri Ranajit Talapatra — Secretary, MGMI, GM (Washery), Coal India Ltd., India

Organisers

Conference (10th AMC)

Chairman, Conference

Shri Manoj Kumar — CMD, Central Mine Planning and Design Institute, India

Co-chairman, Conference

Shri D. B. Sundara Ramam — VP, Tata Steel, India

Chairman, Technical Committee

Amalendu Sinha — Former Director, CSIR-CIMFR, India

Convenor Conference

Shri Prasanta Roy — HoD (Geology), CIL, India

Chairman, Cultural Committee

Shri Samiran Dutta — CMD, Bharat Coking Coal Limited, India

Convenor, Cultural Committee

Shri Awadh Kishore Pandey — GM, MCL, India
Peeyush Kumar — GM/OSD, MoC, India

Exhibition (10th IME)

Chairman, Exhibition

Shri Bhola Singh — CMD, Northern Coalfields Limited, India

Co-chairman Exhibition

Shri Jagdish Prasad Goenka — MD, NMC, India

Convenor Exhibition

Chandra Shekhar Singh — GM, CIL, India

Chairman, Buyer Seller Meet

Shri V. K. Arora — Chief Mentor, KCT Bros. Ltd., India

Technical Committee

Amalendu Sinha — Chairman, Former Director, CSIR-CIMFR, India
Sri T. K. Nag — Former CMD, NCL, India

Bhabesh Chandra Sarkar	Professor (HAG), IIT(ISM), India
Anupendu Gupta	Former DDG, GSI, India
Sumantra Bhattacharya	Emeritus Fellow, IIT(ISM), India
Prabhat Kumar Mandal	Chief Scientist, CSIR-CIMFR, India
Sri Chiranjib Patra	General Manager (UMD/CED), CMPDI, India
M. P. Roy	Chief Scientist, CSIR-CIMFR, India
Sri Ranjit Datta	Former Director (SG), GSI, India
Ajay Mandal	Professor, IIT(ISM), India
Rajesh Mukherjee	Chief Geologist, Tata Steel Ltd., India
Biswajit Samanta	Professor, IIT Kharagpur, India

MGMI Council for 2022–2023

President

B. Veera Reddy	CMD, CCL & Director (Technical), Coal India Limited, India

Immediate Past Presidents

P. M. Prasad	Chairman, Coal India Limited, India
Anil Kumar Jha	Former Chairman, Coal India Limited, India

Vice-presidents

J. V. Dattatreyulu	Former, Director (Operations), SCCL, India
Thomas Cherian	Managing Director, Essel Mining & Industries Limited, India
Bhola Singh	CMD, Northern Coalfields Limited, India
Jagdish Prasad Goenka	Managing Partner, Nanda Millar Company, India

Honorary Secretary

Ranajit Talapatra	GM (WS), CIL, India

Immediate Past Secretaries

Rajiw Lochan	Former Advisor & Head/GM (Clean Energy) CBM Department, CMPDI, India
Prasanta Roy	HOD (Geology), CIL, India

Members

Chandra Shekhar Singh (Hony Jt Secretary)	GM, CIL, India
Prabhat Kumar Mandal (Hony Treasurer)	Professor CSIR-CIMFR, India
Ajay Kumar Singh (Hony Editor)	Former Scientist, CSIR-CIMFR, India
Virendra Kumar Arora	Chief Mentor (Coal), KCT & Bros, India
Jai Prakash Barnwal	Former Chief Scientist, RRL, India
Shri Bhaskar Chakraborti	Former, Dy. DG, GSI, India
Ashis Bhattacharjee	Professor, IIT, Kharagpur, India
Anup Biswas	Former Deputy Director General, DGMS, India
Pravat Ranjan Mandal	Former Advisor (Projects), Ministry of Coal, GoI, India
Awadh Kishore Pandey	GM (Mining/HOD), MCL, India
Nityanand Gautam	Former Advisor, UNDP, India
Netai Chandra Dey	Professor (HAG), IIEST Shibpur, India
G. P. Karmakar	Former Professor, IIT Kharagpur, India
Tapas Kumar Nag	Former CMD, NCL, India
Peeyush Kumar	Director Technical, Ministry of Coal, GoI, India
Bhabesh Chandra Sarkar	Professor (HAG), IIT(ISM) Dhanbad, India
Kalyan Sen	Former Director, CSIR-CIMFR, India
Amalendu Sinha	Former Director, CSIR-CIMFR, India

Supported by

Sl No.	Supported by	Logo
1	Ministry of Coal, Govt. of India Shastri Bhawan, Dr. Rajendra Prasad Road, New Delhi - 110001	
2	Ministry of Steel, Govt. of India Udyog Bhavan New Delhi - 110011	
3	Defence Research and Development Organisation (DRDO), Govt. of India DRDO Bhawan, Rajaji Marg, New Delhi-110011	Defence Research and Development Organisation (DRDO)
4	Coal India Limited (A Maharatna Company) Coal Bhawan Premise No-04 MAR, Plot No-AF-III, Action Area-1A, Newtown, Rajarhat, Kolkata-700156	COAL INDIA LIMITED A MAHARATNA COMPANY

Contents

Exploration, Estimation, Geostatistics and Hydrogeology

Geostatistics in Exploration and Mining

Bhabesh C. Sarkar(✉)

Department of Applied Geology, IIT (ISM), Dhanbad, Jharkhand 826004, India
bhabesh@iitism.ac.in

Abstract. Geostatistics in exploration and mining, over the last eighty years, has witnessed its progress starting with linear kriging, and then advancing to nonlinear kriging, non-parametric kriging, conditional simulation, multi-point statistics and machine learning geostatistics. Corresponding to these advancements, it was felt necessary that a geostatistical modelling study needed a proper connection between geology and geostatistics. The manner of integrating geology with geostatistics can be accomplished by carrying out geostatistical modelling in combination with geology of deposit, mineralisation controls, and zonation of the deposit into several geological 'domains'. An adequate understanding of deposit geology is, thus, an essential pre-requisite to a geostatistical modelling study. It goes without saying that geology controls the distribution of geologically complex grades, hence it is crucial that geology direct estimation modelling rather than estimation modelling direct geology. An improved understanding of type and characteristics of mineralization is achieved only through a suitable integration of deposit geology with relevant geostatistical technique(s).

Keywords: Regionalised variable · Semi-variogram · Block Kriging · Conditional Simulation · Machine learning

1 Introduction

Exploration-Mining life cycle deals with the key activities in mineral industry for exploration and production of raw materials. On the basis of the 'Theory of Regionalized Variables', geostatistics focus on practical application of spatially correlated random variables. Over the years, role of geostatistics has become increasingly important in exploration and mining for quantitative assessment of mineral grade and tonnage distributions. Combined with univariate and multivariate statistical techniques, geostatistics offer a wide-ranging geomathematical methodologies to analyze, model, estimate and measure uncertainties for various forms of spatial features, namely grades, widths and tonnages of mineralized bodies. Development of geostatistics emerged in France starting with the early studies of Matheron (1962). Initially, the methodology was developed on the basis of original work concerning optimal allocation of optimum weights to adjoining sample values for assessing block grade values in gold mines of South Africa (Krige 1962). Developing of any of the workable methods were headed by any of the valid statistical theory (Sichel 1952). Subsequently, it was G. Matheron (1962, 1971) who

A. Sinha et al. (Eds.): AMC 2023, SPEES, pp. 3–15, 2023.
https://doi.org/10.1007/978-3-031-46966-4_1

developed the *'Regionalized Variables' theory'*. Spatial position for the first time was given due consideration and concepts of *'auto-correlation*' and *'auto-covariance'* were connected to a new statistical tool by the name of *'semi-variogram'*. On the basis of this theoretical foundation, a set of apt methods were developed, which are familiar by the common name *'Kriging'* to estimate values at any unsampled point or block locations from a set of observed sample values at regularly or irregularly grid. Geostatistics, thus, provide an important dimension in which sample-to-sample correlation constitute the important foundation for modelling, characterization and evaluation of orebodies in exploration and mining.

2 Why Geostatistics

Geostatistics constitute a set of probabilistic methods that provide spatial models for correlated random variables. Key objective of Geostatistics in exploration and mining is to reliably estimate grade, bulk density and tonnage on a block-by-block framework together with their spatial distributions in a mineral deposit leading to an appropriate mineral inventory (Rossi and Deutsch 2014). Among the methods used in the development of mineral inventory, geometrical and deterministic methods are faced with inability to assess estimation errors and degree of confidence in estimates. Classical statistics produce degree of confidence in estimates but ignore the spatial correlations among samples. Rolling mean (moving average) method and trend surface analysis consider sample relationships over a dimensional space but not estimation error. The stated deficiencies justify that there is a need for a suitable estimation methodology, which is able to provide estimates with minimum variance. Geostatistics based on the *'Regionalized Variables Theory'* is able to produce such estimates satisfying minimum variance condition.

Geostatistics has proved its successful applications to modelling, evaluation and characterization in various fields of geology, mining, petroleum, environmental and allied branches (Rossi and Deutsch 2014), viz. (i) Geostatistics in Exploration and Mining; (ii) Geostatistics in Petroleum Reservoir characterization; (iii) Geostatistics in Hydrogeology; (iv) Geostatistics in Geotechnical Engineering; (v) Geostatistics in Environmental studies; and (vi) Geostatistics in RS & GIS, among many others. During the last eighty years, decade-wise broad developments in geostatistics can be stated as given below:

- 1940–1950: Autocorrelation Theory
- 1950–1960: Regionalized Variable Theory
- 1960–1970: Linear Geostatistics
- 1970–1980: Non-linear Geostatistics
- 1980–1990: Non-parametric Geostatistics
- 1990–2010: Geostatistical Simulation
- 2000–2010: Multi-Point Geostatistics
- 2010–till date: Trendy Geostatistics
  (AI and ML Geostatistics).

3 Geostatistics in Practice

A meaningful geostatistical modelling study with due consideration to deposit geology and controls of mineralization delivers an adequate block-by-block estimates and a mineral inventory of the deposit being evaluated associated with estimation error of each of the block estimates. Essential dimensions in a geostatistical modelling study should entail the following steps:

3.1 Geological Domaining

Geological domaining with reference to statistical concepts of mineral population and geostatistical characteristics of sample data is an essential prerequisites to provide specifics of mineral deposits for improved understanding of various zones of mineralization. A domain represents a zone of geological identity, complementing the deposit geology.

3.2 Sample Value Compositing

Location of individual drill hole samples at given distances of constant interval leads to compositing of sample values wherein the individual drill hole samples are grouped into segments of constant length along a drill hole, subject to geological homogeneity. The values corresponding to the segments of constant length are the composited weighted values. The location and size of the composites are chosen to coincide with the (i) bench elevation and bench height respectively in case of deposit being suitable to open pit mining or (ii) level elevation and vertical cut height in case of deposit being suitable to underground mining. The sample value composites and their respective centroid XYZ coordinates constitute the *composite value database.*

3.3 Exploratory Data Analysis

Probability modelling of frequency distribution of the composite values of constant support provides a significant insight into mineralisation characteristics of a deposit. The type of probability distribution representing the sample values is estimated through classical statistical modelling. The assumption of a probability model fit to a sample frequency distribution is validated graphically by its probability plot and numerically by its degree of skewness, degree of kurtosis and chi-squared goodness of fit test statistics.

3.4 Semi-variography

Computation of semi-variograms and modelling should show the spatial continuity and data structure, as well as geostatistically confirm any previously modelled or observed geological trends. When using semi-variogram analysis, the practitioner should always look for a geological explanation for the main directions. Such a phase would frequently disclose some subtle controls on mineralization that were not immediately obvious in the geology. Should any anisotropy exists in the domains of the mineral deposit, geostatistical analysis also identifies it and attempts to quantify it. The study should corroborate the choice of using hard or soft domain borders as well as the degree of random variation in each direction, whether brought on by issues with sampling or by intrinsic unpredictability in the data.

Semi-variography can show whether the geological model is acceptable, yet it should not be used as a substitute for geological interpretation. For an indicator semi-variography, the degree and direction of anisotropy may vary with grade; for instance, high-grade veins may have a different orientation from the majority of the mineralization. In addition to providing information for the aggregation of grades into larger block sizes, if necessary, the design and modelling of semi-variograms aid in the definition of the fundamental block size to be used in all block modelling procedures.

3.5 Block Model

Preceding to estimating grade, a specific volume of the mineralization that needs to be estimated must be physically decided upon in three dimensions using the geological model and/or the domain model. A block model is often defined as a group of orthogonal cuboids that are either of same size or are subsets of a designated parent block size. Although the method is typically semi-automatic, the blocks must typically be generated within the framework of geological boundaries for each elevation level of operation. Usually, this is provided by a collection of surfaces or a three-dimensional enclosed solid. These are created either by triangulating a collection of points and strings into a digital terrain model, more often known as a surface, or by wire-framing strings of points on a section or plan.

3.6 Block Kriging

Subsequent stage entails estimating block values along with their corresponding variances employing a specific geostatistical estimation procedure, known as kriging. By projecting sample data from several transverse sections, a homogenous mineralized zone is then taken into account and divided into a number of regularly spaced horizontal sections. On each of the horizontal sections, mineralized limits are then defined in light of relevant geological and mining factors. The length at which drill hole samples are composited, which typically equals bench height (in the case of an open pit operation) or vertical lift (in the event of an underground operation), determines the spacing of the horizontal cross-sections. Each of the slices, which are horizontal, has a mineralized boundary marked on it.

Ideally, the intended bench height or cut height should be assumed to represent the height of the block. At least one-fourth of the typical drill spacing should be represented by the other two dimensions. Since the choice of equipment depends on the amount of material it can handle in terms of tonnes, the daily production is another significant contributing factor. The individual slices, when divided into smaller grids, generate a collection of blocks with constant Z (elevation) values in the X (easting) and Y (northing) axes. The block arrays are then kriged slice-by-slice to produce kriged estimate, and kriging variance and a slice average. Block estimates are presented, evaluated visually, and a comparison of the values of the block, sample, composite sample, and individual sample is then undertaken for result reconciliation. Once this process gets completed, the estimated values of blocks will get accepted for mine decisions.

3.7 Mineral Inventory and Grade-Tonnage Relations

Each slice from top to bottom with regularly spaced kriged blocks is stacked one on top of the other to create a 3D array of blocks with their kriged mean (KM) and kriging variance (KV) and tonnages per block, which are calculated by dividing the block dimensions by the mineral's bulk density. The mineral inventory, which provides *in situ* stock of minerals, is referred to as such a 3D framework of blocks. Subsequent step is to create a series of grade-tonnage estimations at various probabilistic cutoff grades next to generation of a mineral inventory. A higher tonnage is typically linked to a relatively low grade. Progressively higher grades may be worked out by increasing the degree of selectivity in mining and thus reducing the tonnage. This is known as grade-tonnage relation through as numerical approach is to model the relation statistically. The method involves a step-wise integration of the block grade frequency curve over a range of grades and calculates (i) quantity of ore, metal and waste; (ii) average grade of ore and waste; and (iii) waste-to-ore ratio. Plots of these relations provide grade-tonnage curves. These curves together with the mineral inventory provide a sound basis for mine decisions.

4 Geostatistical Simulation

A spatial model based on semi-variography provides a means to evaluate a deposit, which is partially known from exploration data combined with appropriate interpretation of deposit geology employing kriging algorithm. As a result, deposit models generated consist of uncertainty. However, if the objective is to map spatial heterogeneity and assess model uncertainty, then it calls for application of geostatistical simulation methods to generate equi-probable multiple images. Geostatistical simulations provide a means of modelling of uncertainty in exploration geology for prediction of mineral grade values (Vann et al. 2002; Lantuéjoul 2002). Various simulation algorithms have been applied to optimise drill hole spacing, uncertainty during feasibility and grade control in daily operations (Rossi and Deutsh 2014). Geostatistical simulations reproduce histogram and spatial structure in order to maintain spatial heterogeneity of actual sample data. Simulation outputs provide improved realisations of local variability and a means to quantifying uncertainty (Dimitrakopoulos et al. 2004). Geostatistical simulations provide equi-probable multiple realisations of the true phenomena retaining the histogram and spatial distribution of the original data. A schematic for sequential Gaussian simulation is shown in Fig. 1.

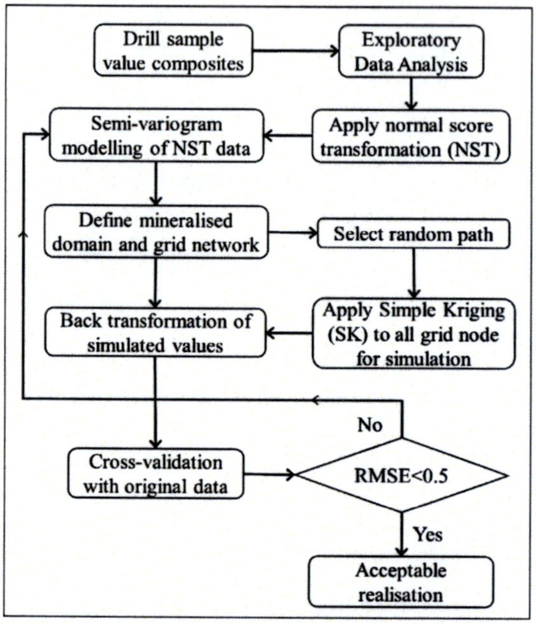

Fig. 1. A workflow for geostatistical simulation

5 Machine Learning Geostatistics

In recent years, importance of machine learning approaches in exploration and estimation modelling has increased due to its flexibility and adaptability. Typically, machine learning algorithms build a mathematical model from training data, or sample data, to make predictions or judgements without needing to be explicitly programmed to do so. It is a technique that makes it feasible to mix multiple data layers and identify key correlations among them in ways that weren't before conceivable. By using sample data from diverse locations, the model is directly trained to capture the spatial distribution. The distribution of grade values in the drilling zone is then predicted using the training model. Main advantage of this strategy is that it does not necessitate intricate mathematical modelling or presumptions on the spatial distribution of sample values. In comparison to traditional shallow neural network (NN) models, modern deep learning provides improved training stability, generalization, and scalability. It performs quite well in a number of classification and approximation problems and is accordingly becoming widely accepted algorithm of choice for highest predictive accuracy. Geostatistics being a parametric approach captures the spatial distribution using a linear function, whereas a NN capture linearity on a local scale and non-linearity on a global scale and yet can retain spatial heterogeneity of the distribution (Zhou et al. 2002). Geostatistics and NN are two strong contenders for mineral grade assessment (Sarkar et al. 2017; Singh et al. 2018, 2022; Sarkar and Ray 2020). A representative structure and operation of a neural network is shown in Fig. 2. An overall workflow of an integrated geological-geostatistical evaluation in practice is shown in Fig. 3.

6 Test Case – An Iron Ore Deposit E

Iron ore Deposit E has a strike of N25°E-S25°W with dip changing between 40° and 50° towards west. The depth of ore bodies varies from 40 m to 120 m. Hematite together with goethite constitute bulk of iron minerals in the ore. These ores are classified as lateritic ore, hard laminated ore, soft laminated ore, blue dust and yellow ochre.

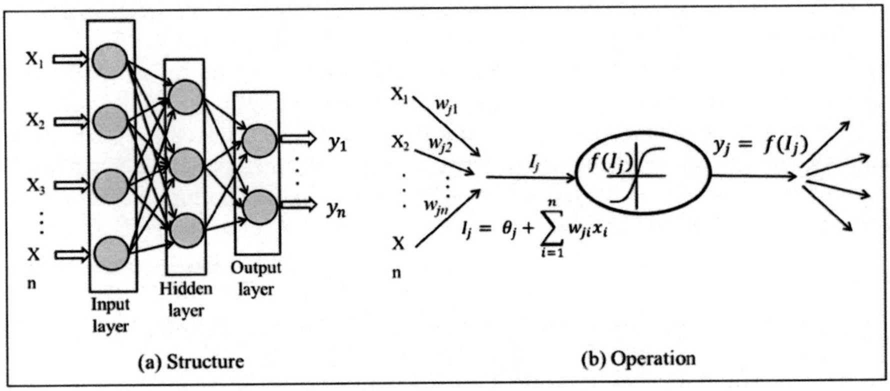

Fig. 2. A representative structure and operation of a neural network

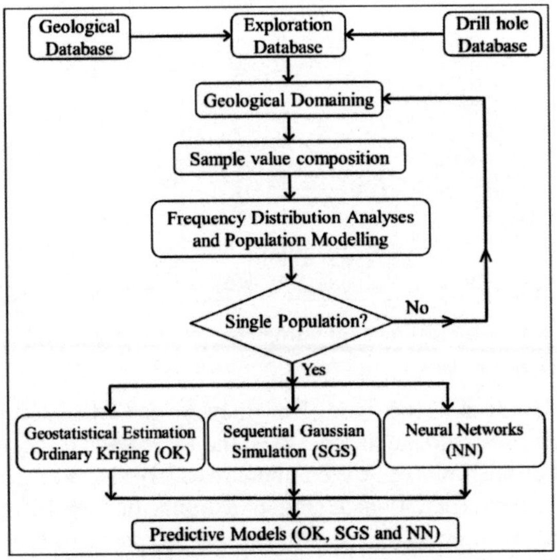

Fig. 3. A workflow of an integrated geological-geostatistical evaluation in practice

With phase-I exploration at 400 m × 400 m grid spaced drilling in 90 drill holes, the mining company estimated iron ore reserves of 445 mt with an average grade of 63.30% Fe at 57% Fe cutoff employing conventional sectional methods.

With phase-II exploration at 200 m × 200 m grid spaced drilling in 85 drill holes, the company estimated a reserve 353 mt with an average grade 62.20% Fe at 55% Fe cutoff. In the present modelling study, a total of 49 drill holes at 200 m × 200 m grid spacing were considered. The assay data consists of Fe value. Using a cutoff grade of 45% Fe, assay data was composited at 1 m interval down the hole that led to a total of 9672 composite samples. 3D spatial coordinates of the composite samples were generated for every constant length composite samples considering the drill midpoint intersection. Classical statistical modelling revealed a negatively skewed 3-parameter lognormal distribution as evidenced by histogram and probability plots (Fig. 4(a) and Fig. 4(b)). A spherical model (Fig. 5) was found to fit the experimental semi-variogram adequately validated through point kriging cross-validation.

A total of 27 horizontal slices have been constructed with mineralised boundaries delineated on each at a vertical interval of 10 m (corresponding to mining bench height) starting from 780 mRL at the top elevation to 520 mRL at the bottom elevation of the iron ore deposit. On each horizontal slice, a block grid of 50 m × 50 m considering the current mining practice have been fitted and blocks have been estimated employing OK within a lateral boundary stretch of 10500E to 11750E and 11450N to 13450N. Each block within individual slices have been assessed producing slice-wise kriged estimate and kriging variance. Choice of the block dimension has been based on current practices for iron ore mining in other adjoining leasehold areas.

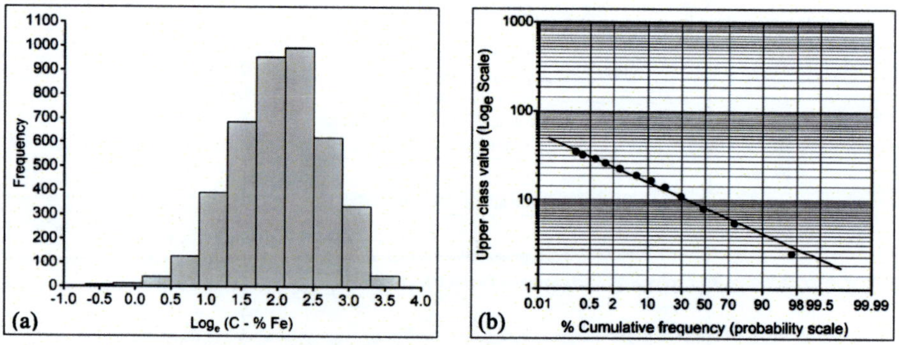

Fig. 4 Log-transformed (a) histogram plot and (b) probability plot

A total of 6,060 blocks were kriged. The global kriged estimate and kriging variance in respect of Fe have been calculated as 60.41% and 17.95 (%)^2 respectively. Central 95% geostatistical confidence limits of Fe is computed as 52.10% (lower) and 68.72% (upper). A representative slice plan at 650 mRL (Fig. 6) exhibits the distribution of high Fe grade blocks in the northern and southern parts, while the central part consists of blocks of mixed grades. The kriging variance maps show discrete patches of varying spatial distributions of block variances uniformly from southern to northern part with increasing kriging variances towards the peripheries of the discrete patches associated with low data density. Individual horizontal slices with kriged blocks were placed vertically below each other starting with 780 mRL slice as the uppermost down to 520 mRL slice as the lowermost. This demonstrates a 3D spatial arrangement of geostatistically estimated

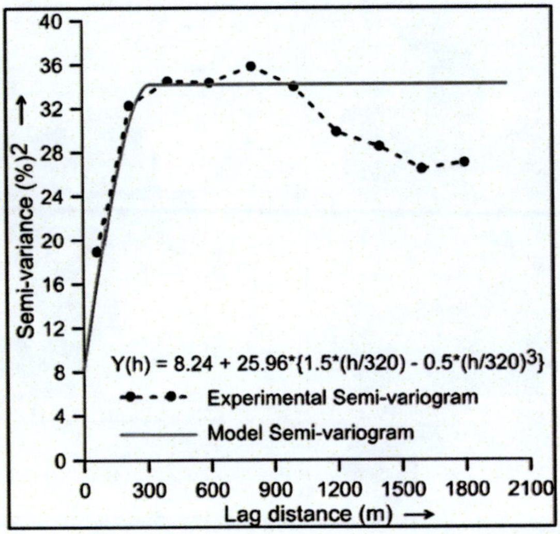

Fig. 5. Experimental semi-variogram with fitted spherical function

blocks with values of kriged estimate and kriging variance. The tonnage estimate of each block has been computed as the product of block volume and average bulk density of 3.00 t/m^3. Total tonnage from top to bottom is calculated as 454.50 mt. The 3D network of kriged blocks established an iron ore inventory providing *in situ* stock of mineral. Grade-tonnage (GT) relations have been established at various probabilistic Fe cutoff grades. The procedure provided estimated (i) tonnages of ore and metal; (ii) average grade of ore; and (iii) waste-to-ore ratio. Plots of GT relations provided GT curves (Fig. 7).

Sequential Gaussian simulation (SGS) were performed using semi-variogram model parameters of the normal score transformed data. The statistical results show that simulated values are very close to original composited Fe values. Out of 50 realisations, five realisations namely #2, #12, #21, #27 and #45 have been chosen because of being close to the statistical parameters of the original data.

The histograms of the selected realisations are well correlated with the histogram of actual values, of which realisation #2 provided closest statistical relationship. The results showed that the simulated values and original values had a fairly similar relationship. The simulated maps of selected realisations at 650 mRL are demonstrated in Fig. 8. The SGS realisations represent a spatial pattern in detail without presenting any smoothing effect and very close to original data. With an objective to capture non-linearity and spatial variability of iron ore grade distribution, an attempt has been made to develop a hybrid GRNN-based model owing to its ability of fast learning and convergence to optimal regression surface. It requires a training dataset and a smoothing factor to model the pattern of ore body distribution.

In the hybrid model, two unsupervised models, namely OK and k-means clustering, are used as pre-processors for feeding data to supervised GRNN model. OK is used to capture the distribution in local scale, whereas k-means clustering is used for dimension

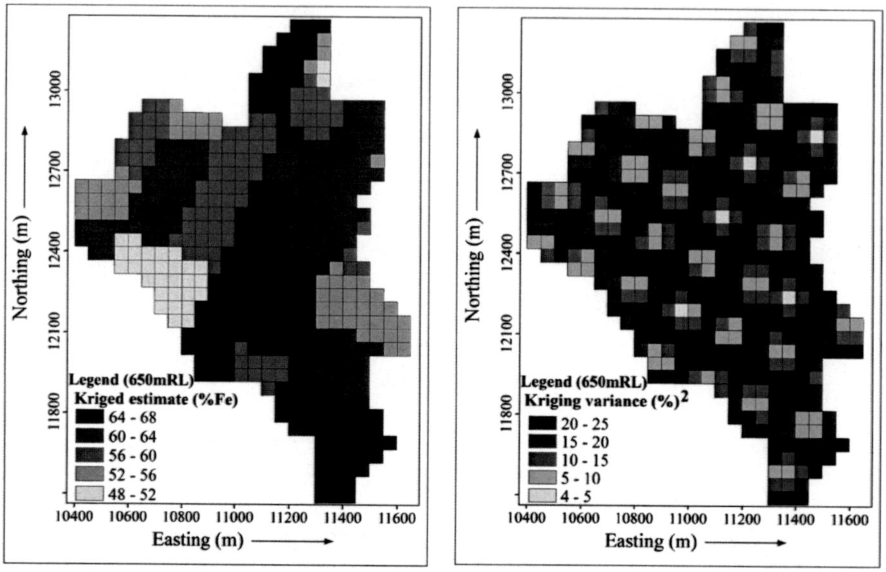

Fig. 6. Kriged estimated map (left) and kriging variance map (right) at 650 mRL

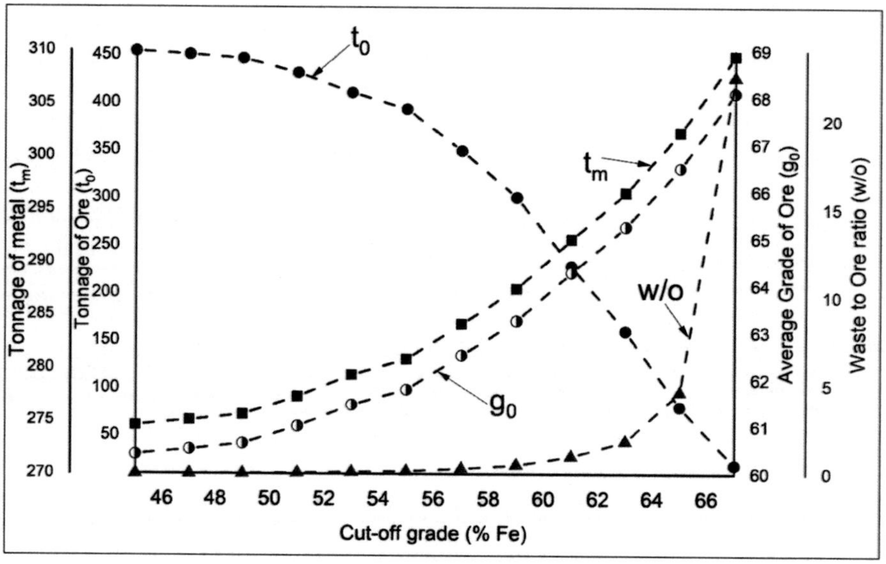

Fig. 7. Grade-tonnage curve for Fe grade

reduction, which aid remarkably to achieve accuracy and speed of the modelling processes. The hybrid model demonstrates to be a powerful data-driven technique to capture the non-linearity of the feature variable distribution. The spatial distribution maps of the predicted Fe values by OK and GRNN have been analysed in the study area by generating

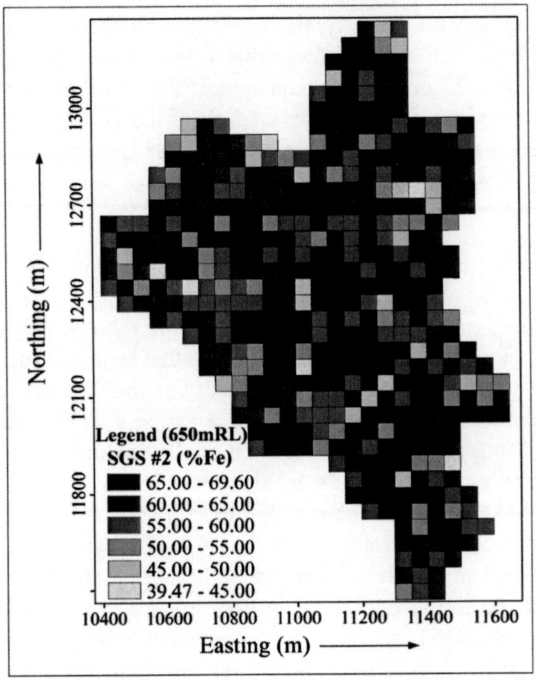

Fig. 8. Predicted Fe values at 650 mRL using SGS.

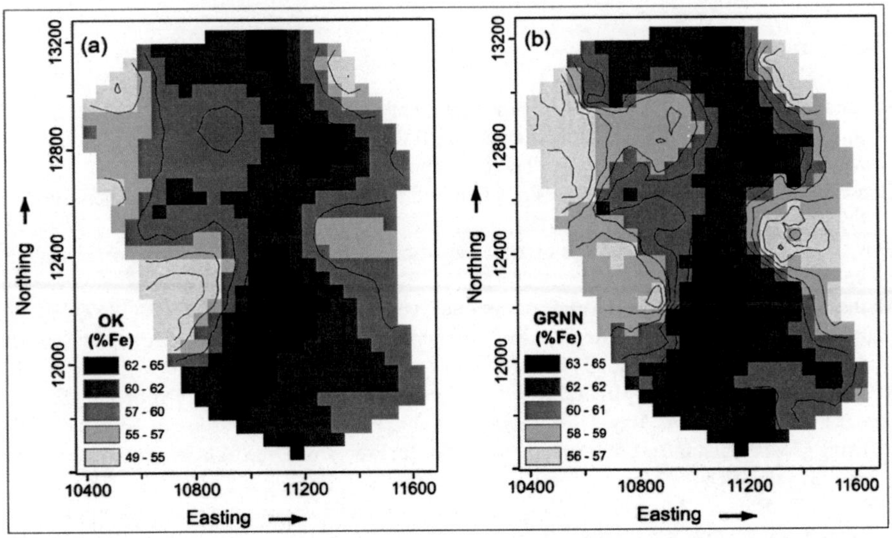

Fig. 9. Fe Grade estimation contours of Fe at 650 mRL (a) OK; and (b) GRNN

Fe grade contours (Fig. 9). Evidently, the spatially distributed estimation maps generated through OK exhibit smoothing effect while those generated by hybrid GRNN retain spatial heterogeneity of the original Fe data values. This is accounted for by the fact that OK involves minimization of variance as a result of the volume-variance relationship, while GRNN, being a nonlinear approximator exhibits similar spatial heterogeneity as with the Fe assay values.

7 Conclusions

Geostatistics is extensively being used and is becoming more widely accepted in exploration and mining. However, there are some issues that require cautious consideration, *viz.* (i) an increase in the mathematical complexity of the theories, and (ii) a decrease in the geological considerations in the estimation process. A meaningful geostatistical model should fulfil four requirements: (i) the model's parameters should be estimable; (ii) it must be operational, or provide an answer to a question; (iii) it must be compatible with data; and (iv) the model's predictions must be cross-validated and verified by experience. Clarification and dissemination of geostatistical concepts to practitioners, and documentation of unsolved issues in evolving geostatistical methods, have had a significant impact on the development of geostatistics in the mineral industry since 1960. It must be emphasized that geology, not estimate models, drives the distribution of geologically complex grades, and that geology should therefore influence estimation modelling. Hence, an adequate geological model is a prerequisite, and there is cause for considerable concern if the geostatistical model disagrees with the geological model.

References

Dimitrakopoulos, R., Li, S., Scott, J.: Quantification of geological uncertainty and risk assessment in coal resource/reserve classification, ACARP Project C11042 Report, vol. I. W. H. Bryan Mining Geology Research Centre, The University of Queensland, 250 p. (2004)

Gandhi, S.M., Sarkar, B.C.: Essentials of Mineral Exploration and Evaluation, p. 406p. Elsevier, USA (2016)

Krige, D.G.: Statistical applications in mine valuation. J. Inst. Mine Surv. South Africa **12**(2, 3), 45–84, 95–136 (1962)

Matheron, G.: Traite de Geostatistique Appliquee, vol. 1. Edition Technip, Paris (1962)

Matheron, G.: The Theory of Regionalised Variables and its Applications. Ecole Nat.Sup. Des Mines de Paris, Booklet No. 5 (1971)

Rossi, M.E., Deutsch, C.V.: Mineral Resource Estimation, p. 332. Springer, Dordrecht (2014)

Sarkar, B.C., Singh, R.K., Ray, D., Kumar, A., Sinha, P.K., Sarkar, V.: Iron ore grade modelling using geostatistics and artificial neural networks. In: Proc. Iron Ore 2017 Conference, AusIMM, pp. 431–437, Feb (2017)

Sarkar, B.C., Ray, D.: Machine learning approach to mineral deposit grade modelling. Indian J. Geosci. **74**(3), 269–274 (2020)

Singh, R.K., Ray, D., Sarkar, B.C.: Recurrent neural network approach to mineral deposit modelling. In: Proc. 4th International Conference on Recent Advances in Information Technology, IEEE Xplore, Dhanbad, pp. 412–416 (2018)

Singh, R.K., Ray, D., Sarkar, B.C.: Mineral deposit grade assessment using a hybrid model of kriging and generalized regression neural network. Neural Comput. Appl. **34**, 10611–10627 (2022)

Sichel, H.S.: New methods in the statistical evaluation of mine sampling data. In: Transactions of the Institutions of Mining and Metallurgy, London, vol. 61, pp. 261–288 (1952)

Lantuejoul, C.: Geostatistical Simulation: Models and Algorithms. Springer-Verlag Berlin, 256p (2002)

Vann, J., Bertoli, O., Jackson, S.: an overview of geostatistical simulation for quantifying risk. In: Proceedings of the Geostatistical Association of Australasia Symposium "Quantifying Risk and Error", March, vol. 6959, pp. 1–12 (2002)

Zhou, Z.H., Wu, J., Tang, W.: Ensembling neural networks: many could be better than all. Artif. Intell. **137**(1), 239–263 (2002)

Prediction of Water Inflow in a Proposed Underground Coal Mine of the Rajgamar Dipside (South of Phulakdih Nala) in Chattisgarh, India

Mrinal Kanti Mukherjee(✉)

Department of Applied Geology, Indian Institute of Technology (Indian School of Mines), Dhanbad 826004, India
mrinal_km67@yahoo.co.in

Abstract. A Hydrogeological investigation has been carried out around Rajgamar-dipside (SOI Toposheet no. 64J/15) to predict water inflow in a proposed underground coal mine. The Groundwater table in the study area, is in general, gently undulating and is shallow with an average elevation of $348 - 355$ mamsl, and occurs at a depth of $0.5 - 6$m below ground level depending upon topography. The groundwater flow-lines converges from north, south and east toward west-central part (Rajgamar) of the study area. Drainage morphometry indicate significant overland flow in the low topography areas leading to excessive soil erosion. Natural seepage rates vary between $4 - 43$ ml/cm^2/min. Solution sinkholes and pocks ranging in diameter of $4 - 10$ cm to $0.5 - 1$ m diameter occur densely within the rocks of Kamthi and Barakar Formations. These solution pits link up laterally and vertically to develop networks of channels that serve as free passage for fast infiltration and subsurface flow of water. Field data and petrophysical properties indicate a possible inrush of groundwater at places through underground solution channels and local fractures within shallow depths (~20 m) and a steady inflow of 4397 gpm to 19834 gpm distributed over a corresponding progressive depth of 10 m to 330 m respectively in a proposed underground coal mine for a gallery whose height is 2.5 m, width is 4 m and length is 1000 m. The study will help mine managers to optimise the pumping rate for dewatering during mining activity.

Keywords: Groundwater · Inflow · Rajgamar-dipside · Porosity · Drainage

1 Introduction

Water is a major consideration in mine planning, be it open cast or underground not only because it is a physical hindrance to the extraction of minerals and fuels, but also because mining activities frequently degrade water quality in important water resources. Water occurrences in underground mines range from sudden water inrushes to steady inflows of Groundwater. Problems that arise are manifolds: Water inrushes and inflows

A. Sinha et al. (Eds.): AMC 2023, SPEES, pp. 16–31, 2023.
https://doi.org/10.1007/978-3-031-46966-4_2

can delay or cause the cessation of mining work together with the loss of life and equipment. The enormous capital investment required by mining operations means that not only is the high cost implication of abandoning equipment, but delays in working are also expensive. Steady inflows into a mine require water to be pumped out because mining cannot proceed under waterlogged conditions. Mining activities can result in the lowering of water levels and pressures in aquifers, thus threatening the supply of good quality water.

The ability to predict water flow rates is an important aid to mine planning. Knowledge of likely water quantities contributes to an assessment of the economic viability of a proposed mine. Forecasted water flow rates are used to determine the optimized pumping capacity required by an operating mine; an accurate forecast results in a reduction in the number of costly delays caused by unexpectedly large flow rates. Predicted water flow patterns are part of the information needed to select the best of several possible approaches to a mining problem on the basis of the operational difficulty and environmental impact of each.

Prediction of water inflow to underground mines involves evaluation of the hydrogeological and geomorphological set up of the area. This include a generalized idea of the geological set up of the area in terms of occurrence and variation of lithotypes and their stratigraphic thickness, structural geometry, petrophysical properties of lithotypes, landform and slope distribution surface drainage characteristics, soil type and soil cover thickness, weathering and erosional patterns, water table altitude and configuration, annual recharge flux from rainfall and other sources, seepage rates and mechanics of groundwater movement.

In this paper, an account of hydrogeological study that has been performed around Rajgamar-dipside, is presented with a purpose to have a first-hand estimation of the inflow of water in a proposed underground coal mine that is planned to be developed in near future.

2 Methodology

Geological mapping along Phulakdih and Kherwa Nala was performed to understand rock exposures, bed orientation, erosional parameters, soil overburden, fractures & joint patterns. Geomorphic studies were undertaken to characterize drainage in terms of drainage morphometry along with erosional parameters and slope of the area surrounding the block. Water table contour maps were constructed using water level data collected through observation wells comprising of dugwells of the area and discharge from natural seepage. The aquifer characteristics comprising of porosity, hydraulic conductivity and texture of the aquifer rocks in block specimens was determined. Porosity was estimated by total saturation method and textural characterization in terms of grain size, grain/matrix ratio, porosity and grain boundary relationships was performed under a transmitted and reflected optical microscope with an automated stepper stage modal point counting facility. Governing equation (Domenico and Schwartz, 1990) is used to calculate water inflow in underground excavations taking into consideration the hydraulic conductivity of the rocks and its variation with depth, in an underground gallery of dimensions 1000 m length, 4 m width and 2.5 m height.

3 Geological Set Up of the Area

3.1 Stratigraphy

The area forms a part of the Gondwana Supergroup that unconformably overlies the basement rocks of the Bilaspur-Raigarh-Surguja Belt. The Basement rocks of the Bilaspur-Raigarh-Surguja Belt is an assemblage of granites and gneisses with metasediments and intrusives spanning between Archaean to Meso Proterozoic age. They however are not exposed in and around the block area, but can be observed far south west outside the study area. The overlying Gondwana Supergroup of rocks in and around the block area is represented by Middle Permian Barakar Formation (290-278 Ma) (Fig. 1). Overlying this is the Upper Permian to Lower Triassic Kamthi Formation (260-240 Ma) (Mukhopadhyay et al. 2010) (Fig. 1). The Barakar Formation is coal bearing and forms a part of the Korba coalfield of Hansdeo-Mand-Raigarh Gondwana basins of Chattisgarh. The stratigraphic succession of the study area is presented in Table 1.

Table 1. Stratigraphic classification of the rocks of the study area

Age	Supergroup	Formation	Lithology
Recent		Alluvium	Sand, silt clay
Carboniferous to Lower Triassic	Gondwana Supergroup	Kamthi Formation	Sandstone, shale, siltstone, thin coal seams (<0.25 m)
		Barakar Formation	Arkosic Sandstone, thin shale, siltstone and thick coal seams (>0.5 m)
		Talchir (?) (not exposed in drill cores)	Green shales, conglomerates, diamictites
Archaean to Meso Proterozoic	Bilaspur –Raigarh -Surguja Belt	Unclassified Metamorphics and intrusives	Granites, Gneisses, unclassified metaediments

3.2 Lithology

Barakar Formation

The Barakar Formation are trough cross bedded medium to thick beds (Fig. 2.a) of gritty to coarse grained arkose in the outcrop, composed of around 70-80% quartz and 20% Feldspars with opaques and minor lithic fragments. The general sorting is moderate and is sub-angular to poorly rounded (Fig. 2.b). The Barakar formation outcrops sporadically in the block area mainly along the Phulakdih nala and is mostly covered under a thin veneer of soil and agricultural land cover in the adjoining areas. The sandstones are often fragile in hand specimens indicating loss in induration at least in the outcrop level. In the boreholes of MRG series in the block area, Barakar sandstones are represented as medium to coarse grained in the lithologs. Coals occur as seams in this Formation.

Kamthi Formation

The Kamthi Formation is characterized by poorly developed cross beds, frequently massive, ferruginous arkosic to sub-arkosic arenites and interbedded red clay and micaceous silt beds (Fig. 2.3). In general Kamthi sandstones are mica rich, angular, moderately sorted quartz arenites and are bound together with ferruginous cement. The Kamthi Formation is of the largest extent in the study area and occurs mostly in the eastern parts. The contact between the Barakars and Kamthi shown in the geological map of the study area (Fig. 1) is based on isolated outcrops and their relationship has been confirmed through traverses made upstream along the Phulakdih nala further to the east of Rajgamar.

3.3 Structure

Bedding: The primary marker horizonof the lithotypes observed in the area under study is the bedding. Individual beds in general, for the Barakar Formation varies in thickness between 10–15 cms to 1.00 m. The beds are cross-stratified (Fig. 2a) with varied degrees of inclination of the foresets. In Kamthi Formation, bedding was recognized in the field mainly by compositional banding and grain size variation. Cross bedding in Kamthi sandstones were not very frequent. Barakar sandstones are arkoses composed of dominantly quartz and relatively less amounts of feldspars as observed in thin sections (Fig.2b)

In general bedding has a shallow dip (Fig. 3). The dip amounts vary from 5° to 20°. However the dip-direction varies over wide ranges. The average dip direction of the beds is towards ENE with a warped bed-form surface that also controls the outcrop pattern (Fig. 1). The contact between the Barakar and Kamthi Formations follow the bed form surface in general. The overall dip of the beds is 15°→ NE to ENE.

Joints: Joints do not occur throughout but occur locally as polygonal cracks that do not extend more than a meter in strike length. Two sets of vertical to sub-vertical joints trending 140° and 80° had been commonly observed in the Kamthi formation (Fig. 4a). Joint frequency ranges between 1.66/m to 1/m. Joints are irregular anastomosing and discontinuous along strike. Joints developed in the sandstones due to uneven loading post lithification. In thin sections of sandstones, uneven loading during compaction and lithification is evident from bent mica flakes that gives a folded appearance (Fig. 4b).

Faults: Faults were not apparently observable in the field. The exploration drill-hole data of the MRG series drill holes of Monnet Ispat Energy Ltd., in the study area, places faults on the basis of apparent displacement of correlatable coal seams intersected in the drill holes. From these drill holes it is interpreted that faults are in general NW-SE trending in the southern sectors of the block area and show curved and branched out extensions towards north to north-east in the central part of the block area. The faults are extensional and vary in throw from 2 m to 40 m. Fault spacing varies from 600 m to 400 m. The fault facing and sense of displacement gives rise to stairways, horsts, angular grabens, bifurcating fault segments and zones of culminations all generally trending NW-SE.

3.4 Weathering Characteristics

The sandstones of the Kamthi Formation and Barakar Formations are considerably porous and is marked by high weathering index due to decomposition of feldspars into kaolinite. This gives rise to the porosity and decreases the consolidation of the rock considerably. Solution sink holes and pocks ranging in diameter of 4–10 cm to 0.5–1 m diameter occur densely within the Kamthi and Barakar Formation of the study area (Fig. 5a). These solution pits link up laterally and vertically to develop networks of channels (Fig. 5b) that serve as free passage for fast infiltration and subsurface flow of water. Soil in the study area is red sandy loam to sandy clay loam (Alfisols). In general soil cover is thin (0–7 m) and at places lateritized.

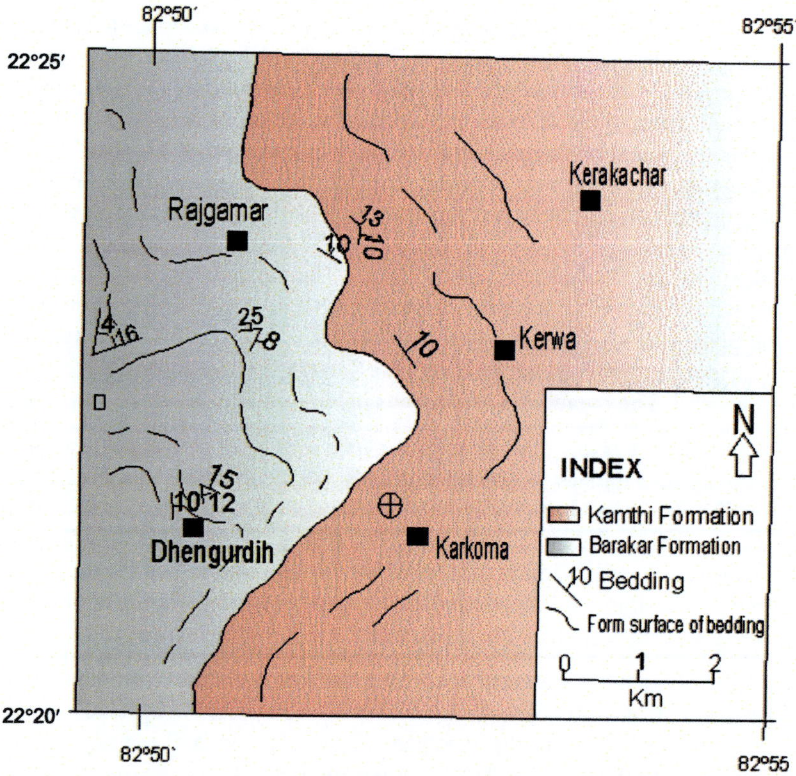

Fig. 1. Geological map of the study area around Rajgamar. (Survey of India, Toposheet no. 64J/15)

Fig. 2. Barakar Formation. (a) Part of trough cross stratification in Barakar sandstones in the Phulakdih Nala. (b) Angular to sub-rounded quartz grains (transmitted light, CPL) with bluish grey to white tones surrounding a feldspar (microcline) clast observed in photomicograph of Barakar sandstone.

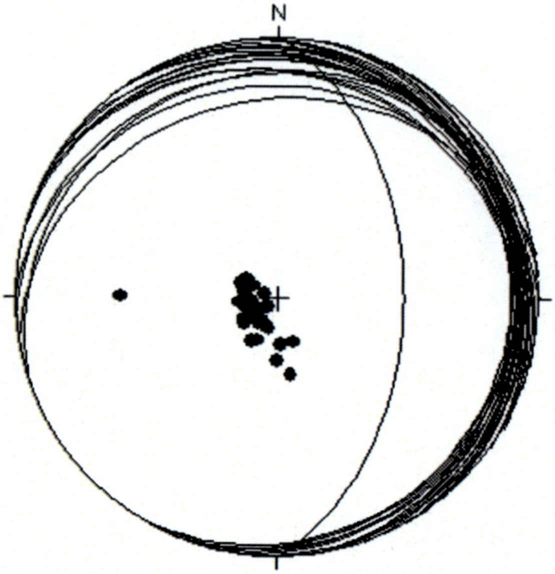

Fig. 3. Equal area projections of bedding (S_0) as great circles and the poles to the bedding planes as solid filled circles.

4 Hydrogeology of the Area

4.1 Surface Drainage

The area receives an annual rainfall of 1050–1000 mm. The area is marked by dendritic drainage pattern of the first order and a semi radial drainage pattern in the second order. The drainage in the block area runs from north east towards west and is opposite to the dip direction of the beds. In the south eastern corner the drainage is directed towards SE (Fig. 6).The morphometric parameters used for analysis of surface drainage and their values determined are indicated in Table 2.

The above data imply that the drainage works considerably hard to minimize the ruggedness. The bifurcation ratio is moderate and implies that there is good accommodation for distribution of water from higher elevation to lower plains through channels. However the central part of the area remains relatively elevated to drain water in a radial pattern. Since the drainage in the block area is directed opposite to the dip of the bedding, the relatively older formations will be increasingly outcropping in the downstream side. The drainage density is 0.63 and is low; indicating thereby that after a heavy precipitation, there will be significant overland flow rather than channel flow. This would increase soil erosion.

4.2 Aquifer Characteristics

The porosity of the representative samples have been analyzed by total saturation method and by petrographic method. The representative micaceous siltstone of Kamthi Formation has a porosity of 12.25% whereas the sandstones show a porosity range between

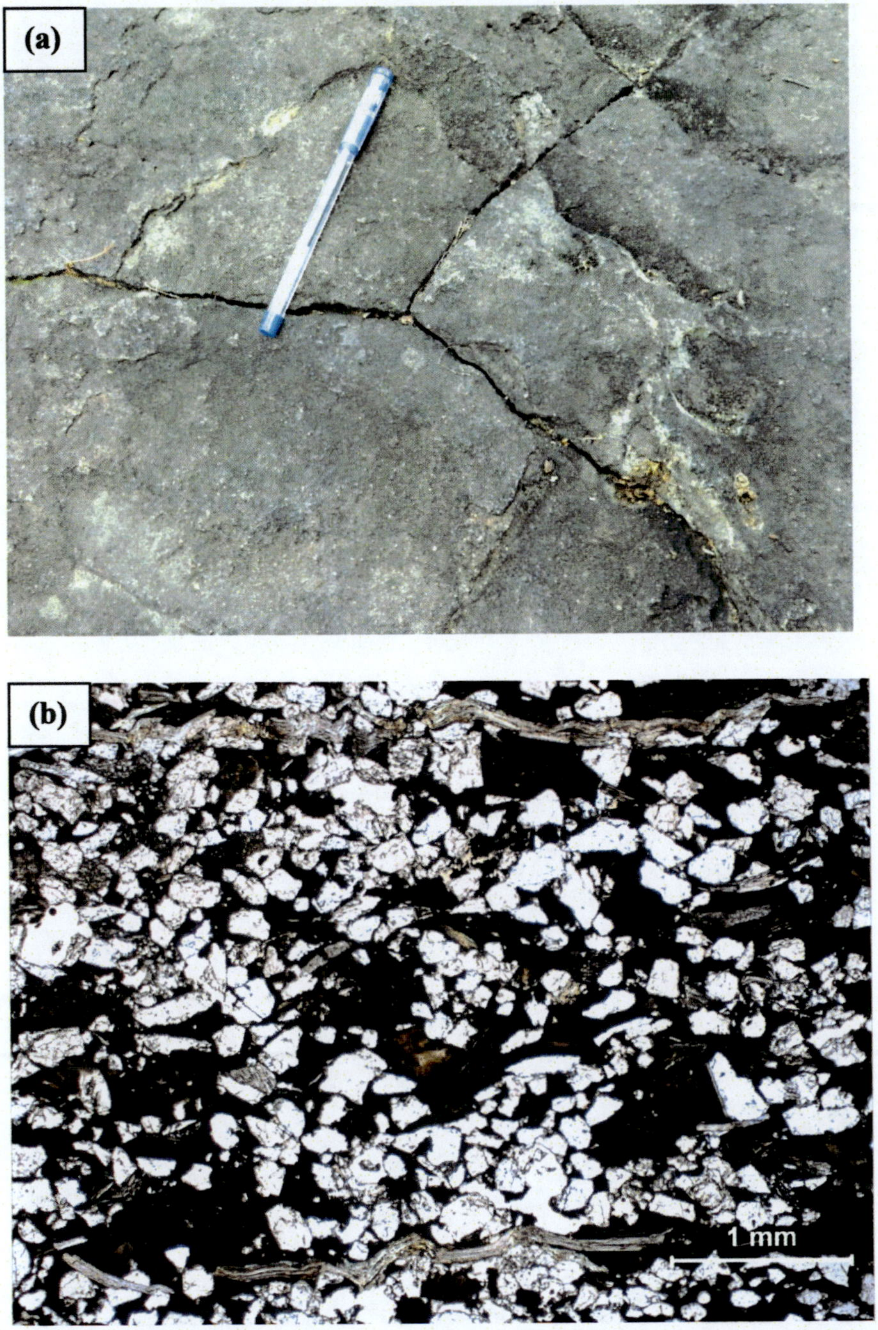

Fig. 4. Kamthi sandstones. (a) Ferruginous coarse sandstone. Looking vertically down on bedding surface facing 30°. Local vertical joint surfaces intersect at various angles. (b) Photomicrograph of Kamthi sandstone. White and gray tones represent quartz and feldspars. Note large films of mica aligned parallel to bedding and folded due to uneven compaction. Intervening dark brown patches are ferruginous cement. PPL

Fig. 5. Weathering patterns (a) Solution sink holes on bedding surface of Kamthi Sandstones in the banks of Phulakdih nala. (b) Solution pocks in Barakar sandstones laterally joining to create more complex networks of solution channels in the northern bank of Phulakdih Nala near Ompur.

Table 2. Morphometric parameters of the study area.

Sl No	Parameters	Values
1	Total area under consideration (A in km^2)	82.935 sq. km
2	Total stream length (L, km)	53 km
3	Drainage density(D_d) (L/A)	0.63
4	Total number of streams (n)	131
5	Stream frequency (streams/sq. km)	1.57
6	Bifurcation ratio (B)	4.9
7	Maximum relief of the area (H)	649m
8	Ruggedness No (H/L)	6.1
9	Soil cover thickness	0-7m

19.28% to 24.72% with a modal porosity of 20%. The epoxy and air bubble filled pore spaces of the sandstones are clearly observable in thin sections of the sandstones (Fig. 7a). Pore spaces are generated with more point-contacts between grain boundaries than line-contacts (Fig. 7b).

The hydraulic conductivity of the semi-weathered Kamthi and Barakar sandstones range between 3.45×10^{-5} m/sec and 4.62×10^{-5}m/s. However, hard and unweathered specimens, have relatively lower hydraulic conductivity of the order of 10^{-6} to 10^{-7}m/s.

The aquifer of the block area is composed of the Kamthi and the Barakar sandstones that together attain a thickness of 350 m. The water table is gently varying with an overall slope towards west (Fig. 8) and therefore follows the topography. The water table is shallow with an average elevation of 348 – 345 m above mean sea level (a msl) and occurs at 0.5 – 6 m below ground level (bgl) depending upon topography. The water table contour map (Fig. 8) shows a converging flow towards west. Discharge rates from Natural seepage points (Fig. 9) were measured in the field that shows a variation between 4 and 43 ml/cm^2/min. The average recharge rate of the block area is 3.9×10^{-9} m/s.

5 Prediction of Water Inflow in the Proposed Underground Coal Mine

5.1 Water Inflow in Underground Excavations

Mine feasibility and environmental evaluations can benefit from predictions of groundwater inflow into a mine excavation, be it open cast or underground. Estimations of mine inflow are made by numerical methods and also by analytical equations. While numerical modeling may be required at advanced stages of mine planning, simple analytical equations for estimating inflow rates can be informative during the initial stages of mine development.

Equations for predicting groundwater inflows in underground excavations are presented in the works of Polubarinova Kochina (1962), Goodman et al. (1965), Verma

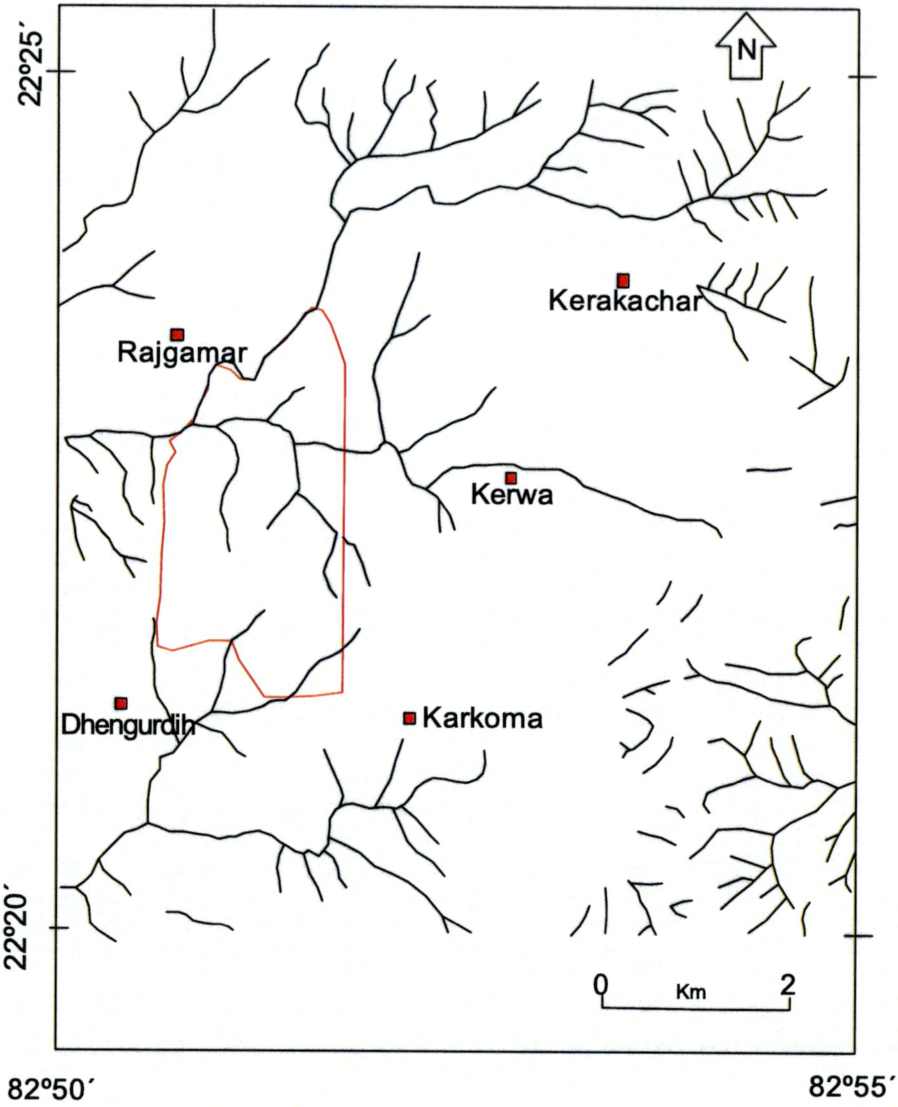

Fig. 6. Drainage map of area around Rajgamar, with the block area outlined.

and Brutasert (1970, 1971), Singh and Atkins (1984), Atkinson et al. (1989), Domenico and Schwartz (1990), Hanna et al. (1994), El Tani (2003). The applicability of a solution depends on the extent to which the real problem under consideration is consistent with assumptions used to derive the mathematical equations. If a solution is judged to be applicable, the accuracy is generally dictated by the appropriateness of the bulk hydraulic conductivity value used to perform the calculations.

Fig. 7. (a) Photomicrograph of Kamthi sandstone. Note that air bubbles occupy a part of the epoxy that filled the matrix pore space during thin-section preparation. The modal volume of porosity in this section was 21.32%. PPL (b) Photomicrograph of representative Barakar sandstone. Note the angularity of the constituent grains of quartz and feldspar that generate inter-granular pore spaces. S is surface contact between two grains and P is grain point contact. The modal volume of porosity is 18.35%.

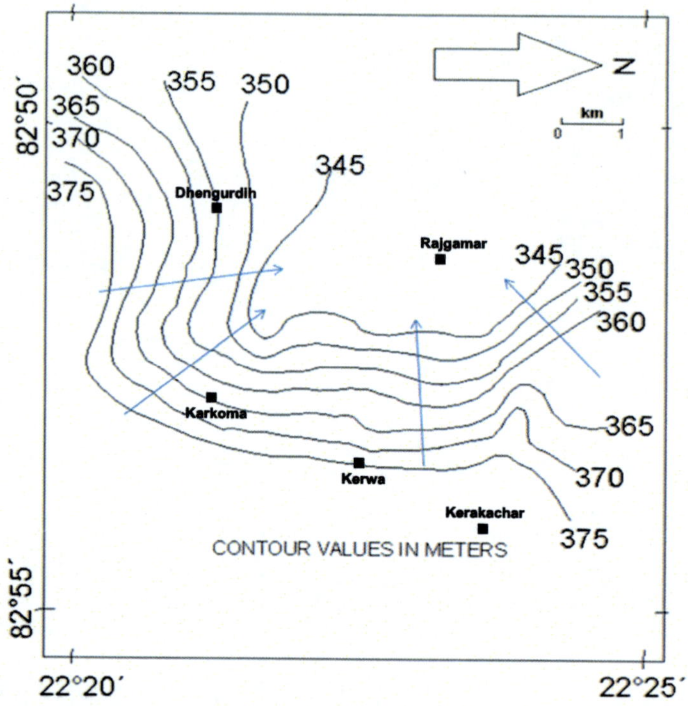

Fig. 8. Water table contour map of the block area with directions of flow lines converging towards west.

The hydraulic conductivity (K) measured in meters per day and reported in meters per second is an important parameter for estimation of groundwater inflows. In general for medium grained sandstones with porosity of 19–20%, a hydraulic conductivity of 3 m/day (3.45×10^{-5} m/s) is a standard value to be used. Porosity of the sandstones of Kamthi and Barakar formations as stated earlier range between 17–24%. Hence a hydraulic conductivity of the sandstones near to the water table is therefore of the order of 3.45×10^{-5} m/s. It is to be understood that hydraulic conductivity of same rock can have a range of values depending on the factors such as weathering, fracturing, solution channels, and depth of burial. The general range of hydraulic conductivity for medium to coarse sandstones is of the order of 10^{-6} to 10^{-7} m/s and can be as less as 10^{-10} m/s.

Analytical equations are generally valid for homogeneous and isotropic aquifers. However rock masses are typically anisotropic and heterogeneous media for which the traditional analytical solutions do not accurately predict underground excavation drainage. However any over estimations predicted by analytical equations are generally to be considered positively as it largely covers the risk factors.

Fig. 9. Natural Seepage of groundwater (seepage points indicated by arrow) that drains into the Phulakdih nala.

5.2 The Governing Equation and Water Inflow

For a horizontal gallery in a fully saturated, semi-infinite homogeneous media, the steady inflow rate is approximated by Domenico and Schwartz (1990) as follows:

$$Q = \frac{2\pi K H_0 L}{2.3 Log(\frac{2H_0}{R})}$$

where Q is the inflow amount in m^3/sec, for the gallery of length L (m), K is the hydraulic conductivity (m/sec), R is the radius (m) of the cylinder having equivalent volume to that of the gallery, L is the length (m) of the gallery, H_0 is the distance (m) between the water table and the centerline of the cylinder having equivalent volume to that of the gallery. A constant hydraulic head is assumed. Table 3 depicts the predicted water inflow in gallons per minute for a gallery whose height is 2.5 m, width is 4m and length is 1000m. In general, the mine inflow amount decreases with increase in depth. This is also illustrated in Fig. 10 where the general trend of decrease provides a linear fit with goodness of fit (R^2) as 0.8852 (t-test on R indicates that R is significant at 5% level of significance). An important point in this regard is the role of fractures and solution channels that can create sudden water inrushes into the underground excavations in addition to the steady inflow. This point is discussed in the Sect. 5.3 and the inrushes are not included in the Table 3.

Table 3. Predicted water inflow in an underground excavation for a proposed coal mine

Depth below water table (H_0) (m)	Hydraulic conductivity (K) (m/sec)	Gallery length (m)	Predicted Inflow (gallons per minute (gpm)	Remarks
10	3.45×10^{-5}	1000	19834	Inrushes are not included
50	1.15×10^{-5}	1000	15840	
90	0.8×10^{-5}	1000	18264	
130	4.5×10^{-6}	1000	13550	
170	3.2×10^{-6}	1000	11871	
210	2.8×10^{-6}	1000	12253	
250	2.0×10^{-6}	1000	10058	
290	1.8×10^{-6}	1000	10169	
330	7.0×10^{-7}	1000	4397	

5.3 Water Inrush in Underground Excavations

Mine water inrushes are caused by factors such as gaping joints, wedge failure during operations, increase in flow path radius during bolting and blasting, solution channels that link up underground to discharge abnormally high amounts of water etc. These are difficult to predict. From the previous documentations of solution channels on the surface, it is expected that fast recharge will take place through solution sinkholes. A margin for such amount of water should always be considered to be added to the amount of inflow. However the inrush problems is expected to decrease with depth as do the inflow.

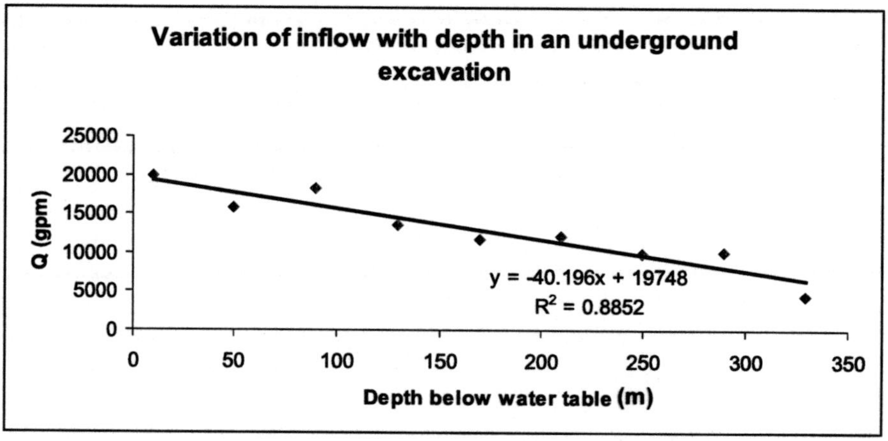

Fig. 10. General trend of the inflow with depth.

6 Conclusions

Hydrogeological evaluation was performed in the study area around Rajgamar dipside (south of Phulakdih nala) for assessing water inflow in a proposed underground coal mine in the study area, in Korba district of Chattisgarh, India. Field studies included geological mapping and lithological data. The field data is integrated to explain lithological disposition and its geometry., The hydrological parameters like landform slope distribution, surface drainage characteristics, soil type and soil cover thickness, weathering and erosional patterns, water table altitude and configuration, annual recharge flux from rainfall and other sources, groundwater seepage rates from natural seepage points were also integrated., The collected rock-sample were studied in the lab to understand textural characterization of the rock types in terms of grain size, grain–matrix ratio, grain-shape, modal volume percentage of mineralogical constituents and determination of porosity and hydraulic conductivity of the rock types in the study area. Based on such studies and assuming steady state flow of groundwater it is concluded that a possible inrush of groundwater at places would occur through underground solution channels and local fractures within shallow depths (~20 m) and a steady inflow in the range of 4397 gpm to 19834 gpm distributed over a corresponding progressive depth of 10m to 330 m respectively would occur in a proposed underground coal mine for a gallery of 1000 m length, 4 m width and 2.5 m height.

References

Atkinson, L.C., Hanna, T.M., Anthony, J.W.: Investigations and implementations of mine dewatering systems. In: World Gold Conference, Singapore (1989)

Domenico, P.A., Schwartz, F.W.: Physical and Chemical Hydrogeology. John Wiley and Sons, New York (1990)

El Tani, M.: Circular tunnel in a semi infinite aquifer. Tunneling Undergr. Space Technol. **18**, 49–55 (2003)

Goodman, R.E., Moye, D.G., Van Schalkwyk, A., Janvandel, I.: Groundwater inflows during tunnel driving. Eng. Geol. **2**, 39–56 (1965)

Hanna, T.M., Elfadil, A.A., Atkinson, L.C.: Use of an analytical solution for preliminary estimation of groundwater inflow to a pit. Min. Eng. **46**(2), 149–152 (1994)

Mukhopadhyay, G., Mukhopadhyay, S.K., Roychoudhury, M., Parui, P.K.: Stratigraphic correlation between different Gondwana Basins of India. J. Geol. Soc. India **76**, 251–266 (2010)

Polubarinova-Kochina, P.Y.: Theory of Groundwater Movement [Translated by R.J.M. De Wiest]. Princeton University, Princeton, New Jersey (1962)

Singh, R.N., Atkins, A.S.: Applications of analytical solutions to simulate some mine inflow problems in underground coal mines. Int. J. Mine Water **3**(4), 1–27 (1984)

Verma, R.D., Brutsaert, W.: Unconfined aquifer seepage by capillary flow theory. J. Hydraulics Div., Proc. Am. Soc. Civ. Eng. **96**(HY6), 1331–1344 (1970)

Verma, R.D., Brutsaert, W.: Unsteady free surface groundwater seepage. J. Hydraulics Div., Proc. Am. Soc. Civ. Eng. **97**(HY8), 1213–1229 (1971)

Reserve Estimation Through Conventional Method and Computer-Aided Software: A Comparative Case Study

Anand S. Sharma[1(✉)], Amar Prakash[1], Sujit K. Mandal[1], K. Prasanth[2], Dilip Kumbhakar[1], and Prerna Jaiswal[1]

[1] CSIR-Central Institute of Mining and Fuel Research, Dhanbad, India
{anandsharma,amarprakash,sujitkm,dkumbhakar,prerna}@cimfr.nic.in
[2] IIT (ISM), Dhanbad, India

Abstract. Reserve estimation, most responsible and irreplaceable task, is carried out at all the stages of a mining project to estimate the quality and quantity of mineral deposits. Accurate reserve evaluation involves a precise assessment of quantity, quality, and sustainability. Various advanced mine planning software are globally available but due to high capital cost most of the mine operators opt for conventional methods for reserve estimation. The present study encompasses a case study of Manoharpur coal block comparing reserve estimation by the cross-sectional method and Minex software based on input data of 238 drill holes and comparison with the actual production. The estimated reserve obtained by the Cross-sectional method and Minex software was 6216695 t and 5619278.33 t respectively. However, the actual production of coal as per the proposed profile was 5251365.86 t. The surface profile of the mine has been obtained by a 3D terrestrial laser scanner. The slight variation in the surface profile, dilution, and cutting-edge outline could be the cumulative parameters affecting the actual production with 7% variation. The Minex software is observed to be more reliable in comparison to the conventional method as the former covers all small deviations in reserve profiles unlike in the latter method.

Keywords: Cross-sectional method · Minex · Reserve · Terrestrial Laser Scanner

1 Introduction

Precise reserve estimation is a strategic factor for mining perspective. Reserve estimation process involved the determination of the amount and type of mineral that can be profitably and lawfully removed from a mineral deposit by mining. The accuracy of the estimate comes up with significant impact on the company's profitability. Governmental entities that must make choices about nonrenewable natural resource policy also place a growing emphasis on accurate mineral reserve estimations (Knudsen et al., 1978). In addition to assessing the quantity and quality of a deposit, estimating coal reserves also takes account the technical and legal ramifications of mining the deposit, beneficiating

A. Sinha et al. (Eds.): AMC 2023, SPEES, pp. 32–44, 2023.
https://doi.org/10.1007/978-3-031-46966-4_3

and selling the finished product (Noble, 1992). Poor resource and reserve estimation practices might have highly negative economic effects. For the accurate reserve estimation the resource information related to the geology, assays, and physical restrictions (topography, property boundaries, prior mining openings, etc.) are essential (Edwards, 2001). Before the development of computers and geostatistical methods, mineral resources and reserves were assessed and estimated manually using, paper, pencils, slide ruler, and calculators to measure the area, volume and grade (Torabi, 2017).Accuracy in estimation of reserve enhanced many folds with the intervene of computer and software. Some of the studies have demonstrated that the advanced mine planning software are useful for the resource and reserve estimation with high level of accuracy (Luke et al., 2014; Pana, 2005). The study carried out to estimate the reserve of Indonesian coal mine by using the nearest neighbor point (NNP), inverse distance weighted (IDW), and kriging method is best in accuracy by providing the smallest error value with smallest root mean square error (RMSE) value of 0.67 (Wijayanto et al., 2021). A comparative study of geometric and geostatistical methods conducted by Afeni et al., (2021) on limestone deposit highlights that the geometric method is effective for tonnage estimation and geostatistical method is effective for the grade estimation.

There are several methods of reserve computation, broadly classified into four namely average factors and area method, mining blocks method, cross-section methods and analytical methods. The classification is based on the criteria for replacing the examined bodies with auxiliary blocks and how averages are computed for variables (Popoff, 1966). Conventional methods for resource estimations are typically performed on plan maps or cross-section maps that slice the deposit into groups of parallel slices. Drill hole location, assay result, and the geological interpretation of the mineralization controls are among the data depicted on the maps. Area averaging, polygonal, cross-sectional and triangular geometric approaches are employed and are based on the geometric weighting of assays (Noble, 1992). The cross-sectional method is the common conventional method used for reserve estimation of various deposits such as limestone (Alpiana et al., 2020), iron ore (Torabi, 2017), coal (Sakdillah, 2020) etc. The volume in the cross-section method is calculated through the produced incision and then the distance between the incisions can be used to calculate the volume (Ganić et al., 2011). For reserve calculation, the mineral deposit is transformed into an analogous geometric body made up of one, several, or an aggregate of close-order solids that best describe the size, shape, and distribution of the variables. This conversion reduces and distorts the original mineral deposit. The selection of appropriate reserve estimation method mainly depends upon geology of the mineral deposit, the exploration technique, the accessibility and dependability of factual data, the intended use of the computations, and the required level of precision (Popoff, 1966). The different techniques used to calculate the coal reserve shows different amount of coal reserves for the same location, this happens due to their different calculation methodology (Wijayanto et al., 2021). The objective of this study is to estimate the reserve for Manoharpur coal block for the financial year (FY) 2021–22 by using the conventional method (cross-sectional) and Minex software and to evaluate the accuracy percentage of both methods by comparing the designed pit profile with the annual survey pit profile.

2 Case Study

The coal mine of M/s Odisha Coal & Power Limited (OCPL) is situated in Manoharpur block, Himgir, Distt. Sundargarh (Fig. 1). The extraction of coal is being following by opencast mining. The mine with a lease period of 30 years is presently running in the third financial year of production i.e., 2021–2022. The coal production of mine during the FY 2019–20 and FY 2020–21 was 1.02 Mt and 2.0 Mt respectively.

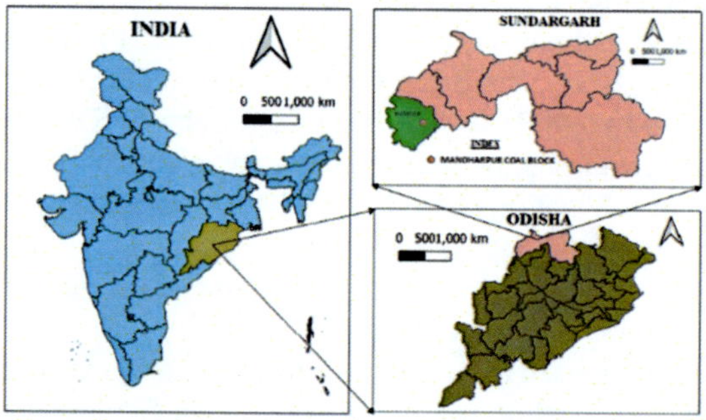

Fig. 1. Location plan of Manoharpur coal block of OCPL

2.1 Regional Geology

Based on the existing borehole data with 100 m × 100 m grid, two major coal horizons viz., Rampura and Lajkura with total of five seams were encountered in the block and six number of local seams were correlated. The Rampur seam is overlain by the Lajkura seam. The average gradient of seams is 1 in 10 with dip direction towards south-west. The present strike length is about 2.40 km and dip-wise length is about 0.7 km (including 1st and 2nd year operations). All the seams are in-crop within the property. Other than infill drilling prior to starting of financial year, the total number of 238 core drilling was done over the lease area. Hence, the borehole density within the lease area comes to 24.338/km^2. The boreholes drilled within the block area is shown in Fig. 2. No faults encountered within the third-year operational area. However, stone/dirt bands were observed intermittently which is having impact on quality and quantity.

2.2 Pit Design for the Financial Year 2021–22

The Manoharpur coal block was proposed to be extracted as single quarry as per the mining plan. The progress of the mine will be towards south-west side of the pit and will continue upto 7th year as per approved mining plan. By looking the demand of coal, the OCPL was planned to produce the 5.5 Mt of the coal for the FY 2021–22. Accordingly, the pit design was created in Minex software by considering the bench parameters as given in the Table 1. As per the proposed excavation boundary for FY 2021–22, the RL of the top bench was considered 288 m and the pit limit RL as 210 m. The benches to be operated for the FY 2021–22 is shown in Fig. 3.

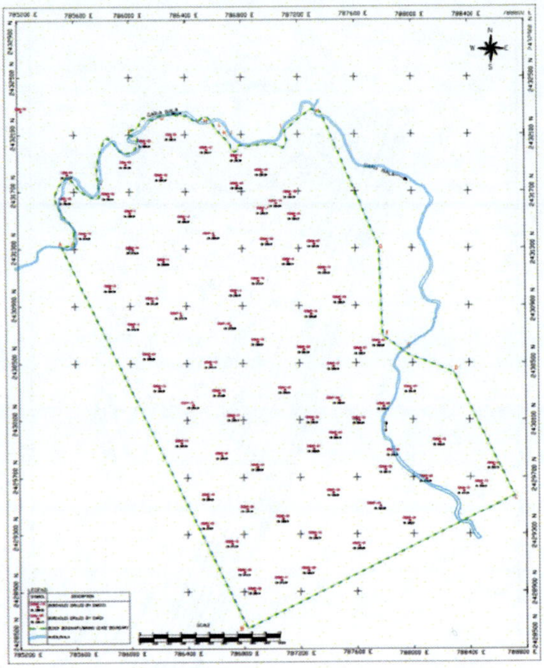

Fig. 2. Boreholes plot within the block area

Table 1. Proposed dimension of the benches for FY 2021–22

Bench Parameters	Topsoil	Hard OB
Bench height (m)	3	10
Bench width (m)	9	30
Bench slope (degrees)	45	70
Overall pit slope	15.13 degrees	

The location of the haul roads was proposed to maintain minimum surface transport equipment density for safety and to minimize the hauling distance. The main entry made as provided in the mining plan with the dimension of 60 m width and gradient of 1 in 20 to approach the coal seams. The soil benches were drawn with 3 m height and 9 m width and having slope of 45°. For more than 250 m RL, benches were designed with 10 m height and 30–40 m width having slope of 70°. In all, there will be 13 benches under operation for the FY 2021–22. Benches were maintained at a uniform width of 30–40 m for the safe movement of vehicles as well as to handle higher volume of coal and overburden. For the benches on the highwall side, width of 10 m and for every 3 benches, width of 20 m was proposed for safety. For the rest of the pit, the benches and

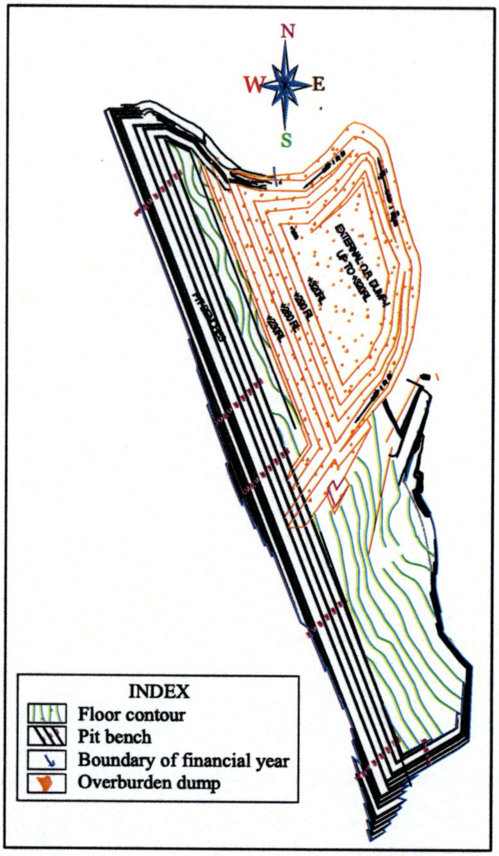

Fig. 3. Pit design for the financial year 2021–22

ramps were drawn to approach the floor of the Rampur II + I seam. The pit design was prepared and incorporated in the Minex software to calculate the reserve (Fig. 4).

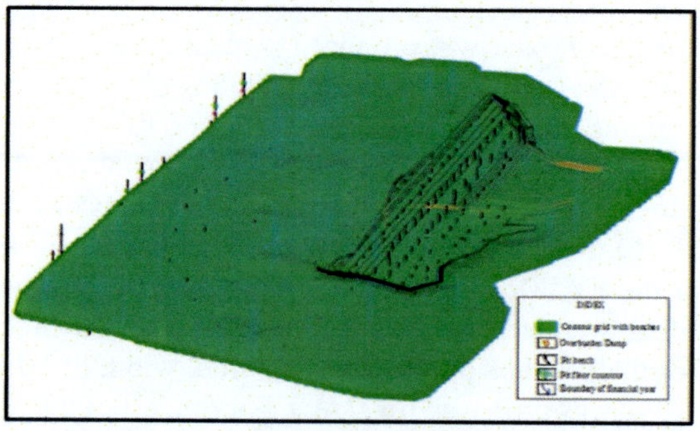

Fig. 4. Grid with benches for the FY 2021–22 by Minex software

2.3 Actual Pit Profile Through Laser Scanner

Terrestrial Laser Scanner (TLS) was used to survey the surface profile of the opencast mine. A number of settings were done to envelope the entire area of the mine. After reconnaissance survey, the stations were located at suitable places to have proper vision of the mine with wide coverage, taking into account the overlapping of scans. Point cloud data were generated by setting the TLS over the established geo-reference stations, following the backsight method. A few pictorial views of the survey conducted by TLS is shown in Fig. 5.

(a)

(b)

Fig. 5. Survey carried by TLS

After precise elimination of noise, a topographical mesh was generated by merging all the point cloud data files with the aid of ATLASCAN software. Three-dimensional view of the opencast mine generated by TLS point cloud data is shown in Fig. 6. This helped in determining the actual profile of the pit including all benches.

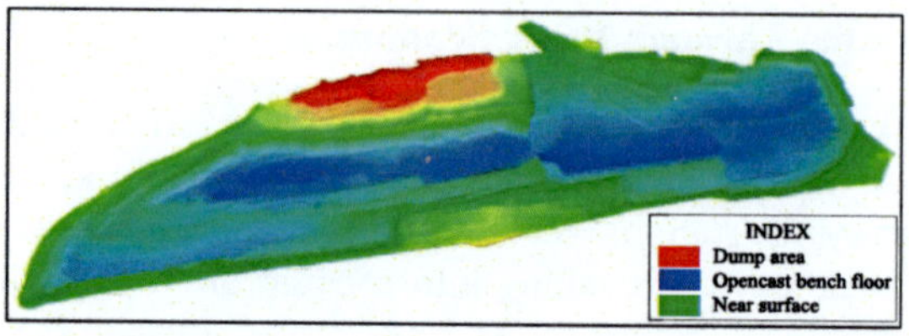

Fig. 6. Three-dimensional view of Manoharpur mine by TLS survey

3 Reserve Estimation

3.1 Cross-Sectional Method

The calculation of a coal reserve is a complex and technical task, consisting of several processes. Understanding of the dimension and qualitative characteristic of the mineral body is necessary for reserve computation. This characteristics can be achieved directly by measurements, chemical analyses and tests or indirectly through assumptions, interpretations, and computations (Popoff, 1966). A total of 53 transverse section lines were developed to divide the coal body into various segments. Depending on the grid interval and borehole locations, the section lines can be spaced at an interval of 50m from N-W to S-E direction (Fig. 7). The designed pit floor and benches for the FY 2021–22 were served as the base profile for the reserve estimation. Out of 53 sections, 48 sections were falling within the pit floor limit, hence only 48 sections were considered for the reserve estimation (Fig. 8&9). The steps followed for the determination of coal reserve is as follows:

The entire pit was separated into sub-blocks along section lines that are half the length of the adjacent sections apart.

i. Each sub-volume block is necessary for reserve calculation. The volume is computed by multiplying the sectional area by the adjacent section's half-distance on each side (i.e., area of influence).
ii. Area of each seam and overburden was measured using AutoCAD software.
iii. To calculate the volume for each seam and overburden the measured area was multiplied by the spacing interval of 50m.
iv. The tonnage for each coal seamwas determined by multiplying the volume by the average density of that coal seam.
v. Thereafter, the suggested losses of each seam were deducted from the estimated tonnage of each seam (Table 2).

Thus, the total quantity of coal to be excavated within the planned block area for FY 2021–22 was calculated as 6216695 t (Fig. 9).

Table 2. Seam wise coal reserves

Seam Name	Volume (m^3)	GCV (kcal/kg)	Grade	Density (kg/m^3)	Coal (tonne)	Mine loss	Coal after mine loss (tonne)
Lajkura Bottom 2 + 1 (LB21)	434574.05	3610	G13	1.73	751813	3%	729259
Lajkura Bottom 2 (LB2)	371530.65	3617	G13	1.73	642748	5%	610611
Lajkura Bottom 1 (LB1)	221228.05	3868	G12	1.69	373875	3%	362659
Local 5 (L5)	51390	3253	G14	1.78	91474	10%	82327
Local 4 (L4)	110778	3521	G13	1.73	191646	10%	172481
Local 3 (L3)	102609.9	3557	G13	1.73	177515	10%	159764
Local 3A (L3A)	83361.8	3697	G13	1.73	144216	10%	129794
Rampur 3 (R3)	819561	3612	G13	1.73	1417841	6%	1332770
Rampur 2 + 1 (R21)	1469444.55	3286	G14	1.78	2615611	3%	2537143
Rampur 2 (R2)	13454.5	3982	G12	1.69	22738	10%	20464
Rampur 1 (R1)	46968.3	3379	G14	1.78	83604	5%	79423
Total					6513081.225		6216695

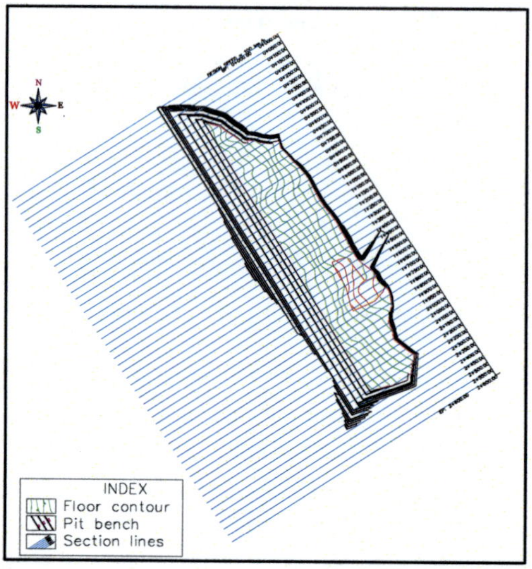

Fig. 7. Plan view of sections for the area estimation.

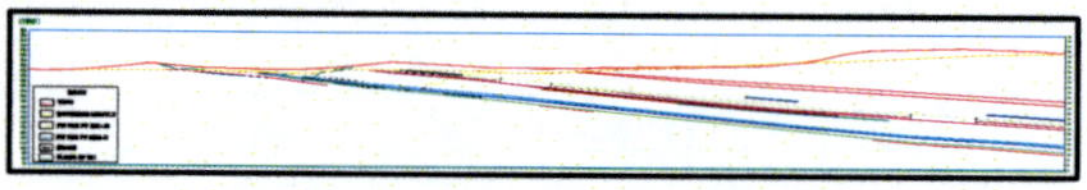

Fig. 8. Single transverse section

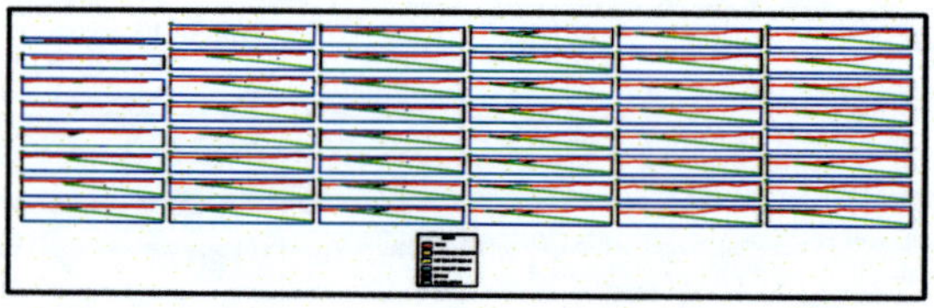

Fig. 9. Transverse sections to compute sectional area

3.2 DTM to DTM method

The maps and reports were evaluated to prepare a database for the reserve estimation and input files were prepared manually. Amount of data required for reserve estimation was large which involved utmost care while feeding as input.

The borehole data which were obtained from geological report went through many transformations to make it in the required format of Minex software for reserve estimation. This database stores all exploration drilling information about drill hole (collar data), complete lithology and assaying data, geotechnical, hydrogeological, geophysical data. Surface contours at 2 m interval were considered for topography. The location of weathered mantle was considered from the borehole data of geological report and accordingly the grid was created by importing the point data in Minex. All the details of the geometry of the block like boundaries, masked polygons were taken into consideration while modeling. The surface contour plan provided in the approved mine plan was considered as a topography surface. As an input data the four main basic files are required for the Minex which is given in the Table 3.

Table 3. Constituents of input data

Collar	Lithology	Assay	Seam
Borehole Id	Borehole Id	Borehole Id	Borehole Id
Easting	From	From	From
Northing	To	To	To
Elevation	Lithology type	Relative Density (RD)	Seam Class
Depth	Colour	GCV	Seam Name
Dip			
Azimuth			
Type			

Based on input data, the borehole model (Fig. 10) and Geological model (Fig. 11) was developed and the reserve within the mining lease area of 652.885 Ha was determined as given in Table 4. The seam L10 was not considered in the reserve calculation due to inadequate data for the griding.

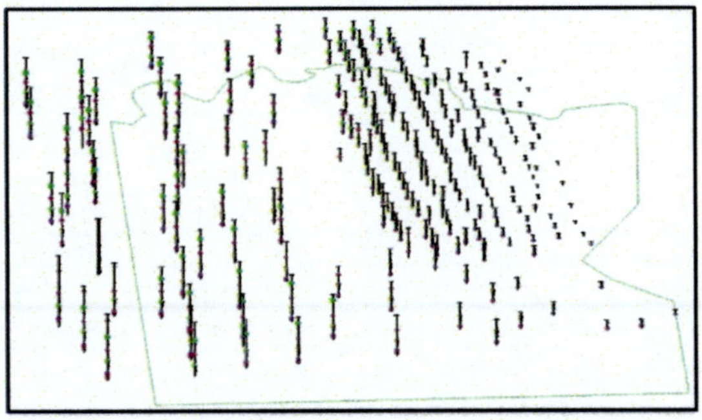

Fig. 10. Borehole modeling done in Minex

Thus, total volume of coal reserve estimated by DTM to DTM method was 56,19,278.33 t.

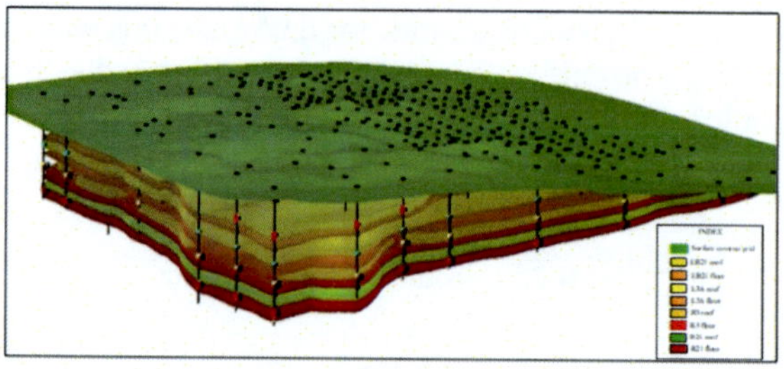

Fig. 11. Seam floor & roof along with topography

Table 4. Mineable reserve for financial year 2021–2022 within the demarked area

Seam Name	Seam Thickness (m)	RD	Coal Area (m^2)	Coal Volume (m^3)	GCV (kcal/kg)	Insitu Tonnes (t)	ROM Tonnes (t)	Waste Volume (m^3)
L9	0		0	0		0	0	5157413.5
L7	0		0	0		0	0	766.49
LJM	0		0	0		0	0	1385.64
L6	1.2	1.77	16836.42	20262.88	3245.6	35959.95	32363.96	221887.48
LB2	1.96	1.71	243192.73	475738.47	3682.16	813446.25	772773.94	1023019.44
LB1	1.1	1.68	299314.13	329185	3994.25	560957.31	544128.56	470439.28
L5	1.36	1.75	23695.7	32289.89	3429.84	56650.91	50985.82	855466.38
L4	1.36	1.74	87299.95	119134.46	3456.88	207170.72	186453.64	1245774.25
L3	1.33	1.71	77322.82	103139.18	3704.1	176873.16	159185.84	2264958
L3A	1.58	1.69	51132.83	80854.41	3716.79	137109.64	123398.67	1129716.38
R3	2.05	1.71	335481.25	687762.31	3735.47	1177740.25	1107075.75	3018693.75
R2	0.7	1.75	468925.47	329039.44	3411.84	577110.31	548254.75	1022997.94
R1	1.9	1.76	484514.75	919897.25	3316.72	1627908.25	1579071	106755.81
L2	0		0	0		0	0	327.33
L1	0		0	0		0	0	10.51
Reserve & OB without considering the parting			2087716.05	3097303.3		5370926.75	5103691.93	16519612.18
Parting reserve & OB							515586.4	289655.3
TOTAL							**5619278.33**	**16229956.8**

4 Result and Discussion

Based on the existing borehole data with 100 m × 10 0m grid, two major coal horizons viz; Rampur and Lajkura with total of five local seams were encountered in Manoharpur coal block. In this FY 2021–22, five numbers of local seams (Local 3A, Local 3, Local 4, Local 5 and Local 6), three numbers of Rampur seams (R1, R2 & R3) and two numbers of Lajkura seams (LB1 & LB2) were considered for reserve estimation. Rampur seam is overlain by Lajkura seam. Rampur seam splits into three major horizons namely Rampur-I, Rampur-II, Rampur-III, in the order of bottom to top. While Lajkura seam is split into two sections namely Lajkura Bottom-I & Lajkura Bottom-II. Rampur-I and II and Lajkura Bottom-I & II is also for the proposed for FY 2021–22. The assumptions considered for the model preparation are as follows:

i. The split parting up to 1m between seam R1 & R2 and LB1 & LB2 was considered as a reserve.
ii. The seam having thickness less than 1m was considered as a waste.
iii. The seams R1, R2, LB1 & LB2 considered as an extractable up to the zero thickness.
iv. The bench width of bench no. at 250 RL is considered 30m.
v. The seam-wise volumes of coal were calculated considering the minimum GCV of 2200 kcal/kg.

The coal reserve computed by conventional method and by DTM to DTM (i.e., surface to surface) was estimated as 6216695 t and 5619278.33 t respectively. Actual excavation of coal for the FY 2021–22 was 5251365.86 t. It was computed through the profile of laser scanner and cross verified through the annual dispatched data of coal. The reserve estimated by the Minex software was observed to be close to actual production with a percentage variation of 7% whereas in cross sectional method it was 18.38%. The percentage variation in reserve determined by Minex may be accountable due to cumulative influence in variation in the surface profile, dilution, and difference in cutting-edge outline.

In sectional area method, sections were prepared at 50 m interval. Any intermediate variation in the geological formation, seam thickness, gradient etc. within the two sections eventually affects the overall reserve resulting into sweeping away from the actual reserve. However, all such variations are taken into account in Minex software and hence the result was found to be in close approximation to actual production. Thus, the reserve estimation by the latter approach proved to be reliable.

5 Conclusion

The accurate reserve estimation is essential for the mining companies for short term and long-term planning. Reserve estimation is time consuming task as it requires correct description of the deposit through exploration. Efforts were made to estimate the reserve accurately for the Manoharpur coal mine before beginning of the FY 2021–22 through the cross-sectional method and DTM to DTM method. The detailed procedure adopted for the computation was presented in this paper. Later, utilizing exploratory borehole data from the Manoharpur coal deposit of OCPL, a comparative investigation was carried out. At the end of financial year, the actual pit profile obtained from the survey by TLS was compared with the predicted pit profile, the slight variation in surface profile was noted which might be a causative factor for the variation in the reserve. The relative difference in tonnage estimation was 18.38% by cross-sectional method whereas it was 7% by Minex software. The Minex software is effective for tonnage estimation for the coal deposit, as shown by the minimal relative difference in tonnage estimation with the actual production. The use of combination of tools i.e. Minex, AtlaScan and Autocad software programs to perform reserve estimating facilities and field validation is an effective tool for mining industry.

Acknowledgement. The authors are thankful to the Odisha Coal and Power Limited (OCPL) management for providing the necessary data for modeling, financial support and encouragement in achieving the desired outputs through Minex software. Authors also express thanks to Director, CSIR-Central Institute of Mining and Fuel Research, Dhanbad, for according permission to present the paper.

References

Edwards, A.C. (ed.): Mineral Resource and Ore Reserve Estimation: the AusIMM Guide to Good Practice. Australasian Institute of Mining and Metallurgy, Carlton Victoria Australia (2001)

Afeni, T.B., Akeju, V.O., Aladejare, A.E.: Geosci. Front. **12**(1), 243–253 (2021)

Alpiana, Ilham, M., Irwan, Matrani, B.F.A.: Quantitative mapping of limestone reserves using cross-section method in Mangkung Village, Western Praya, West Nusa Tenggara. IOP Conf. Ser.: Earth Environ. Sci. 413, 012006 (2020)

Zoran, G., Aleksandar, G., Rade, T., Aleksanda, M.: Podzemn. Rad. **19**, 101–108 (2011)

Knudsen, H.P., Kim, Y.C., Mueller, E.: Min Eng (New York) **30**, 54–58 (1978)

Luke, M., Gideon, C., Fabian, A., Cele, O.: Int. J. Sci. Technol. Res. **3**, 108–115 (2014)

Noble, A.C.: Ore reserve/resource estimation. In: SME Mining Engineering Handbook. Society for Mining, Metallurgy, and Exploration (SME), pp. 344–359. Littleton, Colorado, USA (1992)

Pana, M.: 19th International Mining Congrress, pp. 269–320. Fair Turkey, IMCET2005 (2005)

Popoff, C.C.: Computing Reserves of Mineral Deposits: Principles (1966)

Sakdillah: Int. J. Sci. Technol. Res. **9**, 65–73 (2020)

Torabi, S.A.: Int. J. Eng. Technol. Sci. Res. **4**, 896–903 (2017)

Wijayanto, T.M., Wilopo, W., Indrawan, I.G.B., Sunarko, S.: J. Appl. Geol. **6**, 42–51 (2021)

Non-Gaussian Copula Simulation: A New Approach to Recoverable Reserve Estimation in Indian Open-pit Copper Deposit

K. Dinda[1](✉) and B. Samanta[2]

[1] Advanced Technology Development Centre, Indian Institute of Technology Kharagpur, Kharagpur 721302, India
dinda.krishna@gmail.com
[2] Department of Mining Engineering, Indian Institute of Technology Kharagpur, Kharagpur 721302, India
samantab@mining.iitkgp.ac.in

Abstract. In a prominent open-pit copper mine in India, a non-Gaussian copula-based simulation model is developed to estimate recoverable reserves. This groundbreaking research serves as an investigation into the potential and effectiveness of copula-based simulation models in the field of reserve estimation. The methodology involves comparing two specific selectivity curves: grade-ore tonnage and grade-metal tonnage. These curves are constructed using the copula-based simulation technique, as well as multi-Gaussian kriging, disjunctive kriging, and actual production data derived from blasting operations within the mine. The results indicate that, in the context of estimating recoverable reserves, the copula-based simulation method outperforms both multi-Gaussian kriging and disjunctive kriging.

Keywords: Copula · v-transformed copula · simulation · reserve estimation

1 Introduction

Evaluating and selecting recoverable reserves is a vital aspect of both the feasibility study and long-term planning for mining operations. The strategic planning, encompassing factors like production rate, mine life, cut-off grades, and scheduling, heavily depends on the accurate estimation of the recoverable reserve in the deposit (Ravenscroft, 1992). Challenges often arise in accurately gauging these reserves at the planning stage, especially when faced with limited exploration data collected from widely spaced drill holes. Linear geostatistical methods (Matheron, 1963; Huijbregts and Matheron, 1971; Journel and Huijbregts, 1978) employed at the feasibility stage can prove highly unreliable estimate. Non-linear geostatistical techniques (Matheron, 1976; Verly and Sullivan, 1985; Journel, 1983) may overcome this smoothing issue but remain highly sensitive to the Gaussian assumption. Moreover, their theoretical complexity has hindered widespread acceptance among field engineers. In such situations, simulation than

A. Sinha et al. (Eds.): AMC 2023, SPEES, pp. 45–50, 2023.
https://doi.org/10.1007/978-3-031-46966-4_4

the estimation based techniques are preferable. The main advantage of the simulation techniques over the estimation techniques is that the grade variability in the orebody is faithfully represented by the former. The reproduction of the grade variability is the main factor which influences the recovery of ore. Hence, it can certainly be argued that the results provided by the simulation will not be much different than the actual ore recovery. This study embarks on the development of a non-Gaussian copula-based geostatistical simulation technique for orebody mapping, aiming to effectively synchronize orebody maps. This synchronization facilitates the estimation of minable reserves within a chosen deposit. Copula functions have emerged as a significant tool in spatial modeling for non-Gaussian data. This attention is driven by the innovative manner in which copulas offer a technique to describe the dependence between random variables, transcending conventional methods. The journey of copula-based spatial modeling began with Bárdossy (2006), who first conceptualized the technique for weakly stationary and isotropic random fields. This pioneering work set the stage for subsequent developments. The spatial interpolation algorithm rooted in the copula-based spatial model was then proposed by Bárdossy and Li (2008) and later by Kazianka and Pilz (2010), further expanding the applicability of copula functions in spatial modeling. In recent years, the application of copula-based models has extended into the mining sector, specifically in orebody modeling. Works by Musafer et al. (2017), Dinda and Samanta (2021), and Dinda et al. (2022) exemplify this evolution. The advantages of using copula functions are numerous and substantial:

- Nonlinear Spatial Dependence: Unlike traditional semi-variograms in geostatistics that describe linear dependence, copulas have the ability to describe nonlinear spatial dependencies, providing a more comprehensive understanding of spatial relationships.
- Independence from Marginal: Copulas can capture the pure dependence between variables without being influenced by the marginal distributions, allowing for a cleaner and more accurate representation of dependence structures.
- Flexibility in Modeling Dependence: Whether symmetric or non-symmetric, any spatial dependence can be modeled by copulas, offering an unparalleled flexibility in capturing complex dependence patterns.
- Full Distribution Description: Copulas don't just provide a partial view; they describe the full distribution. This attribute makes them more informative than variograms, offering a richer and more nuanced understanding of the data.

These advantages collectively position copula functions as a robust tool, especially in areas where conventional statistical methods fall short. By allowing for a more intricate understanding of dependence structures and providing a way to model complex non-linear relationships, copulas are paving the way for more sophisticated and accurate spatial modeling, including in vital sectors like mining. In the literature, different types of copulas are available such as: Gumbel Copula (Gumbel, 1960); Frank Copula (Genest, 1987);Vine Copula (Joe and Kurowicka, 2011);Elliptical Copulas (Li and Peng, 2009); V-transformed Copula (Bárdossy and Li, 2008).Due to the observation of a non-symmetric spatial dependence structure in the copper deposit, a non-Gaussian copula-based simulation model, specifically the v-transformed copula, has been utilized for simulating the deposit.

2 Methodology

Copula functions, a concept first brought to light by Sklar in 1959, offer a method to capture the interdependence between random variables while accommodating distinct marginal distributions. An n-copula, or n-dimensional copula, operates within the confines of an n-dimensional unit cube and is characterized by univariate marginal distributions that are uniformly spaced on the interval [0,1]. This unique construction enables the creation of a joint multivariate distribution function, utilizing the given continuous marginal distributions. In doing so, copulas furnish a versatile framework for modeling the relationships and dependencies among various variables. Notable contributions to this field have been made by researchers such as Nelsen (1999), and Bárdossy (2006).

In geostatistics, the core principle is that close measurements are more correlated than distant ones. Traditional methods like semi-variogram analysis capture this but have limitations. Spatial copula models, however, can capture dependencies in multi-dimensional spaces, offering a more nuanced understanding of spatial relationships (Goovaerts, 1997). Different copula models manage spatial dependence in various ways, focusing on the upper or lower tail of the distribution, known as "tail dependency". The Gaussian copula is popular for symmetric spatial relationships, while non-Gaussian copula models can capture the asymmetric tail dependencies. However, some of non-Gaussian copulas may struggle with asymmetrical dependence in higher dimensions.

This approach offers flexibility in capturing asymmetric spatial dependence structures, providing a sophisticated solution for spatial modeling and enhancing the accuracy of simulation. The v-transformed copula model involves the transformation parameters (m, k) as well as the dependence parameter (correlation function). To effectively formulate and infer this model, these specific parameters must be accurately determined. In this study, the maximum likelihood method is employed to optimize these parameters. Using estimated parameters and conditioning data, the v-transformed copula model is applied to simulate an orebody, as described by Dinda and Samanta (2021). Finally, these simulated values are used to calculate the recoverable reserves.

3 Case study

Malanjkhand copper deposit is used for case study purposes. A particular bench, 460 m above mean sea level (AMSL), is used for investigating the model, because detailed blast hole data are available for this bench. A total of 156 composite sample values are available for this bench. Figure 1 displays the location of these composite samples along with a histogram plot. The deposit under study is characterized by a vein-type structure, with the main mineralization occurring within a quartz vein. To accurately define the mineralized portion of the ore body, a detailed computerized wireframe model is created that distinguishes between the ore and waste boundaries. The construction of the wireframe model involves taking a series of longitudinal slices along each borehole section. These slices are subsequently manually digitized to outline the boundary of the orebody. During this manual digitization process, borehole hatches of assay serve as a guiding reference to ascertain the presumed ore boundary. The individually digitized slices are then connected together, forming the comprehensive wireframe model. Figure 2 visually represents the orebody model for the deposit. In terms of simulating the orebody for copper (Cu) concentration, the v-transformed copula model is employed. Prior to implementing the simulation model, the spatial dependence structure must be unravelled. Figures 3 (a) and (b) display the graphs of semivariance and Pearson type correlation against lag distance, respectively, showing a noticeable but weak linear dependence. In contrast, the non-linear relationship, as expressed by Kendall's tau statistics in Fig. 3 (c), is notably stronger. This difference highlights that Kendall's tau is indicative of a distinct spatial dependence, unlike the variogram, which doesn't reveal any structured pattern. This discrepancy can be traced back to the Cu dataset's highly skewed distribution. While it dismantles the structure of the variogram, it leaves the copulas and Kendall's tau correlation unaffected. For validation purposes, 199 blasting blocks are available, and the validation zone is divided into 44 panels. The overall tonnages of ore and metal are calculated by aggregating the tonnages from each individual panel. Further details can be found in the work of Dinda and Samanta (2021). Figure 4 displays the ore and metal tonnages at different cut-off grades for v-transformed copula-based simulation, multi-Gaussian kriging, disjunctive kriging, and blasthole samples. This figure demonstrates that the copula-based simulation model closely corresponds to the actual data.

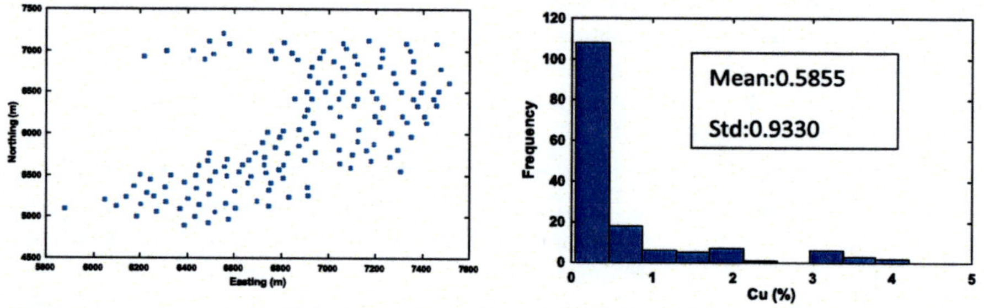

Fig. 1. Location (left) and histogram (right) of the composite samples of 460 m bench AMSL

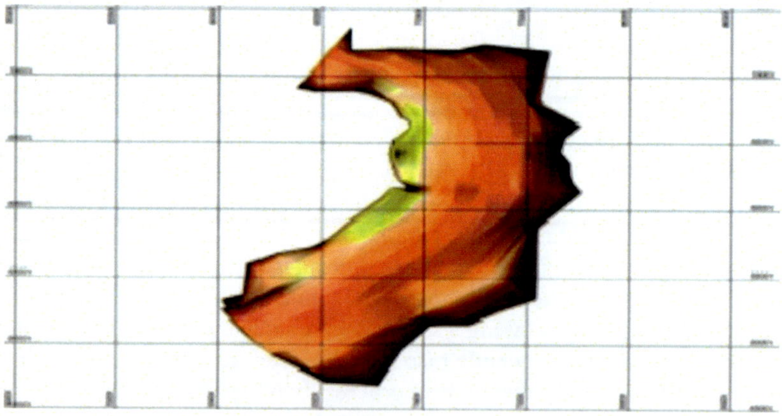

Fig. 2. Wire-frame model of the orebody

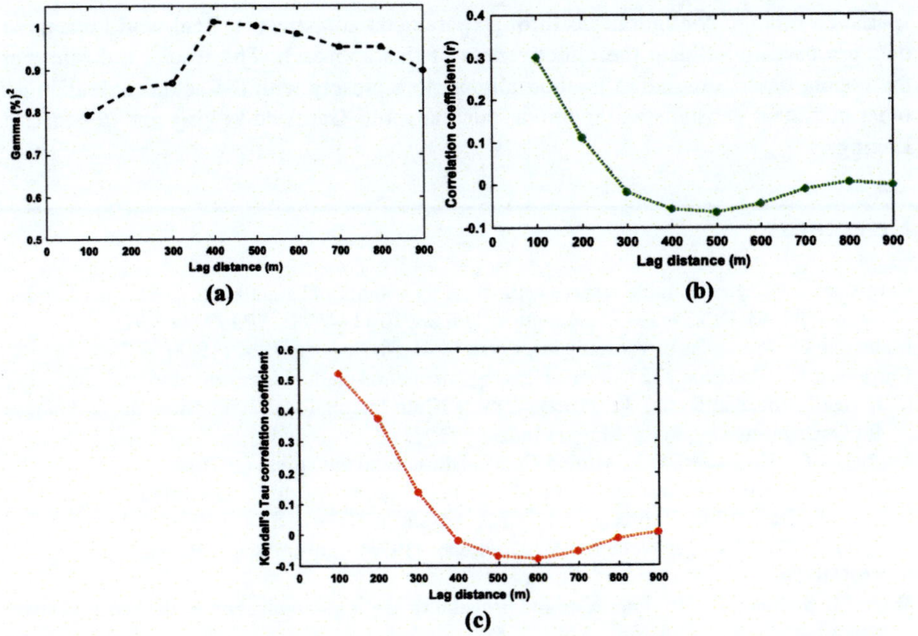

Fig. 3. Semi-variance at different lag distance of spatial dependence, (b) Pearson correlation coefficient of spatial dependence, (c) Kendall's tau correlation coefficient of spatial dependence through bivariate copula function of Cu %

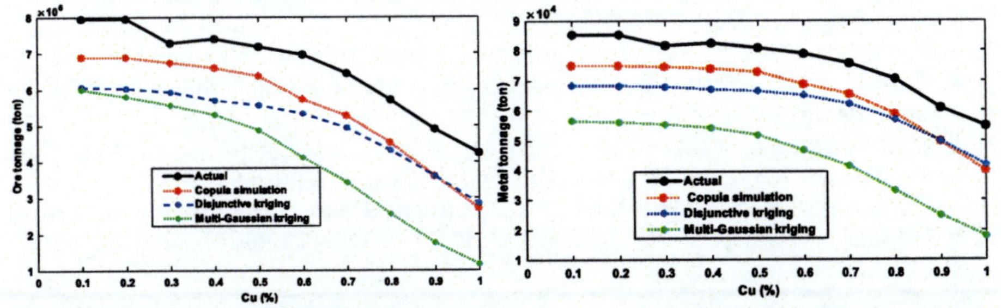

Fig. 4. Grade-ore tonnage (left) and grade-metal tonnage considering SMU size 25 m × 25 m of validation zone of bench 460 m AMSL.

4 Conclusion

In the research conducted so far, a v-transformed copula-based simulation method is utilized to enhance the accuracy of reserve estimation for an Indian copper deposit. This method is chosen in light of the spatial correlation structure of the Cu grade distribution, which shows characteristics of being highly skewed, nonlinear, and asymmetric across a range of quantile values. The v-transformed copula model is considered well-suited for this task due to its capacity to capture a diverse set of asymmetric and nonlinear spatial correlation configurations through parameter adjustments. Real-world blasthole data are used to validate the effectiveness of this approach. The results indicate that the copula-based simulation method aligns more closely with the actual data than do other nonlinear geostatistical methods, such as multi-Gaussian kriging and disjunctive kriging.

References

Ravenscroft, P.J.: Recoverable reserve estimation by conditional simulation. Geol. Soc. London Spec. Publ. **63**(1), 289–298 (1992). https://doi.org/10.1144/GSL.SP.1992.063.01.27

Matheron, G.: Principles of geostatistics. Econ. Geol. **58**(8), 1246–1266 (1963)

Huijbregts, C., Matheron, G.: Universal kriging (an optimal method for estimating and contouring in trend surface analysis). In: Proceedings of Ninth International Symposium on Techniques for Decision-making in the Mineral Industry (1971)

Journel, A.G., Huijbregts, C.J.: Mining Geostatistics. Academic press (1978)

Matheron, G.: A simple substitute for conditional expectation: the disjunctive kriging. In: Guarascio, M., David, M., Huijbregts, C. (eds.) Advanced Geostatistics in the Mining Industry, pp. 221–236. Springer Netherlands, Dordrecht (1976). https://doi.org/10.1007/978-94-010-1470-0_14

Verly, G., Sullivan, J.: Multigaussian and probability krigings–application to the Jerritt Canyon deposit. Min. Eng. **37**(6), 568–574 (1985)

Journel, A.G.: Nonparametric estimation of spatial distributions. J. Int. Assoc. Math. Geol. **15**(3), 445–468 (1983)

Bárdossy, A.: Copula-based geostatistical models for groundwater quality parameters. Water Resour. Res. **42**(11) (2006)

Bárdossy, A., Li, J.: Geostatistical interpolation using copulas. Water Resour. Res. **44**(7) (2008). https://doi.org/10.1029/2007WR006115

Kazianka, H., Pilz, J.: Spatial interpolation using copula-based geostatistical models. In: Atkinson, P., Lloyd, C. (eds.) GeoENV VII—Geostatistics for Environmental Applications, pp. 307–320. Springer, Berlin (2010)

Musafer, G.N., et al.: Spatial pair-copula modeling of grade in ore bodies: a case study. Nat. Resour. Res. **26**, 223–236 (2017). https://doi.org/10.1007/s11053-016-9314-3

Dinda, K., Samanta, B.: Non-Gaussian Copula simulation for estimation of recoverable reserve in an Indian copper deposit. Nat. Resour. Res. **30**, 57–76 (2021). https://doi.org/10.1007/s11053-020-09734-z

Dinda, K., Samanta, B., Chakravarty, D.: A v-transformed copula-based simulation model for lithological classification in an Indian copper deposit. Sci. Rep. **12**(1), 21055 (2022). https://doi.org/10.1038/s41598-022-24233-2

Gumbel, E.J.: Bivariate exponential distributions. J. Am. Stat. Assoc. **55**(292), 698–707 (1960)

Genest, C.: Frank's family of bivariate distributions. Biometrika **74**(3), 549–555 (1987)

Joe, H., Kurowicka, D.: Dependence Modeling: Vine Copula Handbook. World Scientific (2011)

Li, D., Peng, L.: Goodness-of-fit test for tail copulas modeled by elliptical copulas. Statist. Probab. Lett. **79**(8), 1097–1104 (2009)

Sklar, M.: Fonctions de répartition à n dimensions et leursmarges. Annales de l'ISUP **8**(3), 229–231 (1959)

Nelsen, R.B.: An Introduction to Copulas. Springer New York, New York, NY (1999)

Goovaerts, P.: Geostatistics for Natural Resources Evaluation. Oxford University Press on Demand (1997)

Mine Planning and Production

Mass Production Technologies for Underground Coal Mining in India: Status, Challenges, and Prospects

Arvind Kumar Mishra(✉)

CSIR-Central Institute of Mining and Fuel Research, Barwa Road, Dhanbad 826001, Jharkhand, India
director@cimfr.nic.in

Abstract. The current trend in coal production in India shows that underground mining contributed less than 5–8% of the total coal produced. On the other side, the contribution of opencast mining to coal production is at its peak. This trend is not sustainable in the Indian scenario due to environmental issues, coal quality problems, and socio-economic stresses due to opencast mining. The solution lies in the adoption of Mass Production Technology (MPT) in underground mining which can compete with opencast mining in terms of OMS and production rate. Furthermore, Coal India Limited (CIL) has also launched a mission to upscale the coal production from underground mines to 100 million tonnes by 2027–28. Therefore, to advocate MPT and its diffusion in the Indian mining industry, this keynote discusses the existing mining methods and potential MPTs for exploiting deep-seated coal deposits in India. The MPT for underground mining has been defined and the eligible technologies are presented. The focus has been made on Longwall Technology which is one of the finest candidate technologies for MPT in India. Furthermore, the challenges in adoption in the Indian coal mining industry have been elaborated with real cases. Finally, the prospects in R&D, testing, and policymaking for smooth adoption in Indian coal mining have been deliberated. In sum, this keynote can enrich the knowledge of practising mining engineers, mine planners, and policymakers in adopting MPT for underground coal mining in future.

Keywords: Mass production technology · Underground coal mining · Longwall technology · Continuous miner technology · High wall mining · deep mining

1 Introduction

Coal plays a major role in providing affordable electricity and in building infrastructures through its use in steel and cement industries worldwide. In India, coal accounts for 55% of the country's energy needs and will continue to occupy centre stage due to rising energy usage and limited options for other affordable energy sources. On the other side, the growing awareness about the negative effects of coal usage has prompted the government to put stringent environmental regulations in place. Thus, the coal industry is striving hard for increasing production and minimizing negative environmental impacts at affordable costs.

A. Sinha et al. (Eds.): AMC 2023, SPEES, pp. 53–72, 2023.
https://doi.org/10.1007/978-3-031-46966-4_5

The worldwide trend, as shown in Fig. 1, shows that the majority of coal is produced in major countries using underground mining methods. As opposed to the global trend, India is producing about 95% of coal from opencast mining methods [1]. Figure 2 illustrates that coal production from opencast mining is continuing to grow during the last ten years and that from underground mining is declining [2]. This is not the sustainable scenario for India given the limited land resources and high population density in surrounding areas. The hidden costs of rehabilitation & resettlements of the displaced population have reached an unaffordable level.

The opencast mining method of extracting coal has several unwanted environmental implications. This method destroys the landscapes, flora, fauna, soil balance, etc. permanently. The mining operations cause high air, sound, and water pollution in surrounding areas. Blast vibration, air pollution, and water problems cause social distresses in surrounding areas. Further, opencast mining creates coal quality and production issues. The blasting lowers the coal quality due to high dilutions with overburdened rocks and alterations of moisture contents with open storage and transportation. Additionally, stoppage of production in monsoon strains the mining economy.

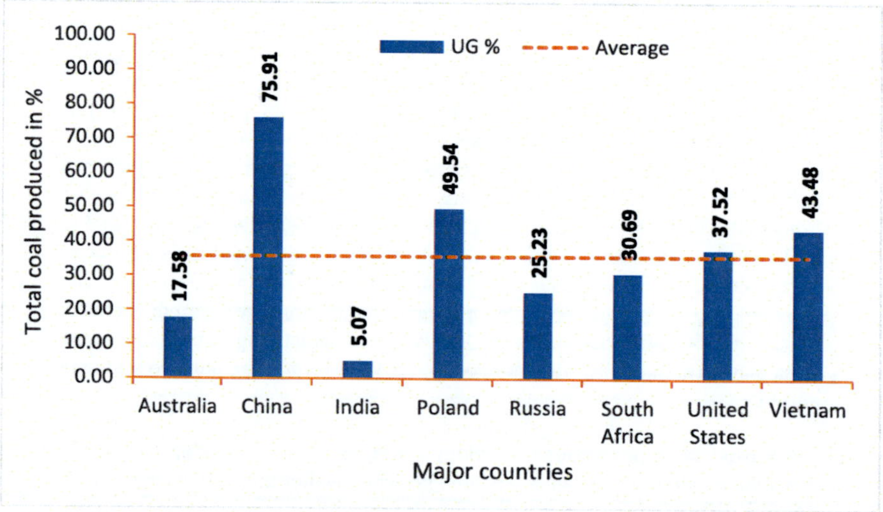

Fig. 1. Share of underground mining in major coal-producing countries [1]

Apart from the negative social and environmental impacts, opencast mining is also reaching the ultimate pit limit of its operation due to the depletion of shallower coal deposits. The transition of mining from opencast to underground is foreseeable and will be intensified in the next 5–10 years from now. Additionally, as per the Indian coal inventory data, about 40% of coal deposits below 300 m of depth are amenable to being extracted using underground mining methods [3], as shown in Fig. 3. Keeping these reasons in view, Coal India Limited (CIL) has unveiled a roadmap for achieving 100 MT production by the financial year 2027–28 in the "Conference on Underground Coal

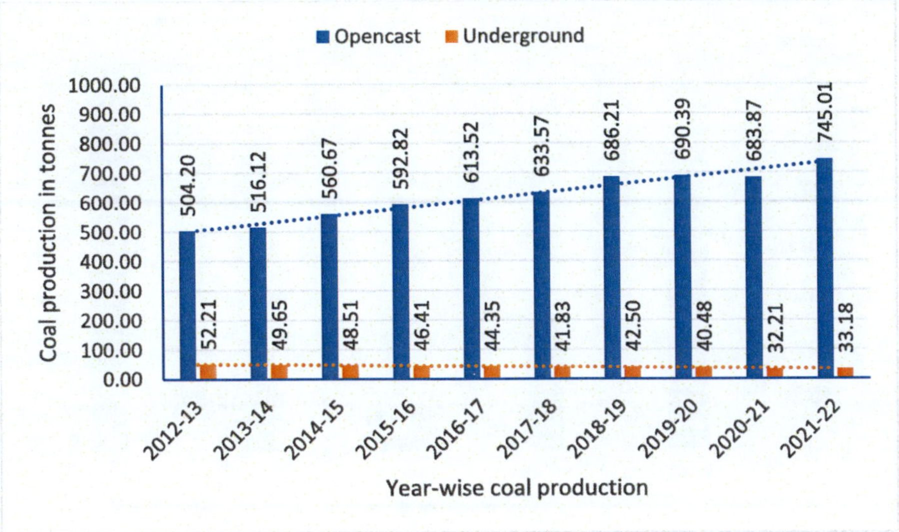

Fig. 2. Underground coal production in India [2]

Mining – Way Forward for a sustainable energy security" held on June 06, 2023, in New Delhi.

Despite the obvious necessity, underground mining technologies have not flourished in line with opencast mining technologies. The reasons may be attributed to the economics and safety of underground mining methods which are inferior to those of opencast mining. The Outputs per Man Shift (OMS) achieved in opencast coal mines are in the range of 10–12 as opposed to very low OMS in the range of 0.70–1.3 in underground coal mines. A majority of coal mines use conventional mining methods of cyclic development and depillaring using SDL/LHD technologies.

To compete with the economics and safety of opencast mining, mass production technologies for underground coal mining must be adopted in the Indian coal industry. The term, mass production, which came from the manufacturing sector, refers to a system of manufacturing that uses a series of standardized processes to create a large number of an item in a relatively short time at lower costs. Here, in coal production, Mass Production Technology (MPT) may refer to the mechanised mining technology capable to produce 1 million tonnes or more coal annually with high efficiency and productivity.

The major candidates of MPT in underground coal mining are longwall, high wall, and continuous miner. However, these methods are mostly applicable to medium-thick (2 m–6 m) coal seams. The Indian coal deposits are of varied nature like thick, thin, inclined, etc. Especially, there is no option other than longwall technology to extract coal deposits below 500 m depth. Therefore, intensive research and development (R&D) for mine design, methods of extraction, machinery customization, ground control, operation safety, regulatory compliance, etc. are required for successful implementations.

This Keynote presents the status of MPTs in Indian coal mines and the case examples of their implementations with the involvement of CSIR-Central Institute of Mining and

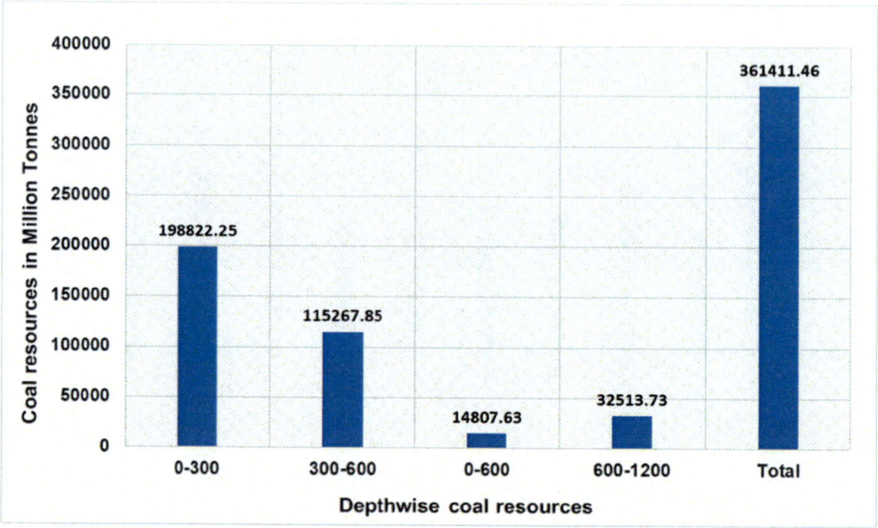

Fig. 3. Depthwise coal resources in India [3]

Fuel Research (CSIR-CIMFR). The challenges of implementing MPTs in underground coal mines with complex geological and mining conditions are explained. To negotiate difficult geological and mining conditions and meet the CIL target of producing 100 MT coal by 2027–28, this Keynote introduces a roadmap for implementing MPTs in Indian underground coal mines.

2 Status of Mass Production Technologies in Indian Coal Mines

Mass production technologies are widely used in Indian underground coal mines with varying degrees of meeting target productivity. Though a few coal mines deploy mass production technology that can truly compete with O/C mining technology in terms of economy and safety minus negative environmental and social distress. CSIR-CIMFR has provided comprehensive R&D, testing, and technical support during the introduction/implementation of mass production technologies in Indian underground coal mines. Some of the successful cases of implementing mass production technologies in Indian underground coal mines are described for longwall, continuous miner, and high wall technologies.

2.1 Longwall Technologies

The longwall is the most suitable candidate for mass production technology given the fact that about 50% of the underground coal production comes from longwall technology and the productivity of longwall is comparable to that of opencast coal mines. India started its first longwall mining project in Moonidih in 1978. For successful implementation of longwall technology, CSIR-CIMFR has carried out several scientific studies to determine the capability of overlying strata, estimation of powered support capacity, prediction of main fall and periodic fall spans, stability assessment of chain/barrier pillars, design of support systems for gate roads, installation chambers and salvage chambers for Adriyala, Moonidih XVI seam, Moonidih XV seam, Jhanjra and proposed longwall projects at Muraidih (BCCL) and Shantikhani (SCCL).

2.1.1 Adriyala Longwall Project (ALP)

Adriyala Longwall Project (ALP) with a high-capacity longwall face has been commissioned by Singareni Colliery Companies Limited (SCCL) in 2014 with a state of the art equipment with, a production capacity of about 3MTPA[4]. The panel and face length of about 2300 m and 250 m respectively, by addressing all the strata problems encountered in the earlier longwall panels. ALP is a successful longwall mine working at more than 362 m of depth from the surface. Some of the works implemented in association with CSIR-CIMFR are shown in Fig. 4.

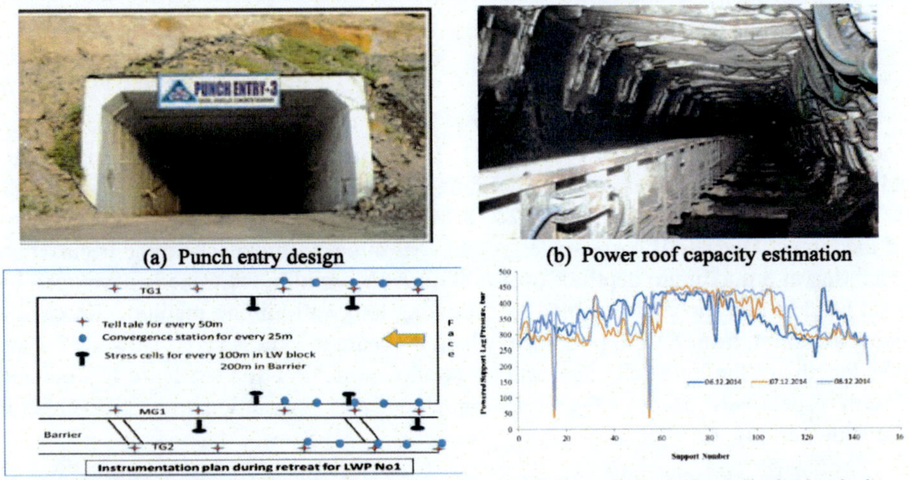

(a) Punch entry design

(b) Power roof capacity estimation

(c) Geotechnical Instrumentation planning

(d) Strata behaviour monitoring/analysis

Fig. 4. Adriyala Longwall Project works done in association with CSIR-CIMFR[4]

2.1.2 Jhanjra Project Colliery (JPC)

Jhanjra Project Colliery comes under Eastern Coal Fields Limited; a subsidiary of Coal India.

Limited is the pioneer in the field of Longwall Technology. Powered support long wall technology was introduced to achieve a minimum production of 9.4MT in 6 years with an average guaranteed production of at least 1.7MT for the R-VI seam of the Jhanjra Under ground block. M/s CODCO had signed the agreement with M/s ECL and took permission from ECL to assign the scientific investigation and its study to CIMFR. Some of the works implemented in association with CSIR-CIMFR are shown in Fig. 5.

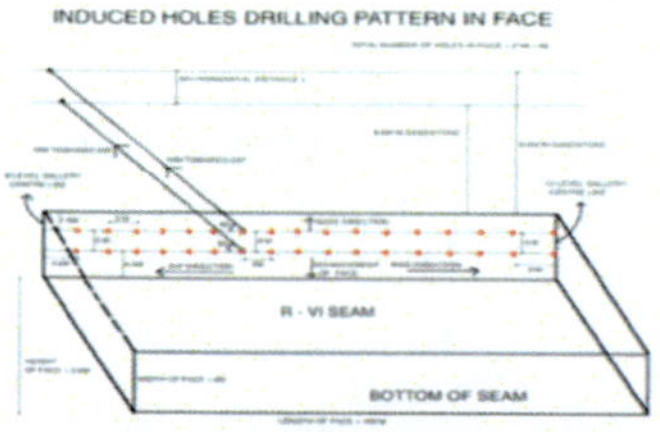

(a) Induced blasting design

(b) Power roof capacity estimation

(c) Geotechnical Instrumentation planning

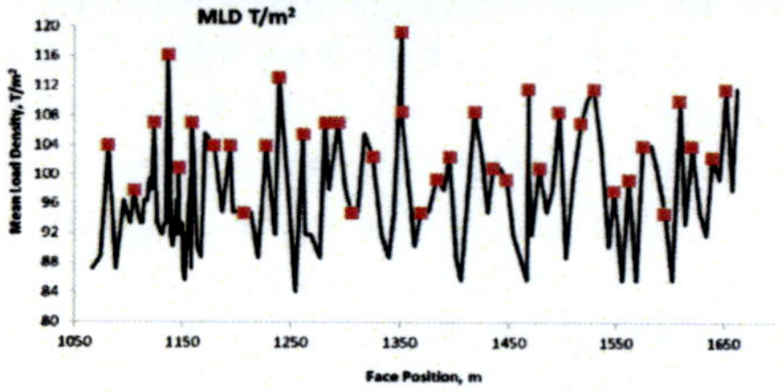

(d) Strata behaviour monitoring/analysis

Fig. 5. Jhanjra Longwall Project works done in association with CSIR-CIMFR

2.1.3 Moonidih Colliery

Moonidih Colliery is the first colliery in India to introduce the fully mechanized self-advancing powered support longwall mining technology way back in August 1978. With a total geological reserve of 1240.8 MT, Moonidih Colliery happens to be one of the largest Degree III gassy underground coal mines in India. The mine is currently operating at a maximum depth of over 600 m. At Moonidih, the mine is developed by road headers and the coal is mined by retreating longwall mining method with caving using a shearer, from XVI Top seam, in which the seam thickness varies between 2–3 m. The lengths of the longwall panel and longwall face in XVI Top seam are 1550 m and 150 m respectively. Some of the works implemented in association with CSIR-CIMFR are shown in Fig. 6.

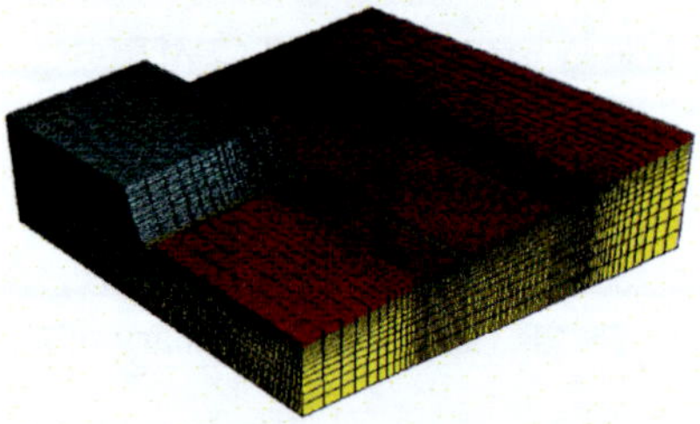

(a) Numerical study of salvage chamber/gate

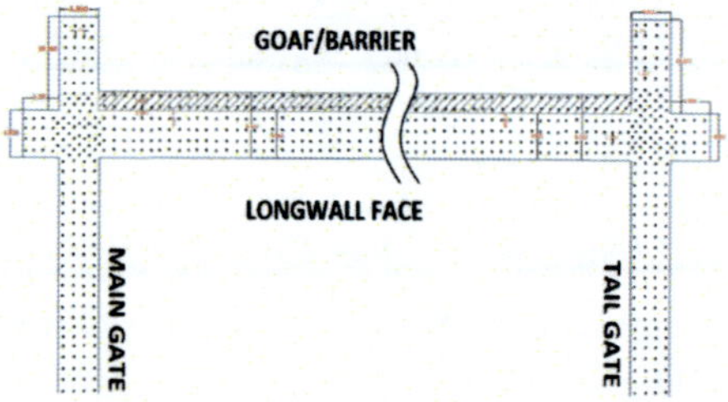

(b) Support system for installation chambers

Fig. 6. Moonidih longwall works done in association with CSIR-CIMFR

2.2 Continuos Miners Technology

Continuous miner technology is a highly flexible extraction method which can fit into the conventional development and depillaring of pillars. This technology is lower in acquisition and operational costs. Indian coal mines are highly suitable to deploy this technology. However, this technology is not economical below 500 m depth due to high costs and low recovery.

2.2.1 Churcha Mine (RO), SECL

Churcha mine (RO), Baikunthpur Area (SECL) has 7 identified coal seams and only workable seam if seam-V of average thickness 3.8 m with roof RMR (unadjusted) of 71. The mine is presently working at a depth of 300 to 420 m. Churcha mine is known for competent immediate roof strata and the presence of dolerite sill above the seam V. This makes a deterrent for the CM technology deployment. However, the technology has been successfully implemented and more than 12 panels have been extracted within the incubation period. Some of the works implemented in association with CSIR-CIMFR are shown in Fig. 7.

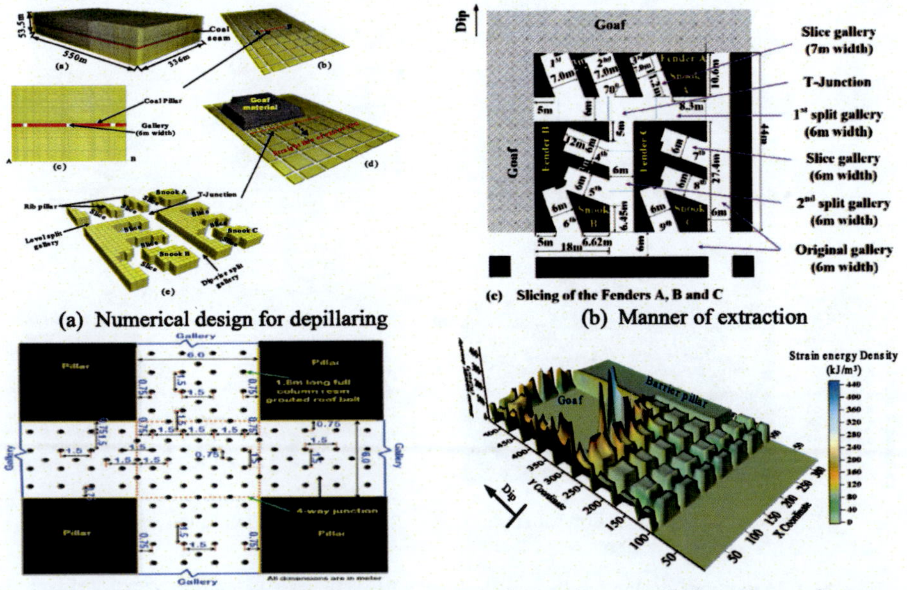

(a) Numerical design for depillaring

(b) Manner of extraction

(c) Support design for roof strata

(d) Coal bump prediction using strain energy

Fig. 7. CM method design for Churcha RO Mine in association with CSIR-CIMFR

2.2.2 Jhanjra Project Colliery

The Jhanjra project Colliery located in Raniganj Coalfield is the largest underground mine of Eastern Coalfields producing 3.4 million tonnes per annum. The Jhanjra block area of 15.10 km^2 is divided into six broad sectors, i.e., Sector-A, B, C, D, E and F. The RVII and RVIIA seams are being worked at 3&4 Incline. The parting thickness between the RVII and RVIIA seams is 31 m. The maximum and minimum depth of covers of RVII and RVIIA seams are 238 m, 134 m and 271 m,174 m respectively. The maximum thickness of the RVII and RVIIA seams at Sector F is 2.39 m and 2.21 m respectively. RVII and RVIIA are recommended to deploy Low Height CM technology. On the hand, the RV seam with a maximum thickness of 4.5 m with a mild gradient in Sector-C of MIC has been designed for normal CM technology. Several panels have been extracted successfully. Some of the works implemented in association with CSIR-CIMFR are shown in Fig. 8.

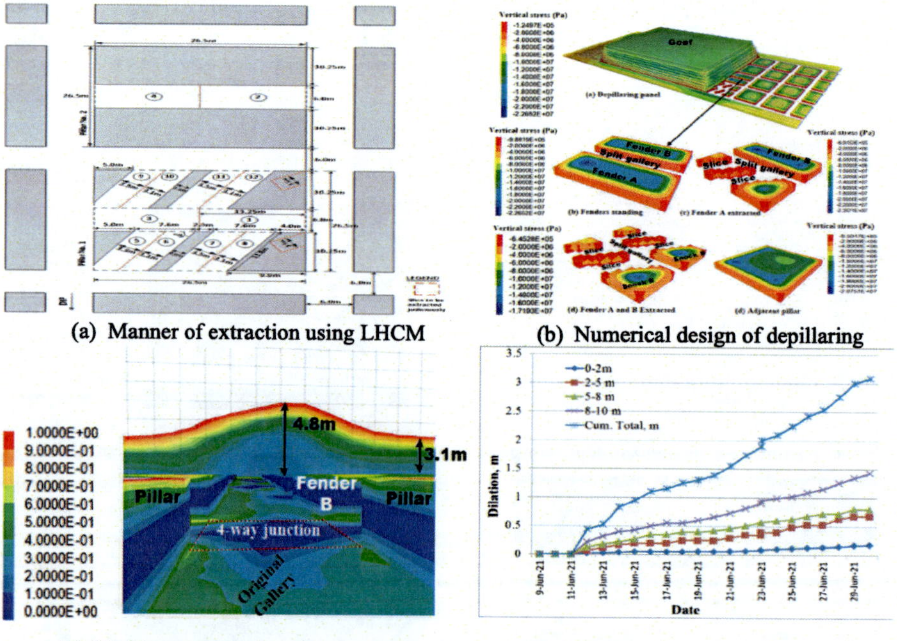

(a) Manner of extraction using LHCM

(b) Numerical design of depillaring

(c) Yield zone at the junction for support design

(d) Roof dilation monitoring in MIC

Fig. 8. CM methods design for Jhanjra Project Colliery in association with CSIR-CIMFR

2.3 Highwall Technology

Extension of an open pit mine is not possible when it reaches the ultimate pit limit. Coal is locked in the high walls of open pit mines. This locked-up coal can be extracted by Highwall Mining Technology. The first high wall mining operation in India was started in December 2010 at OCP-II of Ramagundam Area in SCCL. Other successfully running high wall projects are the Medapalli Opencast Project (MOCP) at SCCL and the Sharda Opencast Project at SECL Quarry SEB & AB at West Bokaro Colliery of Tata Steel Limited. Forthcoming high wall mining projects are Nimcha of Sripur Area in ECL, Narayankuri of Kunustoria Area in ECL, West Barakar quarry of Rajpura Colliery, and Mugma Area in ECL.

2.3.1 West Bokaro Colliery, Tata Steel Limited

Design of Highwall Mining at PCP-3 area of West Bokaro Mine of M/s Tata Steel Limited has been done by CSIR-CIMFR. Three coal seams viz. VII, VI, and V are occurring in descending order at PCP-3 of Quarry AB. All the seams are gently dipping, with 00–80 inclination within the proposed highway sites. The extraction by high wall mining is proposed to be carried out in coal seams VII, VI, and V from bottom to top. The range of thickness of VII is 6.0–8.1 m, VI is 3.0–6.0 m, and V is 2.4–3.3 m. Depths of cover of VII, VI, and V are 15–80 m, 27.3–98 m, and 33–108 m. The parting between seams VII and VI is 12.3–20 m and between VI and V seams is 5.7–10.6 m. Some of the works implemented in association with CSIR-CIMFR are shown in Fig. 9.

(a) Present conditions of Narayankuri OCP (b) Web Cuts in different blocks

(c) Access roadways above high-wall mining (d) Final scheme of extraction

Fig. 9. Highwall mining project works Best Bokaro Colliery

2.3.2 Narayankuri OCP, Kunustoria Area, ECL

At Narayankuri OCP of Mahabir (R) Colliery, the R-VIII B2 seam has been worked out. The decoaled area has been partly backfilled with internal overburden (OB) dump. The OCP had reached within 100 m of the nearby villages and could not be expanded further. So, the mining was stopped due to surface constraints and economic considerations. ECL has proposed to extract the R-VIII B2 seam and R-VII Top (R-VII T) seam within and beyond the Narayankuri OCP by Highwall Mining. Some of the works implemented in association with CSIR-CIMFR are shown in Fig. 10.

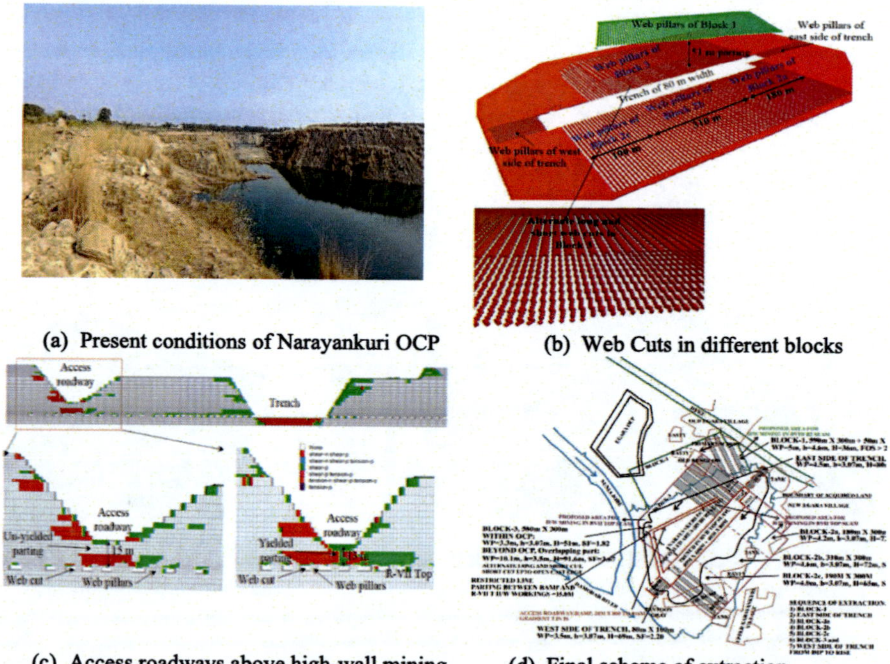

(a) Present conditions of Narayankuri OCP (b) Web Cuts in different blocks

(c) Access roadways above high-wall mining (d) Final scheme of extraction

Fig. 10. Design of Highwall methods at Narainkuri OCP, ECL by CSIR-CIMFR

3 Challenges of Implementing Mass Production Technologies

The recent intensification of mass production technology has resulted in a 3.02% increase in coal production from underground mines [3]. However, the Indian mining industry faces challenges of complex geological factors (depth, surface features, high gradient, etc.), nature of seam thickness, indigenisation of machinery, and regulatory compliances.

3.1 Increasing Depth

The coal deposits occur mainly in the bedded sedimentary rock mass and are found in tabular form with varying degrees of inclinations and thicknesses. The mechanical behaviour of bedded rock in the floor and the roof of coal workings are greatly influenced by the depth from the surface. When any mine openings are made in the underground rock mass, the pre-excavation stresses are redistributed due to gravitational loading, tectonics loads, pressure fluctuations, stress slackening, thermal effects around the openings, etc. The redistribution of stresses at higher depths causes several geo-mechanical issues because of varying geomechanical behaviours of coal-bearing rock mass in underground coal mines. Therefore, mine operators must understand and manage the following basic geomechanical issues for the successful extraction of coal deposits at greater depth. The Key geomechanical challenges in deep mines are as follows.

3.1.1 Rock Mass Classification

The rock mass classification seeks to quantify the properties of the rock mass that may influence its behaviour and to join these numerical values into an index for the rock mass. This index is termed rock mass rating and can be used in excavation stability considerations. Though several rock mass classification systems [5, 6] have been proposed worldwide, CMRI classification for rock mass rating (RMR) has been specifically developed to determine the quality of coal mine roofs in Indian coalfields [7]. The RMR considered layer thickness, uniaxial compressive strength, rock weather ability, groundwater seepage, and structural features for rating.

However, there is no rating for the stresses due to the depth of cover. Rather, they have adjustments for in-situ stresses which may lead to over and under-design of mining structures. Therefore, the RMR must be upgraded by including the depth of cover in its rating system because it is widely used in predicting the stand-up times of exposed spans, support design, cavability, fragmentation, etc.

3.1.2 In-Situ Stresses

At greater depths, in-situ stresses are an important parameter to design stable excavation. These in-situ stresses depend on the Poisson ratio, lateral rock displacement, curvature of the earth's crust, tectonics, rock properties, topography and geological features. Two predominant normal stresses acting on the sub-surface point are given by the following two equations [8, 9];

$$\sigma_v = 0.027H \quad \text{(in MPa)} \tag{1}$$

$$k = (\sigma_H + \sigma_h)/2\sigma_v \tag{2}$$

where, σ_v, σ_H, σ_h, k and H are cover stress, major lateral stress, minor lateral stress, the ratio of the lateral to cover stresses, and depth from the surface. Figure 11 shows the variations of σ_v and k with a depth of mine workings.

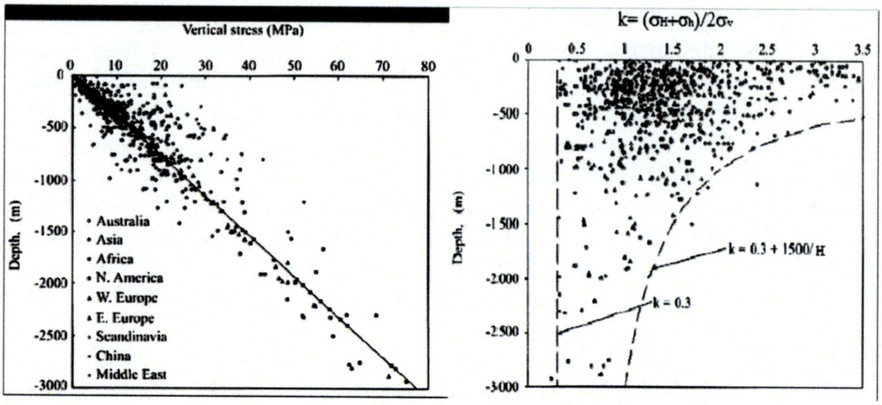

Fig. 11. Depth vs. σ_V and k worldwide [8]

3.1.3 Rockmass Behaviour at Higher Depth

The determination of the compressive strength, tensile strength, elastic modulus etc. of rock mass is a complex problem in designing underground mines. These properties are required to determine the ground convergence around mine excavations, assessment of pillar strength, and the degree of sporadic subsidence. Lab strength of coal or rock expressed by uniaxial compressive strength (UCS) may not be applied directly to the mine structures such as pillars because the corner stresses are uniaxial, the side's biaxial and triaxial at the centre of the pillar. The in-situ strengths are usually less than the UCS due to the presence of geological discontinuity.

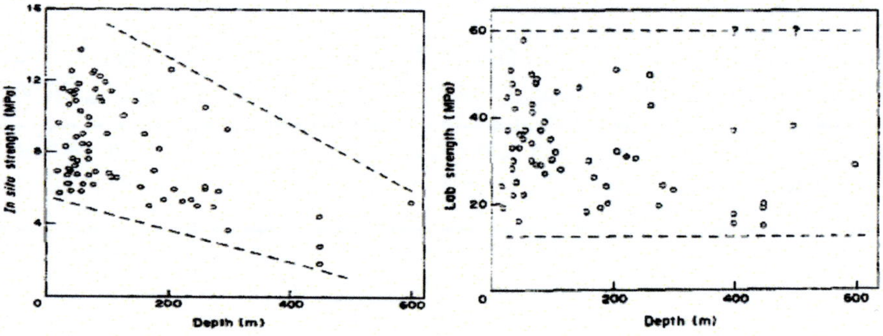

Fig. 12. Variations of in-situ and lab strengths of the tests with depth [10]

This has been confirmed by CIMFR tests, which have conducted extensive in-situ strength tests on 30 cm coal cubes using Schmidt hammers and their corresponding laboratory strength of 2.5 cm cubes [10]. Figure 12 illustrates that the in-situ and lab strengths of coal vary with depth. This variation was observed because the zone of failed coal in a pillar increases with depth, while the lab sample is unaffected by this effect due to its smaller size.

3.1.4 Control of Rock Failures

In deep mines, rock failures around excavation are inevitable due to high stresses. Therefore, the key geomechanical challenge is to control such rock failures during the excavations. This is realized by several types of external supports and rock reinforcement using rock bolting. As discussed in the previous section, deep excavation is constantly prone to dynamic failures of the exposed rock mass. Therefore, support systems must be able to clamp the broken rock integrated, activate frictional resistance in the fractured region, restrict deformation after failure, and yield under high stress and seismic loading environments [11]. An integrated support system consisting of bolts/cables, wire mesh, and steel-rope lacing meets the functional support requirements and can be applied at greater depth. The cone-bolts/D-bolts are yielding bolts and perform well under high stress and rock-burst conditions [12].

3.2 Natures of Seam Thickness

Extraction of thick seams & very thick seam is difficult in comparison to seams up to a thickness of 3–5 m. The concept of a thick seam varies from country to country, the basis for the lower limit of a thick seam being the thickness up to which a seam could be extracted in one lift (pass) with the available equipment and technology [13]. In India, over 60% of coal reserves are thick seams, and a considerable % of coking coal belongs to thick seams. Some thick seam is nearly 30 m thick and one exceptionally thick seam in Singrauli Coalfield is 162 m. Some thick coal seams are given in Table 1 and methods of extracting thick coal seams are given in Table 2.

Table 1. Some of the prominent thick coal seams in India

Jharia coalfields	Seam IX and X, Sudamdih colliery, and Perbelia colliery
Raniganj coalfields	Jambad and Poidih
Singareni collieries	King seam, queen seam, GDK 9,10(Ramagundam), etc
Chinakuri colliery	Disergh seam, ECL
GIDI A mines	KranpuraCoalfields,CCL, Jharkhand
Tipong mines	Assam

Thick seams in India range from flat to steeply inclined, vary in thickness and rate of a dip in the same area, and are, in some places, contiguous and interlaced with numerous stone or inferior coal bands. These thick seams are much more difficult to work on than seams of moderate thickness. The main difficulties include poor recovery, spontaneous combustion, difficult roof control, mechanization, low productivity, etc.

3.3 Mine Mechanization

Mechanization in coal mines for mass production faces several hindrances such as restricted regulations, mismatch of geotechnical factors with machine specification, foreign suppliers, high breakdown, inefficient maintenance, etc. As per the Global Energy

Table 2. Classification of thick seam mining methods concerning the seam thickness [14]

Broad methodology	Extraction	Thickness	Mining methods
Longwall	Single lift	<7.5 m	Sublevel caving
		<9–11 m	Artificial roofing based on sublevel caving
	Multi-lift	<15 m	Stowing in ascending order
Bord & Pillar	Single lift	7–12 m	Blasting gallery
		5–7 m	Cable bolting
			Continuous mining with floor coaling
	Multi-lift	8–10 m	Simultaneous depillaring with stowing
		8–12.5 m	Simultaneous depillaring with underpinning

Monitor data, the Indian coal industry does not give a favour to deploying longwall mining system, as shown in Fig. 13. Though the deployment of continuous mining systems has picked up lately.

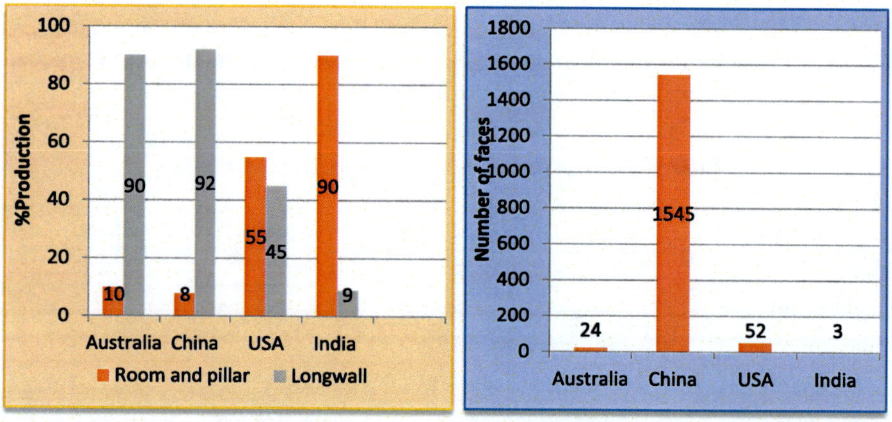

(a) Country-wise longwall production (b) Countrywise longwall faces

Fig. 13. Status of the longwall method worldwide

Further, one example of longwall adoption, as shown in Fig. 14, in India and China depicts the pace of mechanization even though longwall technology has been launched in Moonidih Colliery in India and Meiyukou Colliery in China during the same decade.

The major difficulties of intensive adoption of these mass production technologies are as follows:

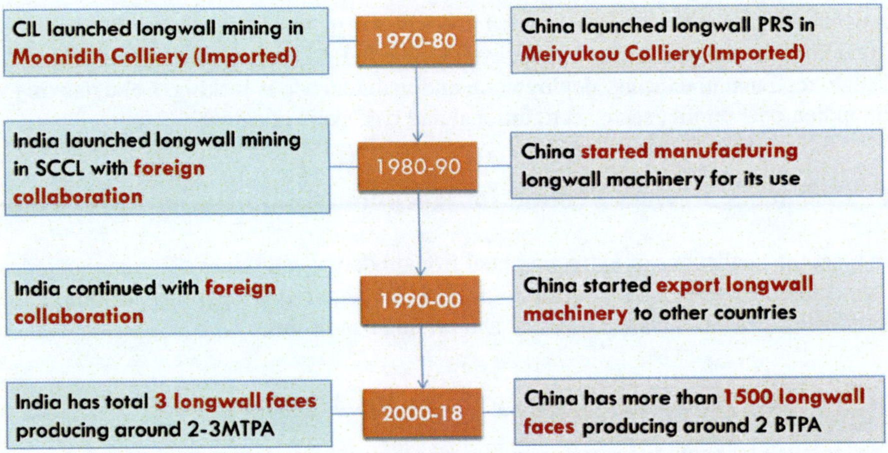

Fig. 14. Adoption of longwall technology in India and China

3.3.1 Lack of Indigenous Manufacturing, Efficient Maintenance, and Testing Facility

Almost the entire face machinery of longwall, high wall, and continuous miner technologies are supplied by foreign suppliers. High breakdown occurs due to a mismatch of geological factors and machine design with 20–28% downtime. The cost of maintenance is very high due to the absence of indigenous manufacturing of spare parts. The internal rate of return (IRR) of all the mines was found to be negative due to high costs and delayed supply. Production & productivity achieved in mines with this high-cost mass production technology is below the project targets. Furthermore, the testing facility to comply with DGMS safety standards does not exist in India and therefore, manufacturers can not deploy mass production technologies without test compliance.

3.3.2 Geomechanical Failures

The cases of longwall failures create a negative perception in the Indian mining industry. However, the reasons for these failures were technical and could be avoided in the next project by including the lessons learnt from them. For example, longwall failures in Patherkhera faces (240T), GDK11 (425T), and Dhemo Main colliery, (370T) were mainly due to inadequate supports and leaking leg circuits. Also, the longwall failures in GDK9 incline(4 × 680 T) of SCCL in 1988, Churcha West (4 × 680 T) of SECL in 1990, and Khottadih (4 × 470 T) of ECL in 1997 occurred due to the influence of intense loadings. The understanding of geomechanical loadings can be assessed and thus failures may be avoided.

3.3.3 Restricted Regulations and Ineffectual Policy

Mechanization in Indian coal mines has not been as intensive as opencast mining. The prime reason may be restricted regulations and ineffectual policies. For, example, the majority of underground coal mines were developed by Bord and Pillar (B&P) method in which the size galleries are restricted to 4.8 m × 3 m which is unfavourable to deploy larger continuous mining machinery. Mission mode R&D for supporting the successful design, test, manufacturing, deployment, and maintenance is lacking. Policymaking in the Indian coal mining sector is traditional and risk-avers in nature.

4 Prospects of Mass Production Technologies in Indian Coal Mines

To meet the challenges of extracting coal from underground mines using mass production technology, the existing technologies may be required to upgrade, customize, and modified as per the Indian geological and mining conditions.

4.1 Potential Technologies for Deep Underground Mining

The longwall technology is time tested method to coal extract deep-seated coal deposits. This technology can be customised to extract coal deposits of varying geological conditions. On the other hand, conventional CM technology in Bord and Pillar mining may not be economical to extract coal deposits below 500 m depth. Also, this technology can not extract coal seams of a thickness of more than 6 m at the moment. Therefore, these two technologies can be customized to suit Indian geological and mining factors.

4.1.1 Customization of Longwall Technology

The longwall technology includes appropriate mining structures, face machinery, transportation system, support system, ventilation network, and many more elements [15]. Mining companies, research agencies, manufacturers, suppliers, etc. should work closely to customize the longwall technology to extract coal from deep, thick, thin, inclined, fractured, etc. coal seams economically and safely. Table 3 provides the summary of longwall variations to meet the various challenges of Indian geological conditions.

Table 3. Variation of Longwall technologies

S.N.	Geological/mining factors	Potential methods
1	Deep-seated coal seams	• Power Support Longwall (PSLW) • Pillarless Longwall with roof cut short-arm beam theory (RSBT) [16]
2	Thick seams (up to 20 m)	• Longwall Top Coal Caving (LTCC) • Split-level Gate roads (LMSG) [17]
3	Developed Pillars at a higher depth	Power Supported Shortwall (PSSW)
4	Thin coal seams (around 2 m)	Low Height Longwall (LHLW)
5	Inclined coal seam (Up to 18°)	Inclined Power Support Longwall (IPSLW)

4.1.2 Continuous Miner for Deep Coal Extraction

Though extraction of developed coal pillars is possible up to a depth of 600 m, a USA's NIOSH study advises the mine operator to not split the pillar during extraction below the depth of 300 m. Some of the prominent methods, such as the Wongawilli method, Yield pillar techniques, etc. are promising methods that can be applied to extract coal at greater depth. Table 4 provides the innovative CM method to extract coal deposits with high recovery.

Table 4. Variation of CM technologies

S.N.	Geological/mining factors	Potential methods
1	Mass production from developed pillars	• CM method
2	Deep-seated coal seams	• Wongawilli method • Yield pillar techniques
3	Thick seams (up to 10 m)	• Multilift slicing method
4	Thin coal seams (around 2–3 m)	Low Height CM (LHCM)

4.2 Indigenous Manufacturing of Mining Machinery

Imported machinery operating in Indian coal mines often has high breakdown records and thus the acquisition and maintenance costs are too high to operate mines economically. The solution lies in the revival of indigenous machine design, testing, manufacturing, and maintenance of mass production technologies.

4.2.1 Designing, Prototyping, and Testing of High-Capacity Longwall Supports

Power roof supports are the most significant component of longwall technology constituting 70% of the total costs. Some of the companies are capable to design and develop prototypes of hydraulic cylinders and other parts of the powered roof support system. For example, CSIR-CIMFR and CSIR-CMERI have attempted to design two-legged powered roof support of high capacity, as shown in Fig. 15. On the other hand, Andhra Pradesh Heavy Machinery & Engineering Limited (APHMEL) has developed 4 × 800T LW PRS under a coal S&T project and tested it at M/s Jessops & Co, Kolkata [18]. However, higher longwall PRS of 1100 tonne capacity is used and no testing facility exists in the country to check compliance with DGMS regulation, ICIS 001/1991 (Indian standard).

Therefore, the need of the hour is to encourage the development of high-capacity PRS through R&D funding. Also, the government should support to establish of a testing facility of at least 2000T capacity for structural crush test of high capacity two-legged powered roof support system for compliance with Indian (ICIS 001/1991), European (EN 1804-2:2001 + A1: 2010), and Chinese standard (MT 312-2000, GB 25974.1-2010, MT 550-1996). These efforts will spur innovation in customizing longwall technology to meet the challenges of Indian coal geological and mining variations.

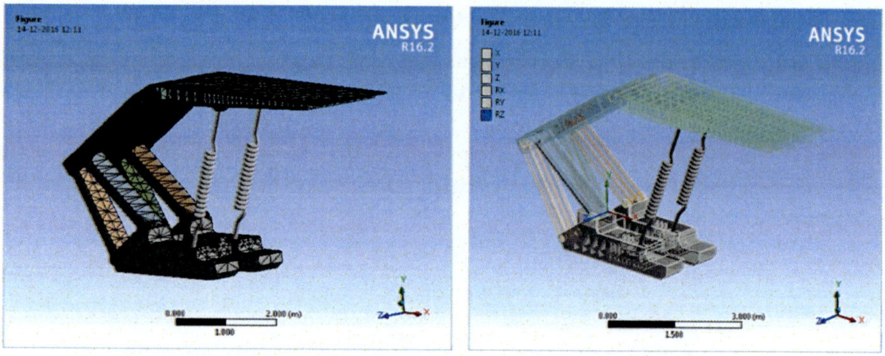

Fig. 15. Power roof support design in collaboration with CSIR-CIMFR and CSIR-CMERI

4.2.2 Designing, Prototyping, and Testing of Specialized Rock Reinforcement Systems

The transition from shallow depth to higher depth will require reliable ground control for safe and productive mass production technology. The reinforcement system using conventional rock bolt assembly may also be needed to be upgraded to tackle dynamic stresses due to rapid excavation at deeper horizons. In this direction, CSIR-CIMFR has initiated several in-house and focused-based research in the design and development of innovative rock bolt components that can solve the myriad problems faced by the industry. One such attempt has been made for designing and development of innovative coupler that can connect two standard rebars to make a longer rock bolt without losing the stiffness and anchorage capacity. Also, the lower bolt can be recovered. The system has been designed, developed, and tested, as shown in Fig. 16 [19].

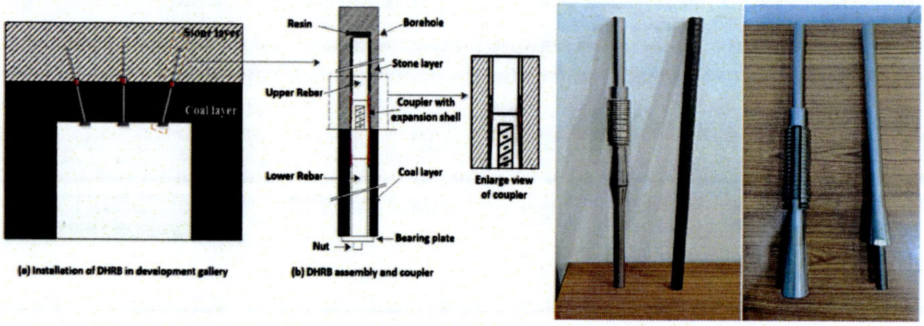

Fig. 16. Rock bolt coupler concepts and 3D printed prototype [19]

This innovation may benefit the extraction of coal in thin seams and also thick seams without procuring any new type of rock bolts. Therefore, the Indian mining industry should promote the design, development, and testing of innovative types of rock reinforcement systems to meet the various challenges of ground control risks in mass production technology.

4.3 R&D Collaboration for Mine Life Cycle

Indian coal producers should engage research organizations right from the planning stage to closure through other phases of mining for customization of mass production technology to suit the local geology and comply with all relevant regulations. Apart from it, the government should make a policy to support indigenous manufacturers with soft loans or financial incentives for manufacturing MPT machinery. Research institutes should conduct research in all areas of MPTs such as design, test, operation, salvage of MPT machinery and mine planning and design for MPT. An academic institute should prepare an exclusive course on MPT for creating a pool of highly skilled multidisciplinary technical manpower. The regulatory agency can facilitate the testing and deployment of MPT technology by framing regulatory provisions based on scientific studies of varying geological and mining conditions of Indian coal deposits.

5 Concluding Remarks

Underground mining of coal deposits using mass production technology can eliminate several issues Indian coal mining faces today such as the destruction of landscape, flora, fauna, water resources, etc. due to pervasive opencast coal mining. Further, the opening of more underground coal mines is expected due to the depletion of shallow coal deposits and a depth of 40% of the total coal inventory below 300 m from the surface. Therefore, Indian coal producers must look for MPTs for their underground coal mining to achieve their desired safety and productivity. There are several critical challenges to implementing MPTs in Indian geological and mining conditions such as increasing depth, nature of seam thickness, high-cost imported machinery, etc. These challenges can be solved if the Indian coal industry can customize the existing MPTs using research and development to suit local geological and mining factors. Indigenisation of mass production technologies should be encouraged through industry-research collaborations, indigenous manufacturing of machinery, and conducive regulatory frameworks. The suggestions made in this keynote can be implemented through R&D funding, manufacturing support, technology demonstrations, and skill development for underground mass production technology. In future, mass production technology will be the mainstream technology to realise sustainable mining goals in the Indian coal industry.

Acknowledgement. The author is sincerely thankful to Dr. Ranjan Kumar and other scientists of CSIR-Central Institute of Mining and Fuel Research Dhanbad for their contributions in preparing this keynote. The views expressed in this paper are the view of the author and do not necessarily reflect those of the institute.

References

1. https://globalenergymonitor.org/projects/global-coal-mine-tracker/summary-tables/
2. Central Mine Planning and Design Institute: Inventory of Indian Coal Inventory. CMPDI, India (2022)
3. Coal Controller Organization: Provisional Coal Statistics. CCO, India (2022)

4. Veera Reddy, B., Nageshwara Rao, K., Sridhar, K., Srikanth, K.: Strata control issues during development and retreat of high capacity longwall face at Adriyala Longwall Project, SCCL. In: DGMS Workshop on "Latest Trends in Strata Control in Longwall Mining with Special Reference to Strata Monitoring Techniques", 20 Jan 2017
5. Terzaghi, K. (1946). Rock Defects and Load on Tunnel Supports. Rock Tunneling with Steel Supports, RV Proctor and TL White
6. Bieniawski, Z.T.: Engineering Rock Mass Classifications: A Complete Manual for Engineers and Geologists in Mining, Civil, and Petroleum Engineering. John Wiley & Sons (1989)
7. Venkateswarlu, V., Ghose, A.K., Raju, N.M.: Rock-mass classification for the design of roof supports—a statistical evaluation of parameters. Min. Sci. Technol. **8**(2), 97–107 (1989)
8. Hoek, E., Brown, E.T.: Underground Excavations in Rock. CRC Press (1980)
9. Aydan, Ö., Kawamoto, T.: The general characteristics of the stress state in the various parts of the earth's crust. In: International Symposium Rock Stress, Kumamoto, pp. 369–373 (1997)
10. Sheorey, P.R.: A theory for In Situ stresses in isotropic and transverseley isotropic rock. Int. J. Rock Mech. Min. Sci. Geomech. Abstracts **31**(1), 23–34 (1994)
11. Wagner, H.: Deep mining: a rock engineering challenge. Rock Mech. Rock Eng. **52**(5), 1417–1446 (2019)
12. Li, C.C., Stjern, G., Myrvang, A.: A review on the performance of conventional and energy-absorbing rockbolts. J. Rock Mech. Geotech. Eng. **6**(4), 315–327 (2014)
13. Bhandari, T., Upadhyay, S., Dash, A., Dewangan, P., Pradhan, M.: Thick coal seam mining challenges and prospects. In: Proceedings of the Current Practices in Mining and Allied Industries, pp. 269–278 (2018)
14. Kumar, R., Singh, A.K., Mishra, A.K., Singh, R.: Underground mining of thick coal seams. Int. J. Min. Sci. Technol. **25**(6), 885–896 (2015)
15. Ralston, J.C., Hargrave, C.O., Dunn, M.T.: Longwall automation: trends, challenges and opportunities. Int. J. Min. Sci. Technol. **27**, 733–739 (2017)
16. Wang, P., Zhao, J., Chugh, Y., Wang, Z.: A novel longwall mining layout approach for extraction of deep coal deposits. Minerals **7**(4), 60 (2017). https://doi.org/10.3390/min7040060
17. Tao, Z., Song, Z., He, M., Meng, Z., Pang, S.: Principles of the roof cut short-arm beam mining method (110 methods) and its mining-induced stress distribution. Int. J. Min. Sci. Technol. **28**(3), 391–396 (2018)
18. https://www.aphmel.com/powered-roof-supports.html
19. Kumar, R., Mandal, P.K., Narayan, A., Das, A.J.: Evaluation of load transfer mechanism under axial loads in a novel coupler of dual height rock bolts. Int. J. Min. Sci. Technol. **31**(2), 225–232 (2021)

Sustainability of Underground Coal Mining in India Vis-a'-Vis Coal Mines Regulations, 2017

Mihir Choudhury(✉)

Phoenix-1105, Siddha Galaxia Ph.-II, Raigachi, Kolkata 700135, India
mihir271259@gmail.com

Abstract. Coal will continue to play dominant role in meeting the growing demand of energy in India. Since nationalization of coal mines, production from underground mines in India has drastically declined over the years while production from opencast mines has grown significantly. Proved and established mass production technologies for underground mining are highly productive and safe all over the world and it must find it's way in underground coal mines of the country, since surface mined coal can not meet the projected demand of coal in future. Further large opencast mines are known for it's nuisance for major environmental damages.

Health, safety and welfare of workmen engaged in belowground workings of coal mines is still an area of great concern in India. Although wages increased considerably, in nationalized coal sector but working environment is still not satisfactory. Age old practice of Bord & Pillar method creates more unsafe conditions in the mine in terms of risk of fire, explosion, inundation etc., requiring stringent statutory obligations, apart from inadequacy of ventilation and lag in roof support, compared to established practices. It's also a fact that cost of coal production by underground operations goes down drastically with quantum jump in output.

This paper looks at the prevailing mining practices in underground mines in India vis-a'-vis new conditions incorporated in Coal Mines Regulations,2017 thereby need for reorganization/planning of large underground coal mines with an aim for large output from a single mine.

1 Introduction

India's overall coal production has seen a quantum jump to 893 million tonnes in FY 2022–23 as compared to 728 MT in FY 2018–19 with a growth of about 22.6%. Priority has been fixed to enhance domestic coal production to reduce the dependence on substitutable coal imports. Thus, coal will continue to play dominant role for growth of India, due to it's availability in abundance. The country is endowed with large deposits of coal more than '360' billion tonnes and it ranks second in world coal production. Presently major share of coal production in India (more than 90%) comes from opencast mines contrary to other major coal producing countries which produce coal from underground mines e.g. neighboring China which ranks first in coal production, produces 90% coal by underground mining, South Africa produces 50% coal from underground. Easy to produce, shallow depth quarriable coal is fast depleting, therefore to sustain growth, country must look for large output from underground mines.

A. Sinha et al. (Eds.): AMC 2023, SPEES, pp. 73–84, 2023.
https://doi.org/10.1007/978-3-031-46966-4_6

Underground coal mining practice in India has remained old, mostly it's Bord and Pillar method of mining with deployment of Side discharge loaders (SDLs) and load haul dumps (LHDs) introduced four decades back which still constitutes significant amount of underground coal production these days. Proved and established technologies which were introduced successfully in developed countries five to six decades back is gradually gaining popularity here. Traditional Bord and Pillar system of mining with deployment of SDL/LHD is neither economic nor it can achieve large output, underground mines producing more than 1000 tonnes of coal per day with the said technology are limited few in the entire country. By introducing SDLs/LHDs, only loading at underground coal faces is mechanized but many operations e.g. drilling, blasting, fumes clearance, dressing, roof supporting, coal clearance etc. remained a bottleneck as it involves many cycles of time consuming labour intensive operations thus it resulted in little gain in productivity. Unlike India other countries have adopted continuous cutting and conveying of coal while deploying minimum manpower in underground operations and replaced above stated unproductive activities.

Further Coal Mines Regulations, 2017 requires stringent statutory obligations for safety and health of workmen engaged in underground coal mines. Since January, 2019 Govt. of India allowed employment of Women in underground mines from 6 AM to 7 PM in technical, supervisory and managerial work who are also to be safeguarded against occupational health, safety and security.

Difficult challenges are ahead when underground coal mining have to be carried out at deeper horizon, presently being operated at comparatively shallow depth. There is need for paradigm shift in planning of underground coal mines these days from low productive and unsafe Bord & Pillar technology to high productive mass production technology for quantum jump in production with due regards to safety and conservation of coal.

2 Present Status and Challenges Ahead

Underground coal mining commenced in British India by adopting labour intensive operations, lease areas in those days were small and hardly any mechanization was introduced for underground coal mining. Coal seams were mostly developed in honeycomb fashion in the past i.e. development of coal seams like a graph paper, driving parallel and perpendicular roadways in coal seams forming pillars to take the load of overlying strata, with utter disregard to conservation of coal as well as safety. Nationalization of all coal mines took place in India in early 70's except a few captive mines of large corporate houses. Even after nationalization, underground mines continued to be operated with very low output, without much of change in technology, except in some of the recently developed mines. Conception of mega mines with large output by proper planning did not take place to reorganise these mines. If country is to mine huge reserve of proven coal deposit at shallow as well as at greater depth, there must be a roadmap for underground coal mining based on the past experiences. Prior to conceiving & planning an underground coal mine, following important aspects needs careful consideration:-

2.1 Reserve

Country is endowed with more than 360 billion tonnes of coal deposit in different coalfields up to a depth of 1200 m. Even if opencast mining is considered up to a depth of about 300 m, less than forty percent of the 360 billion tonnes of coal deposit may be mined by such operation. Further, opencast mining causes major environmental damage in terms of degradation of land, air & water pollution, ground water instability etc. On the other hand, good quality coal deposits can only be mined by underground operations. Therefore, it is high time that underground coal mining must get due importance, for quantum jump in coal production since majority of the country's coal deposit are amenable for underground operations only.

2.2 Conservation

Conservation of coal did not receive much of attention in the past. Developing coal seams by Bord & Pillar method requires no statutory clearances in the country, however it was needed during final extraction. Therefore, many coal seams were developed in the past without any obligation for final extraction. In many occasion, these workings were abandoned resulting in waterlogging of such workings. In some of the mines, depillaring of the workings with caving of overlying strata commenced at shallow depth, without forming proper panels/artificial panels which lead to uncontrolled underground fire in such workings and ultimately all these workings had to be abandoned, presently many such workings remained as waterlogged workings or existing with active fire. As such, these abandoned workings led to the blocking of valuable coal reserves, not only in the coal seams already developed but also other coal seams lying underneath. There is no second opinion that underground coal mining in multi seam workings can best be done from top seam downwards bysystematic mining with due regards to conservation and safety.

Coal Mines Regulations (CMR) 2017 now requires preparation of "Safety Management Plan" by mine management based on identified hazards and risks as a part of submission of "Notice of opening" under Regulation (Reg.) 3 to the DGMS for commencement of mining operation.

Therefore, coal seams may no longer be considered as an abundantly available natural deposits in India which may be mined haphazardly to extract only a small part of the total reserve and leaving rest of the reserve unextracted for ever. It is high time to develop coal seams, with an aim for high percentage of final extraction by proper planning and by adoption of suitable technology.

2.3 Mine Boundary

Prior to nationalization of coal mines in India, underground coal mines were operated mostly with small lease boundaries for each pair of openings, may it be shaft or incline openings. Because of limitation of coal evacuation by rope haulages, dip/rise as well as strike extension of these mines were comparatively small resulting in the conception of small leasehold boundary. After nationalization of coal mines, extraction of coal continued with the same mines with small lease area, which restricted formation of large panels, particularly for present day Longwall mining. Average leasehold area of underground coal mines of the country was around 200 hectares, only a few years ago. In other developed countries, mines are being planned and operated with very large leasehold area. Today, Longwall panels are being planned for a size 3 to 3.5 km long with 300 to 400 m wide face length, which means area of a single longwall panel itself is to the extent of '90' to '140' hectares each. Even for considering formation of '10' such working panels in a coal seam, it would require 900 to 1400 hectares of lease area excluding areas that will be blocked in barriers and main roadways. Thus, present leasehold areas of underground coal mines are too small, to form large panels to get quantum jump in underground coal production.

As per Sub Regulation 3 of Reg.150 of CMR 2017 i.e. Danger of underground inundation clarifies that no working which has approached within a distance of 60 m of any other working whether in the same mine or in any adjoining mine, shall be extended further except with the prior permission in writing of the Chief Inspector and subject to such conditions as he may specify therein. As such, lot of coal reserve will remain blocked or lost in boundaries, forever when leasehold area of the mine is small. Amalgamation of neighboring mines to form large leasehold area is therefore a must for tapping virgin deep seated coal deposits in the country.

In future, mine boundary shall be determined based on major geological disturbances and important surface features only, for achieving large output from a single underground mine.

2.4 Layout

Mine Layout is the lifeline of a mine, every drivage planned for an underground mine is very important apart from suitable location of entries. Concept of mechanization to improve performance of underground mines is so far limited to face mechanization, coal transport by belt conveyors and material transport by rope haulages, apart from planning incline entries or shaft as a means of vertical transport. Mine layout for underground coal mines is an important area which has to be properly addressed in respect of entries, trunk roadways and formation of working panels. Many basic facilities like manriding, use of free steered vehicles etc. also needs consideration during planning of underground mines in the country. Investment for mine development i.e. capital investment for shafts, inclines, boreholes, major underground drivages etc. is not a very costly proposition as compared to total requirement capital for an underground coal project. Underground coal mines, so far have been planned with less than 10% capital investment for mine development. It is an established fact that proper mine planning plays a vital role for increasing production and productivity of underground mines as elaborated in the following paragraphs:

2.4.1 Outlets

Minimum of two openings are essential for any underground mine for intake of fresh air and exhaust of foul air from the underground workings, apart from the statutory provision of two separate means of ingress and egress to the persons employed belowground.

Entries are also meant for transport of coal, men & material and other service facilities. Both vertical shafts as well as Inclines are common in the country, as entries for underground coal mines. Concept of a pair of inclines and an airshaft as entries is now well established for large output from underground mines with belt conveyor as outbye transport of coal. So far incline entries have been considered for transport of coal and material only, thereby restricting it's size, use of the entries for installation of manriding system received little priority.

Now, Reg.93 of CMR 2017 specifies that if travelling distance from incline mouth or pit bottom exceeds 1 km or the travelling is arduous, approved manriding system is to be provided within one year of coming in to force of CMR 2017.

Further, Reg.81 of CMR 2017 i.e. man winding through shaft requires every cage or other means of conveyance of persons shall be provided with an effective means of communication with provision for Audio and visual, data and digital display including a broadcasting system.

In future therefore entries have to be planned with adequate width, height and communication facilities so that a pair of entries may be suitable for accommodating material and coal transport system as well as other service facilities like manriding, communication gazettes apart from adequacy of ventilation.

2.4.2 Trunk Roadways

Till recent past, there was no conception for main roadways in underground coal mines. Most of the underground coal mines had one or two main intake roadways and a main return roadway, to fulfill the need of coal and material transport, which was considered to be adequate to serve as main roadways irrespective of it's size and required number of such roadways. Thus, high resistance of main/trunk roadways caused inadequacy of ventilation in working areas in several occasion, as mine expanded. Trunk roadways are therefore required to be planned not only for coal and material transport, but also for adequacy of ventilation and transport of workmen to avoid arduous travel. Size of main roadways are important, depending on the purpose of it's use e.g. if the same roadway is used for transportation of coal by belt conveyor as well as for manriding at least 6 m wide and more than 3 m in height, should be the size of the main roadway. Moreover, with improved bolting technology strata control for large cross section roadways is no longer an issue, as apprehended earlier.

Further trunk roadways have to fulfil following provisions of new CMR,2017 e.g. Reg.92 requires that where haulage or conveyor system extends to a distance of more than 300 m from the shaft or mine entrance, telecommunication has to be provided.

Similarly, Reg. 93 of CMR,2017 requires travelling roadways are now to be provided with effective telecommunication and also adequate number of digital display boards and communication ports through which any important message or information can be easily transmitted or broadcasted to persons working belowground or passing thereby apart from the provision of manriding.

2.5 Out Bye Transport of Coal

In an underground mine, coal transport system cannot be considered in isolation of the method of wining at the coal face. While planning for face mechanization in underground workings, it is to be ensured that back up facilities like coal evacuation is compatible with face mechanization and there should not be interruptions in coal production due to transport bottleneck. Outbye transport of coal from underground coal face to surface in the past was by in seam rope haulage transport with ultimate evacuation through shafts/Inclines without using belt conveyors which resulted in very low output. There is no such limit for coal production when belt conveyors are used as a means for underground transportation of coal. Presently coal wining in many underground mines of the country is by blasting off the solid in conventional bord and pillar method of mining. However, blasted coal constitutes big coal lumps, when it's directly loaded on to belt conveyor, causes lot of spillage in the entire stretch of belt conveyor roadways. Engagement of considerable manpower, is therefore needed for cleaning of spillage coal and many a time jamming of belt conveyor occurs due to spillage coal, further wear and tear of belt conveyors is also a routine problem. In such situation, use of lump breaker prior to loading of coal on to belt conveyor in underground is definitely a better alternative for smooth functioning as well as for increased life of belt conveyors, which has been introduced off late only with mechanized longwall or continuous miner technology. Unless panel/district production is high, installing lump breaker for small quantum of district production may not be economically viable e.g. in case loading of blasted coal by deploying SDL/LHD at coal faces.

Further following statutory provisions have now been added for safe operation of Belt Conveyors in belowground workings under Reg. 97 of CMR,2017:

- in case number of belt conveyors are used in series sequential control and sequential interlock shall be provided.
- Pull cord switches or any other suitable system shall be installed so that conveyor can be stopped from any place along the entire length.
- The conveyor operator's cabin / place shall be provided with an effective means of communication along with broadcasting facilities through which the operator can communicate to any person present in the conveyor roadway. It shall have facilities of both way communication.
- Audio-visual pre-start warning alarm shall also be provided in the entire length of the roadway conveyor.

3 Important Safety and Health Issues

History of underground coal mining, reveals that workforce were put into most inhuman and dangerous working conditions in the country in older days, even minimum safety and health standards were absent. Fatal or serious accidents were common in underground mines and costs go far beyond those appeared in company's ledger book. Safe and congenial environment at workplace continues to be a problem even today, therefore the endeavor to bring down the hazards, by all means to an acceptable level must continue to reduce the sufferings and agony of the workmen. Following important issues are therefore relevant during planning of underground mines:

3.1 Adequacy of Ventilation

Production, productivity, health and safety in underground mines depend upon environmental conditions belowground, particularly at the place of work. Efficient ventilation system therefore plays a vital role in the performance of underground mines. The workings have to be kept dry at the same time temperature and humidity have to be kept at optimum level, to provide comfortable working conditions. Depth of underground coal mines in the country is increasing day by day. It is the need of the hour, to formulate efficient ventilation system for underground workings to create comfortable working conditions. Most of the existing underground mines, suffer from inadequacy of ventilating air in active working faces. Non erection of ventilation stoppings in time, leakages through stoppings and high resistance roadways are common, causing inadequacy of ventilation in underground active working faces, apart from generation of blasting fumes and dust. In many occasion a single narrow return roadway, whatever may be it's size, has been considered to be adequate as return airway. However, airway resistance increases rapidly with decreasing size. Most of the underground mines were developed in the past with the concept of one or two main intakes and one main return trunk roadway having inadequate size. In many mines, roadway size was restricted as a measure for better roof control but it causes tremendous rise in roadway resistance and underground mines starved due to inadequacy of ventilating air at the working faces. With improved bolting technology, it's a wrong concept to restrict roadway size only because of better roof control, rather aim should be for large cross section main roadways in large leasehold area for adequacy of ventilation. After all, there is need to relax our statutory limit for height and width of roadways in coal seams. Much of the ventilation problems of large underground mines may well be addressed by planning main roadways '20' m^2 in section or beyond, which will also greatly reduce the requirement of total number of such roadways in a mine, at the same time it will provide comfortable working condition. Ofcourse ventilation network analysis and simulation software will always be useful for overcoming ventilation problems of underground mines.

Further following new provisions have been added under Reg.158, 159 and 161 of CMR 2017 for safety and improved ventilation of the district being worked:

- In every ventilating district, two main intake airways shall be provided.
- In case of mine having problems of fire or spontaneous heating in coal seams of degree II or III, Chief Inspector may require construction of explosion proof ventilation stoppings between the main intake and the main return airways.
- Every auxiliary fan, shall have an airduct for conducting the air to and from the face or blind end which shall be maintained within 3 m of the face or blind heading.

3.2 Environment

Unlike drivages in rock, drivages in coal invites hazard due to liberation of noxious and inflammable gasses. Once coal seam is approached by the entries, spontaneous heating of exposed coal surface commences along with chances of liberation of inflammable gases. If the seam is developed in honeycomb fashion, very large surface area of the seam is exposed for spontaneous heating and situation is aggravated due to poor ventilation as it quickens the process of spontaneous heating. Blasting also causes generation of noxious gasses in underground atmosphere. Presently most of the underground coal mines are plagued with poor working environment as it is inherent with the traditional Bord and Pillar method of mining with gradual expansion of workings as well as due to addition of number of working districts.

Continuous telemonitoring of noxious and inflammable gasses, temperature etc. in underground workings will definitely be essential to keep the underground workings safe as well as for ensuring comfortable working conditions. At deeper horizon, with the expansion of workings, it will be difficult to provide comfortable working conditions, unless suitable high capacity fans are planned well in advance, air conditioning of underground workings will be the ultimate choice.

Reg.37 of CMR,2017 i.e. Duties and responsibilities of owner, reveals that Owners shall take all necessary measures to eliminate or minimize the risks to safety and health of persons employed in mines under their control. Also he shall ensure the monitoring, assessment and regular inspection of the working environment to identify the various hazards to which the workers maybe exposed;

Also Owner shall ensure adequate ventilation for all underground workings to which access is permitted;

Further, under Reg.170 of CMR,2017, Chief Inspector may, require in any mine, environmental monitoring devices to continuously record information regarding environmental conditions, to be installed belowground.

3.3 Fire

Unscientific mining in the past, such as selective mining of superior quality coal seams at shallow depth, under inadequate soil cover and poor coal recovery during extraction of in situ coal, was highly attributable to spontaneous heating and thereby occurrence of widespread fire, in various coalfields of the country. This has caused huge loss of coal and blocking of valuable coal reserves in the same seam and underlying seams. Existence of multiseam and multisection workings with multiple openings have intensified the fire and also led to it's spread at fast rate, engulfing a large area and quite a large number of seams. Following important precautionary measures are called for proper mine layout and planning of mine ventilation system:

Development of coal seams in panel system with size of the panel matching with incubation period and rate of extraction.

- Size and shape of the pillars should be such that it affords to long term stability.
- Longwall method of mining is best at greater depth.
- Provision of isolation stoppings around fire area.
- Study of pressure behaviour of sealed off area and pressure balancing if required.
- Peripheral ventilation using smaller capacity fans through large diameter boreholes.
- Inertisation of goaf/fire area by nitrogen flushing.
- Further Reg.140 of CMR,2017 also specifies that management of every mine shall establish a proper organisation for fire fighting in the mine by installing fire stations at surface and also on every main haulage roadway below ground at suitable places in the intake airway near the main shaft with adequate firefighting equipment kept in every such fire station.

- Reg.142 of CMR,2017 also specifies precautions to be taken when a fire exists in a mine that every stopping erected and the pillars containing such stoppings, shall be plastered with fire resistant or fire retardant sealant of adequate thickness using high pressure guns so as to completely fill up the cracks, cavities, crevices, joints, slips, fractures and cleats present in the working in order to completely seal off such area, such that it shall be kept maintained as leak proof.

3.4 Arduous Travel

Time spent by workmen to reach underground working places from surface has increased over the years due to extensive mine workings. The fatigue of workmen is directly related to the gradient and distance of the underground working places in the mine. Long arduous travel is common in several underground mines of the country, due to non existence of suitable man riding system. Introduction of efficient and safe manriding system is need of the day to reduce the unproductive time. It will not only provide effective working hours of the miners but it will also reduce absenteeism, thereby contributing to higher production and productivity of the underground mines. It is high time that future underground coal mines are planned only with suitable manriding system as has been clarified under Reg. 93 of CMR, 2017.

3.5 Surface Constraint

Concept of mining with small leases in the past has led to large human settlement around the mine entries, blocking vital coal reserve in various coalfields of the country, particularly in eastern India. Increase in population over the years in coal bearing areas, has led to the serious surface constraint for final extraction of coal seams. For maximizing extraction from underlying coal seams, surface constraint due to human settlement/villages etc. have to be vacated with proper rehabilitation packages in future. Infrastructures and residential colonies therefore have to be relocated in non coal bearing areas preferably or in an area where mining is not feasible. Expenditure incurred for proper rehabilitation packages, can be made offset by increasing largescale production from underground mines with introduction of mass production technology, as cost of production of coal from underground mines will go down drastically with quantum jump in output. Due to various issues related to possession/diversion of land, Coal India is also identifying mines for application of paste-fill technology for extraction of coal from underground mines as an economically viable alternative where acute surface constraint exists.

3.6 Training

Successful adoption of suitable high productive technology will not only depend on the introduction of mechanization alone, but also with the workmen involved with it. An efficient process can only be managed by trained, skilled and highly motivated employees to sustain good working practices. Therefore, training of workforce, for adoption of technology to enable increased production would be an essential input. Skilled and motivated workforce with proper training in technology as well as creation of improved environmental conditions at workplace, will definitely improve production and productivity of the underground coal mines in the country.

Moreover, Duties and Responsibilities of Owner under Regulation 37 of CMR,2017 now specifies that training of workers engaged in mines is the responsibility of owner. The owner shall ensure that adequate training and re-training programs and comprehensible instructions are provided for workers, at no cost to them, on safety and health matters as well as on the work assigned.

Further manager of every mine under Reg. 110 of 2017 shall, before introducing any machinery or new operation connected with his mine, frame and enforce "code of practice". The codes of practice shall be framed with due regard to the type, size and capacity of the machinery or operation in use and prevailing local conditions. Codes of practices shall provide for 'safe operating procedures' for the machinery or operation to which it relates to and management of every mine, shall hand over copies of such 'code of practices' to concerned officials and persons and ensure effective enforcement thereof.

4 Selection of Method of Mining

Major part of the country's coal deposit, particularly good quality coal seams are extractable by underground operations only, therefore it is high time underground coal mining must get due importance. Coal mining started in British India with Bord & Pillar method of underground mining in honeycomb fashion of development of coal seams. There was no concept of panel barriers in those days, entire operation was manual even coal cutting and out bye clearance of coal. Subsequently, rope haulages and winding through pits gained popularity, blasting was also introduced in due course of time in underground coal mines. After nationalization, basket loading of coal was gradually replaced by mechanizing with SDLs & LHDs for coalface loading and underground transport of coal by Belt conveyors, were also introduced particularly in the nationalized mines of western India. Bord and pillar method of coal mining is still the popular method of coal mining all over the country. Longwall method of mining was also introduced after nationalization of coal mines. Initially Longwall method could not achieve desired result mainly because of following reasons:

i) Inefficient and unmatched coal evacuation system.
ii) Non standard equipment and non availability of quality spares in time.
iii) Cavability problem of overlying massive sandstone roof.
iv) Difficult geo-mining conditions and wrong place of application.
v) Lack of infrastructural facility.
vi) Poor rate of gate road drivage thereby delay in preparation of replacement panels.

Subsequently Continuous miner technology was introduced successfully, with initial setback in Longwall method, considerable success was achieved by both the technologies in recent past.

Off late full scale mechanization of Indian underground mines by joint venture with overseas companies on risk/gain sharing concept has gained popularity with encouraging results particularly with the introduction of continuous miner technology. Consistent coal production on an average '1500' to '2,000' tonnes per day has already been achieved by a single set of Continuous miner depending on seam thickness. Similarly on an average

'2500' to '3000' tonnes per day consistent coal production from a single Longwall face in comparatively thin seam has already been achieved in the country without going for joint venture. At Jhanjra, 6000 tonnes per day coal production on sustained basis has already been achieved from a single longwall face.

There is immense scope for increasing production and productivity of underground mines in the country by adopting said mass production technology. Longwall method of underground coal mining is the best technique all over the world but geology of Indian coalfields prohibited in many occasion, for availability of large uninterrupted properties amenable for longwall mining. However, combination of both the technologies i.e. Continuous miner as well as Longwall technology is likely to be the best option for underground coal mining in India in future.

Identification of potential large coal bearing blocks, in case of existing mines amalgamation of neighboring mines to form large leases, so as to get large minable reserve of coal, is therefore a must. In seam mine boundary shall be avoided unless compelled by geology or important surface features.

Mine planning should be for at least a million tonns annual output if not more by deploying number of continuous minersor by deploying both Continuous miners as well as longwall equipment packages. Production and productivity of underground mines may also be enhanced, by adopting various business models namely 'Mine Developer cum Operator', 'Technology Provider cum Operator' and 'Risk/Gain sharing' with global participation, so as to bring complete system of mining.

5 Cost of Production

Considerable investments are required for extraction of coal from underground coal mines, particularly from deep seated coal deposits. Gestation period is long for exploration, underground mine planning, mine development as well as infrastructure development till the introduction of mechanization. All these requires advance planning and investment made in these respect are to be treated as asset for future.

Present average cost of coal production from underground mines in India with traditional Bord and Pillar method of mining, semi mechanized with SDLs/LHDs and outbye transport of coal by belt conveyor through inclines is more than Rs. 4500/-Per tonne of output and almost all underground mines being operated with this technology are in loss (only 200 to 300 tonnes of coal per day per district is being achieved by deploying SDLs/LHDs). While cost per tonne output for underground mines mechanized with continuous miner technology is less than the sales realization per tonne of coal in most of the mines thereby underground mines are in profit with this technology (1600 to 2000 tonnes of coal per day per district is being achieved with a single Continuous Miner package).

It is a well known fact that cost of coal production goes down drastically with increased output from underground mines. With ever increasing wage cost, adopting fully mechanized underground operations with Longwall/continuous miner technology, thereby enhancing large scale coal production will definitely bring down the cost of operation of underground mines in future.

6 Conclusion

Mine management has so far unable to address the basic problems of underground coal mining in conventional system e.g. adequacy of ventilation, upto date roof support apart from other relevant problems thereby creating unsafe working conditions conducive for occupational diseases. Underground mines have become extensive over the years, arduous travel in several mines thus is an usual phenomena. This added with problem of communication in underground workings and lack of infrastructural facility created poor production and productivity of these mines. Also, underground operations remained labor intensive while average age of workmen is increasing day by day. New provisions have been incorporated in CMR,2017 to address vital safety and health issues of the workmen engaged in prevailing conditions in the underground coal mines.

AS such, underground coal mines suffered from fresh input of quality manpower as well as adequate technological input. There is not much of change in the planning of underground coal mines, mines are still being planned with obsolete technologies in small leasehold area. Adequate attention has not been paid for reorganization of existing mines to liquidate deep seated properties.

Gestation period is long to develop an underground coal mine. It is high time that existing leases are amalgamated to form large property mainly based on geology and important surface features, so as to form large working panels and get surface right of the land with adequate resettlement & rehabilitation schemes. Liquidation of seams standing on pillar, may thereafter be done from top downwards by deploying continuous miners and subsequently these continuous miners can be used for drivage of main roadways/panels in underlying seams. Cost of coal production from underground mines will go down drastically with the increase in quantum of output. Consistent achievement of coal production to the tune of 1500 to 2000 tonnes per day by a single Continuous miner and also consistent production of 3000 to 6000 tonnes per day from a single Longwall face have already been achieved in the country. With these experience and with proper mine planning, underground coal mines in India will definitely be able to achieve more than million tonne annual output economically with Continuous miner/longwall technology from a single mine. There is immense scope of increasing production and productivity of underground mines in the country by engaging ‘Mine Developer cum Operator’ with global participation, so as to bring complete system for large output from underground coal mines.

Dump Slope Stability Rating and Risk Assessment in Opencast Coal Mines: A Case Study

Satya Prakash Sahu[1](✉), Pradeep Kumar Mohanty[1], and Singam Jayanthu[2]

[1] Mahanadi Coalfields Limited, Sambalpur, India
satya.prakash42@gmail.com
[2] National Institute of Technology, Rourkela, India

Abstract. As a part of this paper, detailed risk assessment (identification of hazards, determination of risk score and control measures) was carried out for dump failure in an opencast coal mine. Risk Assessment of dump failure revealed that the hazard i.e. death/injury of unauthorized persons entered near the toe of the OB dump has the highest risk score of 500 whereas the hazard i.e. sliding of external OB dump due to foundation failure has the lowest risk score of 50. Based on individual point ratings for each of the main factors affecting dump stability, the dump stability rating (DSR) was estimated and found to be 400–500 for the mine under consideration for study and the dump stability class was determined as Class II and corresponding category of failure hazard is Low.

Keywords: Hazard · Risk Score · Opencast Project · Dump Stability Rating · Triggered Action Response Plan

1 Introduction

Opencast coal mining methods play a key role in enhancing coal production level which in turn fulfils growth necessities of the country. In case of opencast coal mining, the first step is to remove the overburden economically to expose the underlying coal for excavation and later on needs to be handled safely. Increased level of overburden/waste rock production is also one of the major problems encountered by opencast coal mining as it causes environmental pollution (Agboola et al., 2020). The prime aim of construction of an overburden dump is to provide a stable and safe working surface for waste disposal (Dash, 2019). As opencast coal mining usually involves handling of huge volume of overburden, a portion of it can be backfilled in the de-coaled area of the mine and the remaining portion has to be dumped outside the active quarry in a convenient place considering three significant aspects such as safety, environment and economics.

Several parameters affect the stability of dump slopes (Ulusay et al., 2014; Rajak et al., 2020; Pile, 1991) as depicted in Fig. 1. Now-a-days the stability of overburden dumps is being considered as a foremost problem which needs serious attention in view of (a) occurrence of major accidents due to failure of overburden dumps or failure of a mass of dumped material along the slope of a dump i.e. dump slide, (b) increased volume

A. Sinha et al. (Eds.): AMC 2023, SPEES, pp. 85–98, 2023.
https://doi.org/10.1007/978-3-031-46966-4_7

of overburden which is required to be disposed of due to increase in size of the mine, and (c) scarcity of adequate space for dumping. Selection of the dump site constitutes the primary step for the OB dump design. The dumping site is restricted by factors i.e., open-pit size and location, topographical features of the considered area, surface and ground water flow, working boundaries, rehabilitation requirements and foundation conditions. The information gathered throughout mining exploration, i.e., topography, geology, hydrology and climate etc. be effectively used for the assessment and selection of a dumping site for waste rock generated from mining activities (Nayak, et al., 2020).

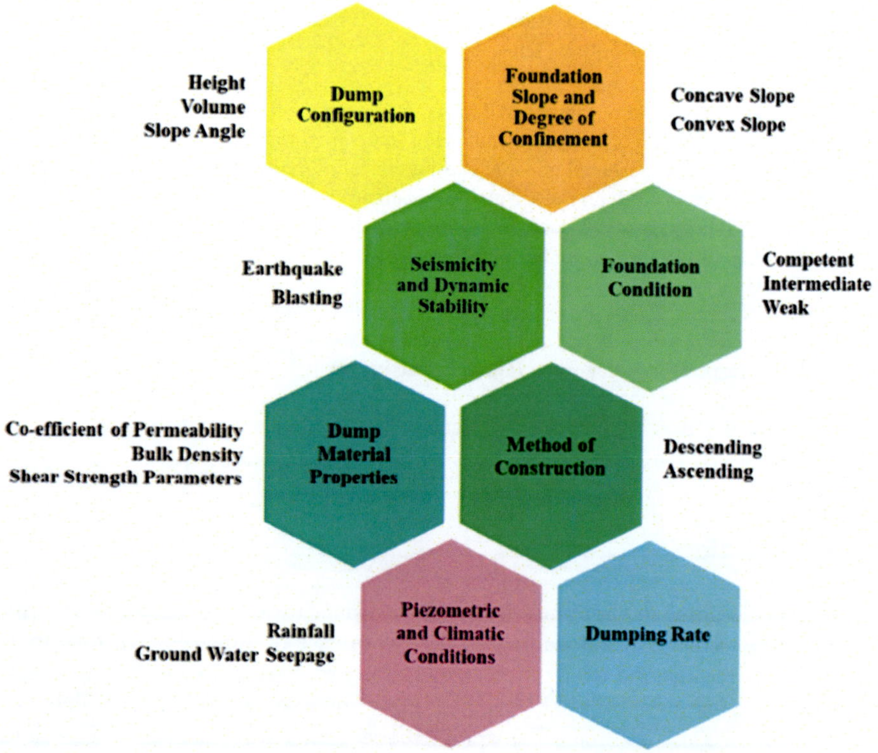

Fig. 1. Parameters affecting dump slope stability

In Indian scenario, majority of large opencast coal mines are facing overburden dump stability problems (dump slide) as it results in loss of production, extra stripping cost to remove land slide material, reforming overburden dump benches, diversion of haul roads, production delays and sometimes loss of human lives and heavy earth moving machinery (Rahul et al., 2015). Dump slides occur due to failure of the floor of the dump and the failure to maintain an appropriate height, volume and slope angle of dump due to non-availability of adequate space. A study undertaken at Western Coalfields Ltd. Shows that the factor of safety of the slope increased logarithmically as the dump angle was reduced while the dump height was remained constant (Kainthola et al., 2011). Another study undertaken at Wardha Valley Coalfields concluded that if an internal dump is made up of several benches rather than a single slope, then its stable increases (Verma et al. 2013). The mineralogical and structural composition of the material which forms the slope also plays a role in its stability (Behera et al., 2016).

Dump failure in opencast mines is being considered as the principal hazard (McQuillan et al., 2018). Opencast mines of India have witnesses several incidents of dump failures, including some fatal accidents. For example, the overburden dump of Kulda OCP, MCL failed on 10.08.2013 which caused trapping and fatalities of 13 persons (HPC Report, 2016). Another accident took place on 29.12.2016, during 2nd shift (2–10 PM) at Dahar Nangi Patch, Rajmahal OCP, Central Coalfields Limited (CCL) which claimed 23 lives (HPC Report, 2016).) Very limited studies had focused on complete risk assessment of the dump failure in opencast mines with in-depth details.

The paper presents the findings of dump slope failure risk assessment in an opencast coal mine. Identification of hazards, determination of risk score has been carried out for dump failure. It has also outlined the control measures of the mining establishment to eliminate or reduce the risks associated with dump failure to improve workplace safety. In addition to these, an overall dump stability rating (DSR) has been calculated for the mine under consideration. Planned actions for dump failure through triggered action response plan (TARP) has also been prepared.

2 Field Study

2.1 Geo-Mining Conditions

Mahanadi Coalfields Limited (MCL), Odisha achieved the highest ever overburden removal 245.96 m^3 in FY 2022–23, registering a growth of 19% as compared to the FY 2021–22 (MCL, 2023). Kulda geological block is located in the north-western part of lb valley coalfield in Sundergarh district in the state of Odisha. It is situated between latitudes 22°01′02″ to 22°03′03″ North and longitudes 83°43′28″ to 83°45′35″ East (Fig. 2). It is a potential block having mineable reserve of about 51.56 MT in present lease area with mine capacity of 21.0 MTY. The mine has 03 nos. of dumps [external dump: 01 no, internal dump: 02 nos.] (Fig. 3) and details of dumps are presented in Table 1.

3 Results and Discussion

3.1 Risk Assessment of Dump Failure

Risk assessment is a process that involves measurement of risk to determine priorities and to enable identification of appropriate level of risk treatment. Now-a-days it has been widely accepted that within mining industries; the various techniques of risk assessment contribute greatly toward improvements in the safety of complex operations as well as equipment. The risk of any hazard is dependent on likelihood and consequence. As per DGMS guidelines, values that can be used to calculate risk score for a hazard is furnished in Table 2.

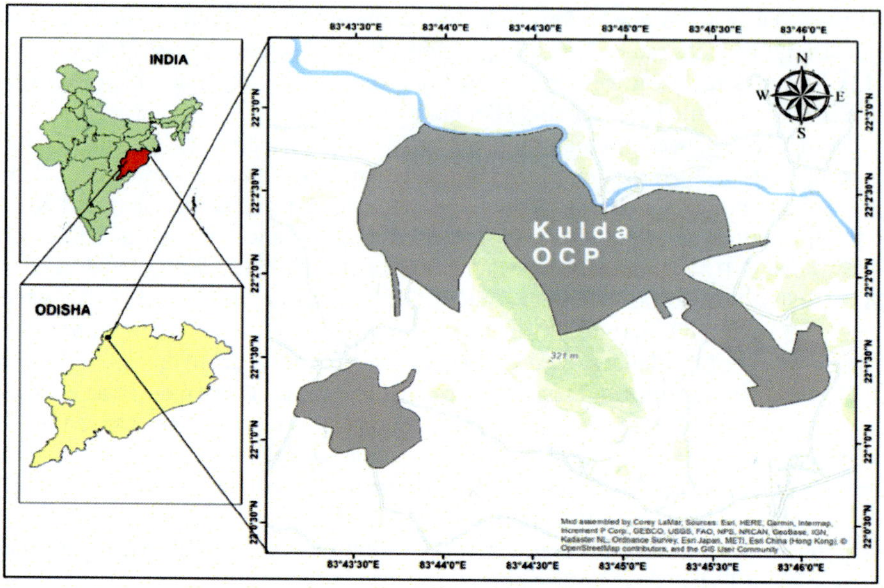

Fig. 2. Study area: Kulda OCP

Table 1. Dump Information of Kulda OCP

Project	Dump Name	Overall Dump Height (m)	Overall Dump Slope	No. of Benches	Height (m)	Bench Slope (degree)
Kulda OCP	External dump	82	1 in 3	5	8	14
	Internal dump 1	85	1 in 8	8	15	14
	Internal dump 2	172	1 in 2	8	29	18

Fig. 3. External dump, internal dump 1 and internal dump 2 of Kulda OCP

Table 2. Scale of consequence, exposure and probability

Scale of consequence		Scale for exposure		Scale for probability	
Several dead	5	Continuous	10	May well be expected	10
One dead	1	Frequent (daily)	5	Quite possible	7
Significant chance of fatality	0.3	Seldom (Weekly)	3	Unusual but possible	3
One permanent disability/ less chance of fatality	0.1	Unusual (Monthly)	2.5	Only remotely possible	2
Many lost time injuries	0.01	Occasional (Yearly)	2	Conceivable but unlikely	1
One lost time injury	0.001	Once in 5 years	1.5	Practically impossible	0.5
Small injury	0.0001	Once in 10 years	0.5	Virtually impossible	0.1
		Once in 100 years	0.02		

The values/scale of consequence, exposure and probability for a particular hazard must be reviewed and agreed by the risk assessment team collectively and consequently risk score is calculated as follows:

$$\text{Risk score} = \text{Likelihood} \times \text{Consequence} \tag{1}$$

Replacing likelihood by exposure and probability Eq. (1) can be re-written as.

$$\text{Risk score} = \text{Probability} \times \text{Exposure} \times \text{Consequence} \tag{2}$$

Further, risks are categorized according to the level of risk i.e., highest to lowest which enables one to establish risk which need immediate attention or may be managed in an on-going manner as given below:

- Risk > 200: High risk requiring immediate action
- Risk < 200 and > 20: Risk requiring management action
- Risk < 20: Low risk but to be reviewed

In Kulda OCP, altogether 07 safety hazards pertaining to dump slope event were identified as given in Table 3. For referring a particular hazard in the risk assessment process each hazard was given unique code such as SFH1 (Slope Failure Hazard 1), SFH2 (Slope Failure Hazard 2), etc. Risk assessment was conducted by assigning the values to consequence, exposure and probability and then risk score for each hazard was calculated using Eq. 2. The mechanism involved in each hazard along with a sample risk assessment sheet is given in Table 4. It has been observed that the hazard SFH6

(Death/injury of unauthorized persons entered near the toe of the OB dump) has the highest risk score of 500 followed by SFH7 (sliding of internal dump due to entry of water) with risk score of 350 whereas the hazard SFH1 (sliding of external OB dump due to foundation failure) has the lowest risk score of 50. Out of 07 identified hazards, 04 nos. of hazards require immediate action and 03 nos. Hazards require management action. The control measure being taken for each hazard is also provided in Table 5.

A fault tree analysis (FTA) is a technique for determining the root causes and the probability of a specified undesired event occurrence. FTA was performed for the accident due to dump failure in opencast mines based on knowledge of procedures, engineering technology and enforcement of statutory provisions. These 03 factors are considered as the contributory factors to accidents pertaining to dump failure (Fig. 4). In our study, OR gate was used because anyone of the causes might contribute to the accidents. Through FTA, 09 number of root causes of dump failure were identified.

3.2 DSR

The DSR system provides a semi-quantitative method for assessing the relative potential of dump stability and recommends the appropriate level of dumping investigation and design. This is based on individual point ratings for each of the main factors affecting dump stability. Each factor is given a point rating based on qualitative and/or quantitative descriptions accounting for the possible range of conditions. An overall DSR is calculated as the sum of the individual ratings for each of the various factors. The maximum possible DSR is 1800 (Pile, 1991). To simplify the rating scheme for possible practical application, four categories or Dump Stability Classes have been defined based on dump stability. Data collected through field investigation were used to calculate dump stability rating for 03 nos. of dumps of Kulda OCP. The rating was estimated for external dump, internal dump 1 and internal dump 2 and found to be 400, 500 and 500 respectively (Table 4). Accordingly the dump stability class was determined as Class II and corresponding category of failure hazard is Low (from Table 3). Therefore, thorough site investigations, sampling, laboratory index testing, routine visual and instrument monitoring may be required.

3.3 Triggered Action Response Plan

TARPs are integral part of hazard management plan which define actions to be taken by site personnel when conditions change from the normal and have the potential to trigger a major hazard event. Triggers and planned actions are reviewed regularly to ensure that they are appropriate. At first level, action may be to confirm that abnormality exists, at second level, preparation for withdrawal of persons, and at final level, actual orderly withdrawal. A detailed TARP for dump failure at Kulda OCP is presented in Table 5.

Table 3. Risk Assessment for dump slope failure

Hazard No.	Hazard	Mechanism	Calculated Risk				Control measures
			CONS	EXPS	PROB	RISK RATING	
1	2	3	4	5	6	7	8
SFH1	Sliding of external OB dump due to foundation failure	√ Dumping of OB done on low lying area or over clay or wet top soil √ Proper garland drain around the external OB dump not provided	01	05	10	50	√ Leveling up of low laying area by dumping blasted material and dozing √ Garland drain of adequate depth and width to be cut for proper drainage of run-off water
SFH2	Sliding of external OB dump due to circular failure	√ Tension cracks developed at the edge of the dump during rainy season, undetected due to poor supervision, causing sliding of OB dump bench, resulting into fatality or injury of persons	01	10	10	100	√ Proper leveling and grading of top and individual OB benches to prevent infiltration of rain water √ Closing of tension crack by dozing
SFH3	Sliding of external OB dump due to Improper bench design	√ Bench height > 30m √ Berm width less than bench height	05	05	10	250	√ Height should not be exceed beyond permissible limit. Berm/bench width shall not be less than the height of the bench
SFH4	Dump slide due to liquefaction of dump mass	√ Accumulation and infiltration of rain water through the discontinuous heaps of dump material causing the dump mass unstable due to hydro-static pressure	05	05	10	250	√ Proper leveling and grading of top and individual OB benches to prevent infiltration of rain water

(*continued*)

Table 3. (*continued*)

Hazard No.	Hazard	Mechanism	Calculated Risk				Control measures
			CONS	EXPS	PROB	RISK RATING	
1	2	3	4	5	6	7	8
SFH5	Dump slide due to fire on dump slope	√ OB bench becomes unstable when rain water enters through the cracks developed due to fire/exhaust smokes	01	10	10	100	√ Carbonaceous material shall not be dumped at the external OB dump. Existing fire shall be covered with top soil
SFH6	Death/injury of unauthorized persons entered near the toe of the OB dump	√ Entry of unauthorized persons at the toe of the external OB Dump for picking of shale/coal pieces due to damage of barbed wire fencing	05	10	10	500	√ Barbed wire fencing has been already provided around the external OB dump to prevent entry of un-authorized persons to the toe of the external OB dump √ Posting of security guards round the clock at the vulnerable points
SFH7	Sliding of internal dump due to entry of water	√ Proper drainage system not provided around the internal dump due to non-supply of hume pipes	05	10	07	350	√ Sufficient number of hume pipes shall be laid down at strategic locations for proper drainage of rain water

Abbreviations: CONS: Consequence; EXPS: Exposure; PROB: Probability; SOP: Safe Operating Procedure. Date of Risk Assessment: xx/xx/xxxx Risk Rating/Score = Cons × Exps × Prob

Table 4. Estimation of dump stability rating for Kulda OCP

Key factors		Range of conditions or description	Kulda OCP External dump	Kulda OCP Internal dump 1	Kulda OCP Internal dump 2
Dump Height	Low	<50 m	Moderate [50]	Moderate [50]	Moderate [50]
	Moderate	50–100 m			
	High	100–200 m			
	Very High	>200 m			
Dump Volume	Small	$<1 \times 10^6$ BCM (bank cubic meters)	Medium [50]	Medium [50]	Medium [50]
	Medium	$<1<10^6-5 \times 10^7$ BCM			
	Large	$>5 \times 10^7$ BCM			
Dump Slope	Flat	<26°	Flat [0]	Flat [0]	Flat [0]
	Moderate	26–35°			
	Steep	>35°			
Foundation Slope	Flat	<10°	Flat [0]	Flat [0]	Flat [0]
	Moderate	10–25°			
	Steep	25–32°			
	Extreme	>32°			
Degree of Confinement	Confined	Concave in plan or section; valley or cross-valley fill, toe buttressed against opposite valley wall; incised gullies that can be used to limit the foundation slope during development	Moderately Confined [50]	Moderately Confined [50]	Moderately Confined [50]
	Moderately Confined	Natural benches or terraces on a slope; even slopes, limited natural topographic diversity; heaped, side hill or broad valley or cross-valley fill			
	Unconfined	A convex slope in plan or section; side-hill or ridge crest fill with no toe confinement, and no gullies or benches to assist any development			

(continued)

Table 4. (*continued*)

Key factors		Range of conditions or description	Kulda OCP External dump	Kulda OCP Internal dump 1	Kulda OCP Internal dump 2
Foundation Type	Competent	Foundation materials as strong as or stronger than dump materials, not subject to adverse pore pressure; no adverse geologic structure	Competent [0]	Competent [0]	Competent [0]
	Intermediate	Intermediate between competent and weak; soils gain strength with consolidation; adverse pore pressures dissipate if loading rate controlled			
	Weak	Limited bearing capacity, soft soils; subject to adverse pore pressure generation upon loading; adverse groundwater conditions, springs or seeps; strength sensitive to shear strain, potentially liquefiable			
Dump Material Quality	High	Strong, durable; less than ~10% fines	Moderate [100]	Moderate [100]	Moderate [100]
	Moderate	Moderately strong, variable durability; 10–25% fines			
	Poor	Predominantly weak rocks of lower durability; greater than ~25% fines, overburden			
Method of Construction	Favourable	Thin lifts (<25 m thick), wide platforms; dumping along contours; ascending construction; wraparounds or terraces	Favourable [0]	Favourable [0]	Favourable [0]
	Mixed	Moderately thick lifts (25–50 m); mixed construction methods			
	Un-favourable	Thick lifts (>50 m), narrow platforms (sliver fills); dumping down the fall line of the slope; descending construction			

(*continued*)

Table 4. (*continued*)

Key factors		Range of conditions or description	Kulda OCP External dump	Kulda OCP Internal dump 1	Kulda OCP Internal dump 2
Piezometric and Climatic Conditions	Favourable	Low piezometric pressures, no seepage in foundation; development of phreatic surface within dump unlikely; limited precipitation; minimal infiltration into dump; no snow or ice layers in dump or foundation	Intermediate [100]	Intermediate [100]	Intermediate [100]
	Intermediate	Moderate piezometric pressures, some seeps in foundation; limited development of phreatic surface within dump possible; moderate precipitation; high infiltration into dump; discontinuous snow or ice lenses in dump or foundation			
	Un-favourable	High piezometric pressures, springs in the foundation; high precipitation; significant potential for the development of phreatic surface or perched water tables in the dump; continuous layers of snow or ice in the dump or foundation			
Dumping Rate	Slow	<25 BCM/m of crest/d; crest advancement rate < 0.1 m/d	Slow (Inactive) [0]	Moderate [100]	Moderate [100]
	Moderate	25–200 BCM/m of crest/d; crest advancement rate 0.1–1.0 m/d			
	High	>200 BCM/m of crest/d; crest advancement > 1.0 m/d			
Seismicity	Low	Seismic Risk Zones 0 and 1	Moderate [50]	Moderate [50]	Moderate [50]

(*continued*)

Table 4. (*continued*)

Key factors		Range of conditions or description	Kulda OCP External dump	Kulda OCP Internal dump 1	Kulda OCP Internal dump 2
	Moderate	Seismic Risk Zones 2 and 3			
	High	Seismic Risk Zones 4 or higher			
Rating			**400**	**500**	**500**

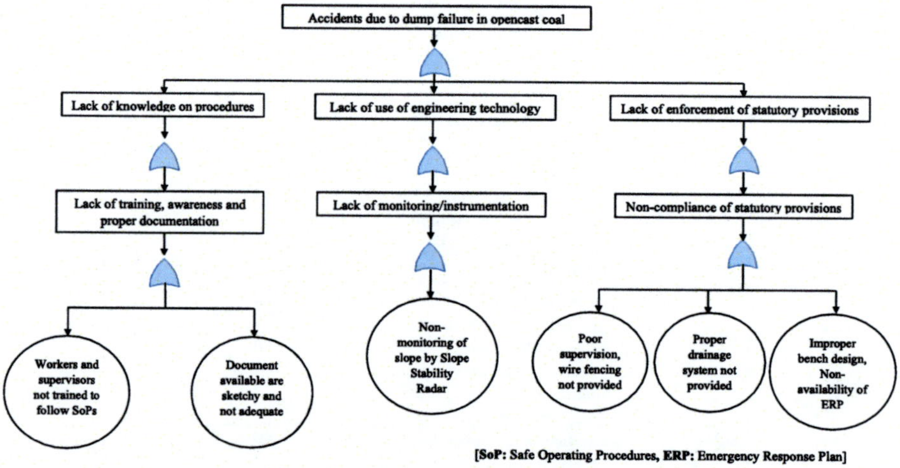

Fig. 4. Fault tree analysis of accidents due to dump failure in opencast mines

Table 5. TARP for dump failure in Kulda OCP

Triggered action response plan Dump Failure	Normal State	Level 1 Triggers	Level 2 Triggers	Level 3 Triggers
	Zero Movement of any pillars	Movement < 10 cm	Movement 10 cm-20 cm	Movement > 20 cm
Persons affected	Action/Response	Action/Response	Action/Response	Action/Response
Survey Officer	Take the RL of each pillar in a week and monitor any movement	√ Take the RL of each pillar in a week and monitor any movement √ Inform to concerned shift in-charge about the change	√ Take the RL of each pillar in a week and monitor any movement √ Maintain close liaison with the concerned supervisors in shift, shift in-charge, manager and safety officer about the change	√ Take the RL of each pillar in a week and monitor any movement √ Maintain close liaison with the concerned supervisors in shift, shift in-charge, manager and safety officer about the change

(*continued*)

Table 5. *(continued)*

Triggered action response plan Dump Failure	Normal State	Level 1 Triggers	Level 2 Triggers	Level 3 Triggers
	Zero Movement of any pillars	Movement < 10 cm	Movement 10 cm-20 cm	Movement > 20 cm
Control room In-charge	√ Continue monitoring as per routine	√ Notify shift in charge about the change &record the detail of the event thereof	√ Considering the severity of alarming situation, information should be passed on to manager, safety officer & rescue station superintendent	√ Ensure evacuation of Man & machinery (if possible) from the possible affected zone
Shift Supervisor/workers	√ Carry out the normal working in accordance with the procedures formed therein	√ If possible, deal with the situation by available means √ Liaise with survey officer about the movement	√ Immediately inform the shift in-charge to stop the on-going work near possible affected zone √ Deliver information to control room √ Start the process of evacuation of man & machinery (if possible) from the affected zone	√ Ensure evacuation of Man & machinery (if possible) from the possible affected zone
Shift In-charge	√ Carry out the regular duty of inspection & supervision	√ Ensure the information received is correct & authentic √ Take measures to control with available means	√ Ensure barring of all related works near the possible affected zone √ Send information to Manager, safety officer, agent, mine in-charge about the situation & aids required	√ Ensure all men deployed are safely evacuated from possible affected zone
Safety Officer	√ Carry out the regular statutory inspections	√ Apprise the correct situation to the manager	√ To co-ordinate all actions of emergency response plan	√ To co-ordinate all actions of emergency response plan
Manager	√ Carry out the regular statutory duties along with managing, direction & control	√ Apprise the correct situation to agent	√ Discuss the situation with mine officials & supervisory staff √ Prepare for level 3 Response √ Apprise the correct situation to agent	√ In addition to level-2 response, Invoke the emergency response team & Rescue recovery team √ Apprise the correct situation to agent, general manager, state administrative authorities, local police station & other higher authorities
Agent	√ Carry out the administrative duties regularly	√ Discuss with manager about the results & findings, initiate any necessary response require thereunder	√ Implement control measures identified by the Hazard Control Plan developed for the purpose	√ Develop a planned strategy to deal with the situation; help can be sought out from nearby projects

4 Conclusion

The following conclusions are drawn from this study.

- Dump stability class was determined as Class II and corresponding category of failure hazard is Low.
- The hazard SFH6 (Death/injury of unauthorized persons entered near the toe of the OB dump) has the highest risk score of 500. Out of 07 identified hazards, 04 nos. of hazards require immediate action and 03 nos. Hazards require management action. Using FTA, 09 number of root causes of dump failure were identified.
- Triggered Action Response Plan (TARP) for dump failure has been suggested and responsibility of the mine management during different trigger levels has been discussed.

References

Agboola, O., et al.: A review on the impact of mining operation: monitoring, assessment and management. Results Eng. **8**, 100181 (2020)

Behera, P.K., Sarkar, K., Singh, A.K., Verma, A.K., Singh, T.N.: Dump slope stability analysis–a case study. J. Geol. Soc. India **88**, 725–735 (2016)

Dash, A.K.: Analysis of accidents due to slope failure in Indian opencast coal mines. Curr. Sci.. Sci. **117**(2), 304–308 (2019)

HPC Report: Report of High Powered Committee on Accident in the Deep Mining Zone of Rajmahal OCP, ECL on 29 December 2016 (2016). https://dgms.gov.in/writereaddata/UploadFile/repor2016ss.pdf. Accessed 24 May 2023

Kainthola, A., Verma, D., Gupte, S.S., Singh, T.N.: A coal mine dump stability analysis—a case study. Geo-materials **1**(01), 1 (2011)

MCL: Annual Report and Accounts 2022–23 (2023). https://www.mahanadicoal.in/Financial/annual_report.php. Accessed 29 July 2023

McQuillan, A., Canbulat, I., Payne, D., Oh, J.: New risk assessment methodology for coal mine excavated slopes. Int. J. Min. Sci. Technol. **28**(4), 583–592 (2018)

Nayak, P.K., Dash, A., Dewangan, P.: Design considerations for waste dumps in Indian opencast coal mines-a critical appraisal. In: Proceedings of 2nd International Conference on Opencast Mining Technology and Sustainability, pp. 19–31 (2020)

Pile, B.C.M.W.R.: Research Committee (BCMWRPRC). Investigation and Design Manual (1991)

Rahul, M.K., Rai, R., Shrivastva, B.K.: Evaluation of dump slope stability of a coal mine using artificial neural network. Geomech. Geophys. Geo-energ. Geo-resour. **1**(3–4), 69–77 (2015). https://doi.org/10.1007/s40948-015-0009-8

Rajak, T.K., Yadu, L., Chouksey, S.K., Dewangan, P.K.: Strength characteristics and stability analysis of GGBS stabilized fly ash-overburden dump. Int. J. Min. Reclam. Environ.Reclam. Environ. **34**(9), 625–648 (2020)

Ulusay, R., Ekmekci, M., Tuncay, E., Hasancebi, N.: Improvement of slope stability based on integrated geotechnical evaluations and hydrogeological conceptualisation at a lignite open pit. Eng. Geol. **181**, 261–280 (2014)

Verma, D., Kainthola, A., Gupte, S.S., Singh, T.N.: A finite element approach of stability analysis of internal dump slope in Wardha valley coal field, India, Maharashtra. Am. J. Min. Metall. **1**(1), 1–6 (2013)

Understanding Runout Behavior of Overburden Dump: Insights for Buffer Zone Design in Open Pit Coal Mines

Anup Tiwari(✉), Bibhuti Bhusan Mandal, and Khanindra Pathak

Department of Mining Engineering, IIT Kharagpur, Kharagpur, India
tiwarianup.at@gmail.com

Abstract. Failure of in-pit overburden dumps in open-pit coal mines triggers the flow of debris toward the active working face, posing threats to personnel and equipment safety. The heterogeneous nature of these dumps and the limited understanding of their engineering behavior contribute to frequent failures. To mitigate the risk of unwarranted failures and enhance safety measures, the establishment of a buffer zone surrounding the periphery of in-pit dumps is necessary, restricting or prohibiting worker and machinery movement. Determining the size of this buffer zone depends on the potential mobility of the failing mass, determined by its runout length and runout area. Although pre-failure deformational behaviors have received significant attention in scientific literature, the need for comprehensive post-failure runout studies has only recently been recognized, driven by the increase in waste dump failure accidents. To address this knowledge gap, a laboratory-scale debris flow flume was developed to investigate the influence of material gradation and slope angle on the runout characteristics of overburden (OB) material. Experimental investigation reveals a strong correlation between runout characteristics and the Fines/Coarse ratio in the debris mixture. Notably, a critical limit of Fines/Coarse exist before and after which the general trend in runout length contradicted each other.

Keywords: Open-pit mining · Mine dump failure · Slope stability · Runout distance · Post-failure study · Buffer zone

1 Introduction

Coal, a vital non-metallic resource in India, plays a crucial role in generating over 75% of the country's power (Ministry of Coal, Govt. of India). Open cast (OC) mining is the preferred method for coal extraction due to its high productivity and lower production costs (Mukherjee and Pahari, 2019). This mining approach involves excavating and removing the overlying strata, resulting in the formation of large overburden dumps (OB dumps). These dumps can be either located within the mine (in-pit) or externally. However, external OB dumps have adverse environmental and social consequences and have higher material handling costs (Dawson, 1994; Robertson, 1985). In-pit dumping offers benefits such as reduced material handling costs and a smaller footprint, but it

A. Sinha et al. (Eds.): AMC 2023, SPEES, pp. 99–109, 2023.
https://doi.org/10.1007/978-3-031-46966-4_8

necessitates careful consideration of safety aspects. The instability or failure of in-pit OB dumps poses risks to personnel and equipment in nearby working areas, resulting in productivity losses and increased mining expenses (Richards et al., 1981; Ulusay et al., 1995). Despite pre-failure deformation studies aimed at preventing such failures during dump slope design, many in-pit dumps still experience failures (Kainthola et al., 2011; Poulsen et al., 2014), as their heterogeneous nature complicates their engineering behavior (Koner and Chakravarty, 2010.

Given the possibility of OB dump failures, establishing a buffer zone around the periphery of in-pit dumps becomes imperative. This buffer zone aims to restrict the movement of personnel and machinery, ensuring additional safety in the event of unforeseen slope failures. The extent of this buffer zone is determined by estimating the mobility of the failing mass, specifically the probable runout distance, which indicates the likelihood of the failing mass reaching a specific location (McDougall, 2017). While Regulation 108 (5) of the (Indian) Coal Mine Regulations, 2017 specifies a buffer distance of 100 m between the toe of a spoil-bank and certain structures, the scientific basis for determining the appropriate safe distance remains lacking.

To understand the movement characteristics and hazard potential of geological debris flow in OB dumps, the study utilizes the concept that OB dump failure resembles a debris flow, where a mixture of water, sediments, rock, and soil surges downslope due to gravity (Zhang and Tang, 2001; Xu et al., 2007). Physical modeling using a debris flow flume has been a conventional approach to study essential aspects of debris flow at a reduced scale using well-characterized materials (Moberly, 2015; Cheng et al., 2019; Hu et al., 2020; Ghasemi et al., 2020; Hu et al., 2021). These experiments do not aim to precisely replicate natural failure but rather facilitate systematic study (Iverson, 2015). Previous quantitative studies on debris flow focused on sediment concentration, size, water content, and slope (Moberly, 2015; Cheng et al., 2019; Hu et al., 2020; Ghasemi, 2020; Hu et al., 2021), but they did not accurately represent the heterogeneous material found in mine OB dumps. Consequently, this study aims to conduct experimental investigations to explore the impact of material gradation and slope angle on the mobility of realistic mine OB dump debris. The findings will establish a reference framework for future research, focusing on understanding the guiding principles for creating buffer zones around in-pit dumps.

2 Material and Method

2.1 Experimental Setup

Hunter et al., 2003, compiled information on 56 OB dump failures, drawing from reports by Dawson (1994), Golder Associates Limited, 1995, and Siddle et al., 1996. Among these failures, seven exhibited confined travel paths, indicating the highest mobility. To replicate confined flow conditions, we designed a 3 m-long flow flume for our experimental investigations (Tiwari et al., 2023). The experimental setup comprises six main components (Fig. 1): (a) Flow flume with a discharge chamber, (b) Supporting metal frame with a hydraulic jack, (c) Deposit plane, (d) LiDAR and LDR sensors with a data acquisition system, (e) Two high-speed cameras, and (f) One digital camera. The flow flume measures 3 m × 0.3 m × 0.3 m, while the discharge chamber's dimensions are 0.3 m × 0.3 m × 0.5 m. To minimize slippage and enhance bed particle shearing, we affixed an anti-slip mat with a studded structure to the base of the flume.At the top of the flow flume, six VL53L0X LiDAR sensors are positioned to measure the height difference (ΔH) between the initial target point (flow flume base) and the instantaneous target point (top of the flowing debris mass), providing the flow depth. Additionally, seven LDR sensor modules are installed at the bottom (Fig. 1) to calculate the velocity of the flow front.

2.2 Material

The experimental debris mix utilized in this study was collected from the overburden (OB) dumps of open-cast projects (OCP) in the Raniganj Coalfields of West Bengal, India. Lithographic data obtained from seven different OCPs in this region revealed that the overburden primarily consists of soil/alluvium, clay, sandstone, and shale. To prepare a well-characterized debris mix, grain size analysis was conducted on fourteen different samples following ASTM D6913/D6913M-17, and their classification was carried out using the MIT soil classification system. In the subsequent sections of this research article, the term "coarse proportion" will refer to particles ranging from −16 mm to +2 mm, while the term "fine proportion" will denote particles smaller than 2 mm. Additionally, any change in the quantity of fine particles, the quantity of coarse particles, and water content will be referred to as a change with respect to the total experimental debris mass 'Q'. For the preparation of the experimental debris mix, several parameters were varied, including: (a) Quantity of fine particles, (b) Quantity of coarse particles, and (c) Gradation of coarse particles (i.e., d_{50}). Three different gradations of coarse particles (Fig. 2) were employed, with d_{50} values of 9.6 mm, 8 mm, and 6.8 mm, while the gradation of fine particles remained constant across all experimental debris mixes. The total mass of the experimental debris mix, denoted as 'Q', was maintained at 30 kg. The experimental scheme implemented in this study is summarized in Table 1.

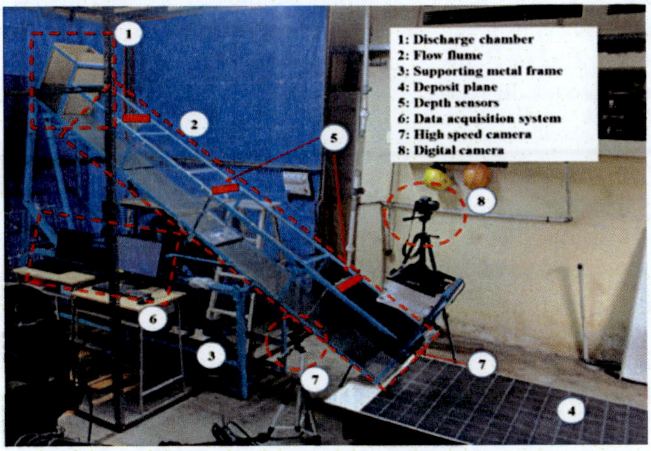

Fig. 1. Designed experimental setup

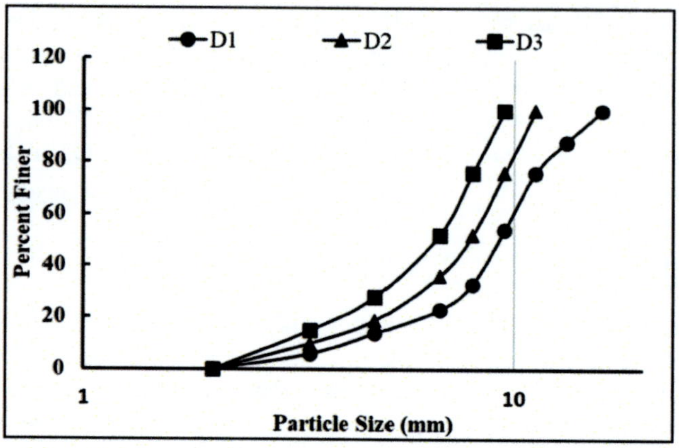

Fig. 2. Particle size distribution curve of three coarse samples

2.3 Method

For each experiment, the discharge chamber gate remained closed initially, and the flume walls were treated with an anti-static spray. Before preparing the samples, the sensors underwent a performance check. The experimental debris mix was then prepared in a bucket, ensuring that the mixing time for all samples did not exceed five minutes. The sensors were triggered by a computer program, while manual triggering was used for the cameras. Subsequently, the prepared samples were poured into the discharge chamber from a height of 0.5 m above the gate, allowing gravity to facilitate the flow. To prevent fines from settling in the mixture, the chamber gate was opened within 10 s after pouring the samples. Readings were recorded to measure the runout length, representing the distance between the deposit front and the flume's toe. This procedure was repeated for all 27 experimental conditions as outlined in Table 1. Each experimental condition discussed in this paper was replicated three times, and the average of these three repetitions was considered the representative value of the response variable. The final values of the runout characteristics and mobility ratio were reported as averages based on these three results.

3 Result

As the length of the flow flume remains constant throughout all experimental conditions, the term "runout length" in this study refers to the distance between the toe of the flow channel and the leading edge of the deposit on the deposit plane. As discussed in the previous section, the total mass of debris mixture (Q) is kept constant in all the experimental condition, the change in fines proportion is represented in terms of 'Q'. In Figs. 3, 4, and 5, we can observe the variations in runout length as the material composition and slope angle change. The results are presented in two sections. In the first section, we investigate how runout characteristics are affected when the d_{50} (median grain size) of the coarse portion of the debris mix is kept constant, while the proportion of fine particles in the experimental mix is varied. In the second section, we maintain a constant proportion of fine particles in the experimental mix while varying the d_{50} of

Table 1. Experimental scheme followed in this study

Material Code	Coarse particles (−16 mm to + 2 mm)		Fine particles (−2 mm)	Water content	Experimental condition at 30° flume slope 'S_1'	Experimental condition at 35° flume slope 'S_2'	Experimental condition at 40° flume slope 'S_3'
	d50	% of total debris mass (by weight)	% of total debris mass (by weight)	% of total debris mass			
G_1	$D_1 = 9.6$	80	$FC_1 = 10$	10	D1- FC1 -S1	D_1- FC_1 -S_2	D_1- FC_1 -S_3
G_2	$D_2 = 8$	80	$FC_1 = 10$	10	D_2- FC_1 -S_1	D_2- FC_1 -S_2	D_2- FC_1 -S_3
G_3	$D_3 = 6.8$	80	$FC_1 = 10$	10	D_3- FC_1 -S_1	D_3- FC_1 -S_2	D_3- FC_1 -S_3
G_4	$D_1 = 9.6$	70	$FC_2 = 20$	10	D_1- FC_2 -S_1	D_1- FC_2 -S_2	D_1- FC_2 -S_3
G_5	$D_2 = 8$	70	$FC_2 = 20$	10	D_2- FC_2 -S_1	D_2- FC_2 -S_2	D_2- FC_2 -S_3
G_6	$D_3 = 6.8$	70	$FC_2 = 20$	10	D_3- FC_2 -S_1	D_3- FC_2 -S_2	D_3- FC_2 -S_3
G_7	$D_1 = 9.6$	60	$FC_3 = 30$	10	D_1- FC_3 -S_1	D_1- FC_3 -S_2	D_1- FC_3 -S_3
G_8	$D_2 = 8$	60	$FC_3 = 30$	10	D_2- FC_3 -S_1	D_2- FC_3 -S_2	D_2- FC_3 -S_3
G_9	$D_3 = 6.8$	60	$FC_3 = 30$	10	D_3- FC_3 -S_1	D_3- FC_3 -S_2	D_3- FC_3 -S_3

the coarse proportion.This approach allows us to analyze the individual contributions of d_{50} and the proportion of fine particles to the overall runout characteristics. By studying these variations, we gain valuable insights into the behavior of the debris flow and the key factors influencing its runout patterns.

3.1 Variation in the Runout Characteristics at Constant d_{50} of Coarse Proportion and Varying Fines Proportion

The laboratory experiments were conducted at flume slopes of 30°, 35°, and 40°, with the d_{50} of the coarse proportion held constant at 6.8 mm, 8 mm, and 9.6 mm for each slope. The results revealed that the runout length was consistently highest for the debris mixture with 0.2Q fines proportion for all three slope angles. However, when the d_{50} was maintained at 6.8 mm and 8 mm, the minimum runout length was observed for the experimental debris mix with 0.3Q fines proportion. On the other hand, for a constant

d_{50} of 9.6 mm, the minimum runout length was observed for the experimental debris mix with 0.1Q fines proportion.

At 30° slope, the runout lengths (in cm) for debris mixtures with 0.1Q, 0.2Q, and 0.3Q fines proportions were 21 cm, 25 cm, and 13 cm, respectively, when the d_{50} was 6.8 mm. As the d_{50} increased to 8 mm, the same trend in runout length was observed. However,when the d_{50} was further increased to 9.6 mm, the runout lengths for the debris mixtures with 0.1Q and 0.2Q fines proportions were 31 cm and 50 cm, respectively. Similarly, at 35° slope, therunout lengths for the debris mixtures with 0.1Q, 0.2Q, and 0.3Q fines proportions were 40 cm, 50 cm, and 37 cm, respectively, when the d_{50} was 6.8 mm. When the d_{50} increased to 8 mm, the trend in runout lengths remained consistent. Inthe same way as we observed in 30° flume slope, at a d_{50} of 9.6 mm, the runout lengths for the debris mixtures with 0.1Q and 0.2Q fines proportions were 54 cm and 68 cm, respectively. At 40° slope, the runout lengths for the debris mixtures with 0.1Q, 0.2Q, and 0.3Q fines proportions were 55 cm, 65 cm, and 48 cm, respectively, when the d_{50} was 6.8 mm. The same trend persisted as the d_{50} increased to 8 mm, and at a d_{50} of 9.6 mm, the runout lengths for the debris mixtures with 0.1Q and 0.2Q fines proportions were 66 cm and 78 cm, respectively. These findings indicate a consistent relationship between the fines proportion, d_{50} of the coarse portion, and the resulting runout lengths at different slope angles (Figs. 3, 4 and 5).

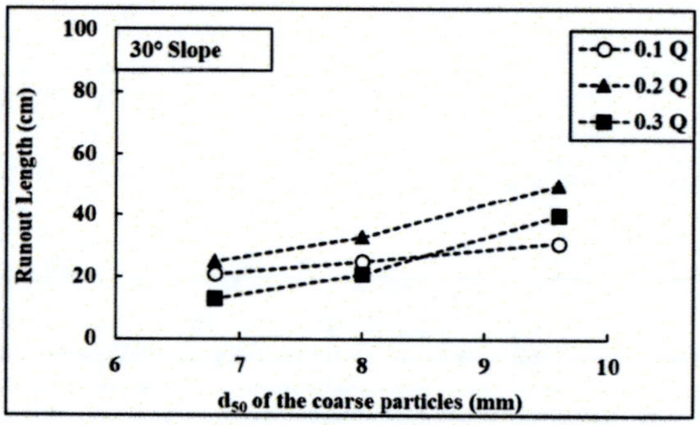

Fig. 3. Runout lengthat 30° flume slope

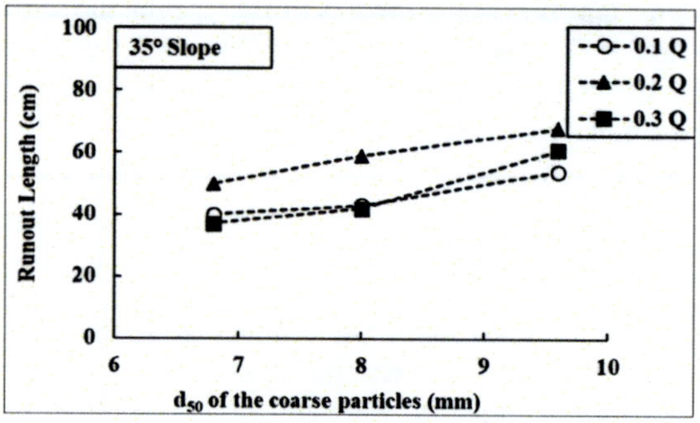

Fig. 4. Runout length at 35° flume slope

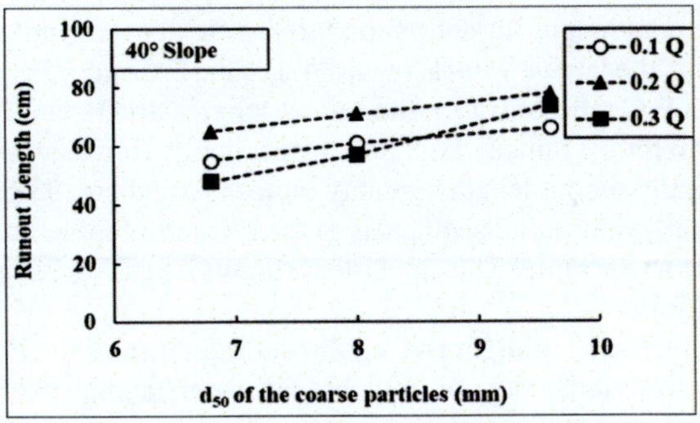

Fig. 5. Runout length at 40° flume slope

3.2 Variation in the Runout Characteristics at Constant Fines Proportion and Varying d_{50} of Coarse Proportion

When the quantity is fixed at 0.1Q, 0.2Q, and 0.3Q fines proportion, the runout length consistently increases with an increase in the d_{50} of coarse particles. Notably, the rate of change in runout length is relatively less pronounced at 0.1Q fines proportion compared to other fines proportions. This implies that the presence of a smaller quantity of fines (0.1Q) has a lesser effect on the runout length with respect to varying coarse particle sizes. However, at 0.2Q fines proportion, the runout length is observed to be the highest, indicating maximum mobility of the debris mix. The runout lengths increase as the d_{50} of coarse particles increases from 6.8 mm to 8 mm, and then to 9.6 mm, suggesting that a fines proportion of 0.2Q allows for optimal mobility of the debris flow. Interestingly, the most significant change in runout length is observed at 0.3Q fines proportion when the d_{50} of coarse particles increases from 8 mm to 9.6 mm. This sharp transition suggests that at a higher fines proportion (0.3Q), even a slight variation in the d_{50} of coarse particles can considerably affect the flow behavior and runout characteristics.

3.3 Effect of F/C Ratio on Mobility Characteristics OB Dump Debris

To comprehend the flow behavior of OB dump, a new index $\frac{Fines}{Coarse}$ ratio is suggested by Tiwari et al., 2023, which bears a strong correlation to runout characteristics. The ($\frac{Fines}{Coarse}$*ratio* or $\frac{F}{C}$*or*α) is expressed as;

$$\frac{F}{C} = \frac{Totalmass\ offines(= -2\ \text{mm})(\%of\ Q)}{Totalmass\ of\ coarse(-16\ \text{mm}\ to\ +2\ \text{mm})(\%of\ Q)} \quad (1)$$

Based on Figs. 6, 7, and 8, a distinct trend can be observed regarding the impact of the F/C (fines-to-coarse) ratio on debris flow runout length. Initially, as the F/C ratio increases, the runout length shows an upward trend. However, after reaching a critical value of 0.286, the runout length starts to decrease. This consistent pattern is observed across all slope angles.

Notably, the runout length for debris mixtures with a coarse particle d_{50} of 9.6 mm consistently exhibits the largest values, regardless of the F/C ratio. For example, at a 30° flume slope and an F/C ratio of 0.125, the runout lengths are 31 cm, 25 cm, and 21 cm for d_{50} values of 9.6 mm, 8 mm, and 6.8 mm, respectively. However, when the F/C ratio increases to 0.286, the runout lengths notably increase to 50 cm, 33 cm, and 25 cm for the corresponding d_{50} values. Subsequently, as the F/C ratio further increases to 0.5, the runout length starts to decrease. This consistent runout length variation trend is evident at all the slope angles.

Overall, these findings highlight the significant influence of the F/C ratio on debris flow runout length, providing valuable insights for understanding and predicting debris flow behavior in different scenarios.

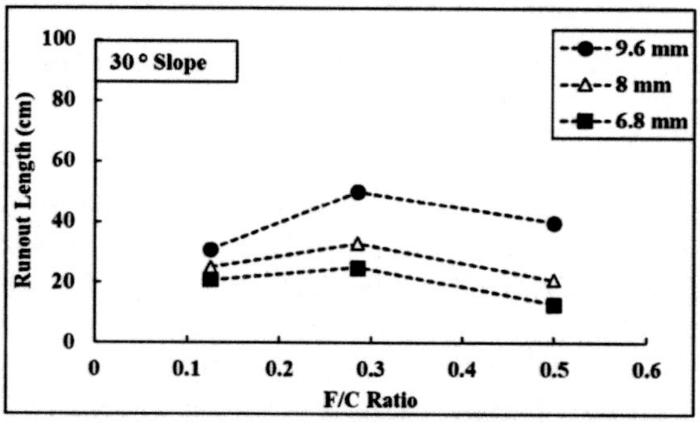

Fig. 6. Effect of F/C on runout length (35°)

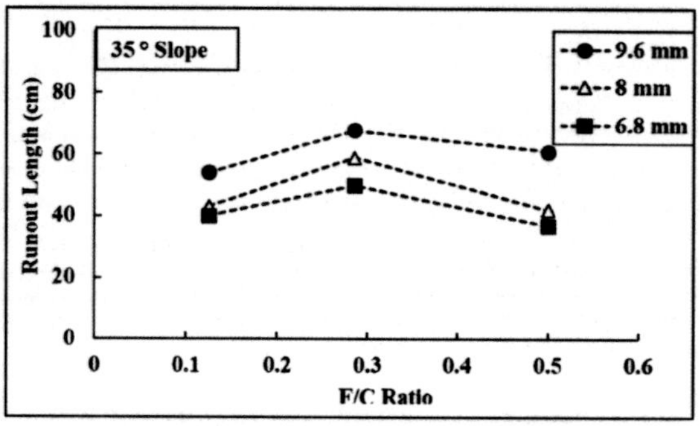

Fig. 7. Effect of F/C on runout length (35°)

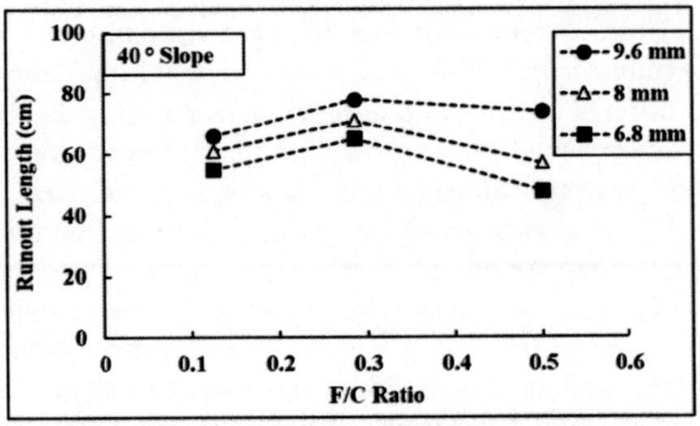

Fig. 8. Effect of F/C on runout length (40°)

4 Discussion

Particle Interactions: Fines act as a lubricating agent between the larger coarse particles, reducing inter-particle friction. This lubrication effect allows the larger particles to move more freely within the flow, leading to increased flow mobility and longer runout distances. Conversely, a decrease in fines can result in higher interlocking among particles, leading to greater internal resistance and reduced flow mobility.

Flow Consistency and Viscosity: The presence of fines alters the flow consistency and viscosity. Higher fines content increases the flow's overall viscosity, making it more cohesive and stickier. This cohesive nature can lead to a more sustained and prolonged flow, resulting in longer runout lengths. Conversely, a lower fines content can lead to a less viscous and more fluid-like flow with reduced runout distances.

Debris Mixture Stability: The fines play a significant role in stabilizing the overall debris mixture. With an appropriate fines content, the debris mix tends to maintain a balanced state, facilitating efficient flow dynamics. However, excessive fines content can lead to the formation of cohesive plugs or dams within the flow, hindering the flow's progression and altering its runout characteristics.

5 Conclusion

The main focus of this paper is to gain experimental insights into the impact of experimental mix composition and slope angle on the runout characteristics of overburden (OB) dump material. To achieve this, nine different experimental debris mixes were prepared by varying the material composition, and tested at three distinct slope angles (see Table 1). The key conclusions drawn from this study are summarized as follows:

1) **Influence of Particle Size and Slope Angle:** The findings highlight the significant influence of particle size distribution and slope angle on the runout length of debris flow. As the particle size increases, the runout distances also increase, emphasizing the importance of considering particle size distribution in accurately predicting debris mix mobility. Similarly, higher slope angles facilitate particle mobility, leading to longer runout lengths.

2) **Prominence of Fines Proportion:** The fines proportion plays a prominent role in determining the runout length. For a given water content in the debris mix, there exists a critical value of fines proportion that maximizes mobility. Beyond this limit, the mobility starts to decrease. Changes in fines proportion have a more pronounced effect on debris mixture mobility compared to variations in particle size. This underscores the significance of considering fines content when assessing flow dynamics and debris flow extent in an overburden dump.
3) **Threshold Limits:** The study identifies a threshold limit of Fines/Coarse (F/C or α) at 0.286. The general trend in the runout length differs depending on the F/C ratio. When the F/C ratio is below 0.286, a positive correlation is observed between runout length and F/C ratio. Conversely, when the F/C ratio exceeds 0.286, a negative correlation is noted between runout length and F/C ratio.
4) **Practical Implications for Debris Flow Assessment:** The quantified relationships between runout length, particle size, and slope angle provide valuable insights for researchers, engineers, and practitioners involved in assessing and mitigating debris flow in overburden dumps. The practical implications can be categorized as follows:
   a) **Slope Stability:** The fines content can significantly impact slope stability in areas prone to debris flow. An increase in fines can modify the shear strength of the debris mixture and affect overall slope stability. Understanding this influence is vital for assessing the potential for slope failure and predicting runout distances during debris flow events.
   b) **Runout Distance Prediction:** The effect of fines on runout distance can vary depending on other factors such as slope angle, flow velocity, and material gradation. Hence, accurately quantifying the role of fines in predicting runout distances is essential for hazard assessment, land use planning, and designing effective mitigation measures.

References

Mukherjee, S., Pahari, D.P.: Underground and opencast coal mining methods in India: a comparative assessment. Space and Culture, India **7**(1), 39–55 (2019)

Dawson, R.F: Mine OB geotechnics (1994)

Robertson, A.M.: Mine OB disposal: an update on geotechnical and geohydrological aspects. In: Geotechnical and Geohydrological Aspects of OB Management, pp. 31–48. Routledge (1985)

Richards, B.G., Coulthard, M.A., Toh, C.T.: Analysis of slope stability at Goonyella Mine. Can. Geotech. J. **18**(2), 179–194 (1981)

Ulusay, R., Arikan, F., Yoleri, M.F., Çağlan, D.: Engineering geological characterization of coal mine OB material and an evaluation in the context of back-analysis of spoil pile instabilities in a strip mine, SW Turkey. Eng. Geol. **40**(1–2), 77–101 (1995)

Kainthola, A., Verma, D., Gupte, S.S., Singh, T.N.: A coal mine dump stability analysis—a case study. Geomaterials **1**(01), 1 (2011)

Poulsen, B., Khanal, M., Rao, A.M., Adhikary, D., Balusu, R.: Mine overburden dump failure: a case study. Geotech. Geol. Eng. **32**(2), 297–309 (2014)

Koner, R., Chakravarty, D.: Stability study of the mine overburden dumps slope: a micromechanical approach. Studia Geotechnica et Mechanica **32**(1), 35–57 (2010)

McDougall, S.: 2014 Canadian Geotechnical Colloquium: landslide runout analysis—current practice and challenges. Can. Geotech. J. **54**(5), 605–620 (2017)

Zhang, L., Tang, K.: Mine Debris Flow. CA, Beijing (2001)

Xu, Y., He, F., Chen, H.: Mine debris flow and its distribution in Northwestern China. J. Mt. Sci. **25**, 729–736 (2007)

Moberly, D.: Laboratory experiments investigating entrainment by debris flows. Doctoral dissertation, University of Minnesota (2015)

Cheng, Y.M., Fung, W.H.I., Li, L., Li, N.: Laboratory and field tests and distinct element analysis of dry granular flows and segregation processes. Nat. Hazard. **19**(1), 181–199 (2019)

Hu, Y.X., Li, H.B., Qi, S.C., Fan, G., Zhou, J.W.: Granular effects on depositional processes of debris avalanches. KSCE J. Civ. Eng. **24**(4), 1116–1127 (2020)

Ghasemi, A.: A Study of Particle Entrainment in Two Common Particle-Fluid Flows in Nature: Bedload Transport in Rivers and Debris Flows in Upland Regions. Doctoral dissertation, University of Minnesota (2020)

Hu, Y.X., Li, H.B., Lu, G.D., Fan, G., Zhou, J.W.: Influence of size gradation on particle separation and the motion behaviors of debris avalanches. Landslides **18**(5), 1845–1858 (2021)

Take, W.A., Bowman, E., Askarinejad, A.: Physical modelling of landslides. Int. J. Phys. Model. Geotech. **20**(4), 175–176 (2020)

Iverson, R.M.: Scaling and design of landslide and debris-flow experiments. Geomorphology **244**, 9–20 (2015)

Golder Associates Limited: Mined Rock and overburden piles: runout characteristics of debris from dump failures in mountainous terrain. Stage 2: analysis, modeling and prediction. Interim Report, Report No. 932-1493. Prepared in association with O. Hungr Geotechnical Research Ltd. British Columbia Mine OB Rock Pile Research Committee and CANMET. Contract No. 23440-0-9198-X86 (1995)

Siddle, H.J., Wright, M.D., Hutchinson, J.N.: Rapid failures of colliery spoil heaps in the South Wales Coalfield. Q. J. Eng. Geol.Hydrogeol. **29**(2), 103–132 (1996)

Tiwari, A., Mandal, B.B., Pathak, K.: Influence of material composition on post-failure behavior of overburden dumps in opencast coal mines. Min. Metall. Explor. **40**, 1199–1209 (2023)

Bridging Efficiency and Safety: A Case Study of Extraction of Thick Coal Seam Having Low Incubation Period Using Continuous Miner

Prabhat Kumar Mandal(✉), Rana Bhattacharjee, Nilabjendu Ghosh, Awanindra Pratap Singh, Subhashish Tewari, and Arka Jyoti Das

CSIR-Central Institute of Mining and Fuel Research, Barwa Road, Dhanbad 826001, India
{pkmandal,rana_b,nilabjendu.ghosh,apsingh,tewari_s,arkajyoti}@cimfr.nic.in

Abstract. The extraction of thick coal seams developed on pillars by underground mining has been a chronic problem in India. The problem increases when the incubation period of such seams is very short, as in the case of Raniganj coalfield. Fast extraction, effective management of mining-induced stresses, and early detection of spontaneous heating and fire incidents are crucial during the liquidation of these thick seams. Minimizing coal loss in the goaf and selecting an appropriate support system for the high roof are important considerations. The paper presents a case study of Khottadih Colliery, where Continuous Miner (CM) technology is successfully employed to extract R-VI seam of 5.4 m thickness in a single pass. The R-VI seam has a very low, around 90 days incubation period and is developed along the floor in Bord and Pillar pattern up to a height of 2.8 to 3.0 m. Overlying seams are extracted by an opencast project and backfilled, with a 23 m high overburden dump above the CM panels. The use of CM with a single-pass height extraction of 5.4 m is the first such application in the Indian coal mining industry, exceeding the previous 4.6 m limit, thereby maximizing productivity while minimizing associated risks.

Keywords: Thick coal seam · Low incubation period · Overburden dump · Nevid method · Continuous miner

1 Introduction

Technological advancements in favour of opencast mining has made underground coal production more or less stagnant for decades. However, to meet future coal demand and embrace clean coal technology, focusing on underground mining is essential. In India, coal seams having thickness 4.8 m or more are referred to as "thick seams" (Singh et al., 2001). However, the criteria used in different nations to classify thick coal seams vary (Fig. 1) (Ghose, 1984; Kumar et al., 2015). The basic rule has generally been to fix the lower limit of a thick seam as the thickness up to which a seam could be extracted in a single lift (pass). Notably, about 60% of India's workable coal reserves amenable for underground mining consist of thick coal seams (Fig. 2) (Kumar et al., 2015).

A. Sinha et al. (Eds.): AMC 2023, SPEES, pp. 110–129, 2023.
https://doi.org/10.1007/978-3-031-46966-4_9

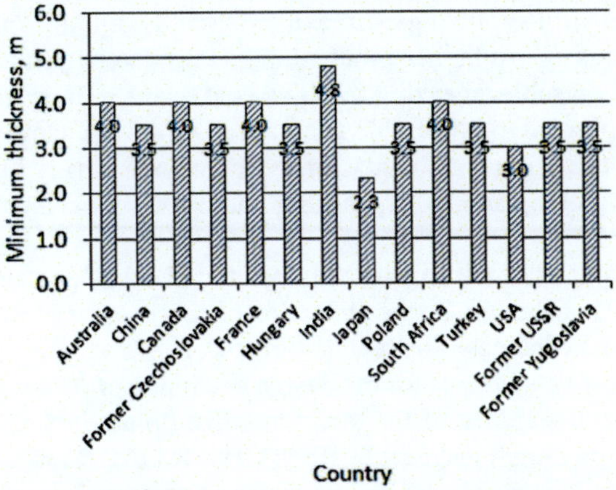

Fig. 1. Minimum thickness norms for thick coal seams in different countries.

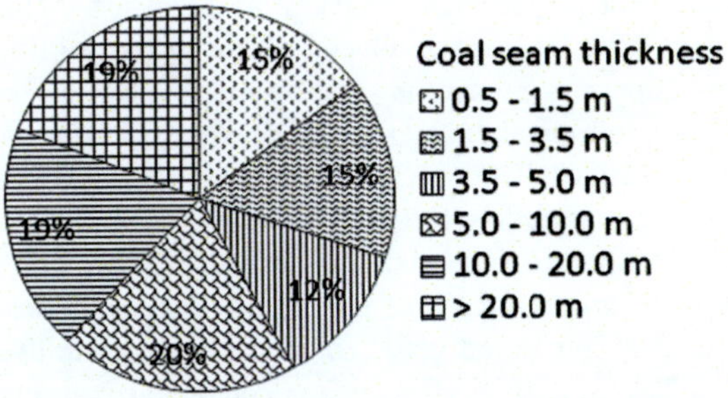

Fig. 2. Distribution of coal seam thickness in India and the corresponding percentage shares for each range.

A huge coal reserve lies locked-up in such thick coal seams in Raniganj Coalfields which is under the command area of Eastern Coalfields Limited (ECL). Majority of these coal seams have been extensively developed in single or multiple sections and are slated for extraction according to ECL's future plans. The thick seams standing on pillars in Raniganj coalfields offer a unique challenge due to their relatively short incubation period. The conventional Bord and Pillar method with stowing is commonly employed for depillaring by Indian mine operators. The blasting gallery method is avoided due to various issues such as poor recovery, ventilation problems, and an increased risk of spontaneous heating and fire. Although the cable bolting method is a viable alternative, it lacks automation and mobility, hindering higher production rates. Top coal caving or other mechanized high-production methods, though established the world over, have not been tried.

These thick seams in Raniganj coalfields are developed along the floor with 2.8–3.0 m gallery height and 4.2–4.8 m gallery width in bord and pillar pattern, an Indian variant of room and pillar (CIMFR GAP Report, 2019). Choosing the appropriate mining method for extracting thick coal seams with optimal recovery and safety is an intricate process (Fettweis et al., 1992). Single-lift extraction of the full thickness of a thick coal seam is advantageous due to higher production and productivity. The extensive use of single-lift working via sublevel caving for thick coal seams in Europe and several other countries can be attributed to favourable geomining conditions. However, in the case of India, where thick coal seams are characterized by comparatively higher strength, the possibilities for employing single-lift working with sublevel caving are limited (Singh et al., 2001).

In India, the majority of thick coal seams are found developed along the floor with galleries around 3.0 m in height. For optimization of coal recovery from these seams, the extraction height can be increased by taking roof coal. However, this pursuit of increased extraction height poses challenges for the design and rating of roof supports, movement of the roof strata, and stability of the structures like pillars including barrier pillars, stooks, or ribs/stumps (Singh and Singh, 1999; CIMFR GAP Report, 2019).

Under Indian geo-mining conditions, the inherent geotechnical problems associated with extracting coal from thick seams are exacerbated. The considerable size of the coal seam, coupled with the overlying roof strata, shallow depth of cover, the dead weight of broken strata if the overlying seams are taken by opencast method and backfilled, and the presence of overburden dump, all contribute to these difficulties. Moreover, the conventional roof support systems may not be suitable for the increased extraction height, leading to higher mining-induced stresses and the risk of pillar overriding (Sheorey, 1992; Singh et al., 1996). The challenges of thick seam extraction include a low percentage of extraction, supporting the high roof, strata control and roof inspection difficulties, sluggish ventilation, potential for violent airblasts due to the collapse of workings in the high void, and an increased likelihood of spontaneous heating due to the substantial amount of coal left in the goaf. To successfully mine these thick seams by underground method, fast extraction, effective management of mining-induced stresses, efficient ground control, and early detection of heating incidents are essential. Minimizing coal loss in the goaf and selecting an appropriate support system for the high roof are critical considerations in this process.

This paper presents one such case study at Khottadih Colliery in Raniganj Coalfield, where shallow cover conditions are prevalent. The study focuses on the implementation of mass production technology using a Continuous Miner (CM) with an increased cutting height to extract a 5.4m thick coal seam in a single pass (CIMFR Report, 1999). The seam has a remarkably short incubation period of just three months and is developed along the floor in a Bord and Pillar pattern. Additionally, the overlying seams are extracted through an opencast project (OCP), and after extraction, they are backfilled. On the surface above the CM panels, there is a substantial height of overburden (OB) dump. The paper describes comprehensive consideration of rock mechanics aspects, suitable extraction methodology, and support patterns which are designed based on field study and numerical simulation.

2 Description of the Study Area

Khottadih Colliery of Eastern Coalfields Limited (ECL) is located in the north-eastern part of Raniganj Coalfields, India. It is a degree II underground coal mine having workings in four seams in ascending order namely R-III/II, R-IV, R-V and R-VI. The extraction of upper seams R-VII (A + B) and R-VII above the R-VI seam are being carried out by Dalurbandh OCP and is backfilled. The backfilled area is overlain by a maximum of 23 m in height of the OB dump. R-VI seam of a thickness of 5.4 m, having variable dip but nearly flat with an average gradient 1 in 10 and depth of cover 81 m to 122.37 m, is developed along the floor in Bord and Pillar pattern. The average pillar size is 30 m × 30 m (centre to centre) and the gallery height and width are 2.8 m to 3.0 m and 4.5 m to 5.0 m respectively. The average coal left n the roof is 2.6 m. The workings are 7 to 8 years old and are supported by 1.5 m long cement grouted roof bolts at a 1.2 m grid pattern. R-VI seam is known for having a very low incubation period of three months only. Earlier, depillaring by caving using conventional supports and with the deployment of LHDs was the adopted method for extraction in the R-VI seam. Depillaring with stowing method was adopted in one panel. For extraction of the thick seam and to address spontaneous heating, 14 panels were extracted using cable bolting-based depillaring method (Mangal, 2021a, 2021b; Mangal and Sarkar, 2021) with caving coupled with an optimized ventilation system.

In the recent past, the mine management has embraced mass production technology by the deployment of Continuous Miner (CM) of increased cutting height for the extraction of developed pillars and roof coal in the R-VI seam in one single-pass of height 5.4 m. The CM panels, B12A, B12B, B14A, B14B and B13A are shown in the mine plan of R-VI seam workings (Fig. 3). Section of various coal seams with parting and working status is shown in Fig. 4.

2.1 Geology

The coal seams are developed in the Raniganj formation. The rock types encountered include coarse to fine-grained sandstones, shaly sandstones, sandy shales, grey shales, etc., with the intermittent shale band and laminae. The rock mass rating (RMR) of the roof of R-VI seam workings is 44.2. The nature of the immediate roof of the R-VI seam is argillaceous shale and medium-grained sandstone and the immediate floor is grey shale and shaly sandstone. The weighted average RQD of 40 m roof rock is 61% whose significance is given in Table 1 (Fig. 5). A 4.5 m upthrow fault (F1) is encountered on the north side of the extracted and proposed CM panels and is traversing from west to east. The strike of the formation in the study area varies from WNW-ESE to NW-SE and dips towards NE. The dip of beds varies from 5° to 7°. The orientation of the face and butt cleats is almost perpendicular to each other and the spacing between the cleats and the length of the cleats are almost uniform everywhere. The face and butt cleat extends in the direction of maximum horizontal principal compressive stress (S_H) and horizontal minimum compressive principal stress (S_h) respectively. Data collected from the roof of the proposed panels reveal that two sets of joints are very prominent (CMPDIL Report 2018). The fracture mechanism for Indian coals is largely devoid of cleat dominance against the case of UK coals (CIMFR Report, 2019). In panel B12 generalized orientation of Joint 1 is N126° i.e. SE-NW and Joint 2 is N213°E. Similar orientations of joints are found in other CM panels (Fig. 6). The average density of Joint 1 and Joint 2 is 2 sets per metre and the lengths of the joints are up to a few metres.

Table 1. Cavability classification in coal measures on the basis of RQD% (Sheorey et al., 1995)

RQD%	Cavability	Proneness to airblasts
0–20	Friable	Not prone
20–40	Well cavable	Not prone
40–60	Cavable	May be prone with a working height greater than 4 m
60–80	Difficult to cave	Moderate to severe airblasts likely
80–100	Very difficult to cave	Very severe airblasts

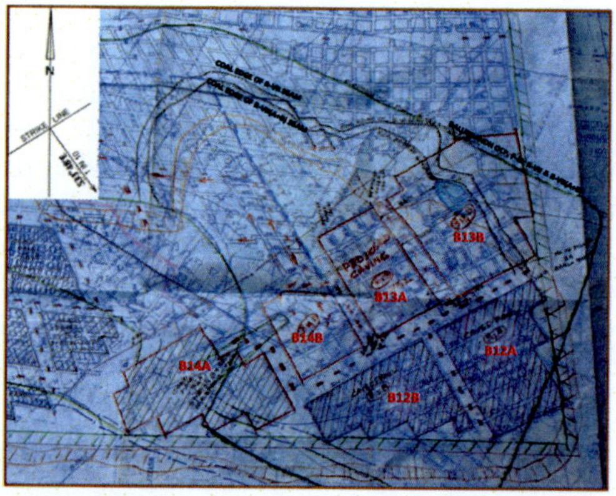

Fig. 3. A part plan of proposed CM panels (sub-sectionalized) in R-VI Seam at Khottadih Colliery, ECL.

3 Floor Bearing Capacity for Suitability of Introduction of CM

A strength assessment of the existing floor was done by Standard Penetration Test (SPT) on four locations at the proposed route of CM in the R-VI seam (Fig. 7). Based on the two criteria, (a) settlement criteria and (b) shear criteria, the bearing capacity of the floor for the proposed CM model was determined to be more than the expected maximum ground pressure on the floor i.e., 27 tonnes/m^2 exerted by the CM. Therefore, the existing floor of the R-VI seam at Khottadih Colliery is found to be safe for the deployment of CM (CIMFR Report, 2016).

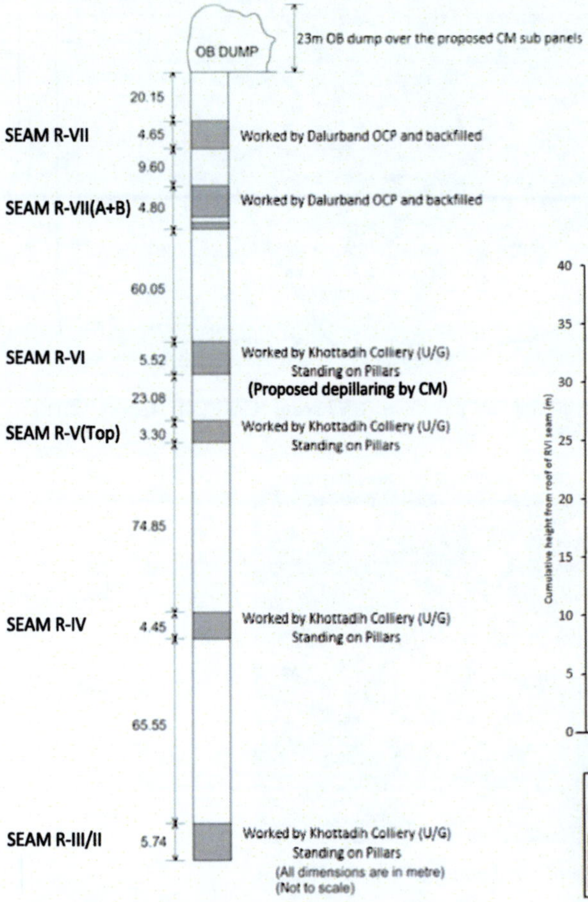

Fig. 4. Section of various coal seams through BH No. DLB-29 near proposed CM panels.

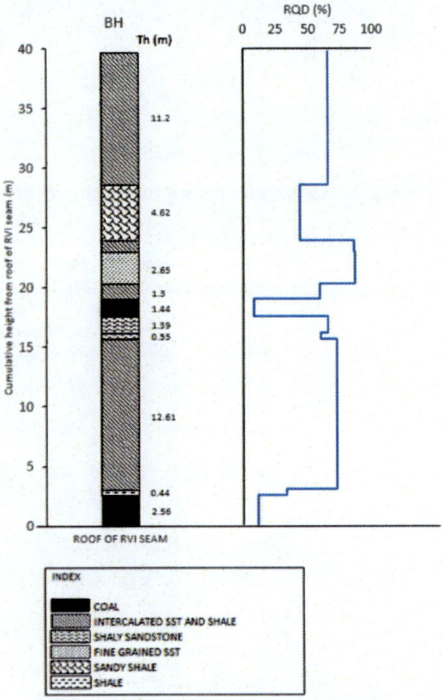

Fig. 5. RQD of roof rock strata of various strata formations above RVI seam in BH No. 5 located at 57L/47D of Khottadih Colliery, ECL

4 Extraction Methodology

The factor of safety (FOS) of the pillars is calculated empirically and by modelling after gallery widening and heightening is found to be less than 2.0. So, the "NEVID" method is suggested for extracting the pillars i.e. with slicing only and not by splitting and slicing which would otherwise disturb the pillar core. Also, the FOS of fenders after the splitting of pillars will be less than 1.0. The method of extraction is so designed that the movement of CM/shuttle cars for depillaring is restricted, to limit the deterioration of the floor. The existing pillars after gallery heightening and widening are extracted as early as possible. Thus, in the present workings of RVI seam, the pillar sizes are such that the pillars are amenable to be extracted by adopting the Nevid method. The panels

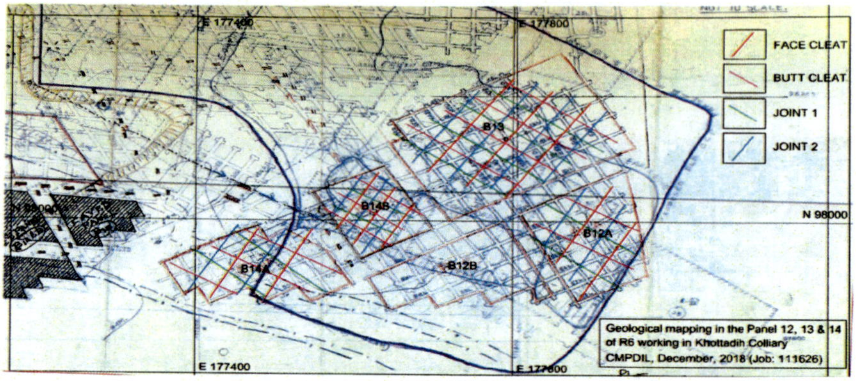

Fig. 6. Geological mapping in the proposed CM panels in R-VI seam (CMPDIL Report, 2018).

Fig. 7. A view of conducting SPT at R-VI seam.

are depillared by the caving method using CM in a straight line of extraction from dip to rise (Fig. 8). The typical layout of the Nevid method of pillar extraction comprises eight number of slices in one pillar of size 24 m × 24 m (corner to corner) with a dimension of a single slice in the pillar is 3.3 m (width) × 14 m (cut-out distance). These slices are to be made by remotely operated CM. During slicing, full seam thickness is extracted in single-lift liquidation. Before the commencement of the depillaring operation, the entire panel is heightened and widened simultaneously up to 5.4 m (i.e. full seam thickness) and 6.0 m respectively using CM equipment. The gallery widening is done by pillar thinning of two adjoining sides of each pillar in an L-shaped manner. The roof and sides of the heightened and widened galleries and junctions are supported immediately after CM operation. The design of the final stump against the junction is important for favourable ground conditions. Depillaring in the panels can be done when Dalurbandh OCP workings on the surface are sufficiently away from the underground workings of the panel under extraction.

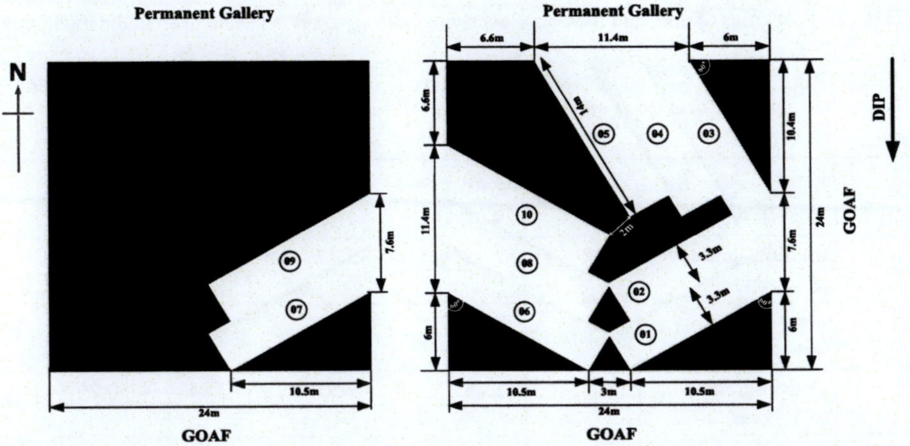

Fig. 8. Proposed pillar cutting design for pillar size 30m x 30m, centre to centre (the extraction line progresses from south to north or dip to rise and the pillar extraction sequence is moving from east to west)

5 Numerical Modelling Study

A 3D numerical modelling by FLAC3D, Itasca (Itasca 2016) is used. Sheorey's rock mass failure criterion is used to convert the intact rock properties to rock mass properties (Sheorey, 1997). The workings are discretized as per the dimensions, in-situ stresses (CMRI GAP Report, 2002) and the material properties (CIMFR Report, 2018) are assigned, collecting the testing results of the rock and coal samples (Table 2). The OB dump on the surface is taken as a broken material and is considered to provide impacts of gravity loading. The stresses σ_{1i} and σ_{3i} (major and minor induced principal stresses) around an excavation are computed from the numerical model and the stability is assessed by estimating the safety factor at different points and drawing the safety factor contours. The geo-mining parameters used in the modelling are given in Table 3.

5.1 Estimation of Factor of Safety of Coal Pillars/Ribs by Numerical Modelling

The pillar strength is estimated by using Sheorey's strength formula (Sheorey, 1992 & 1993). An empirical formula for the estimation of rib/stump strength has been used (CMRI GAP Report, 1987). The numerical modelling results are given in Table 4 and Figs. 9, 10, 11, 12, 13 and 14. If the FOS <1.0 but >0.5, pillar/stumps will stand for a few days. The stump (final snook) against the junction is so designed with F.O.S around 0.45.

5.2 Support Design

The support system is designed based on the rock load height calculated with the help of safety factor contours in the roof of the openings from the results of numerical modelling. The rock load height of a safety factor up to 1.5 is considered for calculating the rock

Table 2. Properties of coal and rock beds overlying/underlying R-VI seam used in the modelling

Bed	Rock type	Density (kg/m^3)	Compressive Strength (MPa)	Tensile Strength (MPa)	Young's Modulus (GPa)	Poisson's Ratio
1	Floor	2321	44.30	4.29	4.94	0.25
2	Coal (R-VI seam)	1470	20.91	1.42	2.00	0.25
3	Medium-grained sandstone	2232	41.75	3.47	2.88	0.25
4	Fine-grained sandstone	2335	39.02	2.84	5.09	0.25
5	Coarse-grained sandstone	2281	41.65	3.55	3.94	0.25
6	Medium-grained sandstone	2302	39.21	3.78	18.19	0.25
7	Intercalation	2198	31.56	2.42	8.00	0.25
8	Intercalation	2276	31.70	5.96	6.92	0.25
9	Broken material	1800	-	-	0.6	0.1

Table 3. Geo-mining parameters of the B12A panel used in the modelling

Average pillar size	25.5 m × 25.5 m (corner to corner)
Developed roadway (Bord) width	5.0 m (average)
Average roadway height	3.0 m
Depth of cover	100 m
Increased height of galleries	5.4 m (designed)
Increased width of galleries	6.0 m
Height of overburden	23.0 m

load. The rock load heights up to block safety factor contours (of value = 1.5) during heightening and widening are shown in Fig. 15 (a-c). Based on the rock density of the immediate roof, the reqFuired rock load to be supported is calculated (CIMFR GAP Report, 2006; CIMFR Report, 2017). The safety factor of the applied support system is given in Table 5. The rock load height, yield zone and support system are given in Figs. 15, 16, 17, 18, 19 and 20.

Table 4. Factor of safety of pillars, ribs and final snook

Sl No	Pillar/Ribs	Effective Width, W_e (m)	Strength (MPa)	Vertical Stress from numerical modelling results (MPa)	F.O.S
1	Pillar after heightening and widening of bords	24.00	7.9	4.4	1.8
2	Rib-1	20.00	6.85	6.45	1.06
3	Rib-2	15.40	5.64	7.09	0.79
4	Rib-3	11.34	4.58	7.72	0.59
5	Rib-4 (Final snook)	9.00	3.97	8.81	0.45

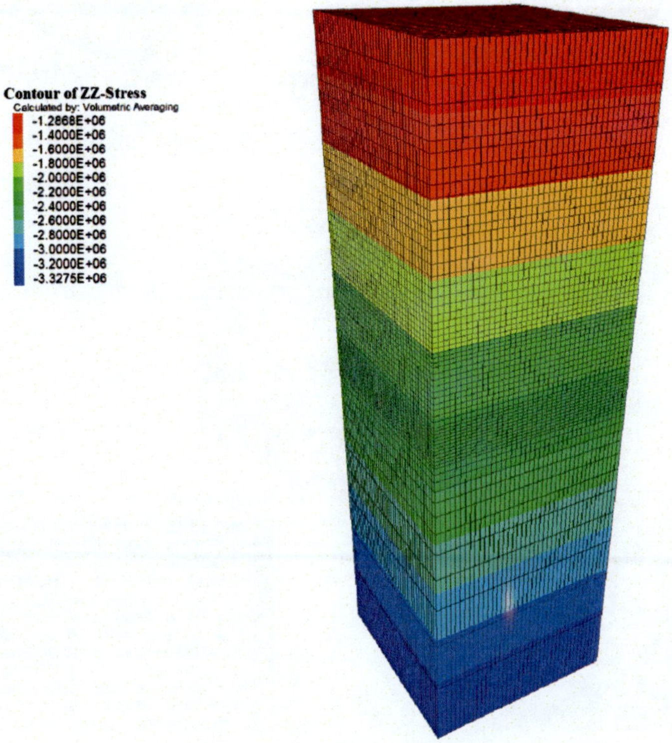

Fig. 9. In-situ virgin vertical stress (Pa) before excavation

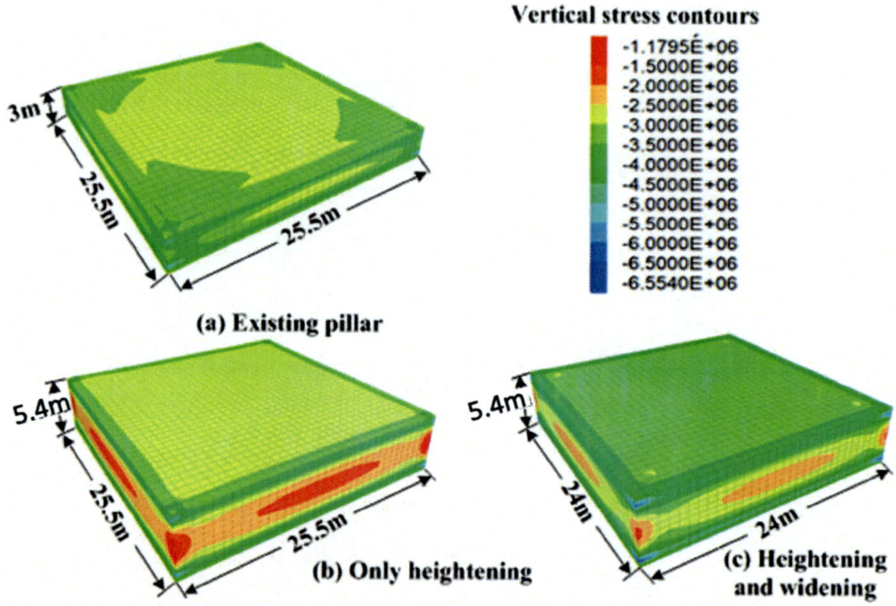

Fig. 10. Vertical stress contours (Pa) on the pillar

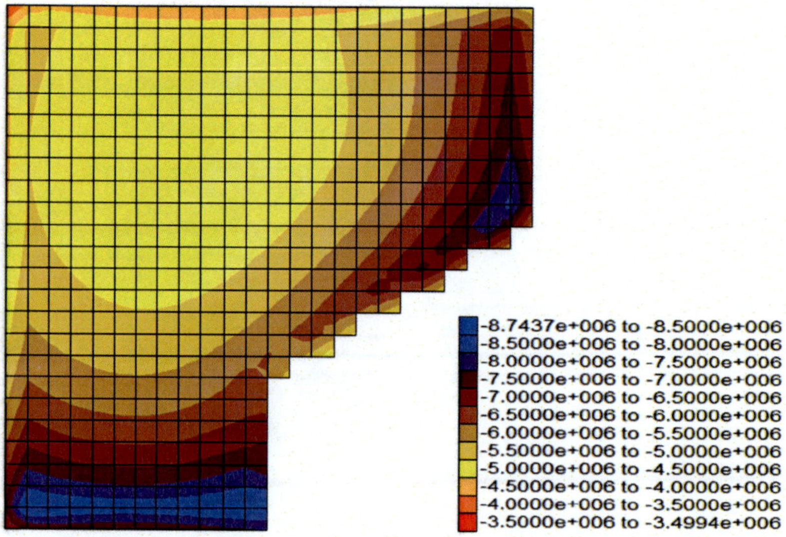

Fig. 11. Vertical Stress contours over Rib-1 after extraction of slices 1 and 2 (Refer to Fig. 8).

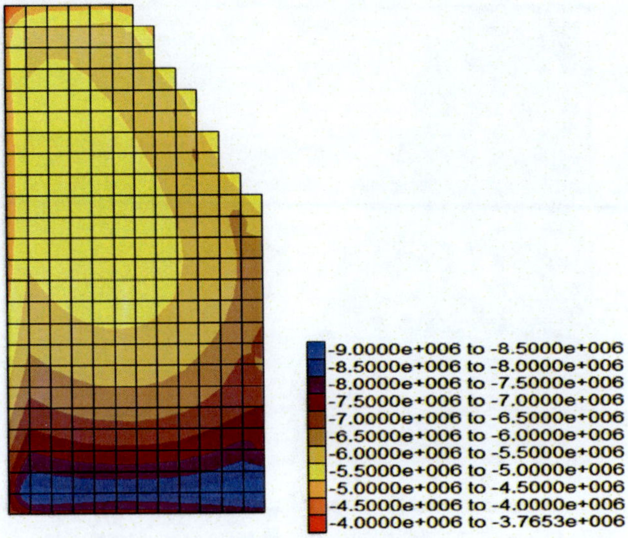

Fig. 12. Vertical Stress contours over Rib-2 after extraction of slices 1, 2, 3, 4 and 5 (Refer to Fig. 8).

6 Instrumentation Plan

Strata monitoring with geotechnical instruments is very essential for successful extraction in the CM panel. The geotechnical instruments used to monitor the strata behaviour during depillaring are given in Fig. 21. In addition to geotechnical instruments, Policeman props are installed against each of the breaker line/goaf edge support.

7 Field Implementation

Till now, five panels have been completed by adopting the Nevid method of extraction using CM in the R-VI seam at Khottadih Colliery. During extraction, caving occurred smoothly in all the panels. Roof falls were generally preceded by bumping sounds. Goaf was found to be well-packed and the fall had not crossed breaker lines (Fig. 22). In the working area, the roof and sides are found to be stable with the designed support (Fig. 23). In five CM panels, 113 pillars were extracted with pillars varying from 30 to 15 pillars per panel depending on the location and geominng condition. All the panels were extracted well within the incubation period of three months. Heating incidents have not occurred so far as was observed during the extraction of earlier panels. Earlier, the cable bolting method was adopted with an average of 12 pillars per panel. The percentage of recovery in the cable bolting method was around 64% while in CM panels in the Nevid method, it is around 74%.

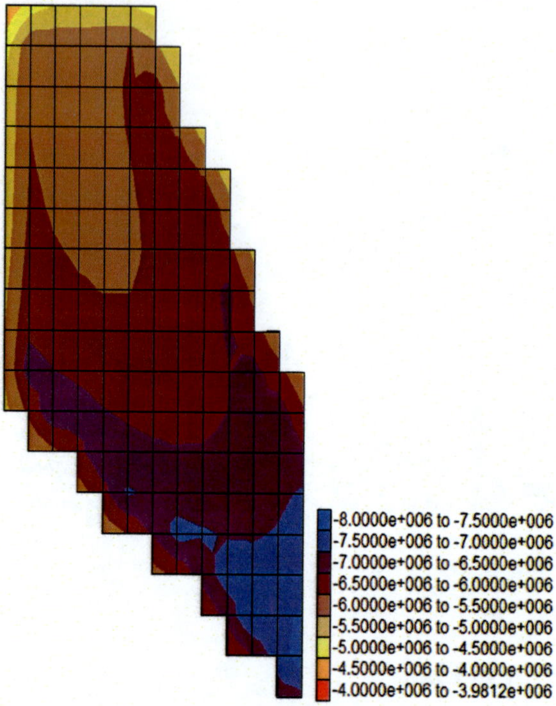

Fig. 13. Vertical Stress contours over Rib-3 after extraction of slices 6 & 8 in addition to earlier extraction, plus 7 & 9 in the next outbye pillar (Refer to Fig. 8).

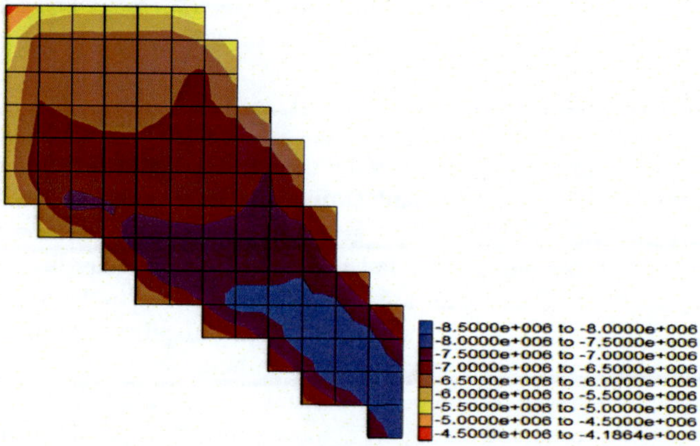

Fig. 14. Vertical Stress contours over Rib-4 after extraction sequences completed in the pillar and CM withdrawn to an outbye safe location

Table 5. Safety factor of applied support system

		Height (m) of safety factor 1.5	Rock load (t/m^2) considering the unit weight of rock as $2.5t/m^3$	Applied support load (t/m^2)	Support safety factor
Heightened and widened	Gallery	2.4	6.0	14.88	2.5
	Junction	3.3	8.25	21.8	2.6

The anchorage capacity/strength of roof bolts of length 2.5 m and bolt diameter 20–22 mm of TMT/MS ribbed bar and grouted with full-column resin capsules is considered as 25 t. The anchorage capacity/strength of full-column resin grouted flexi/bulbed cable bolts is considered to be more than 40 t

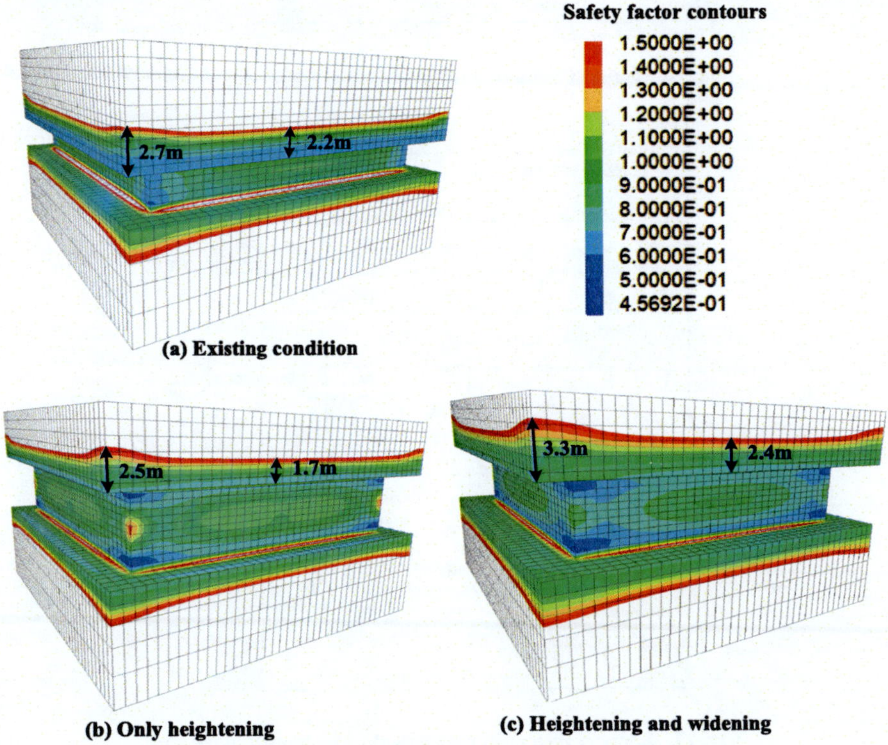

Fig. 15. Rock load height of safety factor contours up to 1.5.

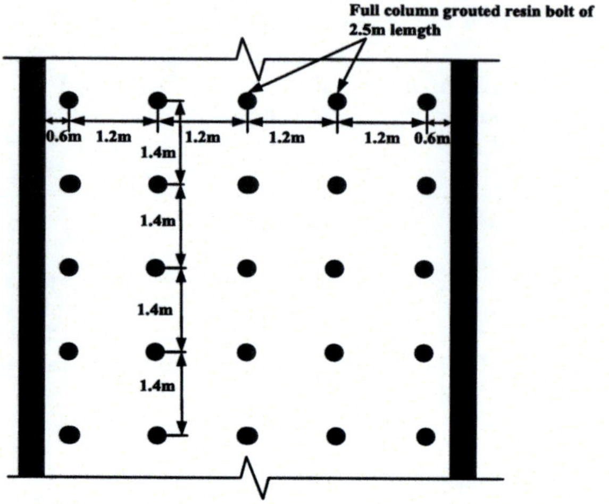

Fig. 16. Support system in heightened and widened gallery.

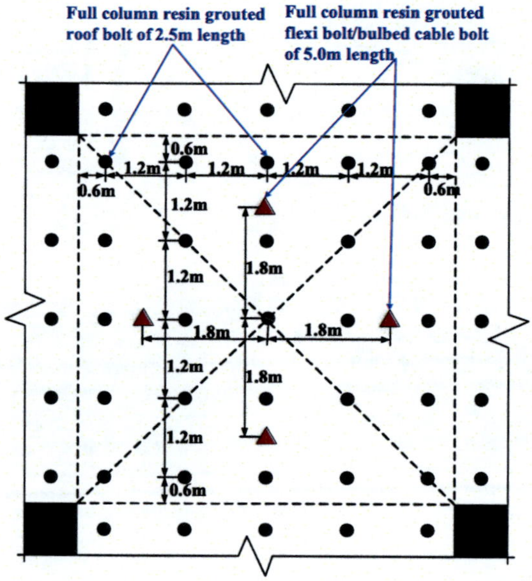

Fig. 17. Support system in heightened and widened junction.

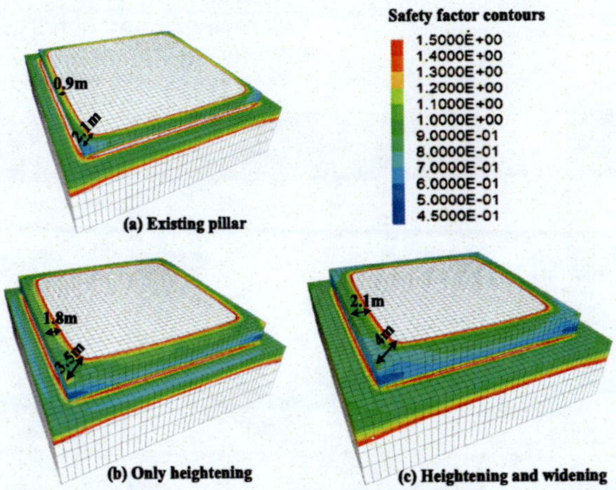

Fig. 18. Yield zone in the pillar with safety factor contours up to 1.5.

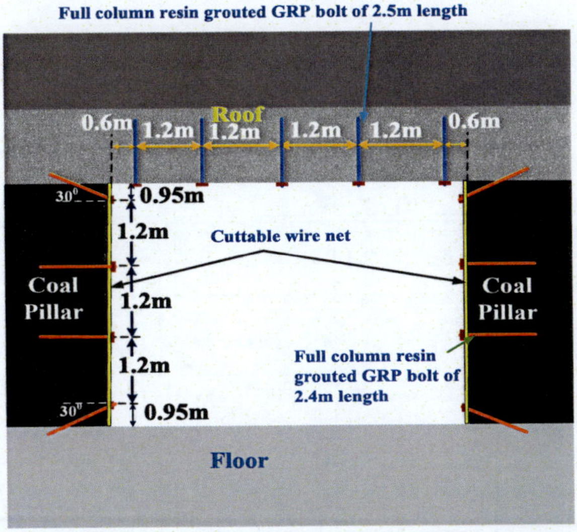

Fig. 19. Side support system in heightened and widened roadways.

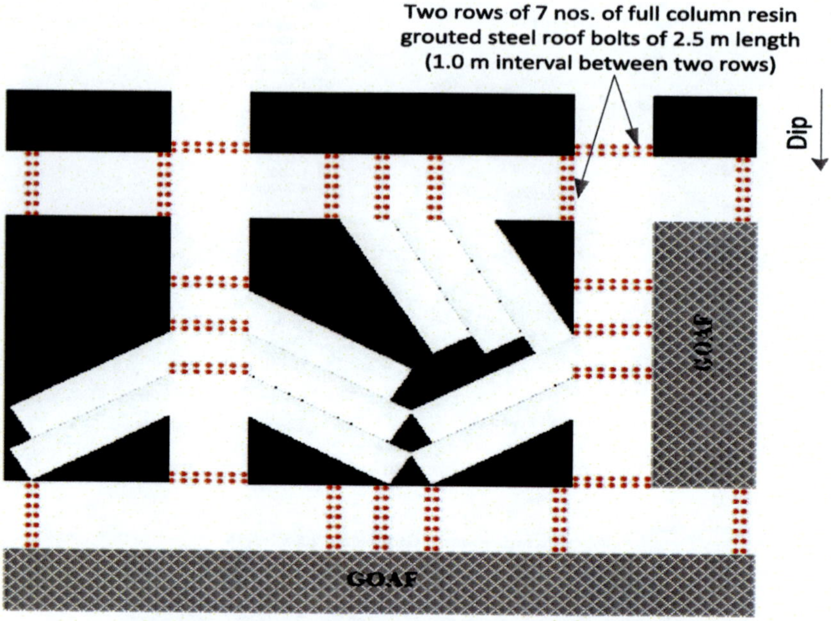

Fig. 20. Breaker line support for pillar size 24m x 24m.

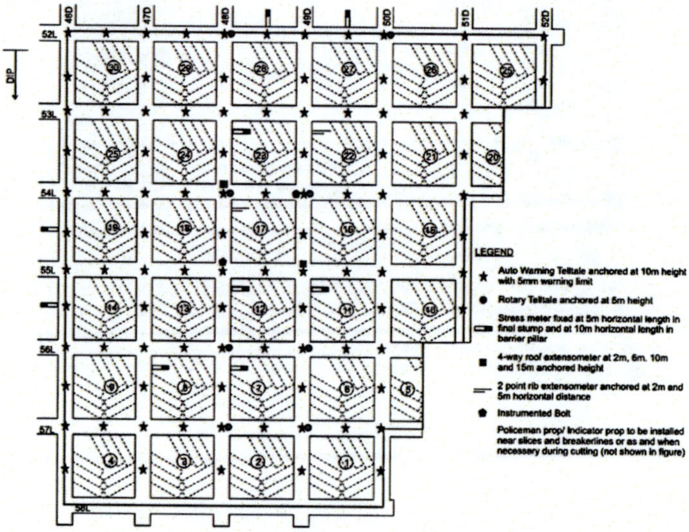

Fig. 21. Instrumentation Plan for CM panel B12A

Fig. 22. Roof fall showing packed goaf at 32D/47L in B14A panel

Fig. 23. Stable roof and sides with the designed support in B14A panel

8 Conclusion

By adopting a fast-paced extraction methodology i.e. Nevid method of extraction using a CM with increased cutting height, it becomes possible to extract the full seam thickness and enhance the number of pillars per panel, even in low incubation coal seams. On a pan-India basis, the single pass extraction height of CM has not exceeded 4.6m. Therefore, CM of single pass height extraction of 5.4m is the first such application in the Indian coal mining industry. Extraction of multiple panels was successfully executed, coping with the deterrents such as equipment size, weight, stability of workings, roof height, floor and face conditions, the low incubation period of the seam, and the presence of an OB dump. The study employed a systematic approach involving detailed field investigations, numerical modelling exercises, stability analysis of underground workings, and comprehensive consideration of rock mechanics aspects, strata control and monitoring plans. The study has significant implications for the coal mining industry, particularly in similar extraction scenarios, and can effectively bridge the gap between efficiency and safety, thereby maximizing productivity while minimizing risks associated with low incubation periods and challenging geological conditions.

Acknowledgements. The authors are thankful to Director, CSIR-CIMFR, Dhanbad, India for his kind permission to publish/present this paper. The authors thankfully acknowledge the support received from Dr. S. K. Singh, Ex. Chief Scientist, CSIR-CIMFR, Dhanbad. The authors are thankful to the CMATPL management for providing all the necessary data and information for this research study. The team expresses deep gratitude to the management of Khottadih Colliery, ECL for providing help related to the proposed area of the CM panel/sub-panels to complete the study. The views communicated in this paper are that of the authors and not necessarily of their organisation they belong.

References

CIMFR GAP Report: Development of Support Guidelines for Depillaring Panels in Indian Coal Mines. Submitted to the Ministry of Coal, Govt. of India (2006)

CIMFR Report: Advice on Load Bearing Capacity in order to decide suitability of introduction of Continuous Miner of Single-Pass height of 5.4 m in R-VI Seam of Khottadih Colliery, ECL: the first time such application in India. CNP/4424/16–17. Submitted to ECL, Burdwan, W.B.-713378 (2016)

CIMFR Report: Advice for Depillaring of R-VI Seam with deployment of Continuous Miner with Increased Height and Width of Extraction at Khottadih Colliery, ECL. CNP/4149/2014–15. Submitted to ECL, Burdwan, W.B.-713378 (2017)

CIMFR Report: Assessment of Physico-mechanical Properties of Rock Core Samples from Khottadih Colliery U/G Borehole, Eastern Coalfields Limited and Advice thereof. SSP/295/2018–2019. Submitted to ECL, Burdwan, W.B.-713378 (2018)

CIMFR GAP Report: Report: To find a methodology of safe liquidation of thick seam of Raniganj coalfields: design, development & showcasing demonstrative trials at Khottadih Colliery, ECL. CSIR-CIMFR, Grant-in-aidProject, GAP/MT/CIL/100/2014–15, Funded from the CIL R&D Board, 125 p. (2019)

CIMFR Report: Advice for Depillaring of B12, B13 And B14 Panels in R-VI Seam with Continuous Miner at Khottadih Colliery of Pandaveswar Area, ECL. CNP/4720/2018–19. Submitted to CMATPL (2019)

CMRI GAP Report: Optimisation of the design of mine pillar parameters and feasibility of extraction of locked-up coal below built-up structures, water logged areas and hard covers. Unpublished S&T project report, CMRI Dhanbad, March 1987, 100p.

CMRI GAP Report: In-situ Stress Measurement in Underground Coal Mines and its application to stability analysis, Submitted to Ministry of coal, Govt. of India (2002)

CMPDIL Report: A Report on Geological Mapping with Rose Diagram for R-VI Seam at Khottadih Colliery, ECL, District- Paschim Bardhaman (W.B.) (2018)

Fettweis, Gunter, B., Fisguss, S.W.: Selection of optimal mining system for underground mining of very thick coal seam. In: Singh, T.N., Dhar, B.B. (eds.) Proceedings of International Symposium on Thick Seam Mining, pp. 159–69. CMRI, Dhanbad, India (1992)

Ghose, A.K.: Underground methods of extraction of thick coal seams—a global survey. Min. Sci. Technol. **2**(1), 17–32 (1984)

Itasca: Itasca Consulting Group Inc, FLAC3D (Fast Lagrangian Analysis of Continua in 3 dimensions). Version 5.0. Minneapolis, MN (2016)

Kumar, R., Singh, A.K., Mishra, A.K., Singh, R.: Underground mining of thick coal seams. Int. J. Min. Sci. Technol. **25**, 885–896 (2015). https://doi.org/10.1016/j.ijmst.2015.09.003

Mangal, A.: Strata stability investigation and convergence monitoring (SSICM) in thick-seam depillaring with caving by cable bolting method. Min. Metall. Explor. **38**(2), 927–944 (2021). https://doi.org/10.1007/s42461-020-00371-x

Mangal, A.: A study of stability investigation of immediate roof for extraction of thick coal seam in India. J. Inst. Eng. (India): Ser. D **102**(1), 203–221 (2021). https://doi.org/10.1007/s40033-021-00264-3

Mangal, A., Sarkar, F.: Underground coal pillar extraction under shale roof and caved goaf of upper lying seam in Raniganj Coalfields. J. Inst. Eng. India Ser. D **102**(1), 223–236 (2021). https://doi.org/10.1007/s40033-021-00254-5

Singh, R., Singh, T.N., Dhar, B.B.: Coal pillar loading for shallow mining conditions. Int. J. Rock Mech. Min. Sci. **33**(8), 757–768 (1996)

Singh, R., Singh, T.N.: Investigation into the behaviour of a support system and roof strata during sublevel caving of a thick coal seam. Geotech. Geol. Eng. **17**, 21–35 (1999)

Singh, R., Mandal, P.K., Singh, A.K., Singh, T.N.: Cable-bolting-based semi-mechanised depillaring of a thick coal seam. Int. J. Rock Mech. Min. Sci. **38**, 245–257 (2001)

Sheorey, P.R.: Pillar strength considering *in situ* stresses. Workshop on Coal Pillar Mechanics and Design, Santa Fe. US. Bur. Mines, I.C. 9315, pp. 122–127 (1992)

Sheorey, P.R.: Design of coal pillar arrays and chain pillars. In: Hudson, J.A., et al. (eds.) Comprehensive Rock Engineering, vol. 2, pp. 631–670. Pergamon, Oxford (1993)

Sheorey, P.R., et al.: Application of the yield pillar technique for successful depillaring under stiff strata. Int. J. Rock Mech. Min. Sci. Geomech. Abstr. **32**, 699–708 (1995)

Sheorey, P.R.: Empirical Rock Failure Criteria. Balkema, Rotterdam (1997)

Performance of Mechanised Depillaring Operations with SDL and Continuous Miner Technology - A Comparative Study

Mithilesh Kumar Mishra[1](✉) and Singam Jayanthu[2]

[1] Department of Mining Engineering, ICFAI University Jharkhand, Ranchi 835222, India
mithileshkumar.m@iujharkhand.edu.in

[2] Department of Mining Engineering, National Institute of Technology, Rourkela 769008, India
sjayanthu@nitrkl.ac.in

Abstract. All countries, who are rich in mineral resources and able to exploit them in judicious and safe manner, achieve faster growth in industrialization and become model of development of modern civilization. India is placed at 3rd position in terms of production. But maximum coal production is from opencast and more importantly the production from underground has a declining trend. To arrest this trend coal companies have started thinking to introduce mass production technology such as continuous miner in underground mines. Due to faster rate of production and advanced safety feature of continuous miner loss of coal is minimized and safety of workers is enhanced along with ensuring conservation. The present study is aimed to analyse overall performance of continuous miner technology and to highlight key points to be taken care of for achieving higher and better utilization of this valuable resource. Cost of production, productivity, and maximum production in the depillaring panel/day of the Side Discharge Loader workings with semi-mechanisation are to the tune of Rs 11000/day, 0.79 and 500 Ton/day respectively whereas with Continuous Miner with full mechanisation the same figures becomes Rs 3800/day, 8.14 and 2000 Ton respectively in typical geomining conditions of depillaring workings in CCL.

Keywords: Mineral resource · Underground mine · Mass Production Technology · Safety · Continuous Miner · Production and productivity

1 Introduction

Indian coal industry has a history of more than 140 years. Today India is the second largest coal producing country in the world after China.India is followed by USA in coal production. Coal is one of the major source of energy in the world. Energy security is very important for all the developing countries. India, being a fast developing country, has to ensure sufficient energy supply to maintain the pace of development.

Coal is the backbone of country's energy supply and India meets about 51%of its energy requirement from coal. (Reference-Installed fuel wise generation capacity electricity report, source: Central Electricity Authority of India).

A. Sinha et al. (Eds.): AMC 2023, SPEES, pp. 130–136, 2023.
https://doi.org/10.1007/978-3-031-46966-4_10

In 1971 when the government nationalised coal industry, 77.45% of the total coal production came from underground and only 22.55% was produced by opencast (Sarkhel & Dey). But today the situation is just reverse and if we take the figure of total production of Coal India in 2023,677.72MT of coal has been produced by opencast whereas only 25.49MT has come from underground (Annual report of CIL). The production of underground & OC since nationalization has been clearly shown in following table:

The data of this table can easily be explained through the following chart (see Fig. 1).

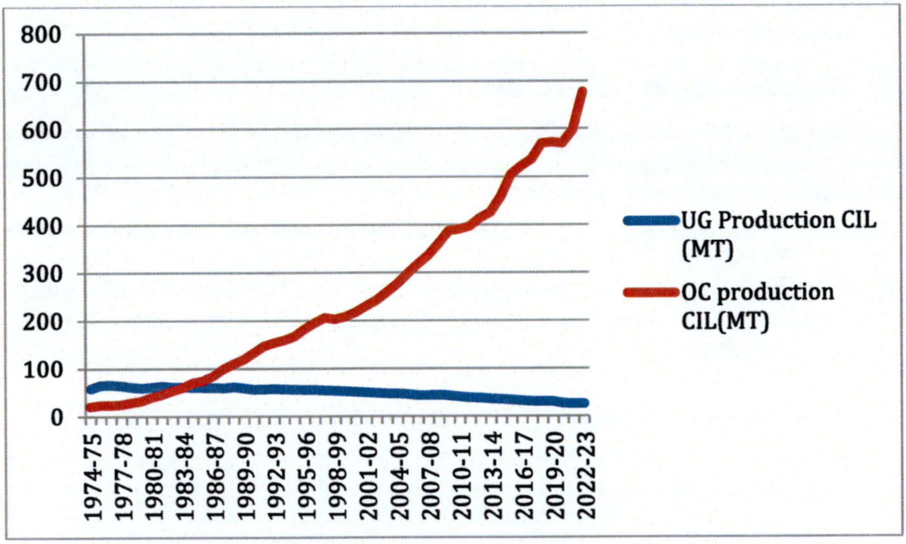

Fig. 1. UG and OC coal Production of CIL since 1974–75

Growing concern from environmentalist, depleting reserve at shallow depth and quality concern has forced coal industry to look forward towards underground and enhancement in coal production from underground was felt necessary. The production from underground comes either from Bord and Pillar method or from Longwall method. In India about 90% of underground production comes from Bord and pillar and the rest comes from Longwall (?). Further in our country a huge reserve is standing on pillar. So it has become necessary to adopt a technique suitable for enhancing the production from Bord and Pillar. Initially Side Discharge Loaders (SDL) and Load Haul Dump (LHD) were introduced and the mines adopted semi mechanisation but the desired output could not be achieved. So, Continuous Miner, a mass production technology suitable for Bord and Pillar was introduced. In India the first Continuous miner was introduced in Anjan Hills Mine of WCL in the year 2002. After that it is being observed that where ever the geo-mining & geo-technical conditions permit continuous miner technology is being introduced. Before introduction of Continuous miner technology, financial feasibility report and scientific study for strata behaviour and its control technique is done. As a huge quantity of coal is standing on pillar CM technology is introduced not only for virgin seams but also for old B&P Mines. In our study we have chosen mines where CM has been introduced in developed mines. Similarly those mines which are having almost similar conditions and using SDL technology has been considered.

This paper is a comparative study of Semi mechanised mines having deployed SDL/LHD with a mine where Continuous Miner is in operation. For comparative study we have taken two mines Dhori Khas where depillaring with SDL and Churi where depillaring with Continuous Miner are in operation. Similarly we have studied data for two different mines of Bankola Area of Eastern Coalfields Limited where SDL and Continuous miner are running.

2 Mining Method

2.1 Semi Mechanisation (SDL/LHD)

In Indian geo mining conditions the most suitable method is Bord and Pillar which consists of two stages. First stage is development in which a series of galleries are driven at right angles to each other and the remnant coal is called pillars which basically support the super incumbent strata. In the second stage these pillars are extracted by forming stook and drivingslices.After drilling and blasting the SDL/LHD loads the blasted coal and as the loading gets faster the production improves. Production further improves when Belt conveyor is installed in association with SDL/LHD. This system is known as semi mechanisation.

2.2 Mass Production Technology (Continuous Miner)

Continuous Miner is a mass production technology and can be used very well in both the stages of Bord and Pillar method.

2.3 Development

In development to achieve high production with safety generally five headings are driven. Generally one continuous miner with two battery hauler or electric or diesel shuttle cars are engaged along with one feeder breaker and one Quadra boom roof bolter capable of drilling on wet and having temporary support system. Feeder breaker normally crushes coal to −200 mm and fed the crushed coal to belt conveyor. An electric equipment load centre is generally provided at a central point and supply power to all sub-assemblies as well as continuous Miner. A LHD is also provided to do auxiliary job such as cleaning for shifting of conveyor and tramming route and face.

2.4 Depillaring

When development is complete, extraction of pillar known as depillaring is started. In this case the cutting drum which is normally 3.3 M wide is sumped in just below the roof level and then the drum starts coming down to the bottom of the pillar. Repeating the same cycle the machine cuts the coal upto permitted cut off distance which vary from case to case depending on the scientific study. When the machine moves forward the gathering apron enters into the cut coal heap and coal is loaded very fast in to the shuttle car or battery hauler. Generally two commonly used methods are Split and Fender Method and Navid Method.

2.5 Split and Fender Method

In split and fender method the pillar is divided in two parts by a dip rise split. This split is supported by roof bolts and then from split we extract the fenders so that it looks like two wings. The original galleries are also heavily supported (see Fig. 2).

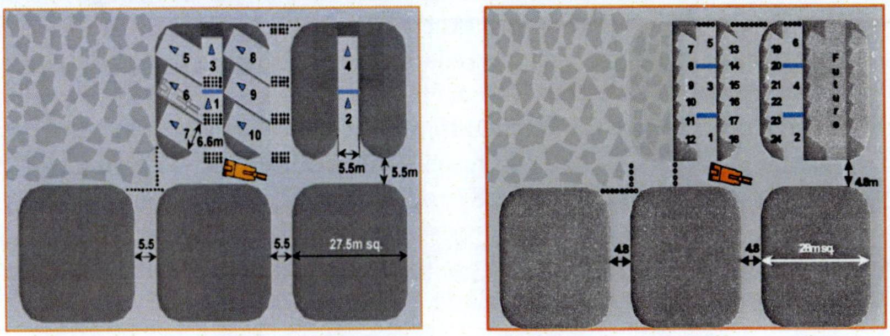

Fig. 2. Split and Fender depillaring systems

2.6 Nevid Method

In Nevid Method the pillars are extracted from original galleries as shown in the adjoining figure. The pillar is cut from three sides and a solid lump of coal is left out in the middle of the pillar. This is very clearly shown in the following figure (see Fig. 3).

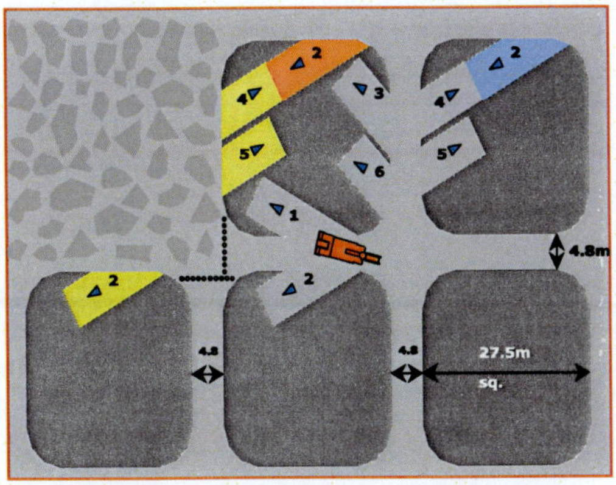

Fig. 3. Nevid method of depillaring

2.7 Comparative Study between SDL and Continuous Miner Technology

For Comparative study the detailed data of Production, Manpower, Productivity, cost of production, Sale value and profi/Loss were collected from different Mines of CCL and

ECL From CCL Churi mines where CM technology has been introduced and Dhori Khas where SDL is in operation were selected and the current data were collected. Similarly from ECL two continuous miner mines namely Sorpi and Nakarkonda Kumardih B and two SDL Mines KumardihA and ShyamSunderPur were chosen and all relevant data were collected.

These data were studied and wasfound tht all the SDL mines are in huge loss and all the continuous miner mines are in profit except Churi which is in loss. But churi mines after introduction of continuous iner has reduced its loss from Rs 11000/Te loss to only Rs 700/Te loss thereby saving a huge amount.

Similarly on productivity front the OMS of SDL mines namely Dhori Khas, SSPur and Kumadih A is 0.79, 0.76 and o.66 respectively whereas the OMS of Churi, Sorpi and Nakarkonda Kumardih B is 8.14, 8.87 and 10.59.

The details of all important parameters such as manpower, OMS, Production, cost of production/Te, Sale price per Te and profit loss of these mines are shown in a table which is on the following Page (see Table 1).

Table 1. .

Name of the Mine	Manpower	OMS	Mechanisation	Total annual Production (Te)	Cost /Te (Rs)	Sale Value (Rs)	Profit/Loss /Te (Rs)
Kumardih A	569	0.66	SDL	93586	35531	8683	(−) 26848
Nakrakonda KumardihB	512	10.59	CM	1328613	2877	5828	(+) 2951
SSPur	969	0.76	SDL + LHD	164653	32002	9561	(−) 22423
Sorpi	180	8.87	CM	496147	2656	6603	(+) 3947
Dhori Khas	604	0.79	SDL	128000	11029	3210	(−) 7829
Churi	337	8.14	CM	706000	3792	3024	(−) 768

On analysis of the data of these mines shown in the table we can easily conclude that whenever the geo-mining and geo-technical condition are suitable a we should adopt mass production technology like CM. CM is successful even in already developed mine. Firstly CM does redevelopment and then depillaring is done.

3 Conclusion

- As continuous miner is a mass production technology it enhances the production of underground mine substantially.
- In churi mine it was found that the loss has reduced very much and thereby mine has earned a notional profit.
- This technology is suitable for redevelopment of mine and then depillaring can be done very successfully.

- With CM High % of extraction to the tune of 70%. Can be achieved Iprovement in production can be achieved if the sub systems are managed in such a way that they operate on full efficiency with zero down time. This can be achieved by adopting preventive maintenance.
- If we compare the cost per unit for a CM district VS SDL /LHD district it is found that feasibility of CM technology is far better than Conventional SDL/LHD district.

References

1. Kumar, N., Sai Kiran, K., Sadasiva Rao, K., Trivendra Reddy, S., Naveen Kumar, S.: Optimization of continuous miner operations in bord and pillar mining. Int. Res. J. Eng. Technol. **7**(6), 3395–3401 (2020)
2. Banerjee, S.: Performance evaluation of continuous miner based underground mine operation system: an OEE based approach. New Trends Prod. Eng. **2**(1), 596–603 (2019)
3. Prasad, M.: Underground coal mining: The way ahead.Projects Monitor, 29 December, 2008-4 January, 2009. [online] http://www.projectsmonitor.com/detailnews.asp?newsid=17688. (2009)
4. Raghwan, V., Ariff Dr, S., Kumar, P.P.: Optimum utilization of continuous miner for improving production in underground coal mines. Int. J. Sci. Res. Publ. **4**(10), (2014)
5. Marepalli, R., Jain, S.C.: Optimum utilization of continuous miner used for pillar extraction in bord and pillar mining method and support requirement in various roof conditions. Int. Res. J. Eng. Technol. **6**(10), 798–803 (2019)
6. AKS University. Deptt. of Mining Engineering (Unpublished), Report on Churi Underground Coal Mine, CCL Strata Monitoring at Panel CMP-W/15 and Geotechnical Assessment for the Redevelopment & Extraction of Pillars in Panel CMP-W/1, On behalf of JMS Mining Pvt Ltd, Kolkata (2019)
7. Mishra, D.P., Sugla, M., Singha, P.: Productivity improvement in underground coal Mines-A case study. J. Sustain. Min. **12**(3), 48–53 (2013)
8. Buddery, P.S., Oldroyd, D.C., Hudson, J.A. (1992/2015-online) Development of a roof and floor classification applicable to collieries, Procc. Of Rock Characterization: ISRM Symposium, Eurock '92, Chester, UK, 14–17 September (1992)
9. Bartels, J.R., Gallagher, S., Ambrose, D.H.: Continuous mining a pilot study of visual attention locations and work position in underground coal mines. Prof. Safety **54** (2009)
10. Best Practices for Continuous Miner Operations. Technical Report January 2009, Research Gate, Contents uploaded by Christopher C. Jobes on 31 July (2015)
11. Modi, J., Bharti, S., Kant, R.: Applicability of continuous miner in room and pillar mining system: higher production and productivity with safety. In: Conference Paper, January2017, Research Gate, Content Uploaded by Sujeet Bharti on 27January 2017 (2017)
12. Samanta, S., Sinha, R.K., Chakraborty, P.B., Kumar, H.: deployment of continuous miner in underground coal mine – a case study of sarpi mine. Res. Square December 21st (2021)
13. Taviti Naidu, P., Krishna Kanth, M.B.V., Keshva Kumar, P., Venkateshwara Rao, M., Ranjith Kumar, S.: Case study of continuous miner along with roof bolter. Int. Res. J. Eng. Tech. **7**(5), 655–657 (2020)
14. Ashok: Study of Depillaring by Continuous Miner in GDK 11 incline
15. Mukhopadhyay, S., Sharma, S.K.: A case study illustration of the layout and support in an indian coal mine with difficult roof condition where continuous miner is in operation. Semantic Scholar (2017)

16. Modi, J., Bharti, S., Kant, R.: Applicability of continuous miner in room and pillar mining system: higher production and productivity with safety. In: International Conference on deep Excavation, Energy Resource and Production Kharagpur (2017)
17. Vegenas, N., Runciman, N., Clemant, S.R.: A methodology for maintenanvce analysis of mining equipment. Int. J. Surf. Min., Reclaimation Environ. **11**(1), 33–40 (1997)
18. Anani, A., Nyaaba, W., Hekmat, A., Córdova, E.A.: Optimizing cut-out distance for maximum coal productivity. Sage Journals **95**(6), 545–559 (2018)
19. Raghwan, V., Ariff Dr, S., Kumar, P.P.: Determining the optimum utilization of continuous miner for improving production in underground coal mines. New Vision Sci. Technol. **9**, 73–86 (2021)
20. Bartels, J.R., Jobes, C.C., Lutz, T., Ducarme, J.P.: Evaluation of work positions used by continuous miner operators in underground coal mines. In: Conference Paper of the Human Factors and Ergonomics Society Annual Meeting, vol. 53, (20), pp. 1622–1626 (2009)

AI/ML Applications and Smart Mining

AI-Based New Tool for Semi-Automatic Interpretation of Reflection Seismic Data

Kalachand Sain(✉)

Wadia Institute of Himalayan Geology, 33 GMS Road, Dehradun 248001, India
kalachandsain7@gmail.com

Abstract. Seismic measurements on the Earth's surface are extensively used to delineate subsurface geologic features for the exploration of geo-resources and understanding geodynamics or seismo-tectonics. Over the decades, seismic attributes (properties or characteristics) extracted from 3D seismic volume have revolutionized interpretation of subsurface geologic environment. It is the high performance computing systems that have allowed processing of voluminous data within a reasonable time, but the interpretation by human analysts still remains tedious, particularly in a complex area. This has necessitated automatizing the process of interpretation to reduce human intervention. We have designed several work flows for merging multiple attributes related to a subsurface structure into a single 'meta-attribute' that has allowed to delineate the 3D configuration of that feature from a large volume of data, even in a complicated region. This artificial neural based approach integrates the intelligence of human assisted machines over a small volume of data, followed by applying the trained system to entire volume for exploring the geologic feature quite accurately at a fast rate. Success stories from a variety of basins across the world including India would be presented.

1 Introduction

Human is one of the most wonderful creation, and the human body is designed to perform several tasks simultaneously. The human brain controls these operations and is considered the most intelligent machine. However, such a smart machine often fails to demonstrate its superiority and efficacy due to some limitations. If you are asked to identify your family member from a crowd, you can do that job within no time. Because, you can identify them using certain qualitative parameters such as white or dark complexion; tall or short height; healthy or weak features; facial characteristics etc. This intelligent machine is a master in qualitative analysis. Again, this task becomes tiresome, if the crowd is very large. Another example, if you are asked to perform certain quantitative tasks like (i) weight or height of each person in the crowd, (ii) how many are boys or girls, (iii) how many are taller than 4 feet, etc., the brain gets exhausted in delivering accurate answers after some time. This is because our intelligent machine or human brain cannot exchange information by sending and receiving signals through the network of our nerve connections continually without any break or rest (Kumar, 2019; Sain and Kumar, 2022).

A. Sinha et al. (Eds.): AMC 2023, SPEES, pp. 139–144, 2023.
https://doi.org/10.1007/978-3-031-46966-4_11

On the other hand, if an engineered machine (calculator or computer), is programed to do such jobs, we get the answers rapidly and easily. An engineered machine may not be so smart or intelligent but has the ability to work continuously and deliver the desired outcome quite accurately. When, they are guided by the most intelligent machine (i.e., human brain), we can obtain optimum solution to a large and complex problem at much ease. This is the reasons why the artificial intelligence or machine learning is getting importance to almost all fields of science, technology, medicines, traffic, games, economy.

The seismic data, acquired on the surface, are rigorously pre-processed and processed using several steps for obtaining improved images of the subsurface, which are being utilised by interpreters for generating encyclopaedic knowledge or information for several purposes with regard to exploration of resources and comprehending geotectonics or seismotectonics. In this context, delineation of subsurface geologic features and their geometry, and variation of physical properties play crucial role for taking certain decision. Seismic method is one of the most suited geophysical techniques that provides accurate subsurface information in terms of structures and properties. Among several interpretation techniques, use of seismic attributes is laudable, as they have the ability to capture geologic responses of the target from the data and guide interpreter for wise judgement with regard to delineation of subsurface geologic features (Sain and Kumar, 2022). Suppose an interpreter is exploring networks of faults from a given volume of seismic data, he/she then looks for certain seismic responses of these features e.g., nature of discontinuity of seismic reflections, which are defined by attributes like similarity and coherency. Interpreter may observe discontinuous structures or faults associated with the low coherency values. However, this attribute may reveal presence of meandering channel complexes also. Thus, an attribute barely responds to a specific geologic feature and thus cannot be used in discriminating geologic target of different kind (Kumar and Sain, 2018; Sain and Kumar, 2022). A single attribute may provide a mixed set of geological information, and thus impedes final interpretation of the geologic target. Seismic attributes, extracted from the seismic data, provide maximum information about a subsurface target or feature from the data volume. But if we rely upon a number of attributes, specific to the faults, we shall be able to distinctly identify the fault system. Now the question arises if a single attribute can delimit the 3D architecture of the fault network. To address this question, we show here that a single hybrid or meta-attribute can serve the purpose, and be used for semi-automatic delimitation of its configuration in the subsurface.

Based on the analogy, as explained in the second paragraph, we have integrated the human intelligence with the machine intelligence to explore the fault system from the entire data space by taking into several attributes, characteristics to the faults, and fusing them into the single meta-attribute using artificial intelligence.

The present work highlights the application of this approach over several prospects in the Taranaki Basin, offshore New Zealand (NZ) and the Indo-Gangetic peripheral foreland basin, onshore India. The output of these works have been validated using available information. Several other examples are available in a book by Sain and Kumar (2022).

2 Application of Seismic Meta-attributes for Delimitation of Faults and Fluid Migration

2.1 Parihaka Prospect for Fault Geometry in Taranaki Basin (TB) off NZ

The 3-D time-migrated seismic data of the Parihaka prospect consists of 1132 inlines and 2904 xlines. The seismic data covers an area of 1028.55 km^2 and was surveyed using 25.0 m × 12.5 m (inl/xl) bin spacing, 4 ms sampling interval, 60-fold coverage and 6.0 s data recording length.

Seismic properties, associated with fault networks are: (i) discontinuities in reflection and sharp changes in dips (Kumar and Sain, 2018; Kumar et al., 2019); and (ii) loss of energy and frequency content (Kumar and Sain, 2018). Hence, seismic attributes like similarity, ridge enhancement filter (REF), thinned fault likelihood (TFL), reference time, and signal-to-noise (S/N) ratio are amalgamated into a single Thin Fault Cube (TFC) meta-attribute to distinctly arrest the fault system from seismic data (Fig. 1).

Fig. 1. 3D view along inline 2248 and xline 4586, co-rendered with the Thinned Fault Cube meta-attribute (TFC), displaying a clear demarcation of thinned fault networks interpreted from seismic data.

Though multiple seismic attributes have been used in delineating faults by several authors (Tingdahl and de Rooij, 2005; Zheng et al., 2014; Kluesner and Brothers, 2016; Huang et al., 2017; Kumar and Mandal, 2017) during the last decade, the present study aims to generate an optimized thinned fault-cube (TFC) meta-attribute for the delineation of thin razor shaped fault system (Fig. 1) from seismic data.

2.2 Indo-Gangetic Peripheral Foreland Basin for Fluid Migration

Fluid Flow through the subsurface strata refers to the passage of fluids from a subsurface source to the surface via hypertonic zones, which can be seen on seismic sections associated with as a variety of geologic features like chimneys, pipes, diapirs, vents, mud volcanoes and related physio-chemical changes in sediments. The fluid flow is identified as a localized and narrow columnar feature, associated with disrupted and chaotic reflection geometry, anomalous amplitude and atypical velocity (Løseth et al., 2009; Cartwright and Santamarina, 2015; Omosanya et al., 2020). The studies pertaining to fluid flow have been of interest because of high-resolution seismic data and apprehension of their ability to delineate features related to hydrocarbon prospects, gas hydrates deposits, magmatic activities, geo-hazards, and paleo-environment.

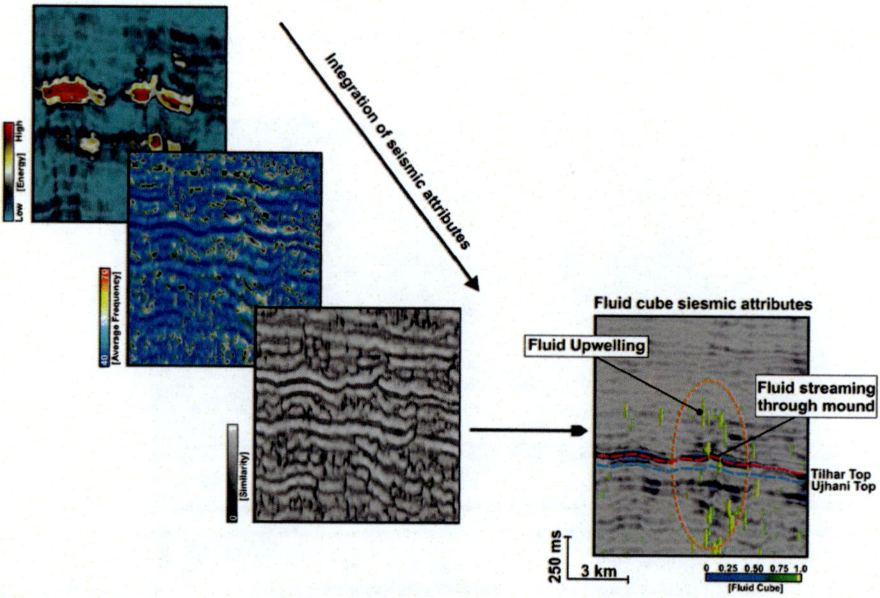

Fig. 2. Amalgamation of different seismic attributes used to generate the Fluid Cube meta-attribute

The Ganga Basin in the central part of Indo-Gangetic-Brahmaputra plains is a peripheral active foreland basin that has been formed on the flexed Indian plate (Singh et al., 1996). This is a product of post-collisional impact between the Indian and Asian plates. Previous research (DGH, 2011, 2017) suggests that subsurface geologic structures in the Ganga valley influence the transport of hydrocarbon, allowing then to flow from the source beds through the reservoir rocks to surface, as revealed by surface geochemical investigations (DGH). This provides an impetus to make a detailed investigation of subsurface fluid flow structures and their structural controls on fluid flow. However, most of the fluid flow studies world-wide are limited to the marine environment. The on-land time migrated 3D seismic reflection data have been used in this study. The data consisting of 643 inlines and 614 crosslines were acquired by ONGC Ltd. Over an area of ~210 km^2 in 2004. The data with a record length of 6 s and sampling rate of 2 ms were processed with a bin size of 40 m by 20 m (inline/crossline).

Different seismic attributes were extracted from the 3D volume and merged into one hybrid attribute, called the Fluid Cube (FC) meta-attribute with a view to visualize the fluid migration in the subsurface.

Fluid flow is observed as vertically focused flow through pipe-like conical vent mostly observed in the central part of the study area (Fig. 2). The high amplitude reflections at the top of pipe may imply that the Ujhani and Tilhar formations are charged with the fluids. The fluid migration routes can be better visualized by co-rendering the amplitude data with the fluid cube meta-attribute (Fig. 3). It is observed that fluid upwelled from the Ujhani-Tilhar central pipe, migrated vertically through the fractures of weaker deformed Tertiary sequences, and finally seeped into the surface being absorbed within the soil particles, which are indicated by the surface geochemical prospecting (DGH).

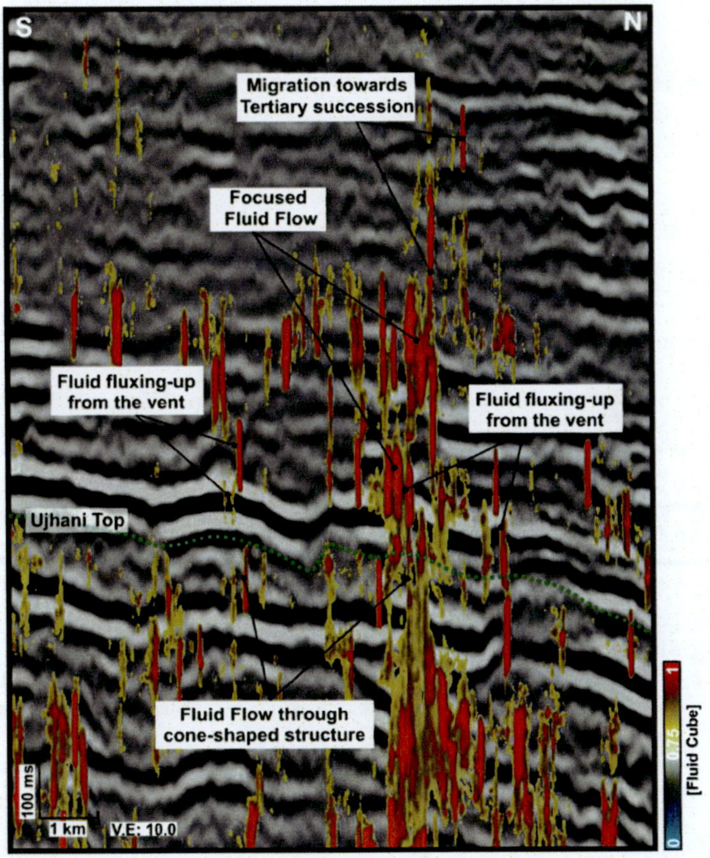

Fig. 3. Co-rendered display (IL 412) of Fluid cube-meta-attribute and the seismic amplitude data showcasing the subsurface fluid upwelling phenomena

3 Conclusion

The present work demonstrates that if the human intelligence is efficiently combined with the engineered machine intelligence, it can augment the interpretation of 3D seismic data quite accurately and at a fast rate. The complex task of delimiting 3D configuration of subsurface geologic features can be achieved from a large volume of data by computing a single attribute, defined as meta-attribute, by fusing several other seismic attributes. This has been demonstrated through two case studies pursued over the Parihaka 3D prospects in the Taranaki basin off New Zealand and in the Indo-Gangetic peripheral foreland basin of the Himalaya. Computation of TFC meta-attribute has distinctly brought out the razor shaped fault system from seismic data volume. Similarly, computation of FC meta-attribute demonstrates the 3D visualisation of fluid migration routes from the source to the surface. This approach is not tedious rather automated and hence accelerates the interpretation of a large volume of 3D seismic data, even in a complex area, with reduced intervention of human analysts. The readers are requested to look into the book by Sain and Kumar (2022) for more case histories in several other basins in the world.

Acknowledgements. The National Data Repository-DGH, India and NZ&PM are acknowledged for providing the data being utilized for this research at SIL-WIHG, Dehradun. KS thanks DST-SERB for providing him with the JC Bose National Fellowship and Dr. Priyadarshi Chinmoy Kumar for being involved in this study.

References

Cartwright, J., Santamarina, C.: Seismic characteristics of fluid escape pipes in sedimentary basins: implications for pipe genesis. Mar. Pet. Geol. **65**, 126–140 (2015)

DGH. Interpretation report of Nelp GV-ONN-2004/1. Directorate General of Hydrocarbon (DGH) (2011)

DGH. NELP-Basin Information docket Category III Ganga Basin. (Unpublished) (2017)

Kumar, P.C., Sain, K.: Attribute amalgamation-aiding interpretation of faults from seismic data: an example from Waitara 3D prospect in Taranaki basin off New Zealand. J. Appl. Geophys. **159**, 52–68 (2018)

Kumar.: A novel approach for enhanced interpretation of 3D seismic data using meta-attributes: examples from taranaki basin, off New Zealand. PnD Thesis, AcSIR-NGRI, Hyderabad, India (2019)

Kumar, P.C., Omosanya, K.O., Alves, T., Sain, K.: A neural network approach for elucidating fluid leakage along hard-linked normal faults. J. Mar. Pet. Geol. **110**, 518–538 (2019)

Kumar, P.C., Kumar, J., Sain, K.: Fluid flow from source to the surface: a case study from the Indo-Gangetic peripheral Foreland Basin. Results Geophys. Sci. **14**(100057), 1–12 (2023)

Løseth, H., Gading, M., Wensaas, L.: Hydrocarbon leakage interpreted on seismic data. Mar. Pet. Geol. **26**(7), 1304–1319 (2009)

Omosanya, K.O., Maia, A.R., Eruteya, O.E.: Seismic, morphologic and scale variabilities of subsurface pipes and vent complexes in a magma-rich margin. Bull. Volcanol. **82**(6), 1–19 (2020)

Sain, K., Kumar, P.C.: Meta-attributes and Artificial Networking: A New Tool in Seismic Interpretation. AGU-John Wiley & Sons (2022)

Singh, I.B., Ansari, A.A., Chandel, R.S., Misra, A.: Neotectonic control on drainage system in Gangetic Plain, Uttar Pradesh. J. Geol. Soc. India **47**(5), 599–609 (1996)

Machine Learning Based Prediction of Spontaneous Combustion Susceptibility of Coal using its Intrinsic Properties: A Safe Smart and Sustainable Mining of Coal Approach

Siddharth Agarwal(✉), Atul Singh, and Pratibha Sharma

Mining Engineering, IIT-ISM, Dhanbad, India
sagarwal@iitism.ac.in

Abstract. Spontaneous combustion and coal mine fire cause widespread damage to accessible coal resources of Jharia coalfields and pose a serious threat to the safety of mine personnel. To study the prediction of spontaneous combustion (SC) susceptibility of coal, 30 samples from various coal seams throughout the coalfields were collected and their intrinsic properties viz. Physical, proximate, ultimate and petrographic determined in the laboratory. The proposed machine learning techniques then were able to classify coal seams based on its intrinsic properties into three SC categories of very risky, risky and low risk. For better visualization of the multi-dimensional data principal component analysis (PCA) projected the obtained clusters along the intrinsic properties as axes in a 2-dimensioanl space. Then wet oxidation potential (WOP) which is a susceptibility index is used to find a relationship with coal intrinsic properties using three machine learning models viz. Support vector machines (SVM), random forest (RF), and elastic net (EN). The clustering results match the laboratory analysis of WOP of each sample and the on-field knowledge. RF algorithm proves to be the best model for the given data to define a relationship between WOP and the intrinsic properties of coal in Jharia region.

Keywords: Coal · Mine Fire · Spontaneous combustion · Clustering · random forest · machine learning

1 Introduction

India has 4th largest reserves of coal; its coal output is projected to reach 1123 million tonnes (MT) by 2023 from the existing levels of around 700 MT (Source: Government of India Ministry of Coal website https://coal.nic.in/content/production-and supplies). India is planning to become a 5 trillion USD economy by the year 2024–2025, for that coal production needs to meet its energy demands in this century and Industry 4.0 standards to be imbibed in its operations. Jharia coalfields have the largest reserves of coal in India with a major problem of spontaneous combustion (SC). SC cause widespread damage to accessible coal resources and pose a serious threat to the safety of mine personnel and

A. Sinha et al. (Eds.): AMC 2023, SPEES, pp. 145–152, 2023.
https://doi.org/10.1007/978-3-031-46966-4_12

their family members (Onifade and Genc, 2019). The overall objective of this research is to develop a computational intelligence based analytical framework for decision-makers by finding the relative importance and underlying relationship of intrinsic properties of coal in predicting the SC of Coal (Sahu et al. 2005) (see Fig. 1).

Fig. 1. Coal fires due to spontaneous combustion at a coal mine in India's Jharkhand state, Source: NPR.org

In addition, classifying coal seams as highly risky, medium risky, less risky and so on for safely conducting mining operations. The idea is to improve the prediction accuracy and finally, developing a methodology to model SC risk in coal seams (Panigrahi et al., 1997; Zhu et al., 2013). This attempt not only performs unsupervised classification correctly matching the actual field results but does so without the susceptibility indices wet oxidation potential (WOP) (Nimaje and Tripathy, 2016). WOP is a parameter for determination of coal's susceptibility to self-oxidation (Panigrahi et al., 1996). Several data mining techniques have been used previously such as fuzzy-c-means clustering, ANN, SVM and linear regression (Namazi et al., 2020; Song et al., 2019; Liu et al., 2019; Zhang et al., 2019; Lei et al., 2019). This the first attempt to combine clustering with high dimensional data projection for SC of coal prediction. The paper aims to bypass the expensive and rigorous laboratory process, expediting the analysis and hence leading to a more accurate decision making (Mohalik et al. 2019).

2 Sample Collection and Experimental Determination of Properties

The samples collected for this study are both from fiery and non-fiery seams within different parts of India. A total of 30 coal samples were collected to cover all coal seams of Jharia and a few others that are prone to spontaneous combustion from other coalfields. The coal seams, operating collieries and the subsidiaries of Coal India Ltd. Operating these mines are such that the first 21 samples were collected from Jharia Coalfields out of which sample 1–14 and 16–21 are from BCCL (Bharat Coking Coal Limited) and sample no15 belongs to TSL (Tata Steel Limited). Samples 22–24 are from ECL (Eastern Coalfields Limited) mines, sample 25–26 from CCL (Central Coalfields Limited), finally sample 27–30, with one each from-SECL (South-Eastern Coalfields

Limited), MCL (Mahanadi Coalfields Limited), WCL (Western Coalfields Limited) and NECL (North-Eastern Coalfields Limited) mines were collected for this study.

The intrinsic properties (Mohalik et al., 2020; Onifade and Genc, 2020) obtained were 15 in number and determined experimentally which included porosity, moisture, volatile matter, ash, fixed-carbon, hydrogen, carbon, total sulphur, calculated oxygen, inertinite, vitrinite and exinite and nitrogen (Saffari et al., 2020; Saffari et al., 2020).

3 Modeling Analysis and Results

This section A describes the results of multidimensional clustering (Fan and Li 2006; Sahu et al., 2012) of data comprising of intrinsic properties of coal seams by using unsupervised learning algorithms.

3.1 Clustering with High-Dimensional Data

The seams are clustered using hierarchical clustering using ward's method of linkage (Freeman, 1994; Jafarzadegan et al., 2019). The method arrives at three number of clusters and the seams that belong to each cluster as depicted in Fig. 2. It builds a hierarchy of clusters such that clusters at one level are combined as clusters at the next level. The formal steps of hierarchical clustering are as follows:

a) Given n data points in m dimensions are to be clustered
b) Start with assuming each data point is a cluster which results in N clusters
c) A similarity metric (distance) is chosen to quantify the separation between the clusters
d) Ward's method in this case gives the best performance (other methods include single linkage, average linkage, complete linkage etc.).
e) Calculate all pair-wise distances between clusters making an n X n matrix. Ward's method uses variance instead of distance metrics as measures of similarity.
f) The most similar pair of points grouped into a single cluster with total C clusters. Distances are calculated in between these newly formed clusters to update the matrix.

The WOP values are the highest for blue cluster which range from 88 to 130. This makes cluster (blue) highly susceptible to spontaneous combustion. The red cluster has seams where WOP ranges from 33 deg. to 80 deg. This makes these seams in cluster (red) moderately susceptible to spontaneous commutation. Finally, the green clustered seams where WOP ranges from 17 to 32 are least prone to spontaneous combustion which is also known through field studies.

Figure 3 shows a projection of the three clusters in this multi-dimensional data, where the axes represent various intrinsic properties of the coal samples. The cluster C1 is represented by a circle, C2 by a cross, and C3 by an inverted triangle. The size of the clusters is defined by the content of porosity in the samples, where a larger size indicates larger porosity. The labels are values of ash in samples, and the color of the samples is based on CPT. The length of the properties as axes denotes the increasing level of influence on samples. It is evident that cluster C3 (inverted triangle) has a much higher influence of moisture, porosity, oxygen, and volatile matter than the other clusters and has less ash content than C1. Cluster C3 represents the riskiest seams for susceptibility to SC whereas C1 is least prone to SC as it has high carbon and ash content, but low porosity, moisture, and volatile matter. Cluster C2 (cross) is moderately risk in terms of propensity for SC.

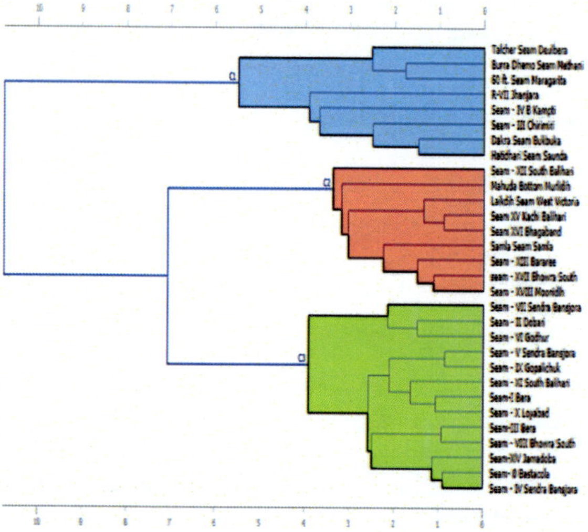

Fig. 2. Dendrogram Hierarchical Clustering by Ward method showing three major clusters marked with seam and colliery names

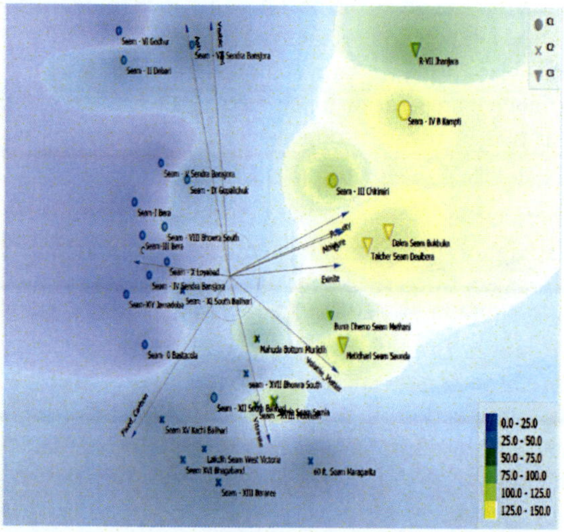

Fig. 3. A 2-D projection of the hierarchical clustering

3.2 Machine Learning Techniques to Model the Relationship Between Intrinsic Properties of Coal and Susceptibility Index Wop

Three machine learning models are tried to find the relationship between major intrinsic variables and WOP. Based on performance measuring criterion such as mean square error (MSE), mean absolute error (MAE) and coefficient of correlation as shown in Table 1. Random forests (Alam et al., 2019; Buja et al., 2008) performs the best amongst all three ML models. RF model has the least MSE, MAE and highest R^2. The best hyper-parameters for the RF model selected after cross validation are listed in Table 2. Figure 4 describes one of the RF trees where each node contains the following information:

i. The variable name and the value used for splitting
ii. The number of samples at each split and the value of mean square error
iii. The value of r-square

The description of hyperparameters used to tune the model are as follows:

- **Number of estimators**: the number of decision trees in the forest. Increasing this hyperparameter generally improves the performance of the model but also increases the computational cost of training and predicting. The optimal value is 6 for this model
- **max_depth**: the maximum depth of each decision tree in the forest. Setting a higher value for max_depth can lead to overfitting while setting it too low can lead to underfitting. The optimal value is 6 for this model
- **max_features**: Minimum number of features random forest considers splitting a node. The optimal value is 4 for this model

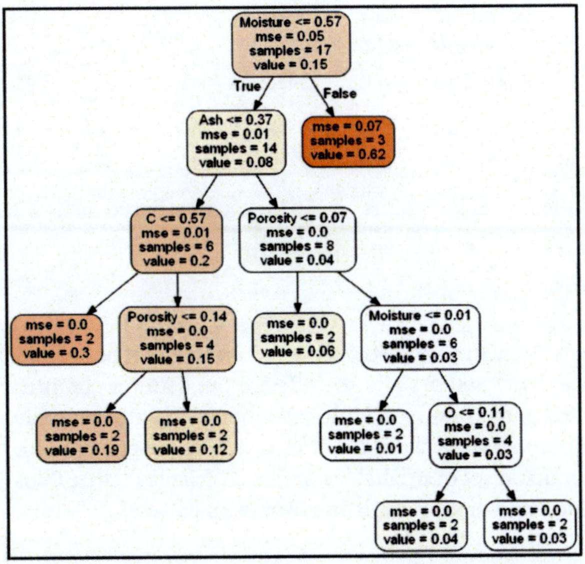

Fig. 4. One of the random forest trees used in estimation of WOP

Table 1. Data performance measures for predicting WOP

Model WOP Train	MSE	RMSE	MAE	R^2
SVM	0.017	0.132	0.110	0.766
Random Forest	**0.008**	**0.090**	**0.066**	**0.857**
Elastic-Net	0.011	0.106	0.088	0.802

Table 2. Best Hyperparameter values of Random Forest for predicting WOP

Random Forest WOP	Values
Number of Trees	6
Number of Attributes considered at each split	4
Maximum depth of individual trees	6
Minimum number of samples in a node before split	3

Table 3. Hierarchical and K-means clusters characteristics

	C2	C1	C3
SPONTANEOUS COMBUSTION TENDENCY	Highly Combustible	Medium Susceptibility	Least Susceptible to SC
INTRINSIC PROPERTIES	High (Moisture, Porosity, Volatile Matter, Oxygen)	Less Porosity and less Ash	High (Ash, C, and Fixed Carbon)
RISK ASSESSMENT	Very Risky	Medium Risky	Less Risky

Table 3 rates the risk of the seams within the clusters based on its intrinsic properties.

4 Discussion

The study suggests that using only 30 data samples might not be optimal for some regression algorithms, and the authors aim to collect more data for comprehensive modeling of Jharia Coalfields in the future. Additionally, not all possible combinations of models with different input variables were developed due to longer training times. The goal was to get a generalized set of predictors or input variables rather than the best possible combination, which may not be realistic given the data size.

5 Conclusion

Both clustering methods confirm that Hatidhari Seam Saunda in CCL, Kamti Seam – IV B in WCL, Talcher Seam Deulbera in MCL, Margarita 60 ft. Seam operated in NECL are most prone to SC. These results match the actual susceptibility condition of coal in these areas. It could be concluded that from the above observations that intrinsic priorities are enough to classify or predict the spontaneous heating susceptibility more accurately than crossing point temperature and differential thermal analysis, for the coal samples studied in this investigation.

i. Hierarchical clustering suggests three clusters in the data representing high, medium, and low risks in susceptibility to spontaneous combustion.
ii. The best hyperparameter values provided for RF can serve as a reference point for researchers or industry personnel when modeling their own data.
iii. Based on the results, mining companies can develop a plan to monitor the intrinsic properties of the coal seam for spontaneous combustion management and prompt an action response accordingly, which will in turn:
    a. Increase production without causing fires.
    b. Manage any incidence of spontaneous combustion effectively.
    c. Minimize any indirect consequences of spontaneous combustion of coal.

The current submission extends the previous work by ensuring sustainable development of the Indian coal sector by identifying the risk of spontaneous combustion in coal seams by their intrinsic properties. Understanding the intricate relationship between the intrinsic properties of coal through machine learning techniques. More experiments will be conducted to determine other SI, such as the Russian U- index, and differential scanning calorimetry (DSC). Data from various sensors for the temperature, gas, when analyzed by AI or ML algorithms can serve as harbingers of spontaneous combustion of coal and thus increasing the much-needed production of coal in a sustainable manner.

References

Alam, M.Z., Rahman, M.S., Rahman, M.S.: A random forest based predictor for medical data classification using feature ranking. Inform. Med. Unlocked **15**, 100180 (2019)

Buja, A., Swayne, D.F., Littman, M.L., Dean, N., Hofmann, H., Chen, L.: Data visualization with multidimensional scaling. J. Comput. Graph. Stat. **17**(2), 444–472 (2008)

Fan, J., Li, R.: Statistical challenges with high dimensionality: Feature selection in knowledge discovery. arXiv preprint math/0602133 (2006)

Freeman, L.: Displaying hierarchical clusters. INSNA Connections **17**(2), 46–52 (1994)

Jafarzadegan, M., Safi-Esfahani, F., Beheshti, Z.: Combining hierarchical clustering approaches using the PCA method. Expert Syst. Appl. **137**, 1–10 (2019)

Lei, C., et al.: A comparison of random forest and support vector machine approaches to predict coal spontaneous combustion in gob. Fuel **239**, 297–311 (2019)

Liu, N., Huang, S., Xie, X.: Prediction of spontaneous combustion tendency grade of sulfide ore based on decision tree combined classifier. In: Journal of Physics: Conference Series, vol. 1302, No. 4, p. 042060. IOP Publishing (2019)

Mohalik, N.K., Lester, E., Lowndes, I.S.: Development of a petrographic technique to assess the spontaneous combustion susceptibility of Indian coals. Int. J. Coal Prep. Util. **40**(3), 186–209 (2020)

Mohalik, N.K., Mishra, D., Ray, S.K., Varma, N.K., Khan, A.M., Sahay, N.: Laboratory investigation to assess spontaneous combustion/fire during extraction of thick coal seam. J. Inst. Eng. (India): Ser. D **100**(2), 229–242 (2019)

Namazi, S., Brankovic, L., Moghtaderi, B., Zanganeh, J.: Comparative study of data mining techniques for predicting explosions in coal mines. In: 2020 10th International Conference on Cloud Computing, Data Science & Engineering (Confluence), pp. 738–743. IEEE (2020)

Nimaje, D.S., Tripathy, D.P.: Characterization of some Indian coals to assess their liability to spontaneous combustion. Fuel **163**, 139–147 (2016)

Onifade, M., Genc, B.: A review of spontaneous combustion studies–South African context. Int. J. Min. Reclam. Environ. **33**(8), 527–547 (2019)

Onifade, M., Genc, B.: A review of research on spontaneous combustion of coal. Int. J. Min. Sci. Technol. **30**(3), 303–311 (2020)

Panigrahi, D.C., Ojha, A., Saxena, N.C., Kejriwal, B.K.: A study of coal oxygen interaction by using Russian U-index and its correlation with basic constituents of coal with particular reference to Jharia coalfield (1997)

Saffari, A., Ataei, M., Sereshki, F.: Studying relationship between coal intrinsic characteristics in spontaneous combustion of coal potential using crossing point temperature test method. J. Min. Environ. **11**(1), 315–333 (2020)

Saffari, A., Sereshki, F., Ataei, M.: Effect of maceral content on tendency of spontaneous coal combustion using the R70 method. Int. J. Min. Geo-Eng. **54**(2), 93–99 (2020)

Sahu, H.B., Panigrahi, D.C., Mishra, N.M.: Assessment of spontaneous heating susceptibility of coal seams by experimental techniques-a comparative study (2005)

Sahu, H.B., Mahapatra, S.S., Panigrahi, D.C.: Fuzzy c-means clustering approach for classification of Indian coal seams with respect to their spontaneous combustion susceptibility. Fuel Process. Technol. **104**, 115–120 (2012)

Song, Y., et al.: Prediction of gas and coal spontaneous combustion co-existing disaster through the chaotic characteristic analysis of gas indexes in goaf gas extraction. Process Saf. Environ. Prot. **129**, 8–16 (2019)

Zhang, J., Zhang, H., Ren, T., Wei, J., Liang, Y.: Proactive inertisation in longwall goaf for coal spontaneous combustion control-a CFD approach. Saf. Sci. **113**, 445–460 (2019)

Zhu, H.Q., Song, Z.Y., Tan, B., Hao, Y.Z.: Numerical investigation and theoretical prediction of self-ignition characteristics of coarse coal stockpiles. J. Loss Prev. Process Ind. **26**(1), 236–244 (2013)

Machine Learning Applications: Coal Fire Prediction Using Graham's Ratio

Surendra Kumar Dogra[1], Jitendra Pramanik[1](✉), Singam Jayanthu[1], and Abhaya Kumar Samal[2]

[1] Department of Mining Engineering, National Institute of Technology, Rourkela, Odisha, India
bidun35@gmail.com, sjayanthu@nitrkl.ac.in

[2] Department of CSE, Trident Academy of Technology, Bhubaneswar, Odisha, India
abhaya@tat.ac.in

Abstract. As the mining industry has evolved, there has been a rise in the number of incidents at mining sites caused by active flames in underground coal mines. It is a complex process that jeopardizes the miner's life and property. The harm inflicted by fire threats in recent history has drawn the attention of the authorities toward implementing measures that will lessen the risk of natural disasters. It is impossible to exaggerate the significance of this finding in terms of its potential to reduce the risk associated with coal's ability to self-heal. This study intends to show evidence that machine learning may be successfully utilized to produce accurate forecasts regarding self-healing, and its primary objective is to accomplish this goal. This research led to the development of AdaBoost classification procedures, which were then used to create methods for anticipating the occurrence of spontaneous fires in UG coal mines. The effectiveness of different ML models is compared to one another using quality metric criteria.

Keywords: Adaboost Classification · Machine Learning · Graham's Ratio · Coal Mining

1 Introduction

There has been a daily increase in the number of fatalities that have occurred in coal mines due to the presence of toxic gases. Their families experienced hardship due to their passing since they were the sole breadwinners for their family [1]. Many lives and possessions were lost as the coalmine disaster disrupted everyday living [2, 3]. Since of India's high coal consumption and production, and since many coal mines are connected to high-gas mines, gas mishaps constitute a significant problem; hence, it is vital to explore and find a solution to this problem for India's coal business. Evaluation of the safety of the mine's gas atmosphere has long been an essential component of mine safety management. Coal mines in India must monitor various factors, including gas concentrations, carbon monoxide concentrations, carbon dioxide concentrations, oxygen concentrations, dust concentrations, wind speed, humidity, temperature, and power state, amongst other factors. Outbursts, gas accumulation, and explosions are all things

A. Sinha et al. (Eds.): AMC 2023, SPEES, pp. 153–164, 2023.
https://doi.org/10.1007/978-3-031-46966-4_13

that can be efficiently avoided in a coal mine, provided the relevant data are monitored automatically and the gas safety state is ascertained as quickly as possible. The study has important theoretical implications and real-world applications for preventing gas accidents [4] and promoting the coal industry's safe and sustainable growth. In the past, research on coal's reactivity with oxygen and attempts to anticipate coal's spontaneous combustion in gaseous environments were primarily conducted in the air [5–13]. On the other hand, it has progressively been clear, as a result of research carried out in more recent times, that the coal-oxygen interaction that occurs in the UG environment is significantly distinct from the one that takes place in the air. One significant distinction is the low-oxygen environments that commonly characterize the coal-oxygen interaction that occurs deep beneath. This is one of the main differences. Mine methane is a type of concurrent product in the process of coal creation, and its emission is a continual process during the mining of coal [14]. Coal mines often produce methane as a byproduct. The methane emission near a mine is hazardous since it can lead to explosions, fires, suffocation, and other mishaps [15]. On the other hand, the deposition of methane in goaf is an occurrence that cannot be avoided in the longwall face. After computational calculations and field tests, Ren and coworkers [16, 17] found that methane buildup persists in the deep goaf of the longwall face. The longwall face goaf is always open and closed during mining coal. As the mining distance rises, methane output decreases, lowering goaf oxygen [18]. Different researchers have found that coal oxidation is proportional to gas medium O2 content. Methane reduces partial oxygen pressure and slows coal-oxygen interaction [19]. Methane's effect on index gases produced during low-temperature coal oxidation needs further study. Because the effective prediction index cannot be estimated in a methane-induced lean-oxygen environment, it cannot be used to evaluate spontaneous combustion in high-gassy mining environments.

This study extensively researched and statistically examined the changing pattern of many indicator gases during coal oxidation at low temperatures under low-oxygen conditions. This is especially important in mines when gas levels are high to avoid the risk of goaf igniting on its own.

2 Related Work

A growing number of practical applications can be carried out with the help of an artificially intelligent algorithm in the engineering field [20–24]. Many attempts have also been made on spontaneous coal combustion [25, 26], gas explosions [27–30], and other similar phenomena. This is all thanks to the enormous development of artificial intelligence (AI). In addition, there are different clever approaches to evaluating the safety of coal mines, which can objectively quantify hazards and provide a solution to the dilemma described before.

Using the ant colony method, he [31] developed a neural network security model. Ant colony optimization trains the model's neural network weights. Using the quantum genetic algorithm, Li [32] optimized the neural net model for the threshold to avoid neural networks from falling into the local minimum. Zhang [33] refined factorization-machine-supported neural networks (FNNs). The updated FNN's fuzzy neurons' improved errors and decision-making have strengthened the system. Wang [34], who combined the grey correlation method with a novel grey correlation degree methodology, proposed the dynamic resolution coefficient to improve its accuracy. Using rough set theory, Bayesian networks, and support vector machines, Li [35] built a multilayer safety forecasting indicator system and a multilevel coal mine risk growth prediction model. Accident causes were identified by Li [35] using Bayesian Network analysis. Together, they created a more comprehensive and accurate safety scenario predicting indicator system with many levels.

Sun [36] created a complete coal mine worker safety risk evaluation model using Fuzzy TOPSIS and integration operator. Dai [37] suggested using a fuzzy synthetic assessment model and an algorithm to assign weights to gas density data. This permitted application use of the data. Peng [38] introduced linguistically intuitive fuzzy numbers to represent evaluation data.

3 Materials and Methods

3.1 Dataset

It uses the data on the mine atmosphere acquired in the GDK-11 inclined shaft. Graham's ratio (GR) was utilized to classify the likelihood of a fire breaking out in an underground coal mine. The equation for Graham's ratio is given as:

$$\mathrm{GR} = \frac{CO}{\Delta O2}$$

As can be seen in Table 1 classifies the dangers into two distinct degrees. The provided dataset includes information on gases such as H2, O2, N2, CH4, CO, and CO2. The observed gas concentration levels are depicted in Figs. 1 and 2 (Fig. 3).

Table 1. Risk Categorization

Index	Label
0	Normal
1	Critical

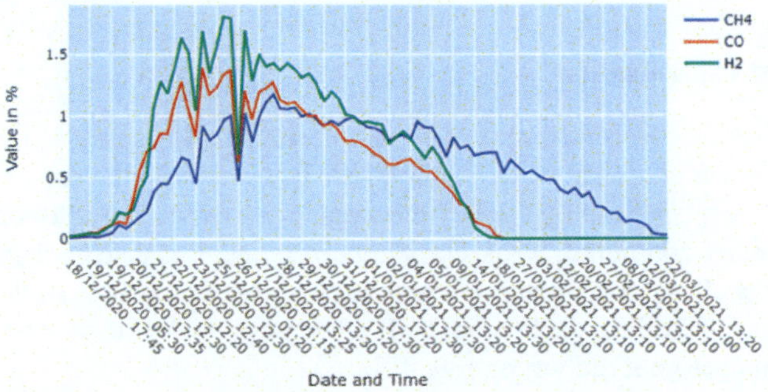

Fig. 1. Recorded values of methane, carbon monoxide, and hydrogen at GDK-11 incline

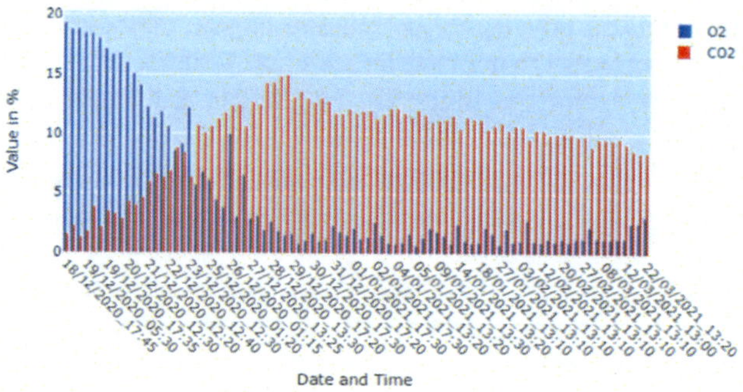

Fig. 2. Variation in the concentration of CO2 and O2 at GDK-11 incline

	H2	O2	N2	CH4	CO	CO2
count	85.000000	85.000000	85.000000	85.000000	85.000000	85.000000
mean	0.616372	4.958683	82.565026	0.581598	0.494216	9.765461
std	0.610058	5.819024	3.758401	0.351774	0.472651	3.302021
min	0.000000	0.677000	74.909800	0.011100	0.000000	1.357000
25%	0.000100	1.237200	78.756400	0.251300	0.000000	8.837800
50%	0.511500	1.928100	82.599400	0.640800	0.489000	10.651000
75%	1.184300	6.529100	86.018600	0.904900	0.884500	11.837800
max	1.807500	19.270400	88.350000	1.176400	1.396600	15.007100

Fig. 3. Descriptive Statistics

3.2 Data Pre-processing

The min-max normalization technique is used to normalize the chosen input data between 0 and 1. When preparing data for a machine language, normalization is frequently used. Normalization aims to convert the numbers in each column of a dataset to a consistent scale without losing or distorting any information. This aids in quickening the learning rate and ultimately results in quicker convergence. Using Eq. 1, the values of a particular input parameter (xi) are changed:

$$Z = \frac{x_i - mean(x)}{[\max(x) - \min(x)]} \tag{1}$$

3.3 Adaboost Classification

AdaBoost is a series-based ensemble method for training and deploying trees. AdaBoost uses boosting to connect a set of weak classifiers in a series so that each weak classifier tries to correct samples misclassified by the weak classifier before it. Boosting creates a robust classifier by combining weak classifiers in series. Because boosting approaches frequently use shallow, non-overfit, but potentially biased models in each decision tree, the term "stump" describes these models. It's been established that one tree's job is to fix what the one before it did wrong [39, 40].

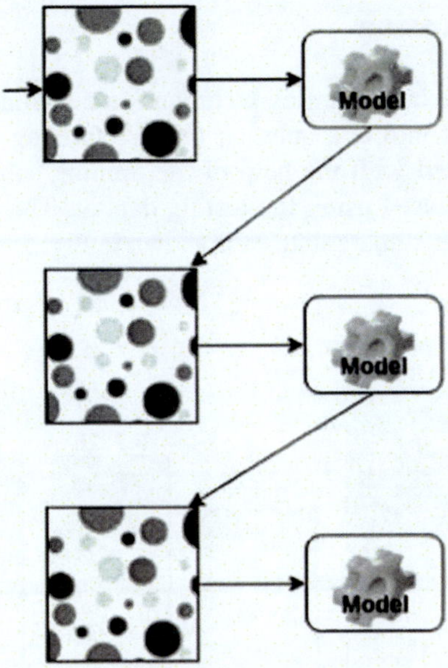

Fig. 4. Basic Schematic of Adaboost classifier implementation

The dataset (85 data points) is used to assess and examine the classifiers' performance. At first, the dataset is divided into two subsets; One subset is used to train the ML classifier. The second subset of data is used as a testing set upon getting the optimum ML model. The performance of the ML classifier on the testing set is assessed via different performance indices. The overall flow of implementation is depicted in Fig. 4.

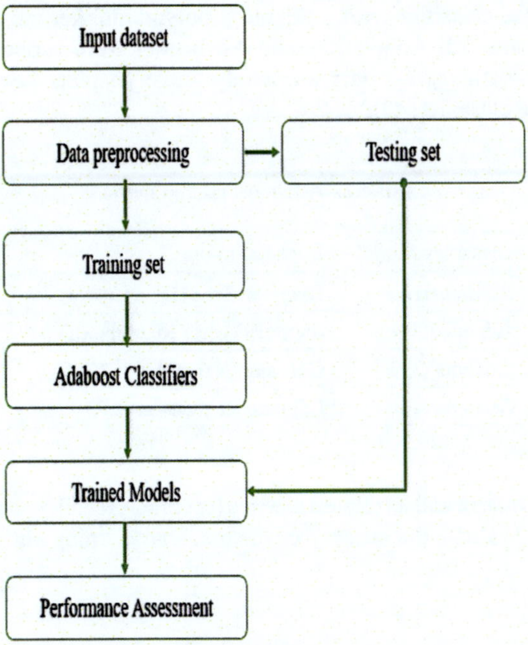

Fig. 5. Implementation flow diagram

4 Result and Discussion

At first, conventional machine learning techniques are applied to the obtained dataset. The dataset is then split into two subsets: one for training and one for testing. The ML algorithms are trained with the help of the training subset. The ML algorithms' effectiveness is then assessed using the testing dataset. The effectiveness of machine learning methods on a benchmark dataset is shown in Fig. 5.

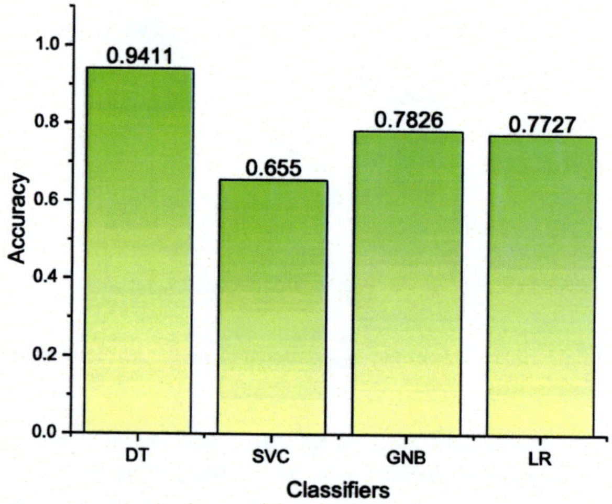

Fig. 6. Accuracy of the conventional machine learning models

In the second step, boosting algorithms are implemented. In Adaboost1, we employed a decision tree base classifier with a degree 3 polynomial kernel. Similarly, the base classifiers for Adaboost2, Adaboost3, and Adaboost4 were support vector, gaussian naïve bayes, and logistic regression, respectively. Table 3 depicts the entire configuration of the various base classifiers (Table 2).

Table 2. Adaboost configuration

Classifier Model	Base Estimator
Adaboost1	Decision Tree (DT)
Adaboost2	Support Vector Machine (SVM)
Adaboost3	Gaussian Naïve Bayes (GNB)
Adaboost4	Logistic Regression (LR)

The confusion matrix and its related metrics (accuracy, recall, precision, and F1score) were employed to evaluate the efficacy of the machine learning algorithm models. The obtained accuracy scores for selected ML models show that AdaBoost classifiers function excellently as predictors. Table 3 presents several additional performances or validation indicators acquired from all the produced models (Fig. 7).

Table 3. Calculated validation indices of different Adaboost models

Models	Accuracy	Precision	F1Score
Adaboost1	0.9545	0.9545	0.9545
Adaboost2	0.7352	0.7352	0.7352
Adaboost3	0.9090	0.9090	0.9090
Adaboost4	0.9117	0.9117	0.9117

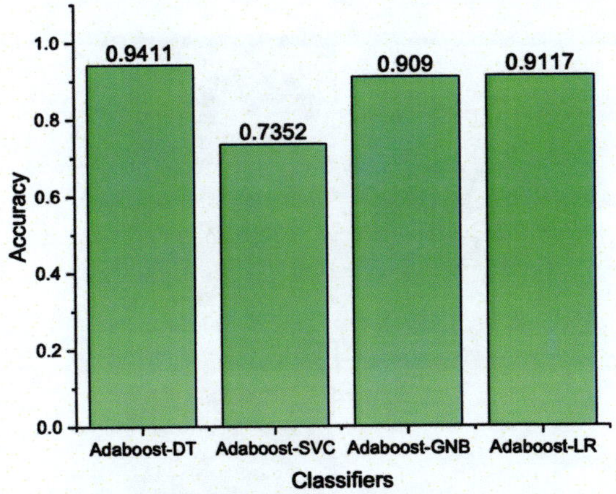

Fig. 7. Accuracy of the Adaboost models

Tree-based ensemble algorithms like Adaboost with a decision tree outperform SVM, GNB, and LR models. The flexibility and power to handle all data may make tree-based ensemble methods better. They handle non-linearity, the complex relationship between dependent and independent variables, and outliers well when other models fail. Ensemble models are famous for collecting even the smallest data features using a small training dataset and a collection of significantly variable geometrical data. In the tested models' confusion matrices, "true positives" are instances where the model correctly identified the coal fire in the test dataset. When a model returns a "true negative," it found no fire. A "false positive" is an instance of "overestimation," where the model wrongly declares the condition "critical" even though there is no fire. In this context, "false negative" means the model missed a coal fire. Correct classifications—including erroneous positives and negatives—are important. Recall measures the percentage of real positives (also known as accurately recognised positives) in a dataset. This parameter determined the model's classification accuracy.

To determine which algorithm is ideal for the current problem, an effort has been made to fit all relevant ones and evaluate their performance. In Fig. 6, we can see the breakdown of test cases by type and the projected outcomes for the specified models. Accurately predicted values appear in the proper columns, while those that did not appear in the false columns (Figs. 8 and 9).

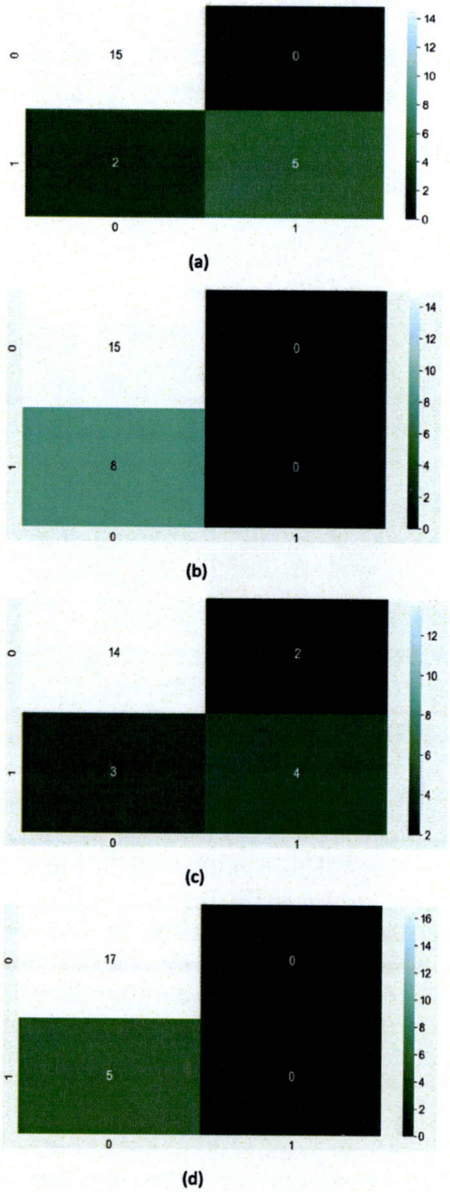

Fig. 8. Confusion matrix: a) DT, b) SVM, c) GNB, d) LR

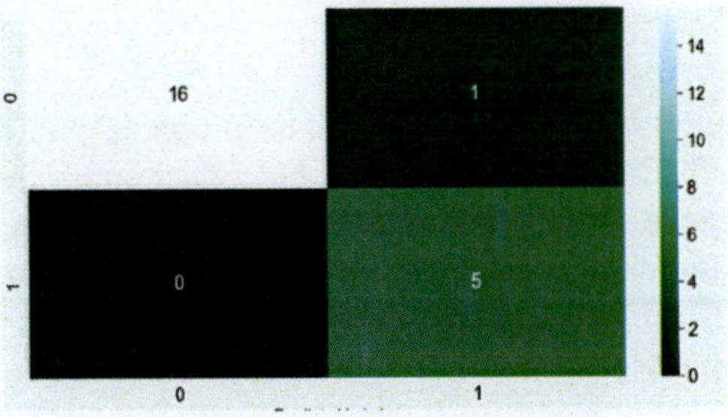

(a)

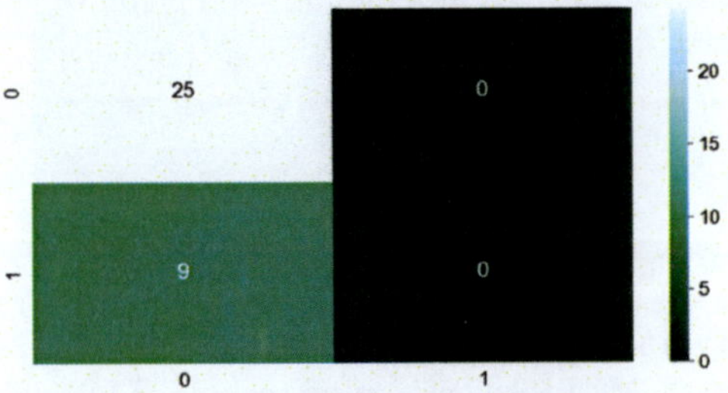

(b)

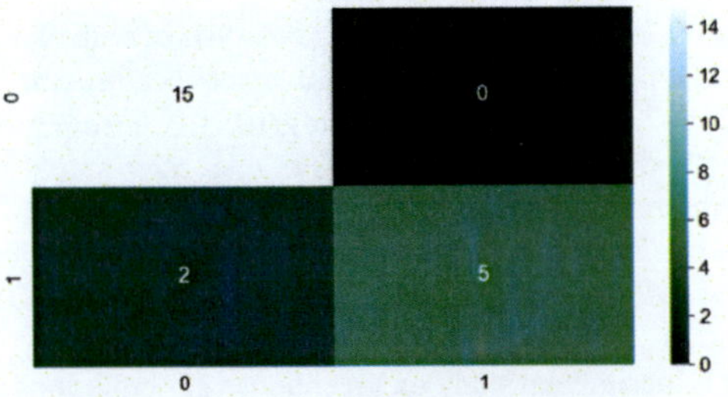

(c)

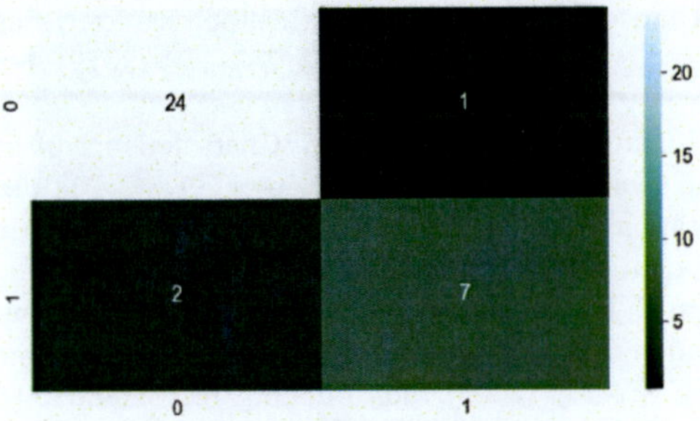

(d)

Fig. 9. Confusion matrix: a) Adaboost-DT, b) Adaboost-SVM, c) Adaboost-GNB, d) Adaboost-LR

5 Conclusion

Risks to machinery, output, and personnel can all be mitigated with good mining environment analysis. Active fire has always been a problem for mining engineers. An accurate forecast system is needed to reduce the danger of unexpected coal fires. Several ML models were created in this research to foretell the likelihood of an active fire in a coal mine. Adaboost-DT was more effective than the other proposed models at predicting the active fire scenario when tested on the dataset gathered from the GDK-11 incline of SCCL. The potential usefulness of the models in improving mining safety is illustrated by analyzing assessment indices such as accuracy, precision, and F1score. Applying ML methods to study and foretell events in a mining setting can benefit significantly from the work presented here.

References

1. Qiao, W., Li, X., Liu, Q.: Systemic approaches to incident analysis in coal mines: comparison of the STAMP, FRAM and '2–4' models. Resour. Policy **63**, 101453 (2019). https://doi.org/10.1016/j.resourpol.2019.101453
2. Liu, Q., Li, X., Hassall, M.: Evolutionary game analysis and stability control scenarios of coal mine safety inspection system in China based on system dynamics. Saf. Sci. **80**, 13–22 (2015). https://doi.org/10.1016/j.ssci.2015.07.005
3. Liu, Q., Li, X., Qiao, W., Meng, X., Li, X., Shi, T.: Analysis of embedded non-safety regulation games in China's two types of coal mines through safety performance disparity, 1980–2014. Resour. Policy **51**, 265–271 (2017). https://doi.org/10.1016/j.resourpol.2017.01.009
4. Cai, C.: Analysis method of gas warning results of coal mine safety monitoring and control system. Ind. Mine Autom. **44**, 15–18 (2018)
5.] Zhu, J., He, N., Li, D.: The relationship between oxygen consumption rate and temperature during coal spontaneous combustion. Saf. Sci. **50**(4), 842–845 (2012). https://doi.org/10.1016/j.ssci.2011.08.023
6. Wang, H., Chen, C.: Experimental study on greenhouse gas emissions caused by spontaneous coal combustion. Energy Fuels **29**(8), 5213–5221 (2015). https://doi.org/10.1021/acs.energyfuels.5b00327
7. Zhang, Y., et al.: Ultrasonic extraction and oxidation characteristics of functional groups during coal spontaneous combustion. Fuel **242**, 287–294 (2019). https://doi.org/10.1016/j.fuel.2019.01.043
8. Zhang, Y., Li, Y., Huang, Y., Li, S., Wang, W.: Characteristics of mass, heat and gaseous products during coal spontaneous combustion using TG/DSC–FTIR technology: the impacts of oxygen concentrations and heating rates. J. Therm. Anal. Calorim. **131**(3), 2963–2974 (2018). https://doi.org/10.1007/s10973-017-6738-x
9. Wang, H., Dlugogorski, B.Z., Kennedy, E.M.: Coal oxidation at low temperatures: oxygen consumption, oxidation products, reaction mechanism and kinetic modelling. Prog. Energy Combust. Sci. **29**(6), 487–513 (2003). https://doi.org/10.1016/S0360-1285(03)00042-X
10. Arisoy, A., Beamish, B.: Reaction kinetics of coal oxidation at low temperatures. Fuel **159**, 412–417 (2015). https://doi.org/10.1016/j.fuel.2015.06.054
11. Ren, X., et al.: Novel sodium silicate/polymer composite gels for the prevention of spontaneous combustion of coal. J. Hazard Mater. **371**, 643–654 (2019). https://doi.org/10.1016/j.jhazmat.2019.03.041

12. Cheng, W., Hu, X., Xie, J., Zhao, Y.: An intelligent gel designed to control the spontaneous combustion of coal: fire prevention and extinguishing properties. Fuel **210**, 826–835 (2017). https://doi.org/10.1016/j.fuel.2017.09.007
13. Hu, X., Cheng ,W,. Shao, Z:. Novel authigenic gas foaming hydrogels for preventing coal spontaneous combustion. e-Polymers **15**(5), 361–368 (2015)
14. Karakurt, I., Aydin, G., Aydiner, K.: Mine ventilation air methane as a sustainable energy source. Renew. Sustain. Energy Rev. **15**(2), 1042–1049 (2011)
15. Zhang, Q., Ma, Q.: Dynamic pressure induced by a methane–air explosion in a coal mine. Process. Saf. Environ. Prot. **93**, 233–239 (2015)
16. Ren, T., Balusu, R.: Proactive goaf inertisation for controlling longwall goaf heatings. Procedia Earth Planet. Sci. **1**(1), 309–315 (2009)
17. Ren, T.: CFD modelling of longwall goaf gas flow to improve gas capture and prevent goaf self-heating. J. Coal Sci. Eng. (China) **15**(3), 225–228 (2009)
18. Qi, G.S., Wang, D.M., Zheng, K.M., Xu, J., Qi, X.Y., Zhong, X.X.: Kinetics characteristics of coal low-temperature oxidation in oxygen-depleted air. J. Loss Prev. Process Ind. **35**, 224–231 (2015)
19. Qu, L.N., Song, D.Z., Tan, B.: Research on the critical temperature and stage characteristics for the spontaneous combustion of different metamorphic degrees of coal. Int. J. Coal Prep. Util. **38**(5), 221–236 (2018)
20. Rassafi, A.A., Ganji, S.S., Pourkhani, H.: Road safety assessment under uncertainty using a multi attribute decision analysis based on Dempster-Shafer theory. KSCE J. Civ. Eng. **22**(8), 3137–3152 (2018)
21. Chen, C.-H., Hwang, F.-J., Kung, H.-Y.: Travel time prediction system based on data clustering for waste collection vehicles. IEICE Trans. Inf. Syst. **E102.D**(7), 1374–1383 (2019)
22. Chen, C.-H.: A cell probe-based method for vehicle speed estimation. IEICE Trans. Fundam. Electron. Commun. Comput. Sci. **E103.A**(1), 265–267 (2020)
23. Chen, C.H., Song, F.Y., Hwang, F.J., Wu, L.: A probability density function generator based on neural networks. Physica A Stat. Mech. Appl. **541** (2020). Article ID 123344
24. Chen, Y., Gu, C., Shao, C., et al.: An approach using adaptive weighted least squares support vector machines coupled with modified ant lion optimizer for dam deformation prediction. Math. Probl. Eng. **2020** (2020). 23 p. Article no. 9434065
25. Zhang, Q., Li, H.G., Li, H.: An improved least squares SVM with adaptive PSO for the prediction of coal spontaneous combustion. Math. Biosci. Eng. **16**(4), 3169–3182 (2019)
26. Zhang, S., Chen, J.Q., Zhang, D.: Gob spontaneous combustion prediction based on neural network. Transducer Microsyst. Technol. **31**(5), 10–12 (2012)
27. Tong, X., Fang, W., Yuan, S., Ma, J., Bai, Y.: Application of Bayesian approach to the assessment of mine gas explosion. J. Loss Prev. Process Ind. **54**, 238–245 (2018)
28. Pejic, L.M., Torrent, J.G., Querol, E., Lebecki, K.: A new simple methodology for evaluation of explosion risk in underground coal mines. J. Loss Prev. Process Ind. **26**, 1524–1529 (2013)
29. Li, M., Wang, H., Wang, D., Shao, Z., He, S.: Risk assessment of gas explosion in coal mines based on fuzzy AHP and Bayesian network. Process Saf. Environ. Prot. **135**, 207–218 (2020)
30. Zhang, Q.: Risk assessment of gas explosion disaster based on random forest model. IOP Conf. Ser. Earth Environ. Sci. **446** (2020). Article no. 22081
31. He, R.J., Du, Y.H.: Application of improved particle swarm-optimization neural network in coalmine safety evaluation. IOP Conf. Ser. Mater. Sci. Eng. **423** (2018). Article no. 12007
32. Li, X., Li, N.W., Yang, Z.: Coal mine safety evaluation model based on quantum genetic algorithm. Comput. Syst. Appl. **21**(7), 101–105 (2012)
33. Zhang, J.A., Li, W.J., Guan, Y.L.: Application of improved FNN in coal mine safety production warning system. Coal Eng. **8**, 168–171 (2013)

34. Wang, D., Liu, L., Zhang, X.M.: The improvement and application of the grey correlation degree method in the evaluation of coal mine intrinsic safety. J. Saf. Sci. Technol. **9**(1), 151–154 (2013)
35. Li, D.W., Li, S., You, M.J.: Research on mine safety situation prediction model: the case of gas risk. In: Proceedings of the 2019 11th International Conference on Wireless Communications and Signal Processing (WCSP), Xi'an, China, pp. 1–6, October 2019
36. Sun, X.D., Zhang, L.X., Qi, H.L.: Coal mine safety production risk assessment model based on fuzzy analytic hierarchy process. Ind. Saf. Environ. Prot. **40**(1), 65–68 (2014)
37. Dai, Z., Liu, J.H., Wang, W.: Collection and fusion of gas density data and its security warning for coal mine. Comput. Engi. Des. **37**, 783–787 (2016)
38. Peng, H., Wang, J., Cheng, P.: A linguistic intuitionistic multi-criteria decision-making method based on the Frank Heronian mean operator and its application in evaluating coal mine safety. Int. J. Mach. Learn. Cybern. **9**(6), 1053–1068 (2018)
39. Yeturu, K.: Chapter 3 - Machine learning algorithms, applications, and practices in data science. In: Srinivasa Rao, A.S.R., Rao, C.R. (eds.) Handbook of Statistics, vol. 43, pp. 81–206. Elsevier (2020). ISSN 0169-7161, ISBN 9780444642110. https://doi.org/10.1016/bs.host.2020.01.002
40. Ding, Y., Zhu, H., Chen, R., Li, R.: An efficient AdaBoost algorithm with the multiple thresholds classification. Appl. Sci. **12**, 5872 (2022). https://doi.org/10.3390/app12125872

An Approach for Implementation of IoT Enables Smart Environmental Monitoring and Strata Monitoring System for Underground Coal Mines

Shankhajit Mitra[1(✉)], S. K. Chaulya[1], Dheeraj Kumar[2], and Ashish Soni[3]

[1] CSIR-Central Institute of Mining and Fuel Research, Dhanbad 826015, JH, India
sm@cimfr.nic.in

[2] Indian Institute of Technology (ISM), Dhanbad 826004, JH, India

[3] Madhav Institute of Technology and Science, Gwalior 474005, M.P., India

Abstract. IoT is the new standard that integrates various technologies and aspects from varied methods. Pervasive computing, sensing technologies, embedded devices, ubiquitous computing, internet protocol and communication technologies are merged for system formation where the digital and real worlds meet.Through the setup of IoT into day-to-day objects, they are modified into smart objects that gather information from the surroundings, controls and interacts with the physical world. Working in the underground coal mines is complex and risk-oriented due to its long and narrow tunnels, complex structures and high probability of various gas concentrations in the coal mines. This leads to many chances of explosion causing the death of workers. Hence, suitable wireless communication is essential for solving the existing issues in the underground mines. Taking all these existing issues of the coal mines into account, IoT based wireless and real-time monitoring system is introduced here to track and monitor these mines for ensuring miners' safety. This paper is mainly focussed on designing and developing of IoT permitted strata condition and a sub-surface mine environment monitoring a web-based system for decision making to predict sub-surface environments.

Keywords: IoT · Sensors · Wireless Networks · Underground coal mines · Miner's Safety

1 Introduction

IoT is the new standard that integrates various technologies and aspects from varied methods. Pervasive computing, sensing technologies, embedded devices, ubiquitous computing, internet protocol and communication technologies are merged for system formation where the digital and real worlds meet. Through the setup of IoT into day-to-day objects, they are modified into smart objects that gather information from the surroundings, controls and interacts with the physical world. The probable count of the interconnected devices and amount of accessible data opens various opportunities for creating services that will bring real profits to the environment, individual citizens,

A. Sinha et al. (Eds.): AMC 2023, SPEES, pp. 165–179, 2023.
https://doi.org/10.1007/978-3-031-46966-4_14

economy and society. Working in the underground coal mines is complex and risk-oriented due to its long and narrow tunnels, complex structures and high probability of various gas concentration in the coal mines. This leads to many chances of explosion causing the death of workers.

As a matter of minimising the intensity associated with the issues, several works relying on IoT has been achieved by the traditional works for communication within the mining industry whenever any unusual activity occurs (Aziz et al. 2020). However, the mining industry is conventional and slow to alterations as a result of infrastructural restrictions in communication, information exchange, data management and storage. Many research works so far on employing Industrial IoT (IIoT) in the mining industry concentrate on particular issues like monitoring ventilation, personal and fleet management, pre-alarm system, analysing accidents and tailing the dam monitoring, whereas the entire IIoT architecture for common mining industry conditions are still missing (Zhou et al. 2017a). Present information associated with Information Technology (IT) in the mining area has been analysed and it finds the main issues of vertical-fragmentation as a result of technological variations of several devices which prevents data distribution, secure information exchange amongst systems and devices and interoperability prevention.

Data processing, interoperability methods and remote controlling progress all the mining stages from searching to reclamation. Moreover, adopting IIoT architecture in this sector affords a safe mining site for the workers, probable mining functions, interoperable surroundings for modern and traditional devices and systems, automation for minimising the intervention of humans, and permits underground surveillance by the convergence of IT and Operational Technology. A system for automatic monitoring is vital for confirming the safety of the engineering construction that happens underground. Present international studies and status of the underground monitoring systems have been summarised from three main aspects such as data acquisition, transmission and processing (Wang et al. 2020). In the data acquisition process, optical-fibre sensors are in process, wireless transmission based on Zigbee has turned to be the trend. In data analysis and processing, several models have evolved endlessly, each of which possess their own disadvantage and advantage. Neural Network (NN) model has been developed recently for this purpose, the integration of NN model and supplementary models explore better accuracy for prediction (Mishra et al. 2019). Likewise, WSN (Wireless Sensor Network) comprising of a diverse array of micro sensors that are interconnected through a wireless communication network for affording much monitoring and sensing services. Initially, distributed system for monitoring is endorsed in the underground coal mine relying on WSN. This has been in accordance with the examination of communication situation corresponding to the underground coal mine (Praveena 2019). Using varied technologies like sensors, communication, microcomputer and automatic detection for understanding the operational functionalities of smart mining area management, this system conquers data warning in real-time. Through industry-camera, it fetches on the remote control and image gathering. It also performed real-time monitoring of security information (Deokar and Wakode 2017).

Similarly, a safe coal mine system has been executed through the usage of the webpage as the medium for data transmission. The device has been introduced for monitoring several parameters within coal mines like light detection, humidity and temperature conditions, fire detection and gas leakage. This sensor system, also termed a large device, has been mounted in coal mines. Every sensor rankings have been automatically afforded into thinking processors that produce various ideas for exploration (Reddy et al. 2021). Besides, laser detecting sensors have been utilised in this model for detecting the existence of light. A notice has also been sent to the mail of the designated person for confirming a proper reply to any probable fire disaster. Vibration sensor has also been utilised for detecting movements happening in mines to minimise unusual events in underground coal mining. A smart monitoring system for coal mines is explored by (Hu et al. 2013). It comprises of a smart server system for the managers and sensor network featured by efficiency and stability. In addition, the duration of nodes in WSN has been computed for representing the efficacy of particular work functionality. Robustness has been the main problem in monitoring mine. For instance, abrupt node departure might lead to sub-network destruction. Because of this, the future works mainly concentrates on the enhancement of robustness related with the suggested sense network. Developing protection of the coal mining for employees through a gas sensor, humidity and temperature sensor, gas sensor, and Arduino persists for tracking mining safety and performing information update for IoT site. Through the use of tools, workers' safety has been confirmed (Prabhu et al. 2019). Also, a system based on block chain appears to be better technology for the mining industry as it might assist in restricting the inclusion and interruption of cyber-attacks as a result of distributed networks and heterogeneous devices. Several open challenges in underground mine communications have been explored. Vulnerabilities such as denial of service attacks and information disclosure have been exhibited during risk and threat analysis on data logger of Safe Advance Goaf Edge Support Systems (SAGES) that comes under information generation of IoT. Hence, it has been advisable for incorporating security through the design strategy from initial development phases. Thus, storage and data transmission could be made secured (Singh et al. 2018).

RFID technology has been selected for communication within mines (Jayasree and Kiruthika 2021; Tajane et al. 2020). Though this research exists in its initial stage, it has already been initiated to find the gaps of the current work by concerning cyber security in IoT, such as vulnerabilities in cyber security, deficiency in standards and structural limitations with the significance on resource and mining industry (Barrie et al. 2017). Proof from the conventional work has high pointed security issues, and there is a lack in the IoT standards. These together possess a threat for mining operations when not mitigated or addressed.

1.1 Digital-Mine System

A Digital-mine is a virtual version of the physical mine. It can also be defined as a simulation of equipment, machinery and production process. It is regarded as the network of methods, models and tools assisting the overall production operation. Digital-system concentrates on monitoring under-ground miner's movement, equipment, strata, environmental conditions in the operating goaf regions, impending hazards, health conditions, flood conditions, dispatch status and mineral production, etc. (Kowalski and Elliott 2020).

1.2 Mine Environment Monitoring

This system has been utilised to monitor the status of humidity, air velocity and temperature in the underground mine through sensors. Due to the harsh environment in the mining tunnels and high accidental risks, there exists a tendency for minimising the presence of humans in tunnels. Accordingly, an inspection robot prototype has been presented, which could assist in maintenance of staff at the inspection time. Recently, the robot has been controlled by an operator by utilising radio. Identified hotspots could also be localized. In addition, its position could be stored as a whole with both kinds of images. Due to the radio-control systems, the operator could remain safe. This kind of robot could be categorised as the mobility monitoring system for the spatial distribution of the underground infrastructure (Szrek et al. 2020).

1.3 Strata Monitoring

This system is comprised of a linear potentiometer along with the piston. The system can measure changes in the convergence and is comprise of four main parts such as convergence, microcontroller, ESP-01 and load-cell sensor. The harsh environment has made it challenging to design reliable and robust networks in the underground mines. To develop a consistent communication system, it is vital to realise the characteristics of an underground mine and various issues faced by wired and wireless communication. Few main properties associated with underground passages impacting the performance of communication systems in the underground mines include refraction, reflection, propagation velocity, noise, path loss and multipath fading. Signal rate offset enhances with the enhancement in moisture content. A communication system based on high frequency requires more attention in comparison to a low-frequency system. In addition, path loss value also enhances with the enhancement in the distance between the transmitter and the receiver. Transmitted signals that are reflected from underground mine walls lead to signal loss. Various mining apparatus and reflectors resulted in signal fluctuation, fading of signal strength and multi-path propagation. Due to temperature change within underground mine, dielectric properties of the medium alters, leading to variation and weakening in propagating signal. Noise created through several types of equipment within underground mine included leads to quality degradation. For operating devices in a harsh environment, it has to be resistant to high moisture content, dust particles and a huge temperature range. Typically used Wi-Fi-based technology for exchanging information affords more delay, high data loss and minimum speed for data transmission. Hence, suitable wireless communication is essential for solving the existing issues of the wired communication in the underground mines. Taking all these existing issues of the coal mines into account, IoT based wireless and real-time monitoring system is introduced here to track and monitor these mines for ensuring miners' safety.

This study is mainly focussed on designing and developing of IoT permitted strata condition and a sub-surface mine environment monitoring a web-based system for decision making to predict sub-surface environments.

2 Material and Methods

The present study focussed on varied parameters such as monitoring the environmental condition and strata monitoring. The utilised approaches and systems are discussed in this section. Software is introduced through the usage of unified observance and predicting hazards to improve safety and productivity in underground mines. Integrated software design include varied modules as shown in Fig. 1 that comprise gas and environment monitoring, strata monitoring.

The modules are introduced for encompassing several aspects of mining in one platform. Python language is used for programming. In addition, MySQL is used as backend support. The software runs on the server and creates a web-based view and real-time information that can be read on the browser. By web-based interface, mine management could perceive real-time mine operations. Mine information can also be accessed from any place within the coal mine having work access.

The software operates in the online-mode alone

- In real-time (online mode), live data from various sensors are continuously displayed on the software page. Nodes transfer the data of various areas to Base Station (BS). Then, BS transfers the data to the gateway. Subsequently, this data is transferred to the system by serial-port, where automatic saving is performed in the database. In this operational mode, application software is employed for monitoring, analysing, storing and seeing information obtained from nodes connected with sensors and also afford E-mail alerts and SMS to the administrations of mine. Data is also analysed and stored in DB. The software continuously gets all the data, predicts imminent explosive conditions, it also displays the live data. When the value of the sensor exceeds the threshold limit, an alarm gets generated representing the affected environment zone utilizing varied colour-codes such as yellow, green, red and orange. When a node exists in a critical situation, SMS and E-mail are sent to respective individuals immediately and confirms the miner's safety.
- In the offline mode, stored data can be displayed by simply choosing the file of needed data.
- In both of the operational modes, software assists in printing hard copy (graphical or numerical as preferred).

An administrator could alter the configurations of software such as mine plan's working area, scale unit, router status and locations and several other vital parameters. Sufficient security could be maintained by a password.

2.1 Environment Monitoring

This module tracks the environmental conditions in the sealed-off region of an underground mine. It monitors the specific gas concentrations in air (SO_2/H_2S/O_2/CO) so that any disaster in the mining area can be avoided. Methane gas concentration is also monitored, which is the major reason for underground mine explosions. A system having model number: DM-O_2/SO_2/H_2S/CO-1 is utilised to monitor the gas concentration status in the underground coal mines through the usage of sensors, ESP-01 Wi-Fi module and ESP8266 that uploads the wireless network information. This wireless-LAN system

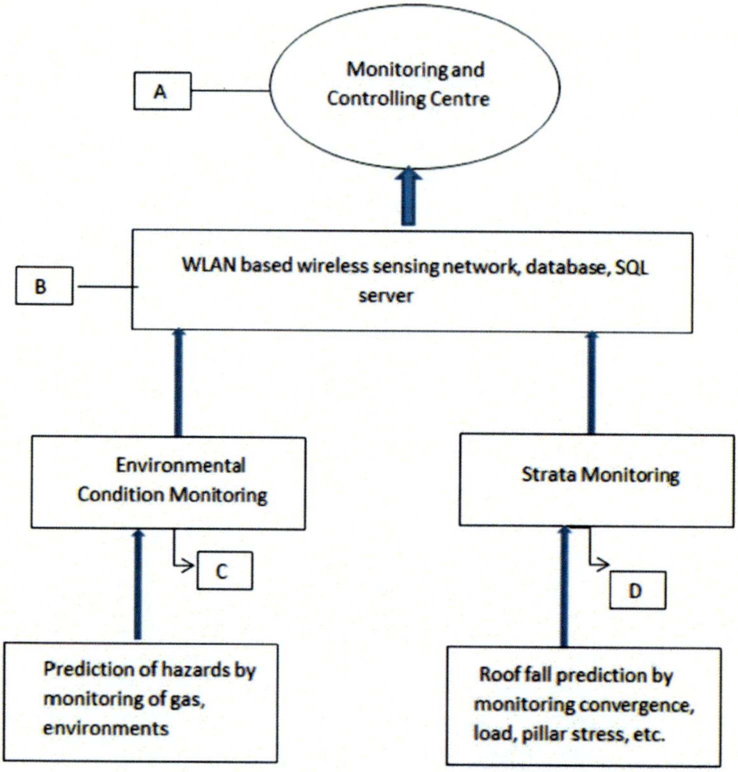

Fig. 1. Various digital mine module

consumes minimum power for data communication for performing digital signal transmission. The sensor module encompasses varied gas sensors like SO_2, H_2S, O_2, CO, CO_2, NO_2 and CH_4 individually in each module that possesses three main components: sensor, ESP-01 and microcontroller. This operates at ISM (Industrial Scientific Medical) frequency band. These sensors persistently need gas concentration and transfer it to the server to monitor underground mine gases in real-time. This also encompasses two major parts such as gas monitoring unit and the power supply unit. The circuit power supply has been supplied from an external source of 12 voltages intrinsically safer power supply. There exists additional system circuitry with the use of a high-quality voltage regulator for regulating the power supply, thereby acquiring a good safety factor. For safety purposes, a fuse is linked at the input system point by constraining the excess current. Further, a current limiting resistor has also been utilized for limiting current in the circuit. The Block diagram of the gas monitoring system is explored in Fig. 2.

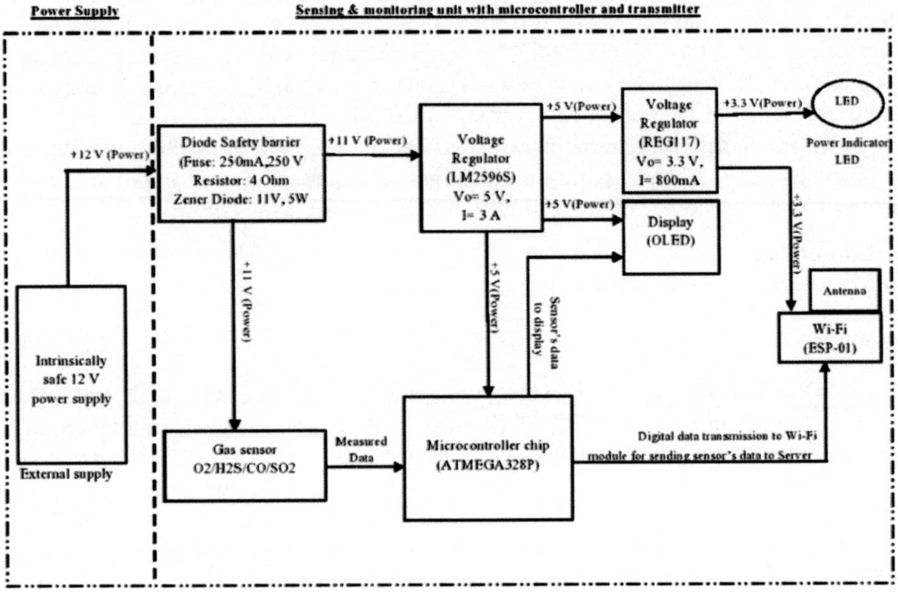

Fig. 2. Block diagram: Gas monitoring system

2.2 Strata Monitoring

Strata monitoring utilises the Wi-Fi network for the wireless transfer of sensor data. This data is continuously retrieved from varied areas in underground operations and transmitted to the control room. The system monitors load and roof convergence. It is controlled through a microcontroller for giving an audio and visual alarm in dangerous situations. The system's software part monitors, analyses and stores information that are received from the nodes connected with sensors and subsequently stored in a database. The software obtains all the data persistently and explores real-time data of each parameter by the graphical interface in a window. It also produces SMS and E-mail alerts comprising of the alarms to supervisors. In case of crossing the threshold limit of the sensor value, the system automatically indicates the impacted zone by varied colours. This situation enables a pop-up window on the control room screen. This module is also integrated with AI and mathematical models to analyse and assist the sensor information in a graphical manner. Hardware circuits comprise of microcontroller, convergence metre, ESP-01 and load cell sensor in addition to the needed electronic components to transfer the data wirelessly. In addition, these are installed in underground mines in varied areas. The load cell is also installed at base support. Moreover, a convergence metre is installed on the roof. This module initiates to monitor and then read the metre and sensor value of the load cell upon switching it ON. It persistently transfers computed values to a corresponding IP address by ESP-01. The software also picks the values of the sensor from the IP address. These values are later checked with a collection of threshold values. Based on this, the software module detects if strata movement exists under safe restriction or not. This accomplishes it by selecting and exploring the alarm or warning level from three levels.

Besides, the threshold power of this device is minimum in comparison to the accessible value suggested in the IEC 60079-0. The temperature range of the operating ambient is −40 °C to 70 °C, and supply voltage has a range of 12 V, circuit's current consumption is 250 mA, and the standard wireless compliance is IEEE 802.11.

Strata monitoring device based on IP possess a linear potentiometer with 225 mm as piston length. It also computes the convergence alteration of a maximum of 225 mm having 1 mm as accuracy. It comprises 4 portions, as discussed below.

a. Convergence
b. Microcontroller – Atmeaga328
c. Load cell sensor
d. ESP-01

Initially, the analogue output of the potentiometer is afforded to the digital converter. Then, it is transmitted to the ESP-01 wireless module. This data is then transferred to the server by a shared WLAN network. Convergence metre has a standalone possessing on-board display that could be utilised by the miners for attaining convergence status. Sensors also fetch data from the underground mine. After this, the data is transferred to MCU, which is then transmitted to software by the server. Values are matched in contradiction to the threshold values based on the software that produces three audio and visual signals for warning and generates reports by relying on the imminent danger to the respective mine management by E-mail or SMS. It monitors the load and roof-convergence in real time. Data recording is performed location wise in the database. The maximum-input voltage corresponding to the device from the internal battery is about 12 V. Later on, voltage regulates down to about 3.3 V and 5 V through the REG1117 regulator and LM2596S. Hence, component analysis for strata monitoring is undertaken at the corresponding needed voltage. In addition, Fig. 3 explores the circuit diagram matching to strata monitoring.

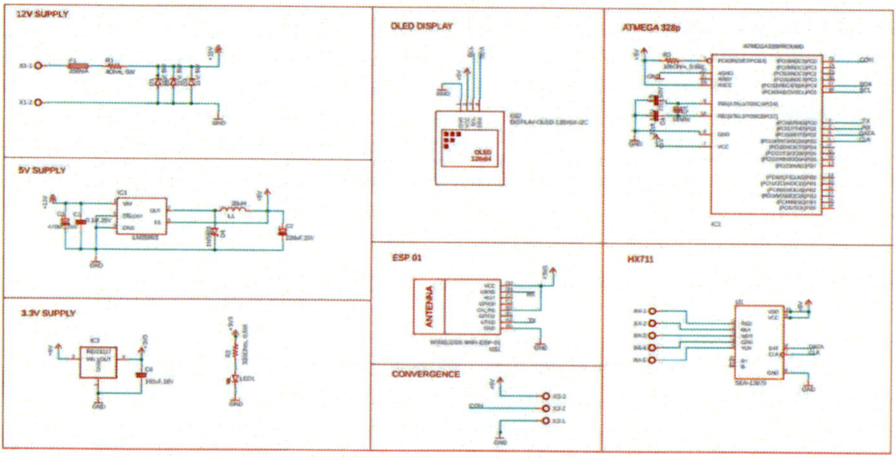

Fig. 3. Circuit diagram: Strata monitoring device

3 Results and Discussion

The varying outputs of each module, with monitoring phase and ensuring safety measures of miners during the mining process are illustrated in this section. The results are determined by executing application software module by module based on requirement objectives. The modules in application software perform task execution for Strata monitoring and Mining Environmental monitoring phases.

Digital Mine Software, designed to facilitate higher productivity and maximize Mine safety by using an integration of sensor technology, computing power, and Wi-Fi-based network technology. The Software consists of various components and modules. The present section enumerated modules outcomes, aiding in the Environmental condition determination and Strata monitoring phases. The software generates effective results, helping the Mining managers or individual application users to gain tracking information and demographic information regarding environmental status inside the mining environment and strata monitoring information. The outcomes are fetched through the execution of Python language in the front end and MySQL in back end support. Generally, the outcomes of the module are obtained during software execution on the server-side. The web-based screenshots and the resulting real-time data were viewed upon browser. The accessibility of such software is explored in any location within the mining site. The mining-related information was depicted through graph representation, map-based interface reports.

In the application software, through each module, the software receives all data continuously, does the prediction of environmental conditions, their monitoring process, Strata data monitoring process. The user enters into the application software after the authentication process through Username and password. After the successful login process, the home page appears in the software. The dashboard of the software consists of various components, as illustrated below (Fig. 4, 5, 6 and 7).

Fig. 4. Properties of Dashboard

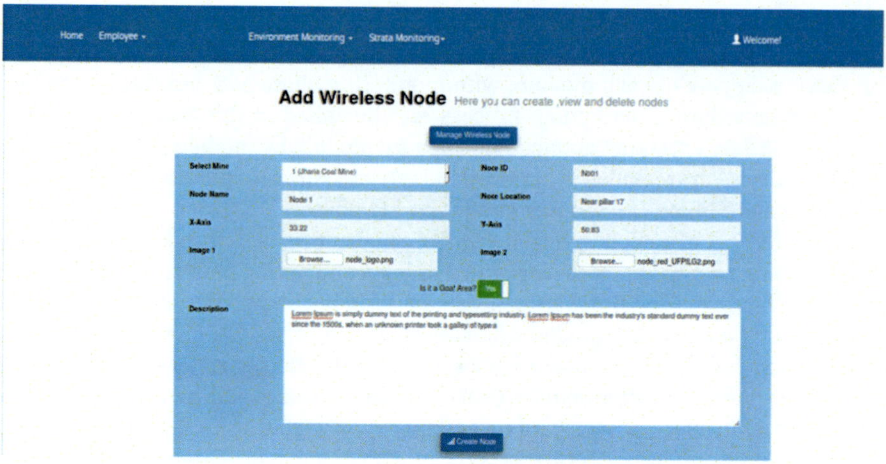

Fig. 5. Add Wireless Node Option

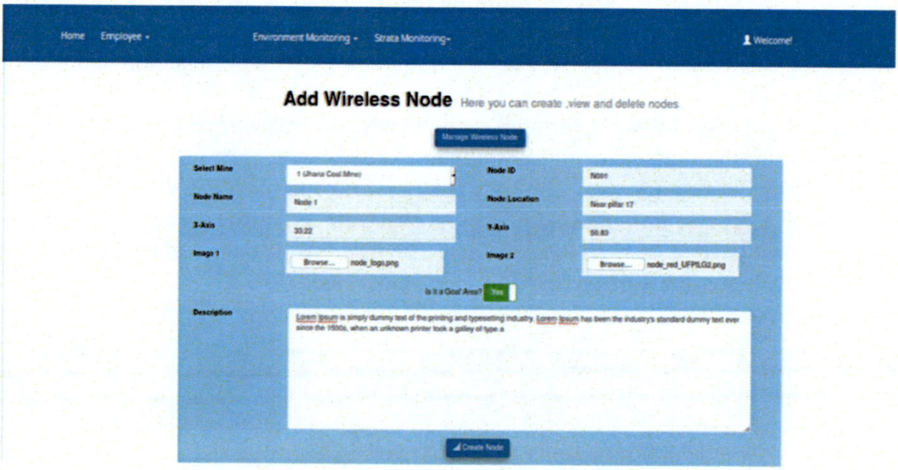

Fig. 6. Wireless Node Details in Data Form

1. System Login process, through authentication by Valid User-name and User password
2. Then the user would be directed to the default home page after the login phases.
3. Traverse to Environmental monitoring option → Select Manage Wireless node, seen as the first label → After clicking the option, select add Wireless-node button on that manage page.
4. Move to Monitoring option → Select Environmental Monitoring option → Select Add Wireless-node option from the respective navigation bar.
5. The selection will yield you an appropriate Wireless node page. All the details related to the Wireless node are filled up in the page. After this, X-axis or then Y-axis is clicked. The selection adds the exact node location in the specified place.

Fig. 7. Addition of wireless node in specific location of Mine Map

6. After the process, the map is opened, such that the wireless node is located in a specific physical address. After node location in map, the escape key is pressed, which exits map.
7. For any node creation, create node-button is pressed for any wireless node addition to the application database. After the addition of a wireless node in the database, the next page directs the user to manage the Wireless node. (Manage Wireless-node option).

3.1 Strata Monitoring System

The strata monitoring analysis and strata monitoring control were explored through built sensors attached to miners or in certain locations. The analysis regarding the strata details may engage managers or users to formulate the prevention strategies or the measures in the prior period. The strata monitoring module would estimate the roof fall of the location through prior monitoring of strata conditions in goaf sites or working areas (underground mines) by employing the IoT enables-sensors. The application sensors collect real-time information from those underground mines. Periodically, Sensor values were ensured in accordance with threshold values. The data is traversed to the micro-controller unit. The unit then does communication to the targeted destination (software) by using the server. Different reports and various audio-visual warnings are generated through the application. In the module, various operations aids to determine strata monitoring information are obtained. The operational outcomes with respect to each feature of the strata monitoring module are depicted as the results. The application system utilises two categories of sensors Load cell and roof-convergence sensor. These all sensors were fitted for measuring out strata conditions, including bed separation of mining roof data, roof load in specifying point, roof convergence and stress upon pillars. The other results that can be determined in the module are rock-mechanic parameters such as convergence, dilation load and stress of the site roof. These conditions are monitored for secure strata control for miners. Before the allocation of employees or employee creation, the site ought to generate the following things. In order to do the process, the following steps are performed (Fig. 8, 9, 10 and 11).

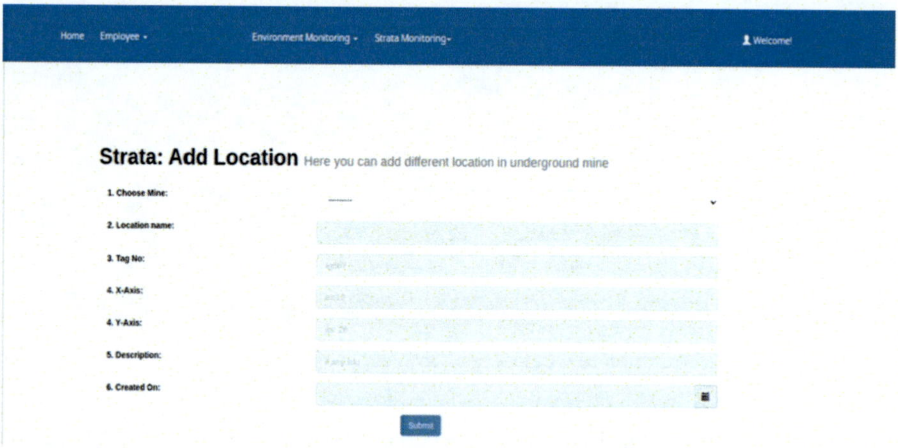

Fig. 8. Location addition in mining sites

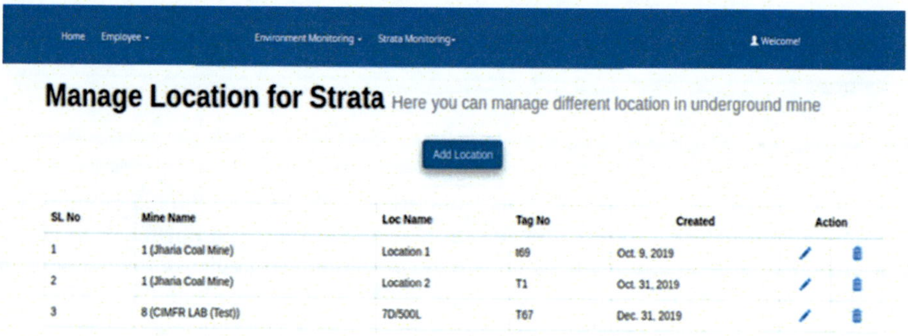

Fig. 9. Strata Location Management

Click on the Home icon → click on monitoring option → then select Add location option from drop-down menu options.

After the "Add location" option is clicked, the window prompts for a new data form. The form consists of various labels such as Choose Mine, name of the location, Tag number, X-axis notation, Y-axis notation, Created date input and the description of the location. After the input values are all given, the submit button is clicked. The submit

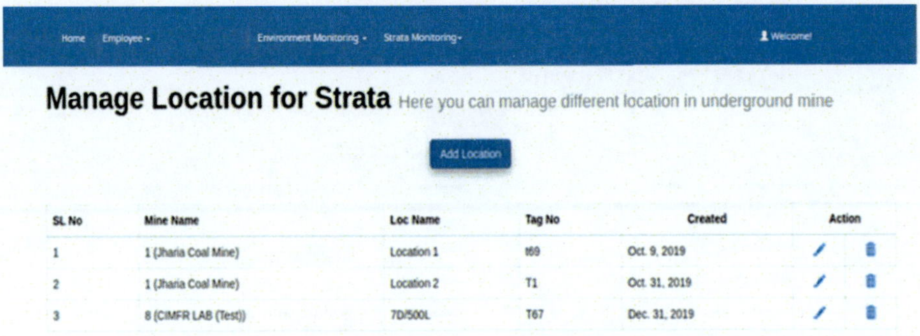

Fig. 10. Deletion of added Location (through trash icon)

button saves the data onto the database. In similar to location addition, the location of strata can be manipulated or managed through the Manage-location option. Similarly same as discussed above, click on Home menu --> followed by selecting monitoring option → after the selection click "Manage location" option.

The below Fig. 10, depicts the Prompting screen as soon as the Manage location: option is selected. The window displays the list of all the strata locations, been added before. The Location details are viewed as in tabular representation, with Mine name, Location name, Tag no, action to be taken and the creation date.

Through the figure, it is clearly revealed that the particular location entry could be added or deleted from the tabular lists. The information can be added through "Add location" in the topmost space. The Entry of location could be deleted through clicking down the Trash icon. In similar to the above feature, the sensor could be managed or added to the location site through performing steps such as Click Home menu → Select Monitoring option → from the drop-down options, select "Add Sensor" option. As soon as the option is clicked, the new data form is shown as below.

The above Fig. 11, represents the data form of "Add Sensor" option. The data form consists of several fields, such as mine name, Level-1 message, level-2 message, level-3 message, Audio play-type, Name of a sensor, Tag number, IP address of sensor, Time-interval Level-1 colour, level-2 colour and level-3 colour, location of the sensor and the Sensor units. These all the fields need to be given for inputs by the user. With respect to the conditions monitoring, various messages are displayed in sensors. The Warning messages are defined in the sensor addition step. Level-1, Level-2 and level-3 messages can be defined as per the severity conditions. The respective colour for respective message levels is denoted in green, yellow and red colour. The Audio could be defined by converting the text message to audio signals. The feature may alert the miner more predominantly, and good significance is based on the severity. The prior information safeguards the miners from hazardous occurrences.

The study outcomes provide results information regarding different research modules such analysing the environmental monitoring conditions, and results of strata monitoring phase. The results of the software modules, and other parameters, ensure to generate a

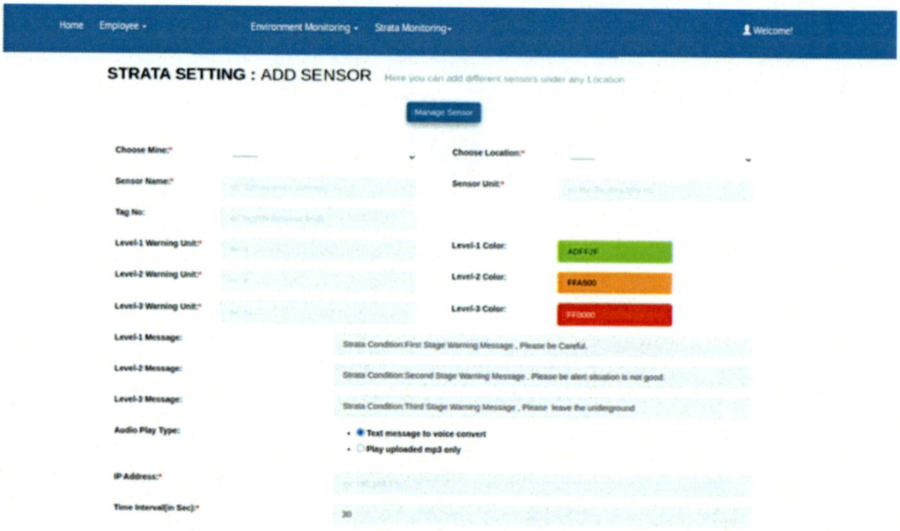

Fig. 11. Adding and managing of sensor in strata monitoring

comfortable and secure Mine working environment. It facilitates safeguarding the miners. The outputs of application software, utilised to store, analyse, view, and monitor information obtained from respective nodes, interconnected to sensors inside the workplace. By using features, it sends email alerts, messages, and Visual alert messages to those mine station administrators. The outputs results were saved within the database. The Software incorporated integrated artificial intelligence and a mathematical model that estimates sensor data changes. If the Sensor data variation is not within range of threshold value, a pop-up window prompts on the monitoring screen in the control room. The research implementation provided a graphical representation of the entire data obtained from sensors depicting all the module options in terms of tabular format and graphs. The results phase yields out a defined dashboard for real-time data display consisting of entire parameters in one window, aids in supervising the status of mining activity and employee safety measures by all higher-level executives. The experimental implementation results are all depicted in the study, illustrating the outcomes of the two phases of Environmental Monitoring Phase and Strata Monitoring-Phase.

4 Conclusion

The complexity of the work environment in the mining sector has made it dangerous. Thus, workers in these mines are susceptible to various risk factors during work that might lead to serious injury or loss that might impact the miners and the mining area. Consequently, it is important to protect these miners as they face threats due to various factors like a gas explosion, fire accident, roof fall, etc. These miners must be earlier made known about the hazardous problems they are going to face. Accordingly, IoT enabled monitoring and smart tracking for the underground coal mines have been developed. Protecting the miners and underground coal mine have been attempted by several traditional researchers. But, most of the studies have not concentrated on India. On the contrary, though few works have focused on securing coal mines in India, they lacked in terms of efficient real-time monitoring and tracking. This research affords some suggestions so that it could be considered in the near future. The enhanced technique will also improve the distance of packet delivery while minimising packet loss. Further work can be carried out by experimental analysis with several network parameters such as network lifetime, communication cost, effective coverage area and energy consumption. In the mine monitoring system, robustness is a vital challenge. This study can also be further enhanced by including an enhanced framework by using other advanced sensors to check the dangers in the underground coal mine. In addition, more empowered IoT frameworks could be developed for further use.

References

Aziz, A., Schelén, O., Bodin, U.: A study on industrial IoT for the mining industry: synthesized architecture and open research directions. IoT **1**(2), 529–550 (2020)

Zhou, C., Damiano, N., Whisner, B., Reyes, M.: Industrial Internet of Things: (IIoT) applications in underground coal mines. Min. Eng. **69**(12), 50 (2017)

Wang, L., et al.: Automatic monitoring system in underground engineering construction: review and prospect. Adv. Civ. Eng. **2020**, 1–16 (2020)

Mishra, P.K., Kumar, S., Pratik, Kumar, M., Kumar, J.: IoT based multimode sensing platform for underground coal mines. Wirel. Pers. Commun. **108**, 1227–1242 (2019)

Praveena, R.: Coal mine monitoring and alert system with data acquisition (2019)

Deokar, S.R., Wakode, J.S.: Coal mine safety monitoring and alerting system. Int. Res. J. Eng. Technol. **4**(3), 2146–2149 (2017)

Reddy, N.V.K., Varma, G.M., Visalakshi, P.: Design of IoT based coal mine safety system using Arduino UNO. Ann. Rom. Soc. Cell Biol., 5663–5670 (2021)

Hu, S., Tang, C., Yu, R., Liu, F., Wang, X.: Intelligent coal mine monitoring system based on the Internet of Things. In: 2013 3rd International Conference on Consumer Electronics, Communications and Networks, pp. 380–384. IEEE, November 2013

Prabhu, D., Nikhil, V.N., Kumar, J.S.: IoT based coal mining safety for workers using Arduino. IJESC **9**(10) (2019)

Singh, A., Kumar, D., Hötzel, J.: IoT based information and communication system for enhancing underground mines safety and productivity: genesis, taxonomy and open issues. Ad Hoc Netw. **78**, 115–129 (2018)

Tajane, P.S., Shelke, S.B., Sadgir, S.B., Shelke, A.N.: IoT mining tracking & worker safety helmet. Int. Res. J. Eng. Technol. (IRJET) **7**(4) (2020)

Barrie, G., Whyte, A., Bell, J.: IoT security: challenges and solutions for mining. In: Proceedings of the Second International Conference on Internet of Things, Data and Cloud Computing, pp. 1–9, March 2017

Kowalski, R., Elliott, B.: Is your workforce prepared for the digital mine? Eng. Min. J. **221**(10), 58–59 (2020)

Szrek, J., Wodecki, J., Błażej, R., Zimroz, R.: An inspection robot for belt conveyor maintenance in underground mine—infrared thermography for overheated idlers detection. Appl. Sci. **10**(14), 4984 (2020)

Vision Enhancement System for Foggy Weather in Opencast Mines

Swades Kumar Chaulya(✉), Girendra Mohan Prasad, Monika Choudhary, Naresh Kumar, Virendra Kumar, Vikash Kumar, and Abhishek Chowhury

CSIR-Central Institute of Mining and Fuel Research, Barwa Road, Dhanbad 826001, India
chaulyask@cimfr.nic.in

Abstract. The presence of fog and smog in the environment hinders the visibility of heavy earth-moving machinery operators in opencast mines. The paper describes an intelligent driving assistance system to enhance the drivers' visibility in opencast mines during harsh weather conditions. The system integrates different devices like proximity radar, infrared cameras, global navigation satellite systems (GNSS), wireless devices, anti-collision laser lights, and image processing algorithms (image dehazing, image stitching, and object detection using artificial intelligence technique) for assisting the drivers to navigate in dense foggy conditions. The system aids the operators in driving during dense fog environments, generates an audio-visual alarm while facing any obstruction, and mitigates colloidal vehicle accidents. The final processed output of cameras, GNSS-based navigation using 3D geo-tagged digital mine map, and proximity radar are displayed on the dashboard screen fitted in forward-facing of driver's seat. The system incorporates artificial intelligence, image processing techniques, and advanced analytics for dehazing foggy video footage, proximity warning, object detection, and fleet management.

1 Introduction

The presence of fog leads to a lack of visibility, which prevents drivers from getting a clear view of the road, which causes accidents and vehicle collisions in opencast mines and transportation roads. Generally, the production rate of opencast mines in hilly reason is very low due to the diminished visibility for the period of rainy and winter seasons. Rain and fog can make opencast mining operations extremely dangerous for drivers practically working blind. While operating heavy earth-moving machinery, the operator shall be aware of surrounding areas to avoid accidents and damage (Saurabh et al. 2016). Due to poor visibility conditions during foggy conditions, opencast mining operations are suspended in the hilly regions. Therefore, it is necessary to improve the visibility of dumpers/shovel operators using latest technologies so that opencast mining activities may be sustained throughout foggy climates to meet the rising demand for minerals (Chatterjee and Chaulya 2019).

Therefore, there is a need for an integrated vision enhancement system to continue opencast mining operations during foggy climate. In addition, the driver assistance system is required for effective visualization of transportation path as well as

A. Sinha et al. (Eds.): AMC 2023, SPEES, pp. 180–192, 2023.
https://doi.org/10.1007/978-3-031-46966-4_15

loading/unloading areas with the help of thermal imaging and infrared cameras, global navigation satellite system (GNSS), proximity radio detection and ranging (radar), roadside edge detection arrangements using flashlight and self-regulating heating cable, light-emitting diode (LED) fitted vest, etc.

Considering the requirement of mining industry, a vision enhancement system has been developed to carry on opencast mining activities in foggy weather. The paper enumerates description of the system and its different modules.

2 System Description

Vision enhancement system (VES) is a smart driver assistance arrangement for proper visualization of transportation path and working faces in foggy weather with the help of thermal imaging and infrared cameras, proximity radar, GNSS, wireless devices, rugged computer integrated with graphical processing unit (GPU), laser light, touch screen, roadside edge detection arrangements using flasher light and self-regulating heating cable, etc. (Fig. 1) The system helps in an enduring opencast mining operation in foggy weather, navigation using 3D mine map and GNSS, collision avoidance, optimization of shovel-dumper performance, mine transport surveillance, production monitoring and various reports generation.

The system provides 180° panorama field of view through real-time video stitching and image processing, rearview, real-time position of vehicles by GNSS modules, object detection using radar and GNSS, audio-visual alarm to the driver on object detection, processed and integrated outputs displayed on a touch screen installed on dashboard, and display screen logically split into four quadrants (Fig. 2). The upper half of the screen displays output of the combined algorithms incorporated with the video frames of internet protocol (IP)-based front cameras, and the bottom half is further split into 3 windows for radar output, GNSS-based navigation, and the view of the road behind the HEMM to provide multi-stage maneuvering and safety (Kumari et al. 2022a). The driver can choose any specific window by means of a touch panel to enlarge it on full screen, even though the other applications run in the background and alert messages pop up as and when received. A high-end GPU is utilized to process data of all the cameras, proximity radar, and GNSS information using advanced analytics and artificial intelligence.

The intelligent vision enhancement system helps to carry out opencast mining and transportation operations safely during foggy weather. The integrated system with multi-stage safety features helps the mining machinery operators to see through the fog and provides alerts in case of route deviation, obstruction, and impending collision (Kumari et al. 2022b). Thus, the system facilitates to improve safety, production, and productivity in opencast mines. These facilities make the system techno-economically feasible for installation in opencast mines.

3 Image Processing Software

The flow diagram of the software for the vision enhancement system is depicted in Fig. 3. Video output of the 3 front cameras fitted in front of a dumper is fed to a high-end graphical processing unit (GPU), i.e., NVidia Jetson TX2 processor. The power supply

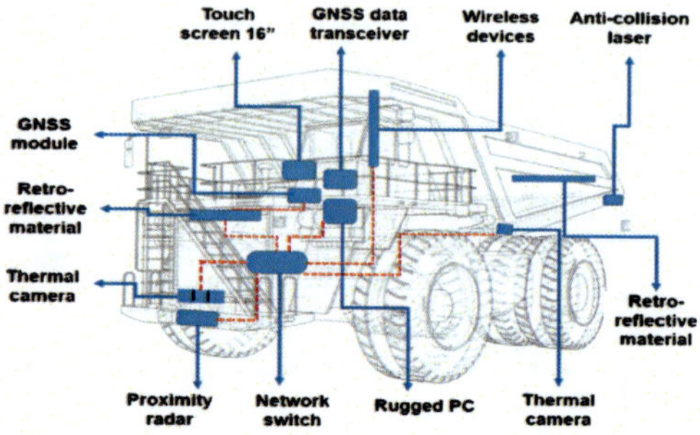

Fig. 1. System deployment architecture for dumper

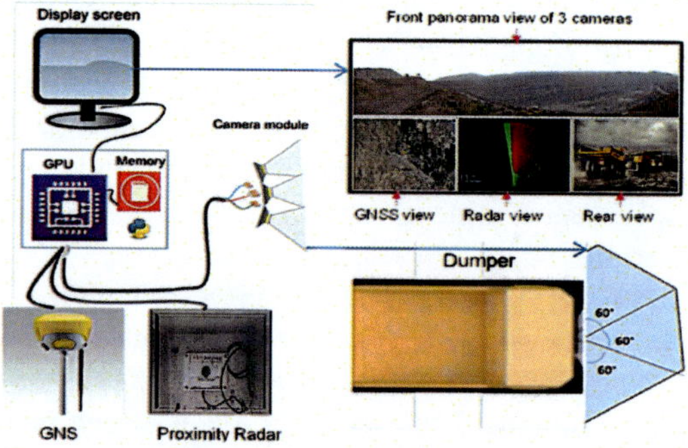

Fig. 2. System function diagram

is given to the cameras from the dumper powertrain. The GPU constantly verifies 'on' or 'off' status of all the 3 cameras. When these cameras are 'on' condition, it proceeds the input images for subsequent processing level. A high-end GPU is installed with the 3 algorithms for processing of videos in real-time, such as (a) video stitching of 3 camera frames (Kumari et al., 2022b), (ii) image improvement (He et al. 2011; Ramya and Rani 2012; Kaur and Sohi 2017; Kim et al. 2018), and (c) object detection using IR camera (Younis et al. 2020). The processed haze fee video and object detection is displayed on a touch screen fixed on the dashboard, so that the HEMM can be operated during foggy weather (Fig. 4).

4 System Modules

4.1 Camera

The system uses 3 thermal/infrared cameras for front (Fig. 5) and 1 for rear end of dumper, which are strategically placed or as per the user requirement to provide a wide field of view for maneuvering dumper and cover blind spot. A GPU is used for image processing. Infrared (IR) illuminators are used to create artificial invisible light to get a better camera output image. Specifications of the camera are as follows:

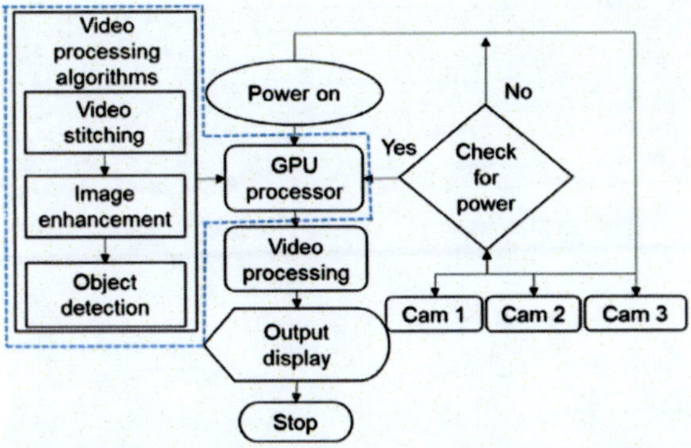

Fig. 3. Flowchart of image processing and object detection algorithm

- Resolution (minimum): 640 × 480 pixels
- Frame rate (minimum): 25 fps
- IP support: Yes (RJ45)
- Recording support: Micro SD: 64 GB or higher
- Video compression: H.264
- Onvif support: Yes
- Field of view: 60° (h) × 45° (v)

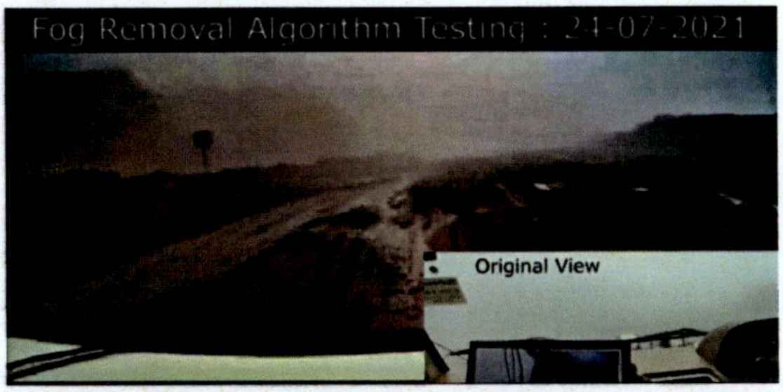

Fig. 4. Display of processed camera image on the touch screen in front of the driver seat

4.2 Proximity Warning Device

Proximity radar (Fig. 6) detects static as well as moving objects, vehicles, virtual target areas (as per as per point 14 of DGMS Circular No. 6, 2020), and system generate an audio-visual alarm for the operators to prevent collision. An algorithm is incorporated into the system which verifies the detected object signal and controls the false alarm. Specifications of the proximity warning device are as follows:

- Sensor: frequency-modulated continuous wave radar of 77 GHz frequency
- License-free band as per Wireless Planning Commission (WPC), Government of India

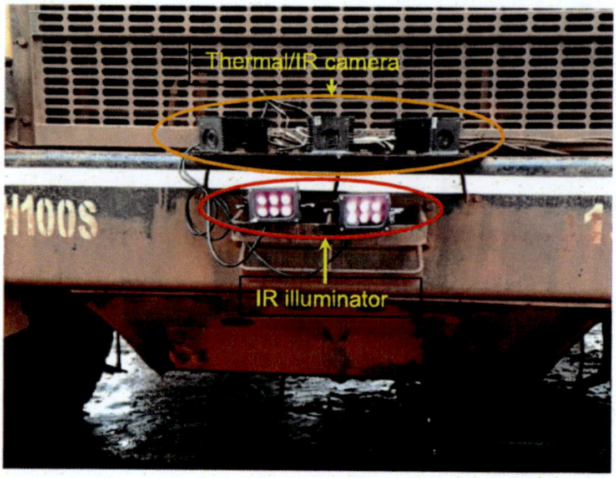

Fig. 5. Thermal/IR cameras installed in front of a dumper along with IR illuminators

- The sensor shall detect objects in a virtual target area as mentioned below:
  - Detection range: up to 80 m (programmable range)
  - Detection width: Adjustable to the width of truck + 0.5 m on both sides
- Warning indication: Audio and visual

Fig. 6. Proximity radar installed in front of a dumper

4.3 Global Navigation Satellite System

Rover-base station based GNSS (Fig. 5) provides 10 cm level accuracy in positioning and attitude. GNSS also provides the location of the surrounding vehicle on the display map as well as it gives warning when any vehicle is approaching less than the predefined safe distance. A rugged computer is used for the software application. Specifications of the GNSS are as follows (Fig. 7):

- Frequency range: GAGAN/GPS/GLONASS
- Mounted on truck
- Accuracy <10 cm
- Real-time (RTK) corrections
- GPS interface: Ethernet/RS232/RS422
- Impedance: 50 Ω
- Azimuth coverage: 360°

Fig. 7. GNSS and wireless device installed on a dumper

4.4 Advanced Analytics with Artificial Intelligence

The system is capable of counting the number of trips made by each dumper from face to dump and tampering alerts (power disconnection or cover tamper). The system provides real-time optimized shovel-truck combination to increase productivity. The advanced analytics with artificial intelligence (Fig. 8) perform the following functions:

- The system provides an object detection alarm.
- Vehicle locations for tracking it on geographic information system (GIS) map or 3D mine map.
- Facilitates monitoring and displaying real-time locations of the vehicle and route undertaken on the GIS map or 3D geo-tagged mine map.

- Provides total distance traveled, total time, time taken per trip or between stops, round trips traveled by a vehicle, and stoppage report.
- Facilitates to monitor route deviations or geo-fence deviations.
- Facilitates to generate different reports on route, distance, number of round trips traveled by vehicles on a daily/weekly/monthly/custom interval.
- Geo-fence: Facilitates to fix lines/points along route or around an area within which a vehicle should travel for a specific period and receive alerts on violation of the defined geo-fenced area.
- Alerts: Provides alert notification when rover or GNSS device is tampered with, or power is disconnected from a vehicle battery, geo-fence violation, route violation, stoppage more than the specified time, etc.
- Display: Displays surrounding vehicle locations on GIS map or 3D mine map.
- Proximity awareness: Calculates and displays distance of the nearest vehicle fitted with GNSS, and gives warning when the two vehicles reach closer than the set safe distance.

Fig. 8. Advanced analytics feature integrated with the system's software

4.5 Navigation in Foggy Weather

The navigation module helps the driver to operate the dumper in harsh weather where the visibility is poor, and route judgment is hampered. The dumper location is determined using GNSS displayed on the 3D geo-tagged mine map to view the dumper location on the virtual mine map and indicate road direction (Fig. 9).

4.6 Wireless Networking

Wireless networking provides seamless connectivity of the mining equipment such as shovels, and dumpers with the control centre, where all the data from the equipment is

Fig. 9. Navigation through 3D geo-tagged digital map and GNSS

captured, stored and processed (Fig. 10). In addition, this wireless network can also be used to provide voice, data from GNSS, and other communication, as well as to provide a live feed from cameras placed on poles/towers at the strategic locations of an opencast mine, which can be viewed in the control room. However, the view of thermal/IR cameras is displayed on the screen fitted in front of the driver's seat after image processing using an industrial computer which is kept in the driver's cabin so the driver can see a front view of the road on the screen and drive the dumper during foggy weather. Thus, the video of the thermal/IR camera fitted on the HEMM is not required to be transmitted to the control room through the said wireless network. Specifications of the wireless device are as follows:

- 802.11a/n radio
- Speed up to 300 Mbps or better
- Frequency range: 5.72–5.875 GHz
- 10/100/1000 Base-T Ethernet

4.7 Fleet Management Module

The system caters to all the aspects of production and performance monitoring, and reporting for mine in a single unified database and software platform (Fig. 11). The user is allowed to view information at every level (i.e., equipment, section, mine, etc.) of the operation. The system is able to accept various performance indicators like trip count,

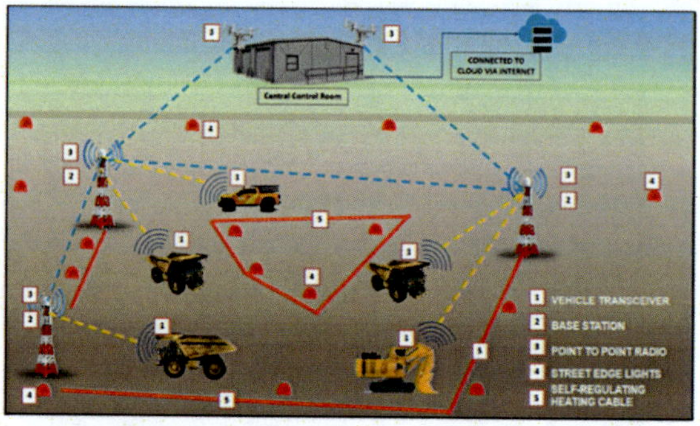

Fig. 10. Schematic diagram of wireless networking throughout the mine

working hours, idle hours, break down hours, maintenance hours, cycle time details, trips per hour and other forms of reporting, HEMM availability/utilization/overall efficiency, operator performance, etc. The system supports exception reports and events such as over speed, low power, low battery, power off, power on, geo-fence enter-exit, tow, power cut, tampering, removing, cover tampering, ignition on, ignition off, motion, trips, stops, distance drove in kilometer. There are various types of notification supported, such as push notifications, web notifications, e-mail notifications, SMS notifications. Other reports such as location history, trip, chart, and summary reports are also available on the web or mobile app. Location history can also be projected on the mine map to provide a visual representation.

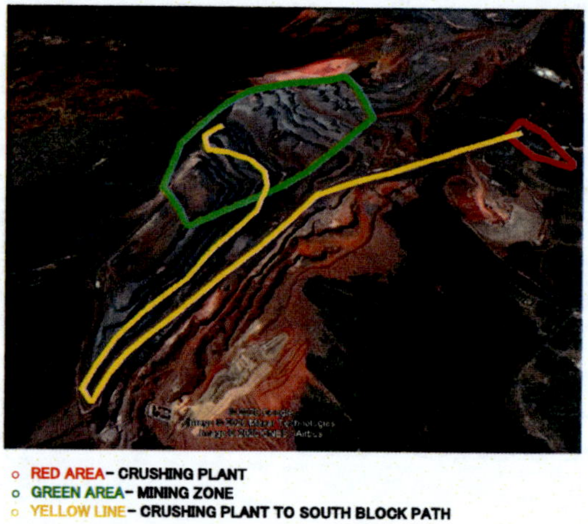

Fig. 11. Geo-fencing of different mining zones for fleet management and trip count

4.8 Anti-collision Laser Light

Two numbers of anti-collision laser lights are fitted on the back of the dumper, which is meant for indicating the form of a red line on the road surface for the approaching fleets from the back of the dumper (Fig. 12). The laser light range is 5–25 m, and the horizontal red line beam width is 5–8 m. Its operating power is 9–24 V DC. It indicates the road surface behind the dumper on which it is fitted during nighttime and in low light foggy condition.

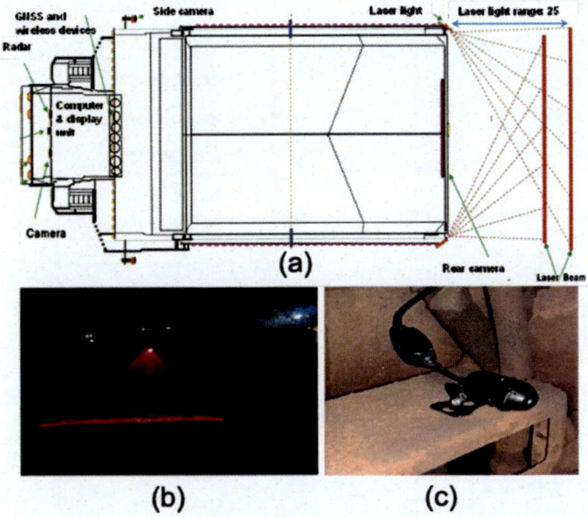

Fig. 12. Laser light for fitting on rear end of a dumper, (a) schematic diagram of laser light installation on a dumper, (b) laser beam at night, (c) laser light fitted on a vehicle rear end

4.9 Other Accessories

The system requires different accessories, such as touch screen, rugged computer, LED fitted vest, street edge lights, self-regulating heating cable, etc. Specifications of these accessories are given in Table 1.

Table 1. Details of accessories of the vision enhancement system

Device name	Photograph	Specifications
Touch screen		• Touch screen size: 16" or more • Mounting: To be mounted on truck/dumper dashboard with the folding arrangement • Interface: VGA / HDMI / Ethernet / RS232 / RS422
Rugged computer		• High-speed, rugged and power-efficient PC including an embedded AI computing device. • Fanless design • 2 Nos. of 64 bit CPU • 8 GB RAM • 32 GB storage capacity • 802.11ac Wi -Fi enabled device
LED. fitted vest		• LED. fittings: 2 horizontal and 2 vertical lines both in the front as well as in the back of the vest • Power supply: 3.7–5 V DC • Battery backup:

5 Conclusions

Mining operation in hilly regions is generally hampered due to the presence of dense fog and poor visibility during rainy and winter seasons. The developed intelligent vision enhancement system helps to continue opencast mining and transportation operations safely during foggy weather. The integrated system with multi-stage safety features helps the mining machinery operators to see through the fog and provides alerts in case of route deviation, obstruction, and impending collision. These facilities make the system techno-economically feasible for installation in opencast mines.

The developed system facilitates multi-layered protection and virtual display of mining activities using innovative techniques to improve visibility of HEMM operators. The system has additional security devices such as proximity radar, anti-collision laser light, etc., apart from visibility enhancement algorithms. The developed image processing algorithm has a significant advantage over the existing image dehazing technologies. The system reduces road accidents by reducing human error. For better safety and easy navigation of dumpers/tippers during foggy weather, augmented reality is being incorporated with the vision enhancement system for long-range virtual visibility of dumper's operator using V2X (vehicle to everything communication), high precision GPS, and digital mine map, along with the processed image of the camera view.

Acknowledgments. The authors are thankful to the Director, CSIR-Central Institute of Mining and Fuel Research, Dhanbad, for granting permission to publish the paper. Further, the authors are grateful to the Ministry of Electronics and Information Technology, Government of India, and National Mineral Development Corporation Limited, Hyderabad, India, to support the R&D project jointly.

References

Chatterjee, D., Chaulya, S.K.: Vision improvement system using image processing technique for adverse weather condition of opencast mines. Int. J. Min. Reclam. Environ. **33**(7), 505–516 (2019). https://doi.org/10.1080/17480930.2018.1496886

He, K., Sun, J., Tan, X.: Single image haze removal using dark channel prior. IEEE Trans. Pattern Anal. Mach. Intell. **33**(12), 2341–2353 (2011)

Kaur, H., Sohi, N.: A study for applications of histogram in image enhancement. Int. J. Eng. Sci. **6**(6), 59–63 (2017)

Kim, K., Kim, S., Kim, K.S.: Effective image enhancement techniques for fog-affected indoor and outdoor images. IET Image Proc. **12**(4), 465–471 (2018)

Kumari, S., et al.: Intelligent driving system at opencast mines during foggy weather. Int. J. Min. Reclam. Environ. **36**(3), 196–217 (2022a). https://doi.org/10.1080/17480930.2021.2009724

Kumari, S., et al.: Real-time video stitching, enhancement and object detection in mines during foggy weather. Concurr. Comput. Pract. Exp. **34**(4), e6631 (2022b). https://doi.org/10.1002/cpe.6631

Ramya, C., Rani, S.S.: A novel method for the contrast enhancement of fog degraded video sequences. Int. J. Comput. Appl. **54**(13), 1–5 (2012)

Saurabh, K., Chaulya, S.K., Prasad, G.M., Ansari, S., Kumar, D.: Assessment of different technologies for improving visibility during foggy weather in mining and transportation sectors. Int. J. Recent Innov. Trends Comput. Commun. **4**(4), 31–40 (2016)
Younis, A., Shixin, L., Jn, S., Hai, Z.: Real-time object detection using pre-trained deep learning models MobileNet-SSD. In: Proceedings of 6th International Conference on Computing and Data Engineering, Sanya, China, pp. 44–48 (2020). https://doi.org/10.1145/3379247.3379264

Interpretable Predictions: Machine Learning Approaches to Understand Slope Stability in the Presence of Joint Networks

Sudhir Kumar Singh(✉) and Debashish Chakravarty

Department of Mining Engineering, Indian Institute of Technology Kharagpur, Kharagpur, West Bengal 721302, India
sudhir_94@iitkgp.ac.in, dc@mining.iitkgp.ernet.in

Abstract. This study leverages machine learning algorithms to predict slope stability in various types of jointed rock masses. Employing a dataset of 40290 cases derived from numerical simulation using the Shear Strength Reduction (SSR) technique, the research analyzes slope stability in the presence of different joint types. The utilized models include Linear, Ridge, Lasso, Random Forest, k Nearest Neighbor, Decision Tree, XGBoost, and AdaBoost regression. The results indicate that XGBoost and Random Forest outperform other models on test datasets, as measured by Mean Absolute Error and R-squared values. Notably, the findings highlight the crucial role of joint networks in slope stability prediction, addressing limitations in previous research. The study also underscores the balance between model performance and interpretability, with complex models providing higher predictive performance at the cost of transparency. Overall, this research presents a robust approach to slope stability prediction, enhancing decision-making in geotechnical engineering and contributing to safer slope design and management.

Keywords: Joint networks · Machine Learning · Predictions · SSR · Rock Slope Stability · Model Interpretability

1 Introduction

The evaluation of slope stability is a critical aspect of geotechnical engineering that assists in preventing disastrous outcomes resulting from unstable slopes. The role of joint networks in the analysis of slope stability has only recently gained significant recognition among engineers and scholars. Joints, or naturally occurring fractures in rocks induced by crustal stresses, alter the mechanical behavior of slopes, making their accurate identification and characterization critical. The stability and strength of a slope can be significantly influenced by these joints. Methods for characterizing joint networks include field mapping, borehole logging, and geophysical surveys. Using these methods, data is collected and analyzed to create a detailed map of the joints (Zheng and Zhao 2004; Wu et al. 2006). Such characterization forms the basis for the accurate analysis of slope stability. Previous studies have highlighted the significance of joint networks in slope stability analysis (Hu et al. 2019; Meng et al. 2019; Wang et al. 2020,

A. Sinha et al. (Eds.): AMC 2023, SPEES, pp. 193–212, 2023.
https://doi.org/10.1007/978-3-031-46966-4_16

2022; Zhou et al. 2020; Shuheng and Yinjun 2021). For instance, researchers have used various joint network models, such as the parallel deterministic joint model, parallel statistical joint model, Cross jointed joint network model, Baecher joint model, Veneziano joint model, and Voronoi joint model, to investigate their impact on the stability of rock slopes (Moradi and Hosseinitoudeshki 2015). However, empirical and analytical methods commonly used for slope stability analysis have limitations, given the assumptions and simplifications involved. The advent of machine learning algorithms and numerical modeling has provided a fresh perspective, enabling a better understanding of the relationship between jointed rock masses and slope stability (Flood and Kartam 1994b, 1994a; Adeli 2001; Lu et al. 2012; Lazarevska et al. 2014; Li and Hao 2016; Chávez et al. 2017).

Machine learning algorithms have shown great promise in solving complex problems across various fields, including geotechnical engineering. The use of these algorithms in slope stability analysis can be highly beneficial, enabling the analysis of large amounts of data and the identification of patterns that might be missed by human analysis. The application of machine learning in geotechnical engineering has been an area of significant interest among researchers (Bolt 1991; Lee and Sterling 1992; Goh et al. 1995; Watson et al. 1995; Lee and Lee 1996; Lee et al. 2003; Benardos and Kaliampakos 2004; Ahmad et al. 2007; Asadi et al. 2011a; Asadi et al. 2011c; Asadi et al. 2011b; Niroumand et al. 2012; Moayedi et al. 2015; Nazir et al. 2015, 2018; Moayedi and Mosallanezhad 2017; Moayedi et al. 2017; Mosallanezhad and Moayedi, 2017; Moayedi and Hayati, 2018, 2019; Moayedi and Jahed Armaghani 2018; Moayedi and Rezaei 2019 (Bolt 1991). Several researchers have explored the use of machine learning algorithms for predicting slope stability, focusing on key input parameters such as cohesion, internal angle of friction (phi), unit weight, height, slope, and pore pressure ratio. Nevertheless, gaps remain in the current research, particularly with regards to class imbalance, data scarcity, and the need to consider various types of discontinuity. Few studies have addressed these factors and their influence on slope stability prediction (Suman et al. 2016; Zhou et al. 2019; Yuan and Moayedi 2020; Amirkiyaei and Ghasemi 2022; Asteris et al. 2022).

This paper seeks to bridge these gaps by considering various types of joint networks and their structural parameters. It aims to expand on the existing body of research on the use of machine learning for slope stability prediction. In the subsequent section we will be seeing data preparation, Generalized Hoek-Brown failure criteria, regression models followed by result and conclusion. The following sections will detail the preparation of the dataset, the regression models, results obtained from a finite element model (FEM) analysis using SSR techniques.

2 Methodology

The methodology for this research is based on the use of a comprehensive dataset of 40,920 instances of artificial slopes. The data has been analyzed using the SSR method of FEM in RocScience's RS2 software. The SSR method is a popular numerical approach for evaluating the factor of safety and the potential failure surface of a slope.

The dataset encompasses a broad variety of joint structures that have been observed in artificial slopes. These structures are exemplified in Fig. 1 through Fig. 5 and include:

- Parallel deterministic joints
- Cross-jointed joints
- Baecher joints
- Veneziano joints
- Voronoi joints

Each of these joint types possesses its own distinct parameters and characteristics that impact the stability of the artificial slopes. It is important to note that the Generalized Hoek-Brown failure criterion has been employed in this study. The Hoek-Brown criterion is an empirical stress criterion that has been widely accepted for estimating the strength of jointed rock masses. This criterion offers an effective way to account for the influence of the inherent characteristics of the rock mass, such as the intact rock strength and the characteristics of the discontinuities or joints within the rock mass. The analysis of the slope section through the SSR method and FEM in RS2 software allows for a thorough understanding of the stability of the slopes, considering the significant role played by the joint networks. This comprehensive methodology, including the simulation of different types of joints and the use of a robust failure criterion, provides a solid foundation for training supervised machine learning models for slope stability prediction.

2.1 Dataset Preparation

Each joint has its own unique set of parameters and a description in the form of figure presented below Fig. 1, Fig. 2, Fig. 3, Fig. 4 and Fig. 5:

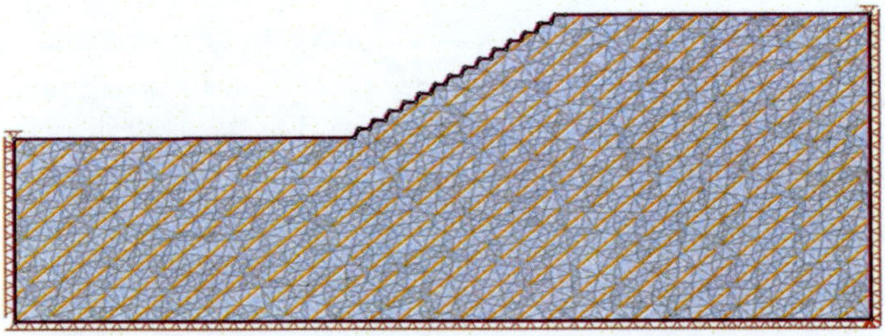

Fig. 1. Representation of parallel deterministic joint network

The required safety levels may vary based on a multitude of factors. These include the accuracy of the input data (Saharan et al. 2020), the scale of the slope, potential repercussions of a failure, uncertainties in the input parameters, the projected lifespan of the structure, anticipated changes in design conditions, and the presence of monitoring systems (Adams 2015). Another significant factor that affects long-term slope stability is time, considering that the strength of rocks can degrade over time due to sustained deformation. As a result, the Factor of Safety (FOS) value could potentially decrease over time (Agarwal 2020).

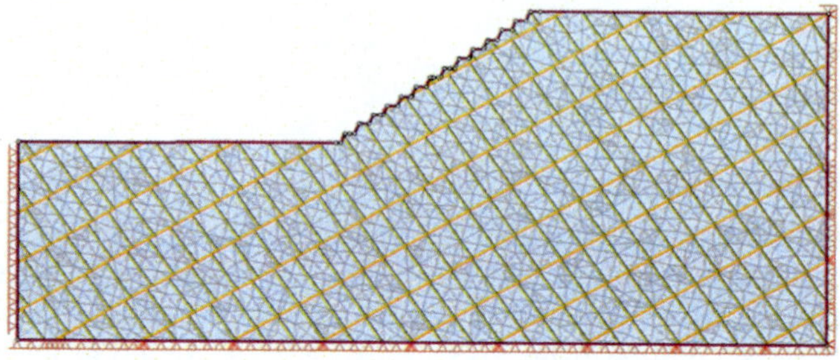

Fig. 2. Representation of cross-jointed joint network

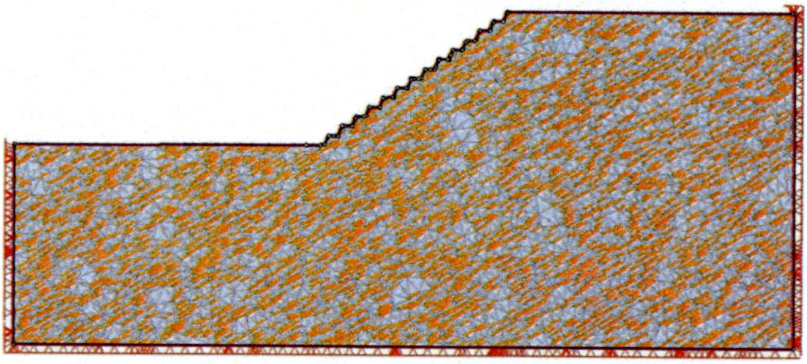

Fig. 3. Representation of baecher joint network

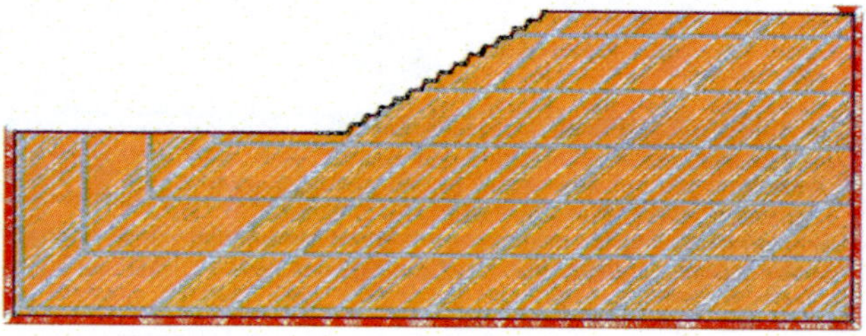

Fig. 4. Representation of veneziano joint network

2.2 Generalized Hoek and Brown Failure Criterion

The Hoek-Brown failure criterion is widely used in rock mechanics to estimate the strength of rocks and rock masses. It was developed by (Hoek and Brown 1988). The failure criterion provides a relationship between the major and minor principal stresses at failure. The generalized Hoek-Brown criterion was later introduced to cover a wide range of rock masses, from heavily jointed to intact.

The generalized Hoek-Brown criterion is given by:

$$\sigma_1 = \sigma_3 + \sigma_{ci}\left[\frac{m_b\sigma_3'}{\sigma_{ci}} + s\right]^a \tag{1}$$

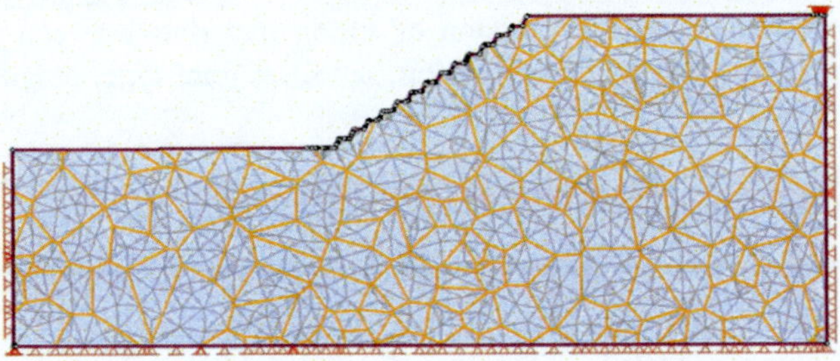

Fig. 5. Representation of voronoi joint network

$$\sigma_3' = \sigma_3 + \sigma_{3i}) \tag{2}$$

$$m_b = m_i exp\left(\frac{GSI - 100}{28 - 14D}\right) \tag{3}$$

$$s = exp\left(\frac{GSI - 100}{9 - 3D}\right) \tag{4}$$

where:

σ_1 and σ_3 are the major and minor principal stresses at failure, respectively.
σ_{ci} is the uniaxial compressive strength of the intact rock pieces.
m_b, s, a are material constants depending on the rock mass properties.
σ_3' is the effective minor principal stress,
defined as the minor principal stress value σ_3
for which the intact rock strength σ_{ci} is defined, plus an additional value σ_{3i}
GSI Geological Strength Index
m_i material constant for the intact rock
D disturbance factor (account for blast damage and stress relaxation)

The FOS in slope stability analysis is defined as the ratio of the available shear strength to the shear stress required to maintain equilibrium. In the presence of joint networks, the shear strength reduction method is commonly used with the Hoek-Brown failure criterion to calculate the FOS.

2.3 Understanding Important Input Features and Output Variables

The input features considered in this study are important for understanding the stability of slopes, in the presence of different types of joint networks and various different external factors. And their effects cannot be ignored. The input features included in this study are as follows:

Geomechanical Parameters - GSI, mi, D, Uniaxial Compressive Strength (UCS)

Slope Geometry - height and overall slope angle but in this model the geometry has been kept constant as it has already been established that height and slope angle have important effect on predicting the slope stability (Singh and Chakravarty 2023).

Parallel Deterministic Joints - cohesion of filling materials between joints, tensile strength of joints, internal angle of friction of joints, joint type, length, inclination, spacing, persistence

Cross Jointed Network - cohesion of filling materials between joints, tensile strength of joints, internal angle of friction of joints, joint type, inclination, spacing, cross inclination, cross spacing

Baecher Joint Network - cohesion of filling materials between joints, tensile strength of joints, internal angle of friction of joints, joint type, length, inclination, intensity type, intensity value

Veneziano Joint Network - cohesion of filling materials between joints, tensile strength of joints, internal angle of friction of joints, joint type, length, inclination, intensity type, intensity value

Voronoi Joint Network - cohesion of filling materials between joints, tensile strength of joints, internal angle of friction of joints, joint type, length, density, regularity

External Factor - pore pressure ratio; rain; horizontal and vertical seismic coefficients

In total there are 23 input parameters used in this study, 3 out of which are categorical variable (joint type, regularity, and intensity type) rest all are numerical variables. There are five types of intensity namely p1 - {joints/area}, p2 - {(sum of lengths)/area}, p3 - {(sum of lengths)/(area0.5)}, p4 - {(sum of lengths2)/area}, p5 - {(sum of lengths)2/area}.

The histograms above in Fig. 6 show the distribution of values in each of the numerical columns in the dataset. The values for all the variables are skewed because most of the variables are associated with different joint parameters and each joint has different set of parameters to describe it so, the other values will be absent. Most of the complex model like Decision Tree, Bagging Algorithms, Boosting Algorithms are not affected by distribution of input features.

The bar plots in Fig. 7 above represent the frequency of each category in the categorical variables of the dataset.

The heatmap in Fig. 8 below shows the correlation matrix for all numerical variables in the dataset, excluding slope angle and height. The observations here are similar to the previous correlation matrix, with GSI, mi, D, and ucs showing a relatively strong positive correlation with FOS.

Collinearity among features in a dataset can affect the performance and interpretation of several types of regression machine learning algorithms.

The Variance Inflation Factor (VIF) Table 1 above shows the VIF values for each input variable after excluding the slope angle and height columns. Here are some observations:

phi_joints has the highest VIF value of 16.45, which is above the usual threshold of 5, indicating high multicollinearity with other input variables. GSI, length, mi, and persistence also have VIF values above 5, indicating potential multicollinearity issues. All other variables have VIF values less than 5, which is generally considered acceptable.

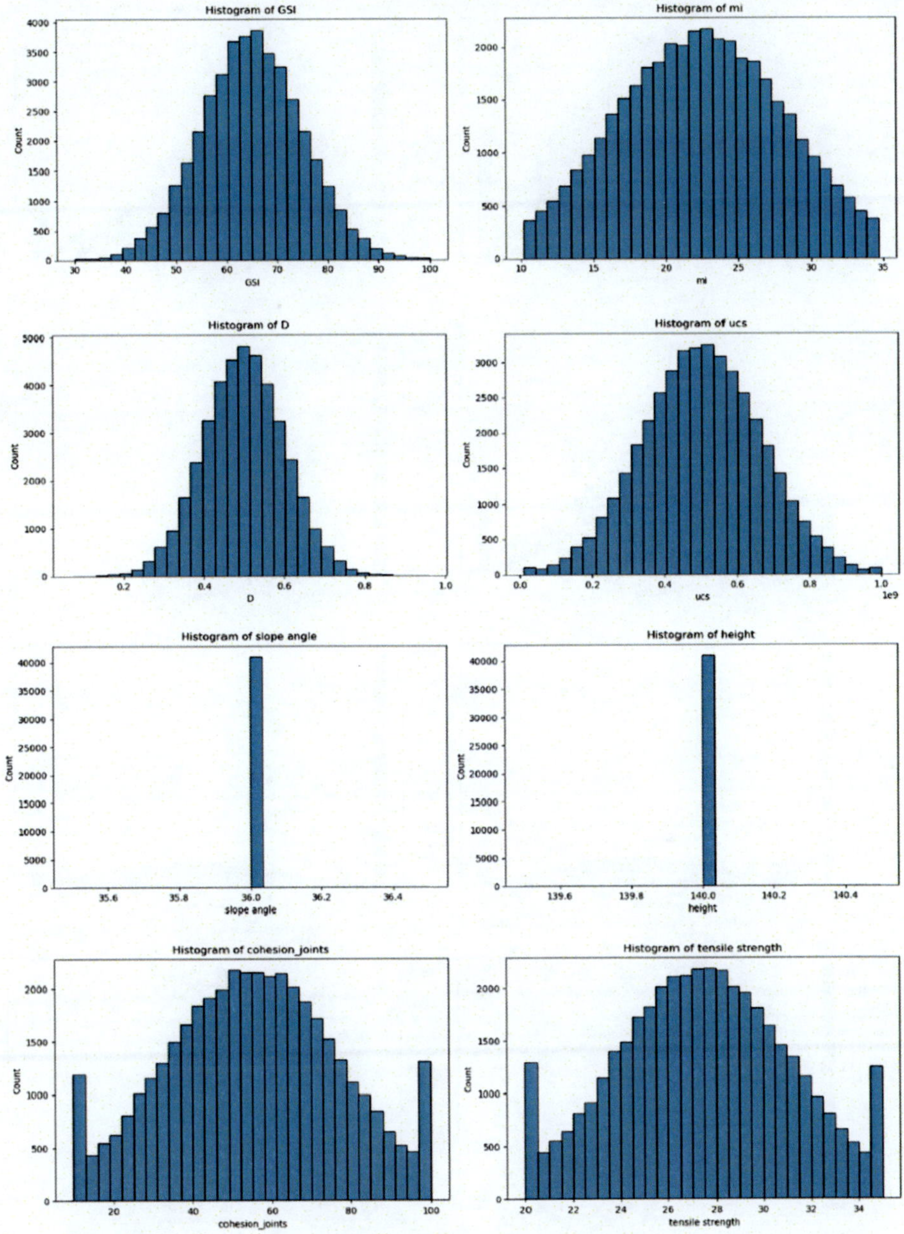

Fig. 6. Histogram of numerical variables

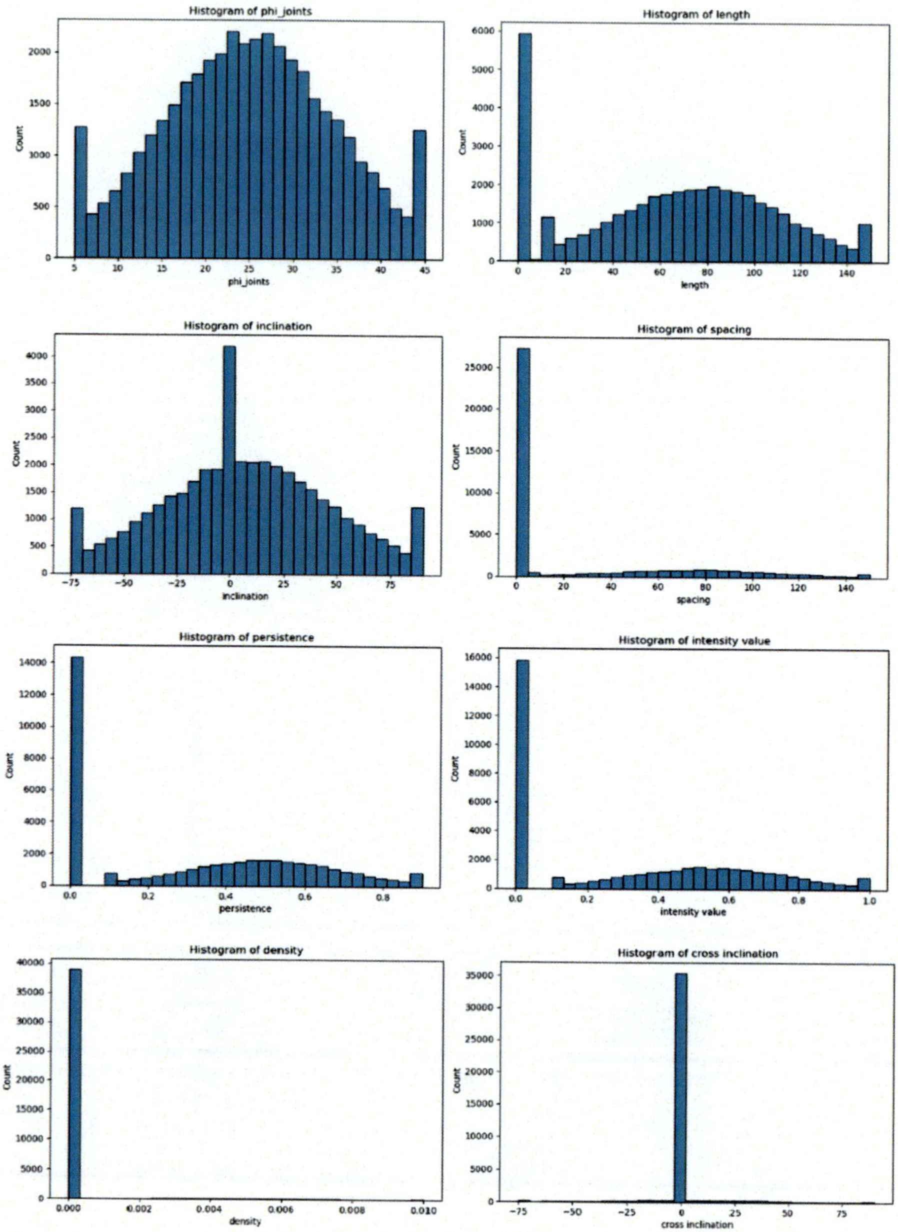

Fig. 6. (*continued*)

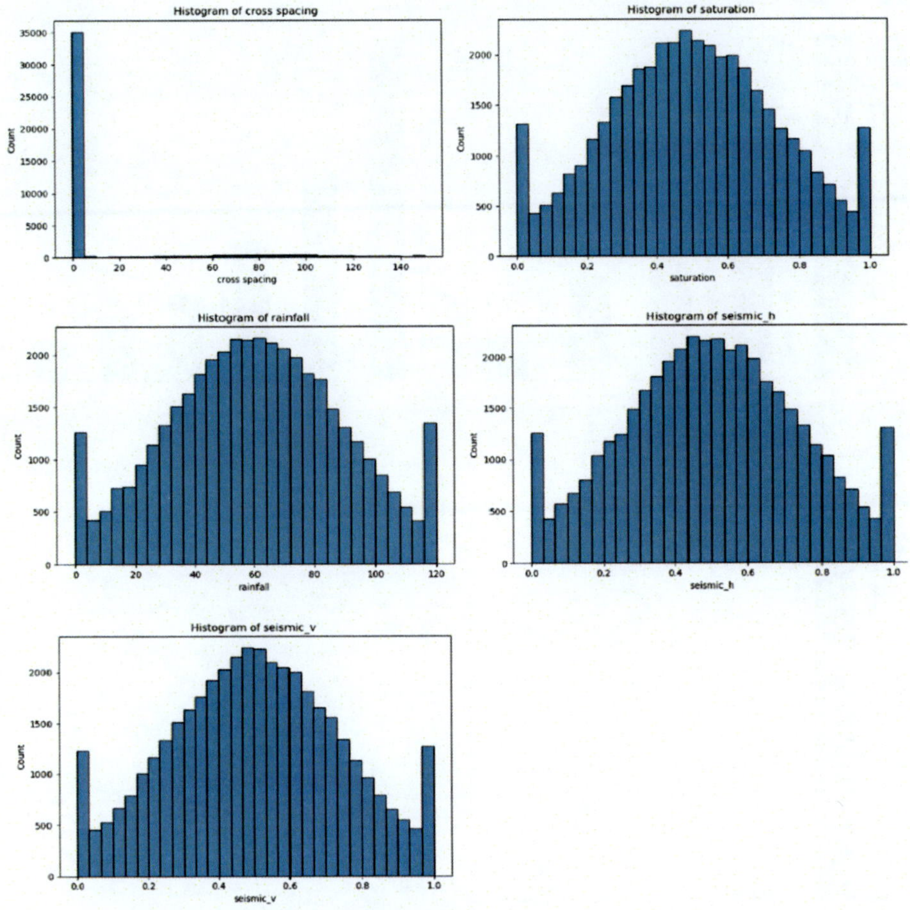

Fig. 6. *(continued)*

2.4 Supervised Regression Algorithms

Machine learning algorithms are powerful tools for modelling the nonlinear and complex relationships among the input features in order to predict the target variable. These algorithms can quickly and efficiently perform calculations that would be time-consuming and computationally expensive using traditional methods, by learning the complex behavior from the available dataset. These techniques have been widely implemented in various engineering fields, including geotechnical engineering. For this study different family of advanced regression algorithms were used and are described as follows:

Linear Regression

It is a type of predictive statistical analysis that attempts to model the relationship between a dependent variable (also known as the target variable) and one or more independent variables (also known as predictors, covariates, or features). The term "linear" refers to

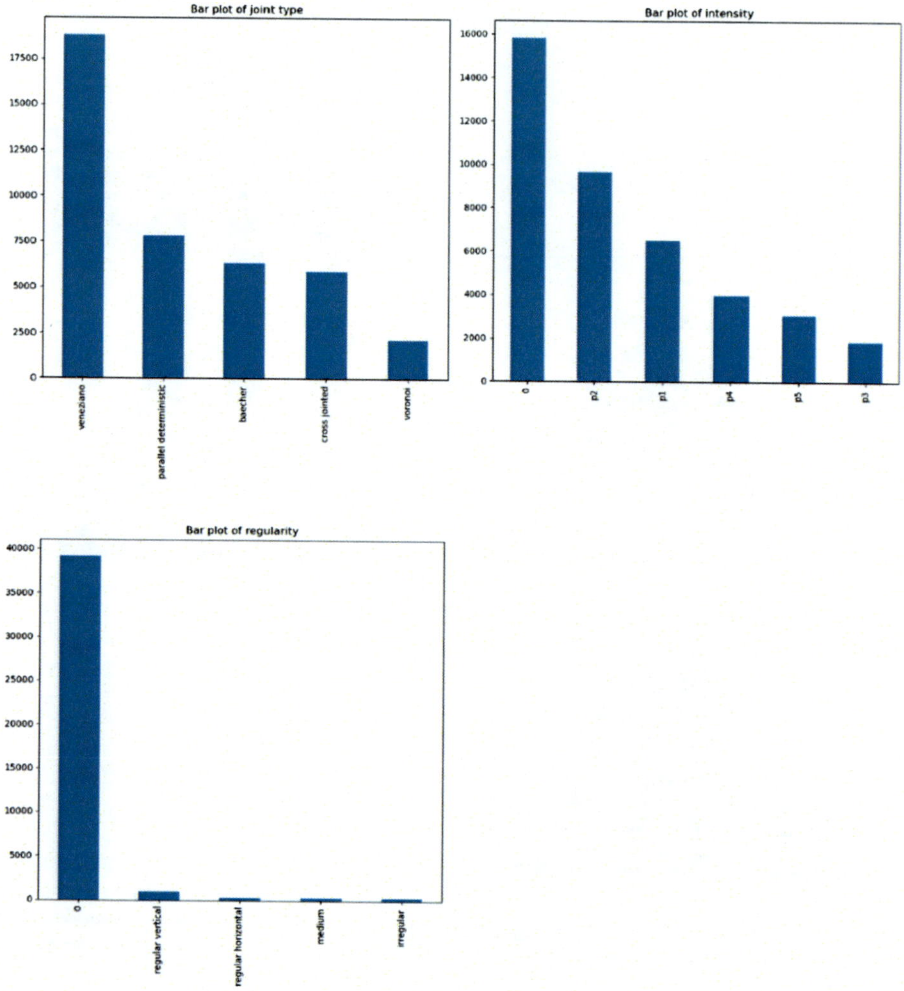

Fig. 7. Bar plot for categorical variable

the fact that the method models relationships using linear predictor functions. The main idea behind linear regression is to fit a straight line (in simple linear regression) or a hyperplane (in multiple linear regression) that best minimizes the sum of the squared residuals. This method of fitting the line or hyperplane is known as the method of least squares.

Here's the basic equation of a simple linear regression:

$$y = \beta_0 + \beta_1 x + \epsilon \tag{5}$$

where:

y is the dependent variable.

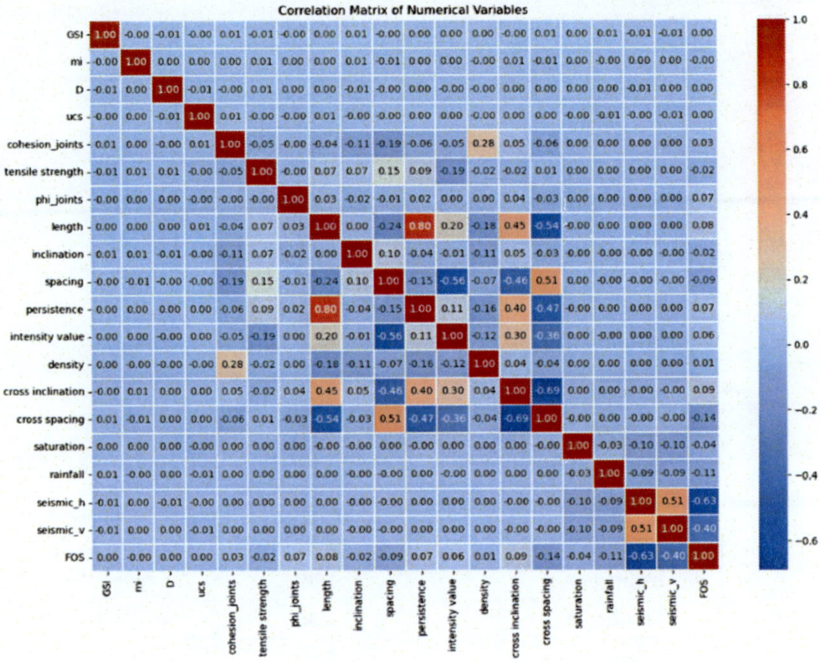

Fig. 8. Correlation Matrix for numerical variables

x is the independent variable.
β_0 is the y-intercept.
β_1 is the slope of the line.
ϵ is the error term (residuals).

The y-intercept, β_0, represents the predicted value of y when $x = 0$. The slope, β_1, represents the predicted increase in y resulting from a one unit increase in x.

In multiple linear regression, the equation extends to:

$$y = \beta_0 + \beta_1 x_1 + \beta_1 x_1 + \ldots \beta_p x_p + \epsilon \quad (6)$$

where:

$x_1, x_2, \ldots\ldots, x_p$ are the independent variables.
$\beta_0, \beta_1, \ldots\ldots, \beta_p$ are the coefficients of the independent variables.

Each coefficient β_i represents the change in the average response y for each one-unit change in the corresponding predictor x_i, assuming all other predictors are held constant.

The coefficients $\beta_0, \beta_1, \ldots\ldots, \beta_p$ in these equations are the parameters of the regression model and are estimated from the data. This is typically done using a method called least squares, which minimizes the sum of the squared residuals. Once the model is trained, it can be used to predict the dependent variable y given new values of the independent variables x. The linearity in linear regression refers to the parameters of the

Table 1. Value of VIF for each independent variable

Input_features	VIF
phi_joints	16.45
GSI	10.13
length	9.84
mi	9.13
persistence	7.64
UCS	3.94
D	3.83
Intensity value	2.92
Cohesion_joints	2.88
Cross spacing	2.78
Spacing	2.58
Cross inclination	2.20
Inclination	1.89
Seismic_v	1.80
Seismic_h	1.80
Tensile strength	1.77
Density	1.19
Saturation	1.04
rainfall	1.03

model, not necessarily the variables themselves. Linear regression models can accommodate non-linear relationships between variables by transforming the data, for example, by creating polynomial or interaction terms.

Ridge Regression (Also Known as L2 Regularization) introduces a small bias into the linear regression by adding a penalty equivalent to the square of the magnitude of the coefficients. This penalty is multiplied by a tuning parameter, lambda (also known as alpha), that determines the strength of the penalty.

The Ridge regression cost function is:

$$Cost = RSS + \lambda \sum_{i=1}^{N} \beta_i^2 \tag{7}$$

where:

RSS is the residual sum of squares,
β_i are the coefficients,
λ is the tuning parameter.

The ridge coefficients minimize a penalized residual sum of squares:

Lasso Regression (Least Absolute Shrinkage and Selection Operator, Also Known as L1 Regularization) works in a similar manner to Ridge Regression, but the penalty is slightly different—it is the sum of the absolute values of the coefficients.

The Lasso regression cost function is:

$$Cost = RSS + \lambda \sum_{i=1}^{N} |\beta_i| \tag{8}$$

where:

RSS is the residual sum of squares,
β_i are the coefficients,
λ is the tuning parameter.

Both Ridge and Lasso regression techniques add a penalty to the cost function, shrinking the coefficients towards zero and discouraging complex models. The key difference lies in how they assign penalty to the coefficients: Ridge Regression uses squared penalty term, while Lasso Regression uses an absolute penalty term. As a result, Lasso Regression can result in zero coefficients i.e., some of the features are completely neglected for the evaluation of output. Thus, Lasso Regression not only helps in reducing overfitting but can help us in feature selection.

Decision Tree

A decision tree is a tree structure that looks like a flowchart, with an internal node representing a feature (or attribute), a branch representing a decision rule, and each leaf node representing a result. The root node is the node at the top of a decision tree. A decision tree divides the data into several sets. To reach a decision, these sets are separated into subsets. The criterion for splitting differs amongst algorithms. Classification and Regression Trees (CART), C4.5 (extension of CART), and Chi-square Automatic Interaction Detection (CHAID) are the three basic forms of decision trees.

Random Forest

The Random Forest technique is an ensemble learning approach that involves the construction of several decision trees during the training phase. The resulting output of this method is determined by either selecting the class that appears most frequently among the trees (in the case of classification) or calculating the average prediction made by the various trees (in the case of regression). Random forests address the issue of decision trees' tendency to overfit to their training data by employing an ensemble approach that combines numerous decision trees through averaging, while also introducing an element of randomization.

XGBoost

XGBoost, also known as Extreme Gradient Boosting, is a Machine Learning technique that belongs to the ensemble methods category and is based on decision trees. It utilizes a gradient boosting framework to enhance its performance. The machine learning model in question is particularly robust when applied to organized or tabular data.

In the context of gradient boosting, the training process involves successively training models, where each subsequent model is trained to rectify the faults caused by its predecessors. The process involves the incremental addition of models until a point is reached where further enhancements are no longer attainable.

K-Nearest Neighbors (KNN)
The KNN technique is a form of instance-based learning that categorizes novel examples by measuring their resemblance to pre-existing instances within the dataset. The KNN algorithm employs a classification approach wherein an object's class assignment is determined through a majority vote from its neighboring objects. Specifically, the object is assigned to the class that is most prevalent among its k nearest neighbors. The variable K typically represents a positive integer of small magnitude.

AdaBoost
AdaBoost, also known as Adaptive Boosting, is an algorithm used for boosting in machine learning. It operates by combining a collection of weak classifiers in order to create a single, more powerful classifier. A weak classifier can be defined as a classifier that exhibits suboptimal performance, yet still outperforms random guessing. The AdaBoost algorithm assigns a numerical weight to each training example, which serves as an indicator of the importance or significance of that example in the training process. This weight is utilized to determine the likelihood or probability of including each example in the training set. Following the training of each weak student, the weights undergo readjustment. Instances that have been misclassified are given more weights, hence increasing their likelihood of being selected in subsequent iterations.

3 Results

Following metrics are utilized for measuring the performance of different regression models:

mae: This represents the Mean Absolute Error, which is a measure of the absolute differences between the true and predicted values.
mse: This represents the Mean Squared Error, which is a measure of the squared differences between the true and predicted values.
rmse: This represents the Root Mean Squared Error, which is the square root of the Mean Squared Error. It measures the standard deviation of the residuals.
mape: This represents the Mean Absolute Percentage Error, which is a measure of prediction accuracy as a percentage. It's the average of the absolute percentage differences between the true and predicted values.
r2: This represents the coefficient of determination, which is a statistical measure that explains how much of the variability of a factor can be caused or explained by its relationship to another factor. It provides a measure of how well the model predictions approximate the true values.

According to Table 2, the Decision Tree model outperforms others on the training set. However, its strong performance doesn't guarantee similar results on unseen data

Table 2. Performance of all models on Training set

model_name	mae	mse	rmse	mape	r2
Linear Regression	0.174	0.06	0.245	15.99	0.821
Ridge Regression	0.177	0.063	0.251	15.429	0.817
Lasso Regression	0.474	0.338	0.581	48.324	0
Decision Tree	0.002	0	0.024	0.232	0.998
Random Forest	0.009	0	0.029	0.807	0.997
KNN	0.047	0.007	0.084	4.152	0.978
XGBoost	0.018	0.001	0.038	1.705	0.995
AdaBoost	0.221	0.074	0.273	19.362	0.784

Table 3. Performance of all models on Test set

model_name	mae	mse	rmse	mape	r2
Linear Regression	0.177	0.063	0.251	15.249	0.817
Ridge Regression	0.177	0.063	0.251	15.435	0.817
Lasso Regression	0.475	0.345	0.588	48.241	0
Decision Tree	0.023	0.004	0.066	2.037	0.987
Random Forest	0.021	0.003	0.056	1.898	0.99
KNN	0.062	0.011	0.109	5.32	0.965
XGBoost	0.021	0.001	0.043	1.927	0.994
AdaBoost	0.221	0.074	0.273	19.362	0.784

due to potential overfitting. Therefore, it's essential to validate the model on a separate test dataset.

Table 3 shows that on the test dataset, XGBoost excels in r2, mse, and rmse, while Random Forest achieves the lowest mae and mape. The optimal model depends on the specific metric of interest, but XGBoost and Random Forest generally perform well. The results in Table 2 and Table 3 clearly indicates that ensemble-based machine learning models is fast and reliable in predicting the stability of slopes in presence of diverse joint networks thus addressing previous drawbacks.

The study evaluates feature importance's using multiple machine learning models. Linear Regression's large 'slope angle' coefficient indicates possible collinearity or scaling issues. Ridge and Lasso Regression, employing regularization, have smaller coefficients, reducing less important features' impact. Tree-based models like Decision Tree, Random Forest, XGBoost, and AdaBoost highlight 'GSI', 'mi', 'D', and 'UCS' as key features, gauged by their frequency of use in data splitting. KNN doesn't define feature importance, as it relies on proximity in feature space. However, we can see from Table 4 that feature importance in XGBoost model uses almost all the features utilized in

Table 4. Feature Importance of each variable according to different models

feature	Linear Regression	Ridge Regression	Lasso Regression	Decision Tree	Random Forest	XGBoost	AdaBoost
GSI	-0.00083398S	-0.000833929	0	0.08867164S	0.088579394	0.003700225	0
mi	-0.001582124	-0.00161515S	0	0.09636476S	0.08991700S	0.004864305	0
D	0.000404662	0.000406355	0	0.0900194	0.089885095	0.005248889	0
ucs	0.000143972	0.000144468	0	0.090697592	0.089694225	0.005136968	0
slope angle	13377189.08	0	0	0	0	0	0
height	-56616374.27	0	0	0	0	0	0
cohesion_joints	0.007315084	0.007252661	0	0.010097182	0.010139456	0.00488065	0
tensile strength	-0.006999584	-0.00701042	0	0.002921022	0.002699205	0.003148806	0
phi_joints	0.030728161	0.030724344	0	0.011156052	0.011303429	0.00687612	0.008684635
length	-0.018465016	-0.018203684	0	0.00833959	0.010563902	0.00408818	0
inclination	-0.016379831	-0.016306413	0	0.026712335	0.027771254	0.006081751	0.008221006
spacing	0.011314633	0.011159542	0	0.0069352	0.00670756	0.007929689	0.00907934
persistence	-0.029234426	-0.028648334	0	0.004768904	0.006051404	0.006296197	0
intensity value	-0.006402588	-0.006528169	0	0.012839781	0.014714491	0.005211013	0
density	-0.005574464	-0.005673839	0	0.000642995	0.00122612	0.004644074	0
cross inclination	-0.014329335	-0.01439542	0	0.000952763	0.001317721	0.002244259	0
cross spacing	-0.030993048	-0.031066158	0	0.021412609	0.011953684	0.043743916	0.007783978
saturation	-0.060272002	-0.060349873	0	0.024012197	0.024004504	0.15516685	0.02413643
rainfall	-0.094859737	-0.094907364	0	0.043648835	0.043608157	0.2581358	0.08120949
seismic_h	-0.299342082	-0.299275028	0	0.397746784	0.397826549	0.37643746	0.799248777
seismic_v	-0.067717416	-0.067805778	0	0.040197965	0.040104447	0.037645765	0.024967514
joint type_cross jointed	-12661456467	-0.063950662	0	0.00187407	0.011674929	0	0.036668829
joint type_parallel deterministic	-14207162840	0.01994341	0	0.000778557	0.000558839	0.003760795	0
joint type_veneziano	0.040008792	0.039290821	0	0.0020687	0.00169113	0.004176626	0
joint type_voronoi	-8045900578	0.011336662	0	0.000556347	0.000472688	0.007798538	0
intensity_p1	-13221861340	-0.008794834	0	0.002544172	0.003102799	0.0081833	0
intensity_p2	-15357800602	0.014467377	0	0.003470805	0.003643258	0.005547983	0
intensity_p3	-7529781246	0.012892866	0	0.001833591	0.001903006	0.003420698	0
intensity_p4	-10734262153	0.010061378	0	0.003099067	0.003226473	0.00525197	0
intensity_p5	-9521314066	0.013041333	0	0.002980738	0.003249902	0.003486295	0
regularity_irregular	-0.007471013	-0.007434839	0	0.000538228	0.000448893	0.000931762	0
regularity_medium	-0.002782393	-0.002808255	0	0.000395478	0.000462624	0.005615772	0
regularity_regular horizontal	-0.005090507	-0.005098612	0	0.000390595	0.000775412	0.005705853	0
regularity_regular vertical	-0.011145334	-0.011096886	0	0.001332032	0.000722439	0.004639479	0

the dataset for making contribution to the final outcome. Feature importance should be contextually interpreted within each model, aiding in feature selection and engineering.

Model interpretability and explainability build trust, ease debugging, comply with regulatory requirements, address ethical considerations, and elucidate influential features which can be easily understood with the help of feature importance. These are crucial in fields where predictions carry significant implications. While complex models yield accurate predictions, their 'black box' nature highlights the importance of interpretable and explainable AI.

4 Discussion

The advantages of machine learning methods in predicting slope stability analysis are multifold, spanning from enhancing predictive accuracy, efficiency, and robustness, to reducing computational requirements, and to improving model interpretability. Given their extensive applicability and versatility, they have the potential to significantly advance the field of slope stability analysis. The ongoing journey of research in slope stability analysis is characterized by constant efforts to address limitations and push the boundaries of our understanding and predictive capabilities. Each successive study builds on its predecessors, illustrating the cumulative and iterative nature of scientific progress.

However, the need for interpretability for instance-based feature importance is something which needs to be addressed in future research and also to make machine learning model more reliable and robust the data should be drawn from different numerical modelling technique like Limit Equilibrium Method (LEM), different failure criterion in FEM like study based on Mohr-Coulomb failure criterion and others. The main idea behind the working of any machine learning algorithm is their working based on available of good training data which comes from previous methods. So, unless the previous methods are producing accurate results our machine learning algorithm won't work.

5 Conclusion

The analysis of the results from various models on the training, testing, and validation datasets offers a profound understanding of the role different features play in predicting slope stability. The incorporation of diverse joint networks in the analysis process has addressed the limitations pointed out in previous studies, reinforcing the importance of joint networks in slope stability prediction. Among the models used, XGBoost and Random Forest seem to perform best across test datasets, as indicated by their lowest mae and highest r2 values. This suggests that these models, given their ability to handle complex interactions between features and provide robust predictions, are well-suited for slope stability prediction. Feature importance analysis further illuminates the role of different features in the model predictions. For the tree-based models, features such as 'GSI', 'mi', 'D', and 'UCS' appear to be the most influential. This information can aid in future feature engineering and selection, potentially improving model performance. This study underscores the importance of model interpretability and explainability in geotechnical engineering. While complex models like XGBoost and Random Forest may provide higher predictive performance, their decision-making process can be hard to interpret. In contrast, simpler models like Linear Regression offer more transparency at the cost of predictive accuracy. Striking a balance between performance and interpretability is crucial depending on the specific requirements of a project. In conclusion, the use of machine learning models, in combination with a detailed understanding of joint networks, provides a powerful tool for slope stability prediction. The study's findings are poised to enhance the decision-making process in geotechnical engineering projects, contributing to safer and more efficient design and management of slopes.

References

Adeli, H.: Neural networks in civil engineering: 1989–2000. Comput. Aided Civ. Infrastruct. Eng. **16**(2), 126–142 (2001)

Ahmad, I., Hesham El Naggar, M., Khan, A.N.: Artificial neural network application to estimate kinematic soil pile interaction response parameters. Soil Dyn. Earthq. Eng. **27**(9), 892–905 (2007). https://doi.org/10.1016/j.soildyn.2006.12.009

Agarwal, A.K.: Annual technical volume of mining engineering division board. Fundam. Slope Stab. **2** (2020)

Moradi, A., Hosseinitoudeshki: The effect of joint models on the stability of rock slopes. Int. J. Geol. Agric. Environ. Sci. **3**(1), 7–10 (2015)

Amirkiyaei, V., Ghasemi, E.: Stability assessment of slopes subjected to circular-type failure using tree-based models. Int. J. Geotech. Eng. **16**(3), 301–311 (2022). https://doi.org/10.1080/19386362.2020.1862538

Asadi, A., Moayedi, H., Huat, B.B.K., et al.: Artificial neural networks approach for electrochemical resistivity of highly organic soil. Int. J. Electrochem. Sci. **6**(4), 1135–1145 (2011a)

Asadi, A., Shariatmadari, N., et al.: Effect of MSW leachate on soil consistency under influence of electrochemical forces induced by soil particles. Int. J. Electrochem. **6**(7), 2344–2351 (2011b)

Asadi, A., Moayedi, H., Huat, B.B., et al.: Prediction of zeta potential for tropical peat in the presence of different cations using artificial neural networks. Int. J. Electrochem. Sci. **6**(4), 1146–1158 (2011c)

Asteris, P.G., et al.: Slope stability classification under seismic conditions using several tree-based intelligent techniques. Appl. Sci. **12**(3) (2022).https://doi.org/10.3390/app12031753

Adams, B.M.: Slope stability acceptance criteria for opencast mine design. In: 12th ANZ Conference on Geomechanics and Human Influence, Wellington, New Zealand (2015)

Benardos, A.G., Kaliampakos, D.C.: Modelling TBM performance with artificial neural networks. Tunn. Undergr. Space Technol. **19**(6), 597–605 (2004). https://doi.org/10.1016/j.tust.2004.02.128

Bolt, G.: Fault models for artificial neural networks. IEEE, Piscataway (1991). (Preprint)

Chávez, C., Equihua, L.N., Dominguez, F.: Stability fem analysis of rock masses modeling pattern of joints. In: XIV International Conference on Computational Plasticity. Fundamentals and Applications COMPLAS 2017, pp. 646–655 (2017)

Flood, I., Kartam, N.: Neural networks in civil engineering. 1. Principles and understanding. J. Comput. Civ. Eng. **8**(2), 131–148 (1994a)

Flood, I., Kartam, N.: Neural networks in civil engineering. 2. Systems and application. J. Comput. Civ. Eng. **8**(2), 149–162 (1994b)

Goh, A.T.C., Wong, K.S., Broms, B.B.: Estimation of lateral wall movements in braced excavations using neural networks. Can. Geotech. J. **32**(6), 1059–1064 (1995). https://doi.org/10.1139/t95-103

Hoek, E., Brown, E.T.: The Hoek-Brown failure criterion - a 1988 update. In: Curran, J.C. (ed.) Proceedings of the 15th Canadian Rock Mechanics Symposium, Toronto, pp. 31–38. Department Civil Engineering, University of Toronto (1988)

Hu, Y., et al.: Study on the process and mechanism of slope failure induced by mining under open pit slope: a case study from Yanqianshan Iron Mine, China. Adv. Civ. Eng., 1–26 (2019). https://doi.org/10.1155/2019/6862936

Lazarevska, M., et al.: Application of artificial neural networks in civil engineering. Teh Vjesn **21**(6), 1353–1359 (2014)

Lee, C., Sterling, R.: Identifying probable failure modes for underground openings using a neural network. Int. J. Rock Mech. Min. Sci. Geomech. Abstr. **29**(1), 49–67 (1992). https://doi.org/10.1016/0148-9062(92)91044-6

Lee, I.-M., Lee, J.-H.: Prediction of pile bearing capacity using artificial neural networks. Comput. Geotech. **18**(3), 189–200 (1996). https://doi.org/10.1016/0266-352X(95)00027-8

Lee, S.J., Lee, S.R., Kim, Y.S.: An approach to estimate unsaturated shear strength using artificial neural network and hyperbolic formulation. Comput. Geotech. **30**(6), 489–503 (2003). https://doi.org/10.1016/S0266-352X(03)00058-2

Li, J., Hao, H.: A review of recent research advances on structural health monitoring in Western Australia. Struct. Monit. Maint. **3**(1), 33–49 (2016). https://doi.org/10.12989/smm.2016.3.1.033

Lu, P.Z., Chen, S.Y., Zheng, Y.J.: Artificial intelligence in civil engineering. Math. Probl. Eng., 1–22 (2012)

Meng, Q.X., et al.: Multiscale strength reduction method for heterogeneous slope using hierarchical FEM/DEM modeling. Comput. Geotech. **115**, 103164 (2019). https://doi.org/10.1016/j.compgeo.2019.103164

Moayedi, H., Hayati, S.: Applicability of a CPT-based neural network solution in predicting load-settlement responses of bored pile. Int. J. Geomech. **18**(6) (2018). https://doi.org/10.1061/(ASCE)GM.1943-5622.0001125

Moayedi, H., Hayati, S.: Artificial intelligence design charts for predicting friction capacity of driven pile in clay. Neural Comput. Appl. **31**(11), 7429–7445 (2019). https://doi.org/10.1007/s00521-018-3555-5

Moayedi, H., Jahed Armaghani, D.: Optimizing an ANN model with ICA for estimating bearing capacity of driven pile in cohesionless soil. Eng. Comput. **34**(2), 347–356 (2018). https://doi.org/10.1007/s00366-017-0545-7

Moayedi, H., Mosallanezhad, M.: Uplift resistance of belled and multi-belled piles in loose sand. Measurement **109**, 346–353 (2017). https://doi.org/10.1016/j.measurement.2017.06.001

Moayedi, H., Mosallanezhad, M., Nazir, R.: Evaluation of maintained load test (MLT) and pile driving analyzer (PDA) in measuring bearing capacity of driven reinforced concrete piles. Soil Mech. Found. Eng. **54**(3), 150–154 (2017). https://doi.org/10.1007/s11204-017-9449-1

Moayedi, H., Nazir, R., Mosallanezhad, M.: Determination of reliable stress and strain distributions along bored piles. Soil Mech. Found. Eng. **51**(6), 285–291 (2015). https://doi.org/10.1007/s11204-015-9291-2

Moayedi, H., Rezaei, A.: An artificial neural network approach for under-reamed piles subjected to uplift forces in dry sand. Neural Comput. Appl. **31**(2), 327–336 (2019). https://doi.org/10.1007/s00521-017-2990-z

Mosallanezhad, M., Moayedi, H.: Developing hybrid artificial neural network model for predicting uplift resistance of screw piles. Arab. J. Geosci. **10**(22), 479 (2017). https://doi.org/10.1007/s12517-017-3285-5

Saharan, M.R., et al.: Factor of safety (FoS) based slope design acceptance criterion: a case study. Ann. Techn. Vol. Min. Eng. Div. Board **2** (2020)

Nazir, R., et al.: Appraisal of reliable skin friction variation in a bored pile. Proc. Inst. Civ. Eng. Geotech. Eng. Geotech. Eng. **168**(1), 75–86 (2015). https://doi.org/10.1680/geng.13.00140

Nazir, R., et al.: Application and design of transition piled embankment with surcharged prefabricated vertical drain intersection over soft ground. Arab. J. Sci. Eng. **43**(4), 1573–1582 (2018). https://doi.org/10.1007/s13369-017-2628-6

Niroumand, H., et al.: Slope stability and sheet pile and contiguous bored pile walls. Electron. J. Geotech. Eng. **17**, 19–27 (2012)

Wu, S.C., Jin, A.B., Gao, Y.T.: Slope stability analysis by strength reduction method based on ubiquitous-joint model. Rock Soil Mech. **4**, 537–542 (2006)

Shuheng, Z., Yinjun, M.: Research on the influence of weak interlayer in open-pit slope on stability. Adv. Civ. Eng. **2021**, 1–9 (2021). https://doi.org/10.1155/2021/4256740

Singh, S.K., Chakravarty, D.: Assessment of slope stability using classification and regression algorithms subjected to internal and external factors. Arch. Min. Sci. **68**(1), 87–102 (2023). https://doi.org/10.24425/ams.2023.144319

Suman, S., et al.: Slope stability analysis using artificial intelligence techniques. Nat. Hazards **84**(2), 727–748 (2016). https://doi.org/10.1007/s11069-016-2454-2

Wang, H., et al.: A statistics-based discrete element modeling method coupled with the strength reduction method for the stability analysis of jointed rock slopes. Eng. Geol. **264**, 105247 (2020). https://doi.org/10.1016/j.enggeo.2019.105247

Wang, Z., et al.: Steep end-slope mining and slope stability of extremely thick inclined coal seam open-pit mine. Math. Probl. Eng. **2022**, 1–9 (2022). https://doi.org/10.1155/2022/3807013

Watson, J.N., et al.: The use of artificial neural networks in pile integrity testing. Civil Comp Press, Edinburgh (1995). (Preprint)

Zheng, Y.R., Zhao, S.Y.: Application of strength reduction FEM in soil and rock slope. Chin. J. Rock Mech. Eng. **19**, 3381–3388 (2004)

Yuan, C., Moayedi, H.: The performance of six neural-evolutionary classification techniques combined with multi-layer perception in two-layered cohesive slope stability analysis and failure recognition. Eng. Comput. **36**(4), 1705–1714 (2020). https://doi.org/10.1007/s00366-019-00791-4

Zhou, Z.H., et al.: Stability of rock slope with bedding intermittent joints based on catastrophe theory. J. China Coal Soc. **45**, 161–172 (2020)

Zhou, J., et al.: Slope stability prediction for circular mode failure using gradient boosting machine approach based on an updated database of case histories. Saf. Sci. **118**, 505–518 (2019). https://doi.org/10.1016/j.ssci.2019.05.046

Machine Learning Techniques for Predicting Health Risk of Mining Vehicle Operators Subjected to Whole-Body Vibration

Amit Sharma(✉), Bibhuti Bhusan Mandal, and Monika Tewari

Department of Mining Engineering, IIT Kharagpur, Kharagpur, India
amitfdh@gmail.com, a.sharma@kgpian.iitkgp.ac.in,
Tewarimonika@iitkgp.ac.in

Abstract. Occupational health and safety in mining operations hold paramount importance, encompassing multifaceted aspects aimed at ensuring the well-being of workers. Prolonged whole-body vibration (WBV) exposure among mining vehicle operators has been associated with a range of adverse health effects, most notably musculoskeletal disorders (MSDs) such as chronic low back pain and lumbar disc herniation. This study aims to develop a predictive model utilizing machine learning techniques, specifically Artificial Neural Networks (ANNs), to forecast the health risk associated with WBV exposure under different operational conditions. The vibration data of the tippers were acquired during its routine operation in an opencast mine in central India. The dataset encompassed various operational factors, including haul road type, operator weight, speed, and payload, under which an ANN model was trained to predict the resulting vibration exposure magnitude. ANN model was developed using Multilayer Perceptron (MLP) architecture to predict the health risk. Backpropagation was used to adjust the weights of the connections between neurons in order to minimize the difference between predicted and actual output. 70% of the total data was used for training and the others for data testing (15%) and validation (15%). The correlation coefficient for the training and test of the optimum network were found to be 97.58% and 97.4%, respectively indicating an excellent correlation between the predicted frequency weighted root mean square acceleration (a_{wrms}) for both training and test stages. AI has revolutionized the mining industry, bringing digitization and automation to the forefront. The predictive model presented here for WBV-related health risks is a powerful tool for mine safety engineering. By harnessing machine learning techniques, it improves risk assessment and enables targeted interventions, improving the quality of life of the mining workforce.

Keywords: ANN · Sustainable mining · MSD · Occupational health · Risk assessment · Whole-body vibration

1 Introduction

In recent years, the field of occupational health has witnessed a growing concern over the detrimental effects of Whole-Body Vibration (WBV) on mining vehicle operators (Bovenzi and Hulshof 1999). Extensive research has established a strong correlation

A. Sinha et al. (Eds.): AMC 2023, SPEES, pp. 213–222, 2023.
https://doi.org/10.1007/978-3-031-46966-4_17

between prolonged exposure to WBV and adverse health effects, particularly musculoskeletal disorders including low back pain (LBP) and lumbar disc herniation (Bovenzi 2010; Bovenzi et al. 2013; Ronchese and Bovenzi 2012; Combs et al. 2018; Mandal and Manwar 2017; Sharma and Mandal 2021). As the mining industry continues to evolve towards sustainability, ensuring the well-being and safety of its workforce becomes of paramount importance. In response to this pressing issue, researchers and practitioners are exploring innovative approaches to predict and mitigate the health risks associated with WBV exposure.

The utilization of Artificial intelligence (AI) and Machine learning (ML) techniques has gained remarkable momentum across various disciplines, ushering in transformative changes in risk predictions and safety enhancement. These technologies have proven their efficacy in a multitude of applications, revolutionizing fields such as finance, healthcare, transportation, and more (World Economic Forum 2019; Chen et al. 2018). By harnessing the power of AI and ML, experts have successfully developed predictive models that analyse complex data patterns, enabling timely risk assessments and informed decision-making. This paper aims to build upon this foundation of cross-disciplinary advancements, delving into the strategic integration of AI and ML in predicting and managing WBV-related risks for mining vehicle operators. Through a comprehensive exploration of existing methodologies and their adaptation to the unique challenges of the mining industry, we seek to contribute to the development of proactive strategies that enhance occupational health and elevate safety standards. The objective of this paper is to determine the efficacy of ANN to predict the WBV exposure of tipper operators under different working conditions.

2 Methodology

2.1 Artificial Neural Networks (ANN)

ANNs are sophisticated computational models inspired by the intricate structure and functioning of the human brain's neural network. These networks are adept at analysing vast datasets, enabling them to perform tasks such as pattern recognition, classification, and regression. ANNs consist of interconnected artificial neurons, organized into layers, which typically include an input layer, one or more hidden layers, and an output layer. Each neuron processes input signals using an activation function and passes the results to neurons in the subsequent layer. The strength of the connections, represented by synaptic weights, plays a crucial role in determining the importance of different features during the learning process. In our study, we have utilized a specific type of ANN called the Multilayer Perceptron (MLP), which is widely applied in various fields. MLPs consist of an input layer that receives relevant data, an output layer for classification or prediction, and one or more hidden layers that enable the network to uncover intricate patterns and relationships within the data. The logsig (sigmoid) activation function, combined with the backpropagation algorithm for weight adjustment during training, empowers MLPs to efficiently model complex interactions and make accurate predictions.

Figure 1 illustrates the architecture of an MLP type ANN designed to predict vibrational health risk levels. The ANN comprises an input layer with n neurons representing different data inputs (x1, x2, x3) and an output layer with a single neuron for making

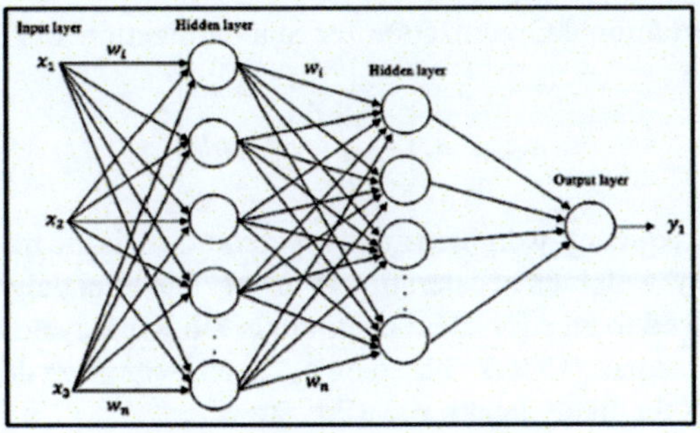

Fig. 1. MLP type ANN model

predictions y1. Neurons are interconnected through weighted connections, and their outputs are determined by an activation function to generate output as follows (Walczak 2012; Bishop 1995; Mandic and Chambers 2001).

$$a_j = f\left(\sum\nolimits_{i=1}^{n} X_i.W_{ij}\right) \tag{1}$$

where, f is the transfer function of neuron i, a_j is the sum of weighted input signal for node j, X_i is the input signal, W_{ij} is the weight or connection strength between node i and current node j and n is the number of input nodes. The feed-forward backpropagation algorithm is utilized to train the network, with the logarithmic sigmoidal (Logsig) function as the chosen activation function, transforming inputs to values between zero and one.

$$Logsig\left(a_j\right) = \frac{1}{1 + \exp(-a_j)} \tag{2}$$

$$MSE = \frac{1}{2}\sum\nolimits_{k=1}^{n}(t_k - y_k)^2 \tag{3}$$

where, t_k is the observed data and y_k is the estimated data. The ANN's performance is assessed using statistical measures like Pearson's Correlation Coefficient and mean square error (MSE). The main goal of this research is to develop an ANN structure capable of predicting vibrational health risk levels by gathering data under various operating conditions and validating the network's predictions against real-world data. The operating conditions include, operators' weight, haul road quality, payload, and speed of tipper.

2.2 ISO 2631-1:1997

The evaluation of the effects of vibration on health, according to ISO 2631-1:1997/Amd 1:2010 is based on calculation of frequency weighted root mean square acceleration (a_{wrms}) for each axis of translational motion on the supporting surface. The standard defines the approach for vibration measurements and frequency weightings to be used in the analysis (International Organization for Standardization 1997).

$$a_{wrms} = \left[\frac{1}{T}\int_0^T a_w^2(t)dt\right]^{\frac{1}{2}} \tag{4}$$

where, a_{wrms} is the frequency-weighted r.m.s acceleration T is the measurement duration $a_w(t)$ is the frequency-weighted acceleration at time t. The total daily vibration exposure for a_{wrms} is normalized to an equivalent continuous 8-h work cycle for risk assessment. Two health guidance zones (HCGZ) based on a_{wrms} and vibration dose value (VDV) are provided to interpret the likely health risk (Fig. 2).

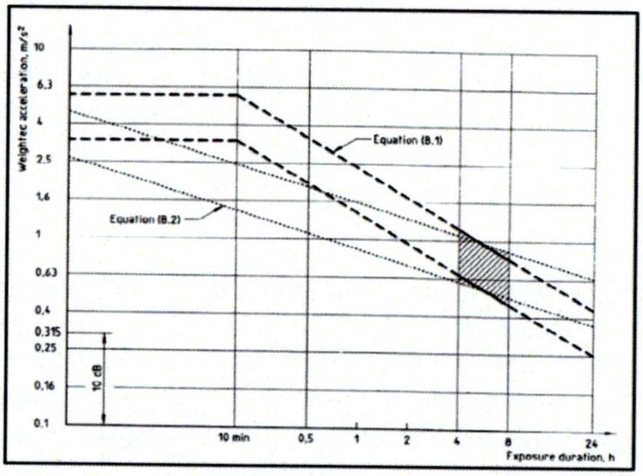

Fig. 2. Health guidance caution zone as per ISO 2631-1:1997

3 Field Study

WBV exposure measurements were recorded from tippers operating in a private sector open-cast coal mine in central India. The block contains an estimated 850 million tonnes (Mt) of mineable reserves, with a corresponding total overburden of approximately 4000 million cubic meters and an overall stripping ratio of 4.5 m^3/t. Open-pit mining operations is conducted using blast, shovel, and dumper methods, while the run-of-the-mine (ROM) coal undergo crushing and washing in a coal handling plant situated within the project area. The role of various operational factors such as the type of haul road, operator's weight, speed, and payload were investigated against the resulting magnitude of the vibration exposure (a_{wrms}).

The type of haul road was classified in two categories based on the rolling resistance (RR) which is a measure of the extra resistance to motion that a haul truck experience (Thompson and Visser 2003). This is done through a qualitative evaluation using a road defect classification system developed by Thompson et al. (2019). This method involves visually assessing the degree and extent of defects on the road, with the most significant defects being potholes, corrugations, rutting, loose material, and fixed stoniness in the wearing course. From the visual observation of a section for different defects, a defect score is computed by multiplying the degree and the extent rating. The summation of defect scores for all the defects gives the roughness defect score (RDS). To estimate rolling resistance, the reference charts relating rolling resistance to roughness defect scores as a function of speeds are used (see Fig. 3). Two haul roads from working bench were chosen for the field study. The haul roads were segregated into two distinct categories, differentiating between RR 2.1 and RR 3.2, based on the rolling resistance derived from the roughness defect scores (Fig. 4). All calculations for rolling resistance were done for a speed of 30 km/h.

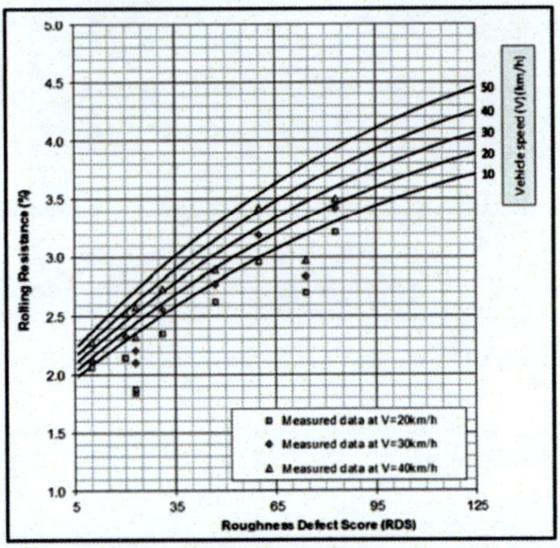

Fig. 3. Plot of rolling resistance vs RDS as a function of speed (Thompson et al. 2019)

During data collection the speed of the tippers was in the range of 15–30 km/hr as stipulated by Directorate General of Mines Safety (DGMS) (circular reference) for Indian haul roads. The speed levels were recorded at 5 km/h intervals which included 15 to 20, 20 to 25, and 25 to 30 km/h. After each loading, the materials were weighed at the weighbridge situated within the coal handling plant mine area. Data regarding the weight of the operators were also collected prior to each cycle of experiments.

4 Vibration Measurement

The vibration measurement followed ISO 2631-1:1997 guidelines. A tri-axial accelerometer (Nor 1286, Norsonic) was used to measure the acceleration transmitted to the seat pad. The sensor was placed at the interface between the seat pad and the ischial tuberosities. Vibration data was collected for the complete cycle of the tippers. The operators were asked to perform their routine work duty cycles which consisted empty hauling to the bench, loading the coal, hauling loaded tipper back to the coal handling plant, and finally unloading.

Fig. 4. Classification of mine haul road with (a) RR 2.1 and (b) 3.2 respectively

5 Results and Discussion

To anticipate the level of health risk arising from whole-body vibration, an artificial neural network was employed. A dataset comprising 199 distinct experimental observations was utilized to analyse the model's predictive capabilities. Input parameters encompassed factors like the haul road quality, speed of tipper, payload, and operator's weight. The output layer was singular, representing the root mean square (a_{wrms}) of vertical acceleration. Since z-axis is found to be the dominant axis in case of tippers plying on haul roads, only vertical axis acceleration was used in the model. The neural network architecture comprised three layers: input, hidden, and output, and was established using MATLAB's 'nnstart' function. A feed-forward back-propagation network was chosen, depicted schematically in Fig. 5. Notably, 70% of the data was allocated for training, with the remainder split between testing (15%) and validation (15%). Transfer functions, specifically 'Logsig' for hidden layers and 'Purlin' for the output layer, were implemented. The 'Logsig' function adjusted weights and biases using the Levenberg-Marquardt algorithm, a rapid hybrid method combining Gauss-Newton and steepest descent techniques. Following network creation, the subsequent phase involved network training. Numerous networks with distinct architectures, varying in neurons and layers, were constructed to identify the optimal configuration. The mean square error served as the criterion for evaluating these networks.

Table 1 displays the superior outcomes derived from different neuron and layer combinations. The selected hidden layers ranged from 2 to 8, balancing complexity, data size, and overfitting risks effectively. Based on the coefficient of determination (R^2)

and mean square error (MSE), the architecture 8-1 (8 neurons in the hidden layer, one neuron in the output layer) emerged as the optimum choice.

Table 1. Comparison of different network topology during training, validation and test stages

Network Topology	Training		Validation		Test	
	MSE	R^2	MSE	R^2	MSE	R^2
I-2	0.005	0.907	0.002	0.953	0.001	0.966
I-3	0.003	0.936	0.002	0.959	0.005	0.921
I-4	0.003	0.930	0.005	0.923	0.003	0.951
I-5	0.004	0.927	0.002	0.941	0.003	0.943
I-6	0.003	0.949	0.002	0.916	0.005	0.929
I-7	0.002	0.957	0.003	0.923	0.002	0.964
I-8	0.002	0.976	0.002	0.967	0.001	0.974

Results indicated a commendable 97.59% correlation for training and 96.71% for validation in this configuration. Regression curves, illustrated in Fig. 6, displayed the strong correspondence between outputs and targets during both training and validation stages. These results underline the robust correlation between the output and target datasets.

The quality of haul roads also holds potential for predicting the overall operational cost of transporting ore and overburden. For instance, in the context of rear dump trucks operating on a 7.3 km, 7% incline with a 91 t payload and 161 t gross vehicle mass (GVM), reducing road rolling resistance from 8% to 4% results in a 29% reduction in capital equipment costs and a 23% reduction in operating costs (Thompson et al. 2019). To assess the model's accuracy, 30 field study data points were randomly compared with estimates from the optimum network. Test results, depicted in Fig. 7, showcased a notable 97.49% correlation between actual and predicted a_{wrms} values. These outcomes substantiate the proposed ANN model's efficacy in predicting vibrational health risk levels, especially in scenarios involving absent or incomplete data. This predicted a_{wrms} values can be further integrated with the duration of exposure as per ISO 2631-1 to derive A(8) values and predict the health effects.

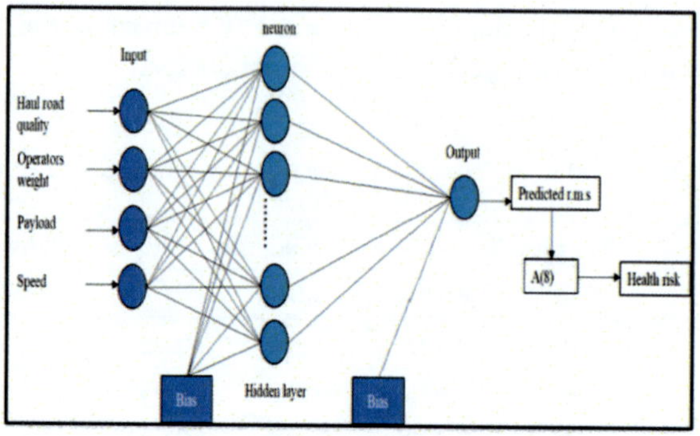

Fig. 5. Architecture of MLP type ANN used

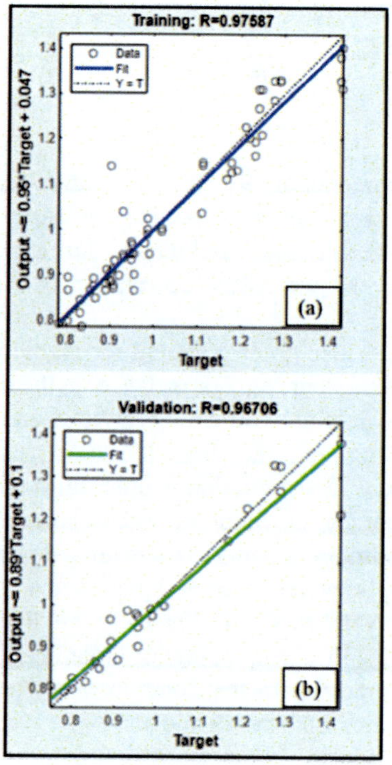

Fig. 6. The regression curves for the optimum network with 8 hidden layers (a) & (b)

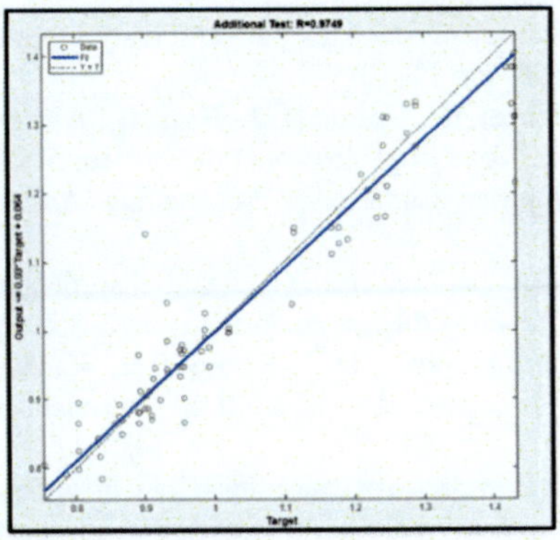

Fig. 7. Test plot for linear regression fit

6 Conclusion

Due to the continuous advancement of technology and the widespread adoption of AI and ML in diverse sectors, it has emerged as a critical element in ensuring global industrial safety. The development of AI-powered prediction tools has enabled the forecasting of potential unsafe events. By learning from past accidents and near misses, these models can anticipate such occurrences and thereby prevent significant hazards. A noteworthy application of AI, as suggested by this study, involves predicting the transmission of WBV to operators of HEMMs. The importance of establishing safe operation limits becomes even more critical due to increasing research indicating the adverse health consequences of WBV exposure (Bovenzi and Hulshof 1999; Bovenzi 2010; Bovenzi et al. 2013; Ronchese and Bovenzi 2012; Combs et al. 2018; Mandal and Manwar 2017; Sharma and Mandal 2021). The predicted a_{wrms} acceleration can serve as a valuable tool for mining engineers, empowering them to enhance safety practices during ore and overburden transportation without the need for repeated field measurements. This predictive capability proves highly effective in establishing safe operational thresholds. The condition of haul roads plays a significant role in determining the extent of WBV exposure experienced by the operator along with operational costs involved. The findings from this investigation shed light on the potential application of ANN in predicting operator vibration exposure. Furthermore, the study opens avenues for future research, wherein a more in-depth analysis can be undertaken to elucidate the specific contributions of each parameter to the levels of vibration to which the operator is subjected. Understanding these individual influences can lead to valuable insights and potentially help in devising strategies to mitigate operator vibration exposure, thus enhancing overall safety and well-being. The outcomes of the neural network training and testing demonstrate significant promise, highlighting the potential benefits of this approach.

References

Bovenzi, M., Hulshof, C.T.J.: An updated review of epidemiologic studies on the relationship between exposure to whole-body vibration and low back pain (1986–1997). Int. Arch. Occup. Environ. Health **72**, 351–365 (1999)

Bovenzi, M.: A longitudinal study of low back pain and daily vibration exposure in professional drivers. Ind. Health **48**, 584–595 (2010)

Bovenzi, M., Schust, M., Menzel, G., Hofmann, J., Hinz, B.: A cohort study of sciatic pain and measures of internal spinal load in professional drivers. In: 5th International Conference on Whole Body Vibration Injuries, Conference Proceedings (Abstracts), Amsterdam, 5–7 June 2013 (2013)

Ronchese, F., Bovenzi, M.: Occupational risks and health disorders in transport drivers. Giornale Italiano di Medicina del Lavoro Ed Ergonomia **34**(3), 352–359 (2012)

Combs, B., Heaton, K., Raju, D., Vance, D.E., Karl Sieber, W.: A descriptive study of musculoskeletal injuries in long-haul truck drivers: a NIOSH national survey. Workplace Health Saf. **66**(10), 475–481 (2018)

Mandal, B.B., Manwar, V.D.: Prevalence of musculoskeletal disorders among heavy earth moving machinery operators exposed to whole-body vibration in opencast mining. Int. J. Community Med. Public Health **4**, 1566–1572 (2017)

Sharma, A., Mandal, B.B.: Attenuation of mechanical vibration during transmission to the human body through mining vehicle seats. Min. Metall. Explor. **38**(3), 1449–1461 (2021)

World Economic Forum: Artificial Intelligence and Machine Learning in Financial Services (2019). https://www.weforum.org/whitepapers/artificial-intelligence-and-machine-learning-in-financial-services

Chen, X., Lin, Z., Ma, H.: Artificial intelligence in healthcare: anticipating challenges in ethical, legal, and social impact. J. Healthc. Eng. (2018). Article no. 4641705. https://doi.org/10.1155/2018/4641705

Walczak, S.: Methodological triangulation using neural networks for business research. Adv. Artif. Neural Syst. (2012). Article no. 251412

Bishop, C.M.: Neural Networks for Pattern Recognition. Oxford University Press, Oxford (1995)

Mandic, D., Chambers, J.: Recurrent Neural Networks for Prediction: Learning Algorithms, Architectures and Stability. Wiley, Hoboken (2001)

International Organization for Standardization: ISO 2631-1 Mechanical Vibration and Shock—Evaluation of Human Exposure to Whole-Body Vibration—Part 1: General Requirements, Geneva, Switzerland. Reference number ISO 2631-1:1997(E) (1997)

Thompson, R.J., Visser, A.: Mine haul road maintenance management systems. J. South. Afr. Inst. Min. Metall. **103**(5), 303–312 (2003)

Thompson, R., Peroni, R., Visser, A.: Mining Haul Roads: Theory and Practice. CRC Press, Boca Raton (2019)

Numerical Modelling and Applications

FLAC 3D Modeling for Excavation Activity in an Underground Marble Quarry in Ambaji (Gujarat - India)

Piero Bettini(✉) and Vidan Ilic

S2R S.r.l., The University of Florence spin off, Viale G. Amendola 24, 50121 Florence, Italy
piero.bettini@gmail.com, v.ilic@s2r-sismosafe.it

Abstract. Dimension stones have been exploited in almost all parts of the world for millennia. In the 20th century, stricter legislation has sometimes limited opencast quarrying with the result that, under favorable geological conditions, underground mining is adopted. The latter gains in popularity also due to environmental and land use positive issues. In the case of the Ambaji marble quarry, the second option was a necessity to augment marble production and to exploit large parts of the concealed marble deposit which, otherwise in case of quarrying, would have required huge investments for the removal of large quantities of useless cap rock with inevitable environmental damage.

This study presents the approach for assessing the stability of underground chambers and pillars through numerical modeling of the marble deposit at Ambaji's Zarivav quarry. To analyze its long-term stability, a detailed investigation was conducted to draw the present and future geometry, the rock mass characterization, and to assess the surrounding rock behaviour. Based on field survey and laboratory tests, numerical simulations were performed using the finite difference code FLAC3D. The geomechanical analysis and the exploitation techniques applied to Zarivav quarry can be a path-finder for other similar projects in mining industry.

Keywords: underground quarrying · marble · underground stability · bolting · numerical modeling · stress-strain state

1 Introduction

D.K. Trivedi & Sons company is the owner of a white marble quarry in Zarivav near Ambaji, a town located to the north of Gujarat near to the Rajasthan State border. The open pit quarry, operating since 1949, increased every year its production, deepening progressively the excavation area. The intense exploitation reduced, over the time, the area available for creating further benches below the present quarry floor with the result that the actual life of the quarry becomes shorter. The other option is to take into consideration the removal of huge volume of useless rock along the western border of the license in order to enlarge the prospective quarrying area at the bottom. This option is found to be a costly hypothesis. To avoid this, a possible alternative option is to enter

A. Sinha et al. (Eds.): AMC 2023, SPEES, pp. 225–241, 2023.
https://doi.org/10.1007/978-3-031-46966-4_18

the marble deposit with underground excavation technique, thus enlarging the potential area to be exploited without the removal of large volumes of rocks with questionable quality (Fig. 1).

Fig. 1. View of the Zarivav quarry.

Underground quarry exploitation and related scientific and practical studies (Gullì et al. 2015, Gullì and Pellegri 2013), have become very common in the Apuan Alps marble district in Italy, gaining in popularity also due to environmental and land use positive issues. In fact, the high initial investment cost in specific cutting machines and the need for highly specialized personnel is supplanted by the subsequent progress of the underground exploitation techniques. Once the quarry has assumed its typical geometry of chambers and pillars, the extraction activity continues underground using the same techniques applied in open pit quarries: overturning of high benches sectioned with diamond wire or chain-saw machines and subsequent sectioning in squared blocks.

2 Geological and Lithological/Structural Features

The exploited marble of Ambaji can be petrographically defined as a white, fine to medium grained pure calcium carbonate marble. It took origin in a remote geological era as primary marine limestone deposits whose lacking of fossils and radiometric dating makes uncertain the age. However, the geological belonging (Prakash and Singh 2012) of this sequence to the Todgarh Formation of Kumbhalgarh Group (Delhi Supergroup) allows to roughly consider as the age of the original deposit as Paleoproterozoic-Mesoproterozoic (1,5–1,0 billion years) which later underwent to several phases of metamorphism that gradually re-organized the texture of the original sediment. Dynamic stresses repeatedly modified the original spatial orientation of all the assemblage of rocky formations underlying and overlying the limestone deposits. As practical consideration, the exploited marble deposit appears set-up as massive body with sub-vertical layering, evident all over the quarry benches, fronts, and underground chambers. The marble formation conformably overlays calc-silicate rocks, dolomite, and calc-biotite schists.

Traces of original sedimentary limestone variations appear in the marble sequence with shading tonalities from white to green/greyish hue. The marble fronts are predominantly massive with a metamorphic layering dipping towards WSW (N230° clockwise $\pm 10°$), with elevate gradient ($70 \pm 5°$ approximately). The prevailing mineralogical composition of this marble is calcium carbonate (95–98%) plus others small components in negligible percentages. Two sub-vertical dark-green dykes of dolerite composition cut the marble formation in different positions, striking NNW-SSE and approximately N-S with local variations. Their thickness ranges in the order of 7–8 m and the stones are harder than marble. Evidence of their continuity in the marble formation has been found in some core-drillings, as well as into the underground chambers.

Several sub-vertical sets of pervasive joints/fractures affect the marble walls all along the perimeter of the open pit; their main orientation is orthogonal to the layering, (bold line in Fig. 2): approximately striking ENE-WSW (see Fig. 3). Some joints are few millimeters open, some wider and filled with loose sediments and/or encrusted by calcium carbonate. In some cases, these joints are grouped in fracture belts few meters wide.

Another set of sub-vertical joints is oriented parallel to the layering, NNW-SSE (Fig. 2); this joint set is less frequent than the previous one, the fractures are few millimeters open and unfilled. The fractures are predominantly dry, no water flow coming from the joints/fractures has been noted all along the quarry wall. Most of the observations carried out inside the new chambers, confirm the expectations: The new mapped fractures follow the main trend of orientation which is ENE-WSW.

3 Geomechanical Data

The Ambaji white marble has been characterized in terms of composition and mechanical properties of the intact rock by several laboratory tests (Politecnico Turin University Laboratory tests according to UNI Standards; Department of Mining Engineering/Faculty of engineering - Jai Narayan Vyas University Jodhpur Laboratory tests, 2019; CISR-CIMFR Dhanbad - Government of India Scientific Institution Laboratory tests according to International Society of Rock Mechanics (ISRM) Standard, 2020). Large field data were gathered to characterize the Ambaji marble geomechanical features and subsequently to define its rock-mass physical-mechanical properties.

The features had:

- Rock-mass distribution that resulted from the presence, spacing and persistence of the discontinues.
- Morphologies and characteristics of discontinues, according to ISRM standard and finalized to the GSI classification (Hoek 1994; Hoek et al. 1995; Hoek and Brown 1980; Hoek et al. 1988; Marinos and Hoek 2000), Bieniawski (1989) RMR rock-mass classification and Burton and Bandis (1990) JRC and JSC parameters.

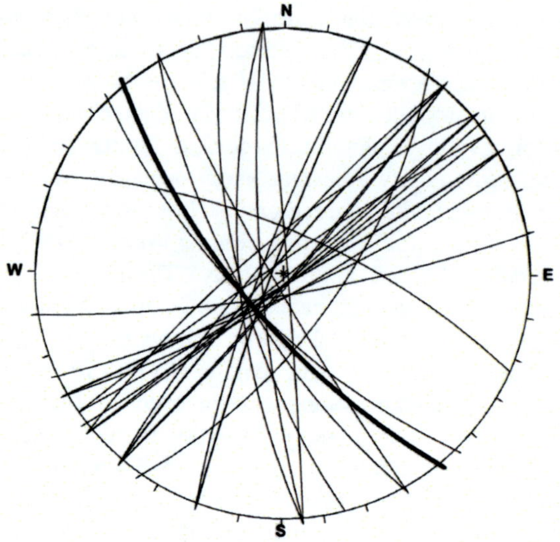

Fig. 2. Major planes plot

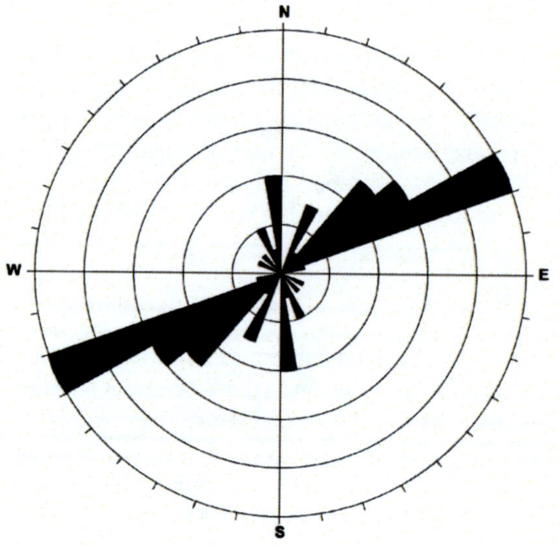

Fig. 3. Rose-diagram.

Thanks to the field survey activities previously carried out (please refer Bettini 2018), it was possible to classify the Ambaji marble into two rock-mass categories: intact and scattered-fractured. Intact and scattered-fractured marble correspond to the overall Ambaji marble formation where the quarrying exploitation has been carried out over the

Table 1. Ambaji White Marble mechanical properties.

Property	Unit	Value
Density	kg/m^3	2880
Imbibition coefficient	%	0.064
Uniaxial comp. strength	MPa	53.3
Bending strength	MPa	8.8
Young Modulus	GPa	63.7
Poisson's ratio	–	0.307
Wear thickness loss	mm	6.9
Impact collapse height	cm	36
Linear thermal expansion coefficient	10^{-6} °C	8.8
Knoop micro hardness	MPa	1382

years. From the geotechnical viewpoint, the Zarivav quarry is characterized predominantly by scattered-fractured rock-mass. Each of these two typologies are characterized by specific features of discontinuity morphology and spacing (Table 2).

According to the Hoek-Brown failure criterion (Hoek et al. 2002) and the empirical estimation of rock-mass modulus (Hoek and Diederichs 2006a, 2006b), the corresponding rock-mass geomechanical properties were computed for each Ambaji marble rock-mass type (intact and scattered-fractured) using RockLab software (Copyright © 2002–2011 Rocscience Inc.).

4 Constitutive Model

Stresses and displacements are calculated in an infinite Hoek-Brown medium subjected to an in-situ stress field. The Hoek-Brown criterion is used for plastic yielding when the minor principal stress, σ_3, is compressive. The criterion is based on a nonlinear relation between major and minor principal stresses, σ_1 and σ_3:

$$\sigma_1 = \sigma_3 + \sigma_{ci}\left[m_b\frac{\sigma_3}{\sigma_{ci}} + s\right]^a \quad (1)$$

where σ_{ci} is the unconfined compressive strength of the intact rock, and m_b, s and a are material constants that can be related to the Geological Strength Index - GSI and rock damage (Hoek et al. 2002) (Table 3).

The Hoek-Brown envelope is extended for σ_3 tensile by a combination of the Mohr-Coulomb envelope (tangent to Hoek-Brown at $\sigma_3 = 0$) and a tensile cutoff at $\sigma_3 = -s\sigma_{ci}/m_b$.

The numerical implementation of the Hoek-Brown model uses a linear approximation (see Fig. 4) where the nonlinear failure surface is continuously approximated by the

Table 2. Ambaji Marble geomechanical input data.

	Intact rock-mass	Scattered fractured rock- mass
Uniaxial comp. strength (MPa)	53.3	53.3
Density (kg/m^3)	2880	2880
Hoek & Brown index m_i (Hoek 2020)	9	9
R.Q.D	100%	83.5%
Discontinues spacing	–	>2 m
Discontinues aperture	–	0.1–0.5 mm
Discontinues persistency	–	10–20 mm
Discontinues roughness	–	rough
Discontinues morphology	–	planar
Discontinues infilling	–	none
Wall rock weathering	–	none
JRC (Barton et al. 1974)	–	12–16
Discontinues hydraulic condition	–	dry
Excavation category (Barton et al. 1974)	C (ESR = 1.6)	C (ESR = 1.6)
Excavation method (Barton et al. 1974)	undisturbed	undisturbed
RMR/GSI	100/100	75/72

Mohr-Coulomb tangent at the current stress level, σ_3. The current tangent Mohr-Coulomb criterion is:

$$\sigma_1 = \sigma_3 N_{\Phi_c} + 2c_c\sqrt{N_{\Phi_c}} \tag{2}$$

where,

$$N_{\Phi_c} = \frac{1 + \sin \Phi_c}{1 - \sin \Phi_c} = \tan^2\left(\frac{\Phi_c}{2} + 45^\circ\right) \tag{3}$$

Table 3. Ambaji Marble mechanical property values.

Rock property	Intact rock mass	Scattered fractured rock-mass
RMR/GSI	–/100	**75/72**
Modulus ratio (MR)	850	**850**
Intact Modulus E_i (Mpa)	45305	**45305**
Hoek & Brown index m_b	9	**3.311**
Hoek & Brown index s	1	**0.0446**
Hoek & Brown index a	0.5	**0.501**
Rock mass cohesion (Mpa)	11.0	**3.9**
Rock mass friction angle (°)	41.8	**35.9**
Rock mass uniaxial comp. strength - sigc (Mpa)	53.3	**11.2**
Rock mass global strength - sigcm (MPa)	49.3	**15.3**
Rock mass tensile strength - sigt (MPa)	−5.92	**−0.72**
Deformation modulus E_{rm}(GPa)	45	**35**

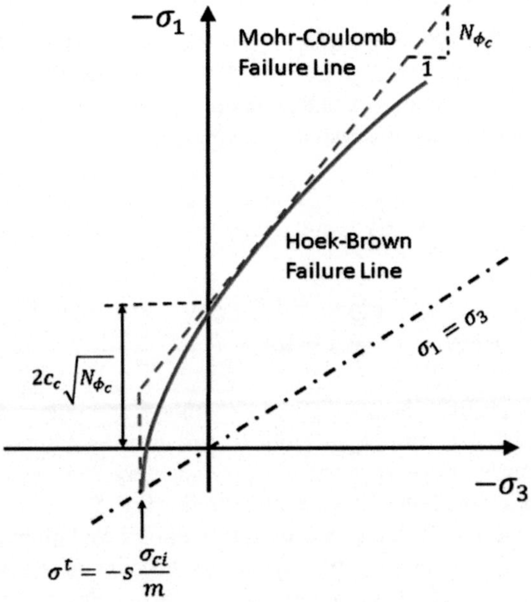

Fig. 4. Hoek-Brown failure criterion.

5 In-Situ Stresses and Boundary Conditions

The model has roller boundaries on all four sides and the bottom, applying the symmetry suitably. The top is left free. Next, the initial conditions are assigned. For this case this is only stresses due to gravity, so gravity is assigned using *model gravity* command. Stresses due to gravity are initialized with the *zone initialize-stresses* command. In-situ stresses are influential factors to assess the stability of an underground structure. Based on a thermo-elastic shell model of the earth, Sheorey (1994) predicted the average horizontal stress. This theory gives the value of the mean horizontal stress as:

$$\sigma_H = \sigma_h \frac{\nu}{1-\nu}\sigma_v + \frac{\beta EG}{1-\nu}(H + 1000) \tag{4}$$

where,

- H is the depth of cover (m)
- σ_v is the vertical in-situ stress (MPa)
- σ_H is the major horizontal in-situ stress (MPa)
- σ_h is the minor horizontal in-situ stress (MPa)
- E is Young's modulus (GPa)
- β is the coefficient of thermal expansion (1/°C)
- G is the geothermal gradient (°C/m)

In the study by Sheorey et al. (2001), this equation is shown to fit in-situ stress measurement data with the values of $\beta = 3 \times 10^{-6}$/°C for marble and G = 0.024 °C/m. After calculation of horizontal in situ stresses it was estimated the average horizontal to vertical ratio about 0.47 so that value is assigned using *zone initialize-stresses ratio* command.

Then, the model is solved to reach initial equilibrium without any excavation. In the further step the open pit geometry is created by removing some *mesh zones*.

6 Development of Underground Excavation and Estimation of the Safety Factor of the Pillars

The next step is to generate the underground chambers and pillars. This is done using the *zone relax excavate* command in order to simulate the actual exploitation. Each excavation step considers the removal of approximately 2 m width marble benches progressively reaching the final quarry configuration (Fig. 5).

The pillar strength is calculated considering Hoek-Brown failure criteria. For example, the rockburst phenomenon generally occurs because of propagation of failure through the rock mass and, in designing against such failures, the mining engineer is more interested in the average strength of the pillar than in the detailed mechanism of failure propagation. Under such circumstances the uniaxial compressive strength of the rock mass, defines the strength at the pillar boundaries only and not in the interior of the pillar. The question that arises from consideration of such problems is how to define the average or "global" rock mass strength that can be used by designers.

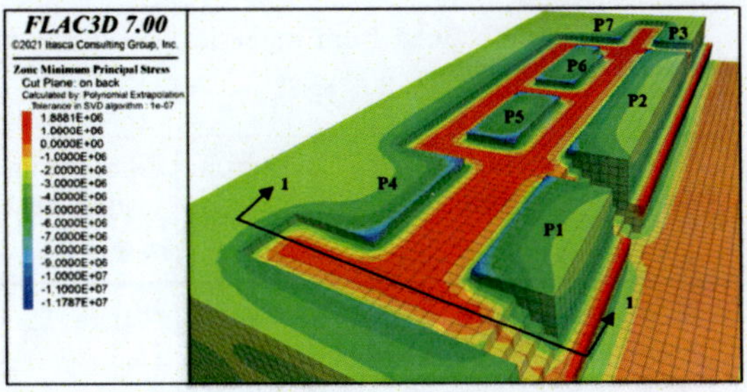

Fig. 5. Major principal stress acting on pillar bottom (approx. 366 msl) [Pa].

The average strength of pillars dependent upon the degree of confinement provided by the rock mass. Because of the highly curvilinear nature of the Hoek-Brown failure envelope at low stress levels, a very small amount of confinement results in a large increase in the strength of the rock mass. Unless a very detailed analysis is carried out, it is difficult to calculate the average strength in a structure in which the confining stresses can range by significant amounts. Furthermore, if the acting stresses on the pillar borders (i.e. zones with zero degree of confinement) overcome the uniaxial compressive strength (σ_c value by Hoek Brown criterion) the spalling phenomenon may occur.

The problem of fitting an equivalent Mohr failure envelope to the failure envelope defined by the Hoek Brown criterion can be very complicated. There is no direct theoretical link between the two criteria and any fitting process must be iterative. The actual process of curve fitting has been discussed in detail in Hoek et al. (2002) and the following relationship between the global rock mass strength σ_{cm} and the Hoek Brown parameters is given by:

$$\sigma_{cm} = \sigma_{ci} \frac{(m_b + 4s - a(m_b - 8s))(m_b/4 + s)^{a-1}}{2(1 + a)(2 + a)} \tag{5}$$

for the confining stress range:

$$0 < \sigma_3 < \frac{\sigma_{ci}}{4} \tag{6}$$

This stress range was chosen based on experience and was found to work well for a wide variety of practical situations. By using the program RocLab for the generalized Hoek-Brown criterion the value of σ_{cm} (the global strength of the rock mass) is calculated using Eq. 5. The FOS of the marble pillars after the first step (enlargement Area 1, Area 2, Area 3 and Area 4 at 365 a.s.l.) is given in Table 1:

$$FOS = \frac{\sigma_{cm}}{\sigma_{pm}} \tag{7}$$

where,

σ_{pm} = major principal stress
σ_{cm} = global strength of the rock mass (Table 4)

Table 4. FOS of the marble pillars obtained from numerical modeling after the 1st step of excavation.

STEP 1	Side A [m]	Side B [m]	Average pillar stress from FLAC- σ_{pm} [MPa]	Pillar strength for scattered fractured rock- σ_{cm} [MPa]	FOS
Pillar 1	30.00	12.00	4.33	15.3	**3.53**
Pillar 2	82.00	12.00	3.84	15.3	**3.98**
Pillar 3	32.00	12.00	3.56	15.3	**4.29**
Pillar 4	30.00	18.00	4.19	15.3	**3.65**
Pillar 5	30.00	12.00	5.53	15.3	**2.77**
Pillar 6	40.00	12.00	5.38	15.3	**2.84**
Pillar 7	32.00	24.00	3.34	15.3	**4.59**

7 Reinforcement Design

The plasticity state of the main chambers is also examined. Figure 6 shows the contours of minor principal stress inside the chambers, particularly on roof. The average values of minor stress in the outer layer of the roof chamber (approx. 2 m) ranges between 0.5 MPa up to 0.7 MPa. The maximum value reaches 0.93 MPa but only locally in the central section of the span. This implies that the minor principal stress overcomes the tensile strength of fractured rock mass (0.72 MPa) and yield phenomenon could occur, but only on the outer layer of the roof chamber. It is to be noted that positive value indicates tensile stresses. Such indication usually shows that plastic flow is going to happen, but it is possible for an element simply to "sit" on the yield surface without any significant flow taking place.

Those conditions are confirmed by Fig. 7, representing failure state.

The plot of the field of velocity vectors is more difficult to interpret, since both the magnitudes and the nature of the pattern are important. As with gridpoint forces, velocities never decrease precisely to zero. The magnitude of velocity should be viewed in relation to the displacement that would occur if a significant number of steps (e.g., 1000) were to be executed. In this phase of modeling, the current displacements in the system are of the order of 3 mm (see Fig. 8), and the maximum velocity in the velocity plot is 2.5^{-9} m/step (see Fig. 9), and about 16065 steps would produce an additional displacement of 0.00004 m, or approximately 0.04 mm, which is 1.3% of the current displacements of the outer layer of chamber roof. In this case, it can be said that the global system is in equilibrium even if all the velocities seem to be "flowing" in one direction, except the chambers roof that is characterized by yielding condition because of the 1.3% increase ratio.

So, it is evident that the outer layer of chambers roof (approx. 2 m), composed by yield zones as shown by numerical modelling, should be retained by proper reinforcement or support system. Thus, the reinforcement should be able to bear the weight of the yielded (loose) rock mass as an extension to support resistance concept (Fig. 10).

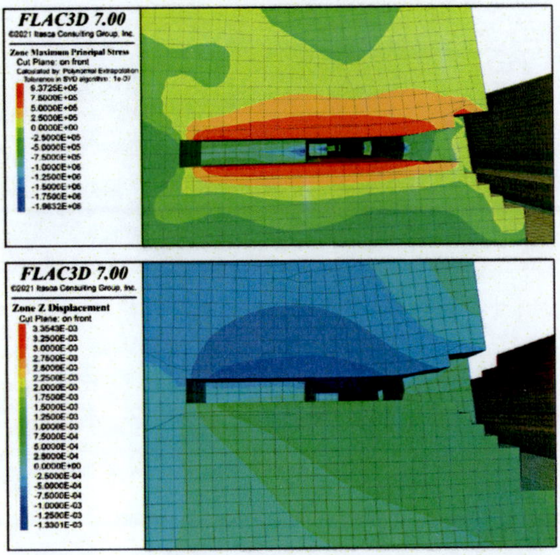

Fig. 6. Minor principal stress [Pa] on chamber roof and vertical displacements [m] after 1° step – section cut 1–1.

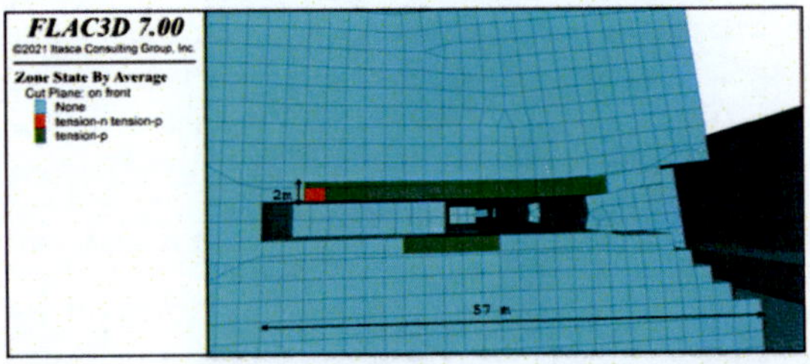

Fig. 7. Zone state of chamber roof after 1° step - section cut 1–1.

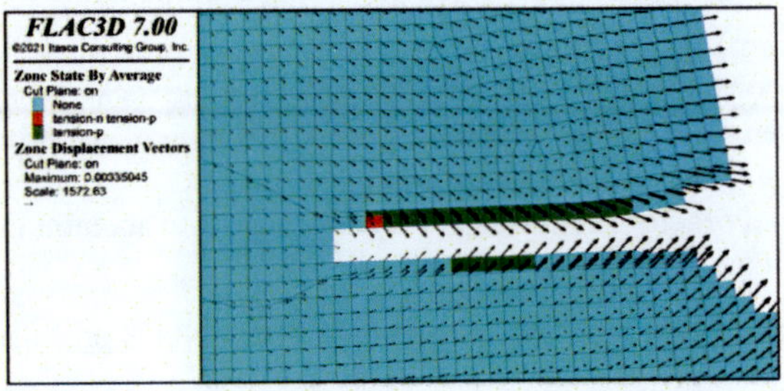

Fig. 8. Zone displacement vectors after 1° step

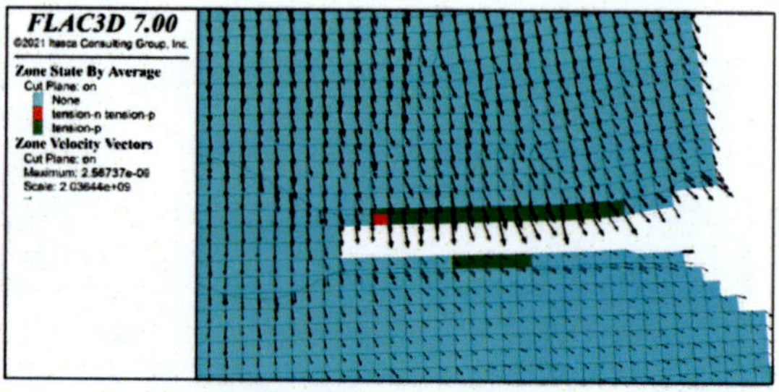

Fig. 9. Zone velocity vectors after 1° step

Based on the rock density, the required rock load to be reinforced is calculated by following equation:

$$P_b = D \cdot h_F \cdot s_x \cdot s_y \tag{8}$$

where,

- P_b is the rock load to be supported (yield block of rock mass) (t);
- D is the rock density (t/m^3);
- h_F is the height of the yielded (loose) roof rock obtained from the FLAC numerical model.
- s_x is the distance (spacing) in x direction between the rockbolts (reinforcement system);
- s_y is the distance (spacing) in y direction between the rockbolts (reinforcement system);

The factor of safety FoS of the reinforcement system is given by next equation:

$$FOS = \frac{A \cdot c}{1/4 \cdot \xi \cdot P_b} \tag{9}$$

where,

- A is the area of the grouted interface (equal to the product of the perimeter of the drill hole (D = 32 mm) by the grouted length in the intact rock mass La = 1.5 m);
- c is least of the cohesion of the scattered fractured marble rock-mass (equal to 3.9 MPa) and tensile strength of grout system (2 MPa for cement grouting or 2.6 for resin grouting);
- ξ is a partial safety factor (in this case 1.5) that takes into account the grout efficiency;
- Pb is the rock load to be supported.

The installation of two types of fully grouted rockbolts is proposed:

- length 3.5 m fully grouted, 32 mm borehole diameter, made of TMT, at least Fe500 grade, cold-rolled 24 mm threaded bar or rebar, grouted with 2 component resin;
- length 3.5 m fully grouted, 32 mm borehole diameter, made of TMT, at least Fe500 grade, cold-rolled 24 mm threaded bar or rebar, grouted with thixotropic cement grout;
- the rockbolts spacing should be 2.2 × 2.2 m; (Table 5)

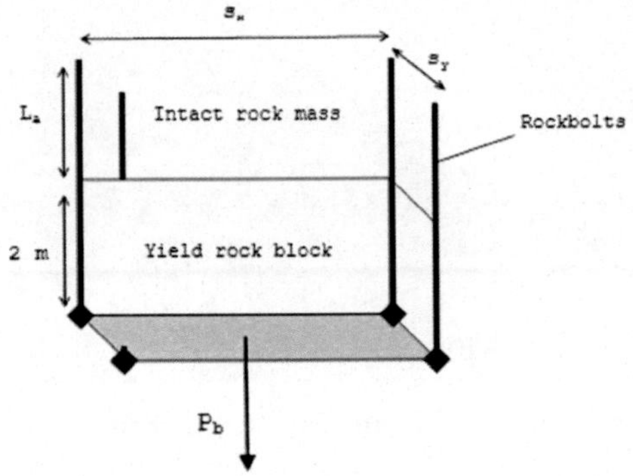

Fig. 10. Support system scheme

Table 5. FoS calculation of reinforcement system (grout material-rock interface).

rock load Pb [kN]	Bar type	Grout system	Compressive strength (28 days) [MPa]	tensile strength (28 days) [MPa]	Bolt grouted strength Sb [kN]	FoS
66.5	Fe500D TMT bars Ø 24 mm	thixotropic cementitious grout	45	2.0	201.0	**3.02**
66.5	Fe500D TMT bars Ø 24 mm	2 component resin systems	60	2.6	261.2	**3.93**

8 Downward Excavation

The simulation is carried out to assess the stability of the marble pillars after the second step of downward excavation (−5.5 m height) plus a third step of downward excavation (−11.0 m height). Figure 11 shows the development of the marble deposit on different levels (i.e., 365, 359.5 and 354 a.s.l.) due to the future envisaged excavation activity. Major principal stress contours are also shown. The average major principal stresses acting on the pillars after downward excavation decrease while the average minor principal stresses increase compared to stress situation after the 1° step of excavation, because the pillars in this phase are slenderer. The maximum vertical displacement is close to 2 mm (Fig. 12).

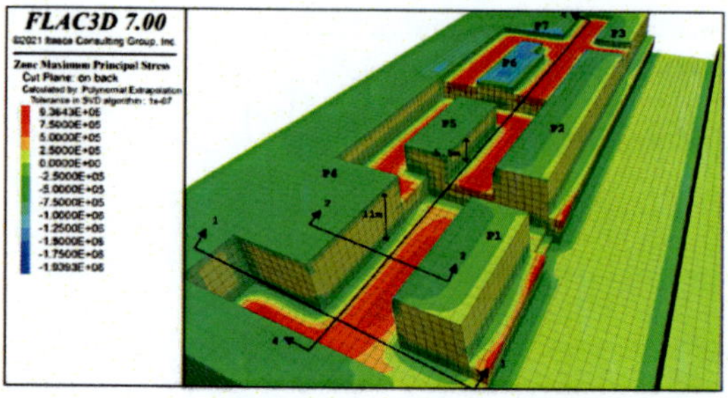

Fig. 11. Major principal stress contours [Pa] after downward excavation.

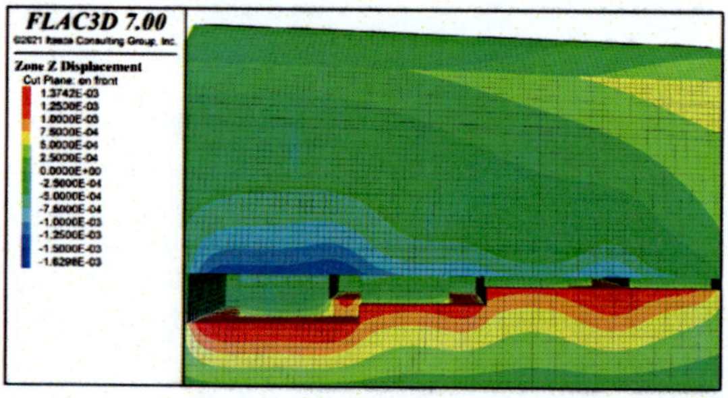

Fig. 12. Vertical displacements [m] after 2° step - section cut 4–4.

9 Conclusions

Based on the detailed field investigations, the evaluation of geo-mechanical properties and the numerical 3D modeling, the following conclusions may be drawn for the underground quarrying at Zarivav Marble mine of Ambaji:

1. The size of the pillars should be kept at least 360 m^2 to guarantee FOS > 2 in each excavation phase. However, smaller pillars may also be made, provided the support system should then be installed (for example steel sets/plates with rockbolts), particularly on the corner pillars. Otherwise, the spalling phenomenon could occur because the maximum major principal stress exceeds the uniaxial compressive strength of the rock mass according to the Hoek-Brown criterion.
2. Stability of the pillar is important to keep stability of the entire underground excavation. To estimate the strength of the pillar, it is considered that the pillars are devoid of large discontinuities but characterized like scattered-fractured rock-mass. The global strength of pillars of 15.3 MPa is estimated by Hoek-Brown formulation. The FOS of pillars is found to be more than 2.0 which is considered as long-term stable.
3. The standard width of the gallery should not be more than 12 m, but 18 m width is also achievable. In the final underground configuration, every chamber is about 16 m high. There are not any significant variations of rock mass condition induced by downward excavation compared to the first phase (i.e. development of underground excavation at 365 a.s.l.).

4. All the chambers, galleries, and junctions to be developed in the marble must be supported with systematic bolting with fully grouted roof bolts of 3.5 m length, 2.2 × 2.2 m spacing. The proposed bolts are fully grouted, 32 mm borehole diameter, made of TMT, at least Fe500 grade, cold-rolled 24 mm threaded bar grouted with 2 component resin. Alternatively, the grouting may be done with thixotropic cement grout. However, the manufacturer specifications of rockbolts and DGMS guidelines should be followed. The FOS of those rockbolts is found to be more than 2.0 which is considered as long-term stable.
5. Generally, steel wire mesh in not required and it should be installed only in roof portions with frequent joints and cracks. Roof bolts must be tested during the installation (according to DGMS guidelines), for example with standard pull-out test, to estimate real bearing capacity. If the recorded bearing capacity is less than the prescribed, then the rockbolt length or spacing should be reviewed. Alternate support system may suitably be adopted.

Acknowledgement. The Consultants are grateful to the owner and the mine management of Zarivav Marble mine of M/s D.K. Trivedi & Sons for the support received in any phase of the activity and particularly to CEO Mr. Kiran Trivedi, Mr. Manan Trivedi, and Mr. Jagdish Joshi. Furthermore, the Consultants are highly thankful to Dr. Gautam Banerjee and Dr. Prabhat Kumar Mandal and other CIMFR scientists for their edifying and precious suggestions and courteous hospitality.

References

Balmer, G.: A general analytical solution for Mohr's envelope. Am. Soc. Test. Mat. **52**, 1260–1271 (1952)

Barton, N.R., Lien, R., Lunde, J.: Engineering classification of rock masses for the design of tunnel support. Rock Mech. **6**(4), 189–239 (1974)

Bettini, P.: Technical feasibility report for Zarivav's marble underground prospective quarry in Ambaji - Gujarat (India) (2018)

Bieniawski, Z.T.: Rock mass classifications in rock engineering. In: Bieniawski, Z.T. (ed.) Exploration for Rock Engineering, pp. 97–106. Balkema, Rotterdam (1976)

Bieniawski, Z.T.: Engineering Rock Mass Classifications. Wiley, New York (1989)

Deere, D.U.: Chapter 1: Geological considerations. In: Stagg, K.G., Zienkiewicz, O.C. (eds.) Rock Mechanics in Engineering Practice, pp. 1–20. Wiley, London (1968)

Gullì, S., Pellegri, M., Marchetti, D.: Mechanical behaviour of Carrara marble rock mass related to geo-structural conditions and in-situ stress. Operative unit of mining engineering (UOIM) ASL Massa Carrara - Italy. In: Proceedings of the 8th South American Congress on Rock Mechanics, November 2015

Gullì, D., Pellegri, M.: Stress analysis on Carrara marble quarries. Operative unit of mining engineering (UOIM) ASL Massa Carrara - Italy. In: Proceedings of the International Symposium on In-situ Rock Stress, 20–22 August 2013, Sendai, JP, pp. 839–855 (2013)

Handley, M.F., Webber, R.C.W.: Examine the criteria for establishing the small span small pillar concept as a safe mining method in deep mines. Research Enterprise at University of Pretoria (Pty) Ltd., GAP 828, January 2002 (2002)

Hammett, R.D., Hoek, E.: Design of large underground caverns for hydroelectric projects, with reference to structurally controlled failure mechanisms. In: Proceedings of the American Society of Civil Engineers International Conference on Recent Developments in Geotechnical Engineering for Hydro Projects, pp. 192–206. ASCE, New York (1981)
Hoek, E.: Uniaxial compressive strength versus Global strength in the Hoek-Brown criterion, Vancouver, 30 March 2005 (2005)
Hoek, E., Brown, E.T.: Empirical strength criterion for rock masses. J. Geotech. Eng. Div. **106**(GT9), 1013–1035 (1980)
Hoek, E., Brown, E.T.: The Hoek-Brown failure criterion - a 1988 update. In rock engineering for underground excavations. In: Curran, J.C. (eds.) Proceedings of the 15th Canadian rock Mechanics Symposium, pp. 31–38. Department of Civil Engineering, University of Toronto, Toronto (1988)
Hoek, E., Diederichs: Empirical estimation of rock mass modulus. Int. J. Rock Mechan. Min. Sci. **43**(2), 203–215 (2006)
Hoek, E., Marinos, P.: Predicting Tunnel Squeezing. Tunnels and Tunnelling International. Part 1 – November 2000, Part 2 – December, 2000 (2000)
Hoek, E., Kaiser, P.K., Bawden, W.F.: Support of Underground Excavations in Hard Rock. Balkema, Rotterdam (1995)
Hoek, E., Carranza-Torres, C.T., Corkum, B.: Hoek-Brown failure criterion-2002 edition. In: Proceedings of the 5th North American Rock Mechanics Symposium, Toronto, Canada, vol. 1, pp. 267–73 (2002)
Hoek, E., Marinos, P., Marinos, V.: Characterization and engineering properties of tectonically undisturbed but lithologically varied sedimentary rock masses. Int. J. Rock Mech. Min. Sci. **42**(2), 277–285 (2005)
Hoek, E., Diederichs, M.: Empirical estimates of rock mass modulus. Int. J. Rock Mech. Min. Sci. **43**, 203–215 (2006)
Iannacchione, A.T.: Pillar Design Issues for Underground Stone Mines, National Institute for Occupational safety and Health, Pittsburgh Research Laboratory, Pittsburgh, PA (1997)
Baryakh, A.A., Gegin, A.S.: Stability assessment of interchamber pillars. J. Min. Sci. **33**(I) (1997)
Itasca (2016): Itasca Consulting Group Inc, FLAC3D (Fast Lagrangian Analysis of Continua in 3 dimensions). Version 7.0. Minneapolis, MN (2019)
Jalote, P.M., Kumar, A., Kumar, V.: Geotechniques applied in the design of the machine hall cavern, Nathpa Jhakri Hydel Project, N.W. Himalaya, India. J. Eng. Geol. (India) **XXV**(1–4), 181–192 (1996)
Marinos, P., Hoek, E.: Estimating the geotechnical properties of heterogeneous rock masses such as flysch. Bull. Eng. Geol. Environ. (IAEG) **60**, 85–92 (2001)
Marinos, P., Hoek, E., Marinos, V.: Variability of the engineering properties of rock masses quantified by the geological strength index: the case of ophiolites with special emphasis on tunnelling. Bull. Eng. Geol. Environ. **65**(2), 129–142 (2006)
Martina, C.D., Maybeeb, W.G.: The strength of hard-rock pillars. Int. J. Rock Mech. Min. Sci. (2000)
Maybee, W.G.: Pillar Design in Hard Brittle Rocks, School of Graduate Studies Laurentian University Sudbury, Ontario, Canada (2000)
Merh, S.S.: Geology of Gujarat, Geological Society of India, 1995 Bangalore, 1 and 3 p. (1995)
Palmstrom, A., Singh, R.: The deformation modulus of rock masses: comparisons between in situ tests and indirect estimates. Tunn. Undergr. Space Technol. **16**, 115–131 (2001)
Prakash, I., Singh, P.K.: Geology and Mineral Resources of Gujarat, Daman and Diu. Geological Survey of India Misc. Publications, no. 30, part XIV, pp. 1–44 (2012)
Salcedo, D.A.: Macizos Rocosos: Caracterizaci6n, Resistencia al Corte y Mecanismos de Rotura. In: Proceedings 25 Aniversario Conferencia Soc. Venezolana de Mecánica del Suelo e Ingeniería de Fundaciones, Caracas, pp. 143–172 (1983)

CSIR-CIMFR report, 2019, on "Advice on feasibility of underground mining of marble with suitable cross-section of gallery opening and pillars at Zarivav Marble Mine", submitted by CSIR-Central Institute of Mining and Fuel Research, Barwa Road, Jharkhand, India (2019)

Sheorey, P.R., Mohan, M.G., Sinha, A.: Influence of elastic constants on the horizontal in situ stresses. Int. J. Rock Mech. Min. Sci. **38**, 1211–1216 (2001)

Stillborg, B.: Professional Users Handbook for Rock Bolting. Trans Tech Publications, Atlas Copco (1986)

York, G., Canbulat, I., Kabeya, K.K., Le Bron, K., Watson, B.P., Willims, S.B.: Develop guidelines for the design of pillar systems for shallow and intermediate depth, tabular, hard rock mines and provide a methodology for assessing hanging wall stability and support requirements for the panels between pillars, CSIR Mining Technology, GAP334, December 1998 (1998)

An Investigation on Optimal and Safe Design of Highwall Mining Panel Beneath Jointed Overburden Strata Using Discrete Element Modelling

Siddhartha Roy[1](✉) and Hemant Agrawal[2]

[1] Mine Ventilation & MT Lab Services Department, R & D Building, CMPDI (HQ), Ranchi 834008, India
siddhartha.roy@coalindia.in

[2] Technical Secretariat Department, 3rd Floor, CMPDI (HQ), Ranchi 834008, India
hemantagrawal@coalindia.in

Abstract. When the mine boundary is reached during surface mining operations, and conventional opencast mining becomes impractical to proceed further, a significant amount of coal reserve is left trapped under open-pit highwall. To overcome these challenges, the Highwall mining method can be seen as a viable option to increase coal production from previously operated opencast mines by extracting the trapped coal reserves and exploit thin seams that were left unmined due to economic limitations. In highwall mining, the safety and stability of a Highwall panel rely heavily on scientifically-designed dimensions of openings, web pillars, and barrier pillars. Pillar dimensions are determined using empirical formulae suggested by various researchers. However, designing highwall mining panels becomes challenging in cases where there is a presence of jointed rock mass in the overlying overburden strata, making it difficult to rely on available empirical formulas. In this study, a 3D discrete element model of a Highwall panel has been prepared incorporating the presence of jointed overburden rock mass using 3DEC™ software. Factor of safety for the overall highwall structure was determined based on observations of different combinations of web pillars, barrier pillars, and opening size. Based on which optimal and safe design of Highwall mining panel has been recommended.

Keywords: Highwall Mining · Stability of web pillars · Stability of barrier pillars · 3D modelling · Discrete Element Method · Jointed rock mass

1 Introduction

India, the second largest coal producer of the world, meeting its 60% energy demand from this fossil fuel (Roy et al., 2022a). The need for coal is increasing unceasingly to meet the demand in various sectors. In order to meet the demand, mining sector is upgrading with high mechanization (Roy et al., 2022b). As mining advances, higher stripping ratios slowly become limiting factors for surface mining operations. It is unavoidable, at some

A. Sinha et al. (Eds.): AMC 2023, SPEES, pp. 242–253, 2023.
https://doi.org/10.1007/978-3-031-46966-4_19

point, for mining operations to attain an economic threshold where the cost of overburden removal exceeds the value of the coal. When stripping ratios lose viability, recovery of coal resources under the highwall can be accomplished by other means. Generally, where the ordinary open-pit mining is tough or impractical in a small-scale due to an uneconomic stripping ratio, the highwall mining systems could be applied to extract the remaining coal reserves (Sasaoka et al., 2016). In India, there are numerous thin coal seams in various coal fields, which are not extractable through the conventional surface mining technique. Highwall mining method is preferred and the only available practical way to extract those thin coal seams (> 1 m) and the coal reserve left over in the highwall of a surface mine. Most of Indian coalfield are surrounded by forest lands, getting permission for mining in such area is a difficult task due to stringent environmental laws. Mine operators might contemplate the extraction of trapped coal of open-pit mine by employing the Highwall mining technology. Highwall mining method is much safer than other surface mining technologies. Occupational safety related organizations like MSHA, NIOSH has admitted that fatality and injury rates are significantly lower than that of other comparable surface mining methods.

Even highwall mining operation is proved to be a safer mining technique, though there is threat of collapsing of roof of entry, web and barrier pillars (Porathur et al., 2017). The biggest threat in a highwall mining operation to both equipment and personnel's safety is ground control (Shen, 2014). The presence of a thin and inferior jointed layer in the immediate roof may cause potential stability issues for the entry roof (Ross et al., 2019). In such cases the safety and stability of a Highwall panel greatly depends on the accurate designing of opening, web pillar and barrier pillar (Fan, 2015). In general, the safe dimensions of highwall mining operation are designed based on the empirical formulas. However, there is limited scope in the empirical method to incorporate the effect of discontinuities in determining the stability of sub-surface structures. The 3D modelling method is an established technique used in the field of geosciences for determination of stresses acting upon some structure, deformation in a structure due to load, discontinuity planes etc. Through 3D modelling of a highwall panel located below jointed overburden, the size of web pillars and barrier pillars can be optimised aiming for high recovery ratio of coal resources. In this paper, 3D discrete element models have been developed in 3DEC™ software to check out the stability of the entry roof and pillars and optimum dimensions of opening, web and barrier pillars.

2 Basics of Highwall Mining

When the stripping ratio of surface mining operations loses profitability, the highwall mining method can be applied to recover the residual coal reserves beneath the final highwall (Fig. 1). In the most general sense, highwall mining is the process of extracting coal reserves from the base of an exposed highwall left by the previous surface mining. Typically, this is done by using the remotely controlled mining methods to horizontally drive a series of nearly parallel entries for a significant distance (Adhikary et al., 2002). A portion of the residual coal reserves are recovered through highwall operations by punching long rectangular entrances into the coal seams (Luo, 2014). Web pillars are left to save the overburden layers from collapsing, and to delineate the entries from each other.

Recent advances in highwall mining systems have enhanced mining safety and increased financial gains, which makes the highwall mining an important coal production method throughout the globe as well as in India (Porathur et al., 2017). Moreover, highwall mining operations are performed by remote control, allowing workers to work in a safe environment free from hazards such as roof falls, gas, dust, flooding, and vehicle movements. Furthermore, due to the limitations of other mining methods, and the tighter environmental control requirements, the highwall mining method should be considered by more coal operators.

Fig. 1. Opening made due to extraction of coal from Highwall

Highwall mining is a combination of surface and underground mining methods (Fig. 2). With the miner's cutter module pushing into the coal seam through a series of push-beams, deep quadrangular entries are driven into the coal seam. As the cutter head moves forward into the coal seam, fully enclosed push-beams are inserted behind the cutter head. A typical mining cycle in highwall mining includes sumping (forward movement) and shearing (vertical movement of the cutter-head to extract the full height of the coal seam). As the cutting cycle continues, the cutter-head is progressively pushed into the coal seam to a depth of 6 m. Fully enclosed push-beams are inserted behind the cutter-head as it moves forward into the coal seam. Push-beams are a series of 6 m long, rectangular, reinforced steel boxes. Cut coal is transported to the launch vehicle via push-beam conveyors. It also supports the hose chain that controls and powers the cutter. Then, the coal drops to the conveyor system, where coal is moved to the centre of the bench for stockpiling and transport. The mining cycle continues for 6 m and the process is repeated until the operator reaches the desired depth. After full coal extraction, the push-beams are removed from the back of the cutter-head (Fiscor, 2002). The current highwall mining techniques can handle coal seam thicknesses ranging from 1 m to 4.8 m with a dip of up to 8 degrees (Luo, 2014). A representative schematic diagram of Highwall mining cutting pattern is shown in Fig. 3.

Fig. 2. Highwall Mining machine extracting coal

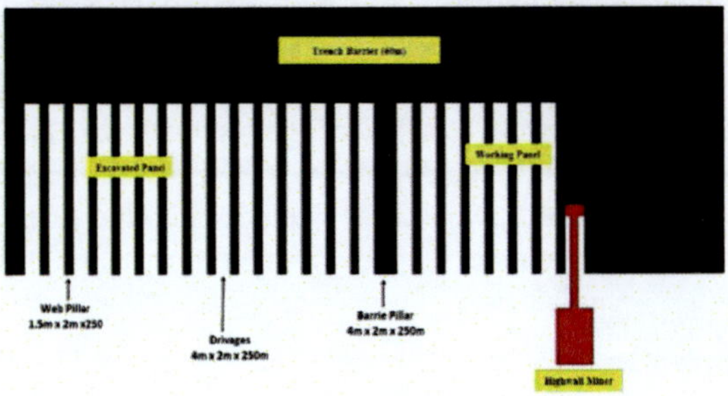

Fig. 3. A schematic diagram of Highwall mining cutting pattern.

3 Advantages of Highwall Mining

Since a highwall mining operation only requires a small bench to obtain entry to the coal seam, highwall mining systems not only produces less environmental troubles to the surrounding environment and land, but also are able to generate considerable economic benefits. The highwall mining system is highly transportable and can be transported from one pit to another in a matter of hours or from one mine to another in a matter of days. The reserve life of opencast coal mines would be greatly increased by the effective highwall mining technologies. Moreover, since the entries are unmanned, many of the safety issues associated with underground mining are eliminated. In general, the highwall mining method is very productive, flexible and cost-effective when used in proper situations. Furthermore, highwall mining technique has ability to selectively extract a high-quality segment of a seam.

4 Advancements in Highwall Mining

Maintaining the coal pillars (web & barrier) and maintaining parallel openings is crucial because the majority of entries don't reach the intended or pre-set depth, the position and direction of the highwall miners must first be known and controlled remotely in order to accomplished this goal. In recent days sensors are used to keep the highwall miner machine in the seam automatically. An array of inclinometers, a ring laser gyroscope, a programmable logic controller, and passive gamma detector systems for the roof and floor are among the sensors used (Ralston et al., 2001). These remote detection systems can be used for the following two things: (1) identifying roof and floor rocks to prevent cutting into them, and (2) figuring out how much pillar width is still left between the previous and current entry to keep the pillar design width.

5 Stability of Highwall Panel

The stability of the highwall structure itself has a significant impact on the safety of a highwall mining operation. When the highwall is first created during contour and open-pit mining operations, it usually remains stable. But when the highwall mining operation causes sizable movements and deformations at its base, instability might happen. The instability of the pillars in the panel section and the inadequate design of the pillar system are the main causes of highwall instability. The presence of discontinuities in the overlying strata may deteriorate the stability of the highwall panels.

In ideal situation (ignoring the discontinuities in the rock mass), the panel design parameters such as opening width, web pillar size and rib pillar size can be obtained by various available empirical formulae.

5.1 Empirical Formulae

A thorough plan should be followed during highwall mining in order to minimise any potential risks. The planner must decide three things while creating a highwall mining layout viz. the width of the web pillars; the distance between web pillars and the width of the barrier pillars. The overburden depth, mining height, and highwall miner opening width affect the design parameters (Zipf, 2005). Additionally, the mine designer should evaluate the pillar strength, applied stress, and pillar factor of safety. Mark-Bieniawski suggested the following formulas for designing a Highwall mining panel.

$$S_{W/B} = S_T\left[0.64 + 0.54\frac{W_{B/W}}{H_{B/W}}\right] \tag{1}$$

$$ST_W = S_V[W_W + H_W]/W_W \tag{2}$$

$$FOS_W = S_W/ST_W \tag{3}$$

$$P_W = N[W_W + H_W] + H_W \tag{4}$$

$$ST_B = S_V[P_W + W_B]/W_B \tag{5}$$

$$FOS_B = S_B/ST_B \tag{6}$$

where,

$S_{W/B}$ = strength of web pillar or barrier pillar

$W_{B/w}$ = width of barrier or web pillar
$H_{B/w}$ = height of barrier or web pillar
S_W = strength of web pillar
ST_W = stress on web pillar
S_V = vertical stress
W_W = width of web pillar
H_W = height of web pillar
FOS_W = factor of safety of web pillar
S_B = strength of barrier pillar
P_W = width of highwall panel
N = numbers of web pillar in a panel
ST_B = stress on barrier pillar
W_B = width of barrier pillar
FOS_B = factor of safety of barrier pillar

These empirical formulae may be used for designing the highwall panels in ideal situation, whereas, in actual situation, a lot of joints are present in the overlying rock mass during highwall mining. In this study, attempts have been made to consider those joints on the overlying strata by developing a 3D Discrete Element Model. The methodology adopted for this study is presented below.

6 Methodology

To conduct the 3D modelling study, Mine-A was selected where there is presence of jointed rock mass in the highwall of opencast. A borehole lithology of Mine-A has been shown in Fig. 4. Physico-mechanical properties of rock and coal sample was collected from records of the mine. The properties of rock, coal and discontinuities were used in the 3D model to build a true representative model. As observed in many different literatures, authors have preliminarily practiced 3D modelling of highwall panel without considering the joints in the rock mass. To compare the results between the above-mentioned model and the model incorporating jointed rock mass, initially a model without the jointed rock mass has been developed. Thereafter, three different 3D models with jointed rock mass (Pattern-I, Pattern-II, and Pattern-III) of different panels size were developed. The model without jointed rock mass was compared with models with jointed rock mass to check the difference in FoS (factor of safety) and other stability parameters.

7 3D Modelling

Dimensions of three different patterns of highwall mining panels are given in Table 1. The average thickness of the seam considered for modelling of highwall mining is 4.5 m. The extraction height for each highwall panel is considered as 4 m (same each pattern). The main purpose of developing three different 3D models was to find out an optimum design of highwall mining panel beneath jointed overburden in which the extraction will be maximum without compromising the safety and stability of the structure. The models were developed by discrete element modelling method with the help of 3DEC software. Firstly, the model was run up to the equilibrium state and then the excavation was done and run until the model reached to the solved stage. Finally, model was run to solve the factor of safety. The principal stresses on web, barrier pillars and on the roof of the opening was checked. The displacement of roof strata at the opening was also checked to know the chance of subsidence. The same procedure of model development and analysis of result was carried out for three different dimensions of opening. The depth of the highwall panel from the surface is 95 m. Whereas, for 3D model, the immediate 3 layers from coal seam i.e., Sandy Shale, MGSST (Medium Grained Sandstone) and Shale comprising the combined thickness of 15 m has been taken. The modelled panel (comprising highwall panel and 15 m as mentioned above) has been imposed with the load of the remaining 80 m rock mass above (0.0027×g × (model height-200 m)). The model is subjected under gravitational force and the boundary fixidity condition applied to fixed for the left, right and bottom boundaries whereas, the top kept free.

Table 1. Dimensions used for development of three different patterns of panel

Pattern	Barrier pillar width	Web pillar width	Width of opening	Extraction height	Length of opening
I	3.5	3	3.5	4	250
II	4	2	4	4	250
III	3	2	4.5	4	250

8 Model Development

The developed model consists of five layers, the bottom to top layer sequence is CGSST, Coal, Sandy Shale, MGSST and Shale. Twelve sets of joint planes dipping in a dip of 76° and dip direction of 279° has been incorporated in the overburden layers of the rock mass. After considering the geological data, physico-mechanical properties of various layers and joints, stress field and fixidity condition into the 3-D geometry model of 3DEC™ have been imposed and then an initial stage of analysis has been done to attain the equilibrium state of the model without any excavation. After the initial stage of analysis, it has been realised that the stresses in the model is stable and represent the condition as it should be in that depth. The excavation was made into the coal seam for a length of 250 m. The dimension of gallery in which the excavation was made is shown in Table 1. The stability of web and barrier pillars and partings was determiner after the excavation was made. Mohr-Coulomb Model is used to simulate the real behaviour of highwall structures. The reason to choose this model is that the Mohr-Coulomb failure criterion is frequently used to assess the state of failure in the study of soil and rock mechanics.

In 3DEC, the material properties that must be defined for a Mohr-Coulomb material include: density, bulk modulus, shear modulus, friction angle, cohesion, dilation angle, and tensile strength. The aforementioned properties of coal, overlying layers and joints

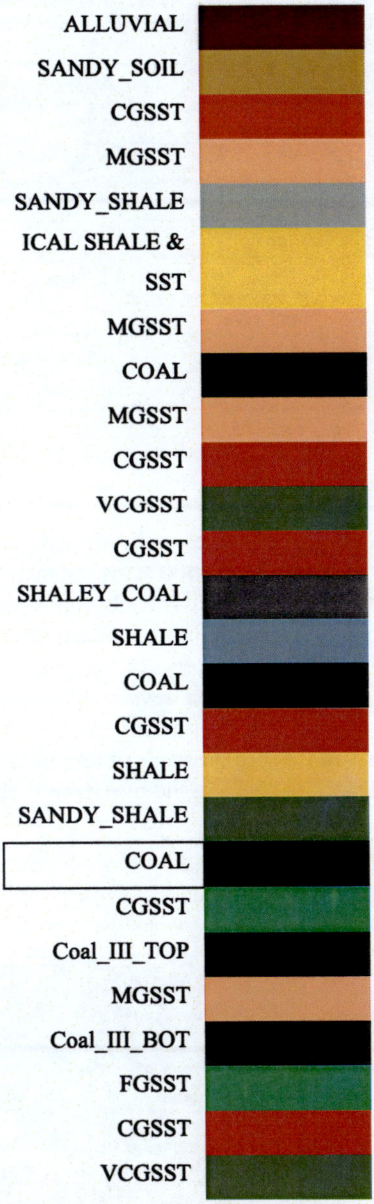

Fig. 4. Lithology of borehole at the Highwall side

were inserted in the model, however, only the properties of coal used in the model development are shown in Table 2. The compressive strength of coal was 17.24 MPa.

Table 2. Physico-mechanical properties of coal used in model

Properties	Value
Density (kg/m^3)	1450
Poisson's ratio	0.25
Young's modulus (pa)	678.61e6
Friction angle (degree)	28.06
Cohesion (pa)	1.08e6
Shear modulus (Mpa)	271.44
Bulk modulus (Mpa)	452.4
Tensile strength (pa)	0.059e6

9 Results and Discussion

Initially each model was run to attain the equilibrium state, it was observed that the unbalance force was near zero at around 1500 steps for each case. Which indicates that the models attain the equilibrium state. It was observed that the vertical stress (σ_{zz}) was increasing with depth. Post extraction of coal from the model, the factor of safety (FoS) of the overall model has been evaluated. As mentioned above, a 3D model without joints in the overlying rock mass has been developed as presented in Fig. 5, the FoS observed was 2.78 with z-displacement in the web and barrier pillars ranging from 0.2–1.6 cm. Although, the FoS and z-displacement observed is under safe limits but this model ignored the most important joints in the overlying rock mass.

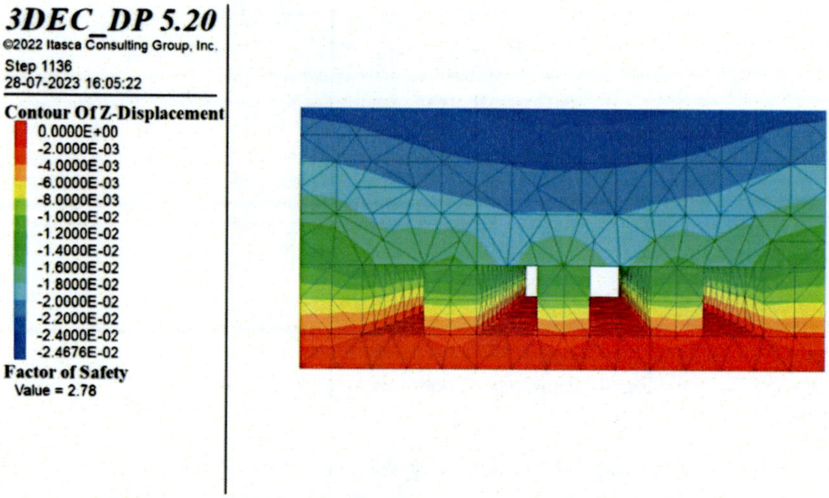

Fig. 5. Highwall panel without joints in the overburden rockmass

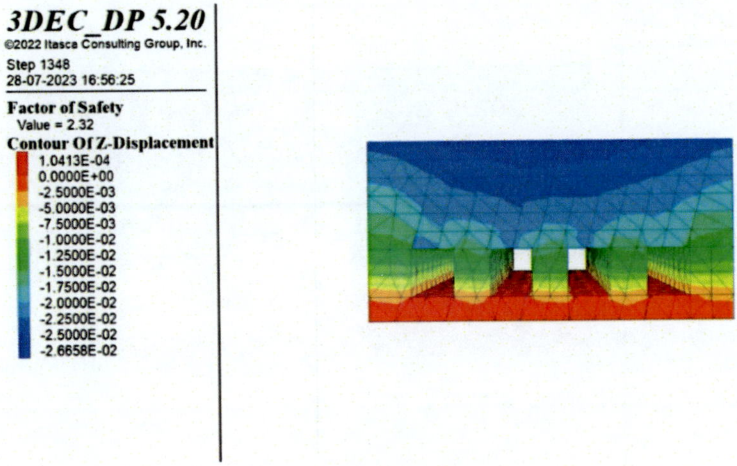

Fig. 6. Pattern-I highwall panel

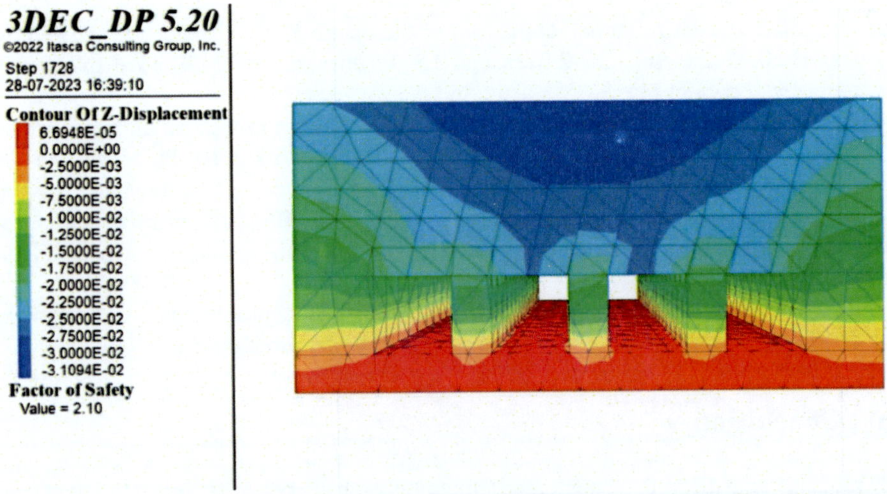

Fig. 7. Pattern-II highwall panel

The models with joints in the overlying rock mass were developed in three different patterns as discussed above. The outcomes of FoS modelling along with Z-displacement are shown in Figs. 6, 7 and 8 for Pattern-I, II, and III respectively. It has been found that the FoS of Pattern-I, Pattern-II and Pattern-III panel was 2.32, 2.10, and 1.93 respectively. Out of the three patterns of cut chosen for the study, the maximum percentage of extraction is obtained in Pattern-III and it's FoS obtained is 1.93, which is above the minimum desired (>1.5) FoS. Therefore, the Pattern-III panel configuration is being considered as the optimum panel configuration, so in this paper we have given emphasis on the 3D modelling outcome of Pattern-III panel configuration. The Z displacement at

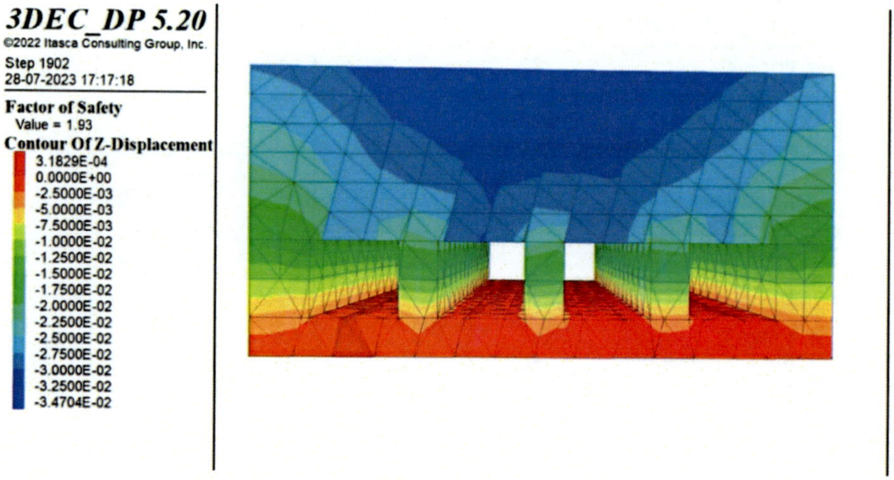

Fig. 8. Pattern-III highwall panel

the web pillar, barrier pillar and at the roof of the opening was also determined and the values were 0.25–2.0 cm, 0.25–2.0 cm, and 2.5–3.0 cm respectively. The Z displacement contours are shown in Fig. 8.

It has been observed that the maximum principal stress in web pillar, barrier pillar and partings between openings ranges between 1×10^5 and 7×10^5 Pa, which is lower than the compressive strength of coal and hence the geometry of the highwall panel is stable. Similarly, the low z-displacement values in web pillar, barrier pillar and at the roof of openings indicate a very less convergence resulting in stable geometry of the highwall panel.

The 3D modelling using 3DEC may thus help the field engineers to design the highwall panels with highest percentage of recovery and productivity.

10 Conclusion

3D models using discrete element modelling method have been developed to investigate on the optimum but stable size of pillars in Highwall mining panel below jointed overburden. The objective of this study was to determine the optimum size of the web pillar, barrier pillar and the openings so that the recovery of coal can be increased without sacrificing the safety and stability of the structure. Initially, a model without joints in the overburden rock mass was developed and then three 3D models incorporating joints in the overburden rock mass was developed to realize the effect of joints in stability parameters. It was found that in case of the joints are overlooked in the model then FoS will be overestimated by nearly 30%. This overestimation may result in deficiencies in effective designing of highwall panel and may result in inaccurate or compromised planning.

After investigating the results of 3 models incorporated with joints, it has been found that web pillar of 2 m × 4 m, barrier pillar of 3 m × 4 m and opening dimension of

4.5 m × 4 m are the optimum. The factor of safety of the Pattern-III model was 1.93. The maximum Z displacement at the roof of the openings was only 3 cm and in case of web pillar and barrier pillar it was 2 cm. Maximum principal stress that was observed around the immediate roof of the highwall panel, web pillar, and barrier pillar was in the range of 1×10^5 to 7×10^5 Pa, which is well below the strength of coal. Through these specifications of cutting, a high percentage of coal may be recovered within the panel without sacrificing the safety.

References

Adhikary, D.P., Shen, B., Fama, M.E.D.: A study of highwall mining panel stability. Int. J. Rock Mech. Min. Sci. **39**, 643–659 (2002)

Fan, M.: Design Programs for Highwall Mining Operations. West Virginia University (2015)

Fiscor, S.: Contour mining puts highwall system to the test. Coal Age 107 (2002)

Luo, Y.: Highwall mining: design methodology, safety and suitability. Coal Energy Res. Bur. Final Rep. (2014)

Porathur, J.L., Pal Roy, P., Baotang, S., Karekal, S.: Highwall Mining – Applicability, Design and Safety, p. 324. Taylor & Francis Group plc (CRC Press), UK (2017)

Ralston, J.C., Hainsworth, D.W., Reid, D.C., Anderson, D.L., McPhee, R.J.: Recent advances in remote coal mining machine sensing, guidance, and teleoperation. Robotica **19**, 513–526 (2001)

Ross, C., Conover, D., Baine, J.: Highwall mining of thick, steeply dipping coal – a case study in geotechnical design and recovery optimization. Int. J. Min. Sci. Technol. **29**, 777–780 (2019)

Roy, S., Mishra, D.P., Bhattacharjee, R.M., Agrawal, H.: Genetic programming for prediction of heat stress hazard in underground coal mine environment. Nat. Hazards **114**, 2527–2543 (2022). https://doi.org/10.1007/s11069-022-05478-6

Roy, S., Mishra, D.P., Bhattacharjee, R.M., Agrawal, H.: Heat stress in underground mines and its control measures: a systematic literature review and retrospective analysis. Mining Metall. Explor. **39**, 357–383 (2022). https://doi.org/10.1007/s42461-021-00532-6

Sasaoka, T., Karian, T., Hamanaka, A., Shimada, H., Matsui, K.: Application of highwall mining system in weak geological condition. Int. J. Coal Sci. Technol. **3**, 311–321 (2016). https://doi.org/10.1007/s40789-016-0121-6

Shen, B.: Highwall mining stability. In: Taishan Academic Forum – Project on Mine Disaster Prevention and Control, pp. 25–37. Atlantis Press (2014)

Zipf, R.K.: Failure mechanics of multiple seam mining interactions (2005)

Numerical Modeling of Split-Set Friction Bolt

Sai Rohith Samala[1], John Loui Porathur[2](✉), Ayan Giri[3], and Vinod Kumar Jagapthal[2]

[1] Western Coal Fields Ltd., Pench Area, Chindwara, Madhya Pradesh 480441, India
[2] CSIR-Central Institute of Mining and Fuel Research, Nagpur Research Centre, 17/C, Telankhedi, Civil Lines, Nagpur, Maharashtra 440001, India
johnlouip@gmail.com
[3] Hindustan Zinc Limited, Zawar Mines, Udaipur, Rajasthan 313901, India

Abstract. Other than grouted type of rock bolts, such as resin or cement encapsulated bolts, friction type of bolts are also getting popular in mining and other underground structures. A split-set friction bolt is one such rock reinforcement. They are cheaper and quicker to install as compared to the grouted type bolts and can provide reinforcement to the rock mass immediately after installation. They are used mostly as secondary rock mass reinforcement, such as for pinning a wire mesh prior to resin encapsulated bolting. However, in very weak rock mass and re-developed roadways through cemented backfills, grouted bolts are difficult to install and therefore split-sets offer a better primary ground support. Basically, a split-set friction bolt is a long hollow 'C' shaped cylinder which is inserted into an undersized hole. The frictional force generated due to the undersized hole create a pressure radially due to which the bolt fits into the rock. In this study a numerical modelling procedure is developed to simulate the mechanism of the split-set. The model is tested by simulating its ultimate pull-out load for different lengths and diameter of the split-set, which is indirectly a measure of its load bearing capacity. In this study pull-out load of different lengths and diameters is investigated using elasto-plastic numerical analysis in a finite difference software viz. FLAC3D. The trends obtained from numerical modelling are validated using field tests.

Keywords: Split-set friction bolt · numerical modeling · pull out test · load bearing capacity

1 Introduction

Ground control and roof protection in underground mines are most necessary for safe operation. Good roof support provides safe ground control. Reinforcing of underground extraction is done a very long ago by various roof bolts under different conditions. A split-set bolt is one of its kind. Split-set rock bolt, which is known as friction rock bolt, was invented by James J. Scott in 1973, who conceived and patented it and was eventually marketed as the 'Split-set'. Friction bolts have been used for over 30 years, but little is known about their actual mechanical behaviour when installed in a rock mass.

A split-set friction rock bolt is a long hollow cylindrical 'C'-shaped rod of different diameters inserted into the lesser diameter hole using a tapered end. Inserting it into the

A. Sinha et al. (Eds.): AMC 2023, SPEES, pp. 254–263, 2023.
https://doi.org/10.1007/978-3-031-46966-4_20

undersized hole forces the 'C' shape to close and develop a radial pressure around the hole in the rock (Davis, 1979, Brady and Brown 2006 Komurlu & Kesimal, 2015). The bolt stays due to the frictional contact between the bolt shank and the rock so it is termed a friction bolt (Komurlu & Kesimal, 2017). The working principle of the friction bolt is explained by Scott (1980), Davis (1979). The amount of frictional contact between the rock and bolt depends on various factors such as the smoothness and consistency of the hole, length of the bolt, diameter of the hole and the bolt, physico-mechanical properties of the rock mass and the bolt material, thickness of the bolt section, etc.

The friction bolt is of different diameters (33 mm to 47 mm) and lengths (0.3 m to 3 m). The friction bolt became popular in many engineering applications especially in the mining industry because it bears load immediately after insertion into the rock, it is cheaper and very quick to install. It offers support resistance to the rock due to the friction at the interface and can prevent rock block movement across and along the bolt (Komurlu & Kesimal, 2015, Hassell & Villaescusa, 2005).

The load bearing capacity of friction bolt is a factor of normal stress at the interface and amount of maximum load required to break the frictional bond at interface between the sidewall and the bolt shank. The load bearing capacity of a split-set depends on the diameter, length, thickness, thickness of the slot, ratio of hole size to bolt size, insertion method and rock mass condition (Tomory et al., 1998). The friction bolts have a good degree of energy absorption capacity as a result of its load and displacement behaviour (Li, 2010, Li et al., 2014, Stacey, 2016, He et al., 2014).

The pull-out test is an effective tool to indirectly measure the load bearing capacity of a bolt. It is a general practice to conduct pull-out test in the field to test the efficacy of the bolting system. The pull-out test is conducted by applying the pull axial force at the collar of the bolt by holding it and increasing the load using hydraulic jack until it slips from the rock. The amount of load at which the bolt slips is said to be the pull-out load, which is a measure of the load bearing capacity of the bolt. Higher the frictional contact surface area at the bolt-rock interface greater will be the pull-out load.

2 Numerical Modelling Procedure

Numerical modelling is done to understand the basic mechanism involved in a split-set installation and a pull-out. Numerical modelling technique is done using a finite difference software viz. FLAC3D of ITASCA Consulting Groups Inc., Minnesota, USA. The elasto-plastic analysis of split-set bolt was earlier done by (Davis, 1979) which stated that the load bearing of a friction bolt is not only by stress distribution due to disturbed medium around the bolt rock interface but also by factor of friction, gravity and free loading condition on the bolt. In the current modelling, the ultimate pull-out load bearing capacity of bolt is evaluated for different bolt diameter of (36 mm, 42 mm and 46 mm) and bolt length (0.3 m, 0.5 m, 0.9 m, 1.2 m, 1.8 m and 2.4 m).

The input parameters used for the numerical modelling study includes in situ stress, rock mass properties and geometry. Numerical modelling is done using Mohr-Coulomb elasto-plastic analysis for rock as well as the bolt material. The pertinent properties of rock mass, rock bolt and the interface are stated in Table 1. The interface between the rock and bolt is considered to be frictional with no tension. The interface is modelled all along the length of the bolt.

Table 1. Material properties used for numerical modelling

Properties	Rock mass	Bolt	Interface
Young's modulus (GPa)	50	200	50
Poisson ratio	0.214	0.26	0.214
Density (kg/m^3)	2970	7850	–
Cohesion (MPa)	5.61	152	–
Friction angle (deg)	39.9	45	10
Tensile strength (MPa)	1.5	250	–

Table 2. Results of some pull-out tests

Bolt dia (m)	Bolt length (m)	Pull out load in field (t)	Pull-out load in model (t)
36	0.9	4	6.48
36	0.9	4.5	6.48
47	0.9	10	8.42
47	1.8	11.4	16.57
47	1.8	15	16.57
47	1.8	14	16.57

Several modelling trails were performed to represent the action of the split-set bolt closer to the actual field condition. Initially a decoupled model is prepared, with the bolt remaining outside the borehole as shown in Fig. 1. A plane of symmetry passing through the centre of the bolt and its slot is used to reduce the size of the model. The cross-sectional view of the grid used to model the rock mass and the bolt is given in Fig. 2(a) and (b) respectively. In the rock mass, the virgin model, loaded with the elastic properties of the rock mass and the in-situ stress is simulated with Mohr-Coulomb elasto-plastic material model. In the next step the borehole is made and the model is brought to equilibrium.

Friction bolt is simulated taking into consideration its actual working principle. In reality when the friction bolt is made to converge into an undersized hole, stresses develop within the bolt when it is driven into the hole. Radial stress acts at the rock-bolt interface, which results in the frictional bonding between the rock and the bolt. This scenario is simulated in the numerical model by shrinking the bolt from its original size to that of the hole size by applying external grid point velocity vectors when the bolt is outside the borehole and inducing the required grid point displacements as shown in Fig. 3. Once the bolt attains the hole size, it is inserted into the hole and allowed to integrate with the rock mass. The coupled model is then allowed to equilibrate; while doing so the

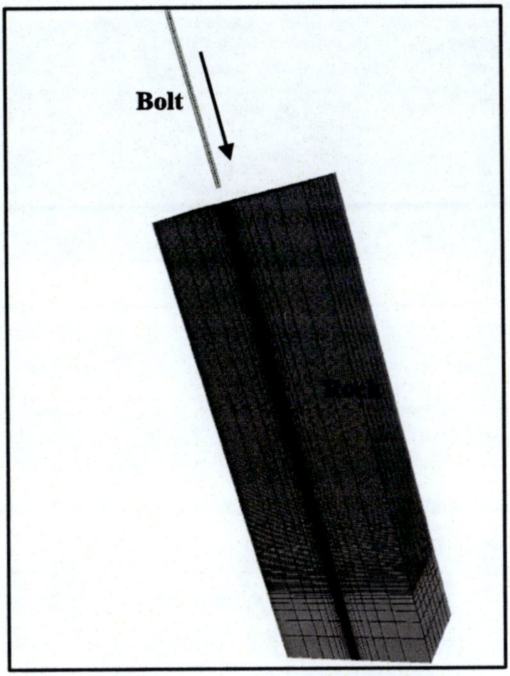

Fig. 1. Modelled rock and the bolt before coupling

stresses developed within the bolt is allowed to be transferred to the rock mass through the interface. The integrated model is shown in Fig. 4. Thereafter the pull-out test is conducted by applying the pull force as velocity vectors on to the bolt front end and the total reaction force (pull-out force) and the bolt displacement are monitored until the bolt slips from the rock.

The numerical modeling revealed, that in all the pull-out tests conducted, the load was found to increase gradually, but later started to flatten when the interface between the rock and bolt started slipping. When the full length of the interface slipped the load curve reaches its maximum and there after fully flattens. A typical loading curve during the pull-out test is shown in Fig. 5.

3 Effect of Bolt Length and Bolt Diameter

Using the numerical modelling procedure explained above a parametric study was conducted to see the effect of bolt length and bolt diameter on ultimate pull-out load bearing capacity of the friction bolt. Figure 6 shows the graph of pull-out load versus bolt length for various bolt diameters. As anticipated the load bearing capacity of the bolt increased with the increase in length of the bolt for all the three diameters of the bolt simulated.

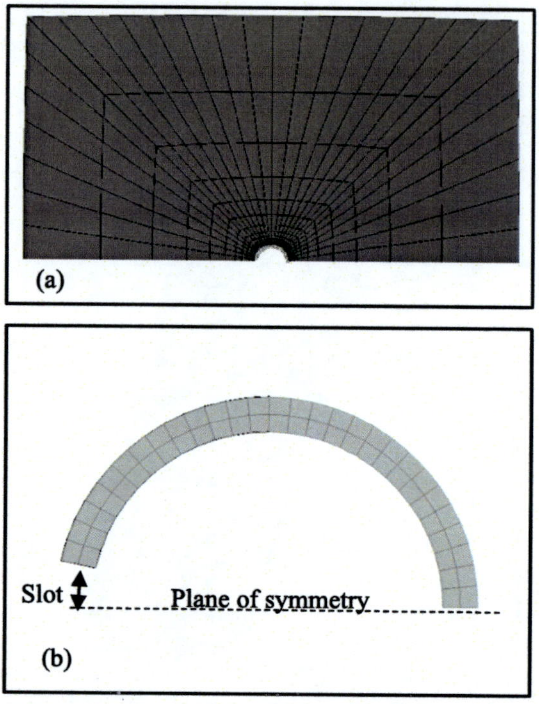

Fig. 2. Cross-sectional view of grid used for modelling (a) rock mass (b) bolt

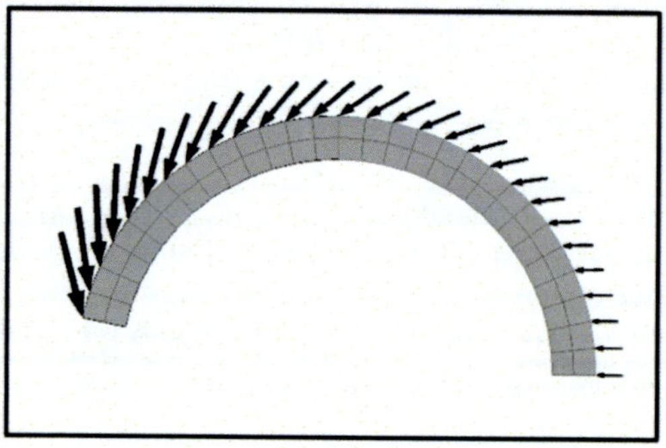

Fig. 3. Compression of bolt by applying angular force.

The effect of bolt diameter was studied using the modelling methodology. Figure 7 shows the pull-out load versus bolt diameter graph. The graph indicates slightly non-linear increasing trend for the pull-out load with respect to the bolt diameter. Similar trend was obtained for all the bolt lengths tested using the numerical model.

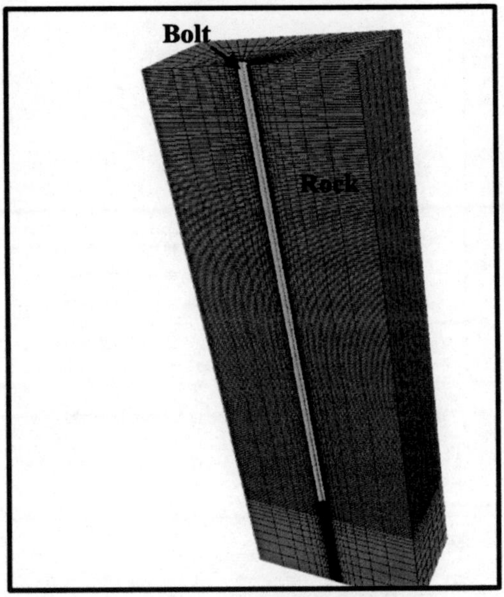

Fig. 4. The coupled model after bolt is inserted into the borehole

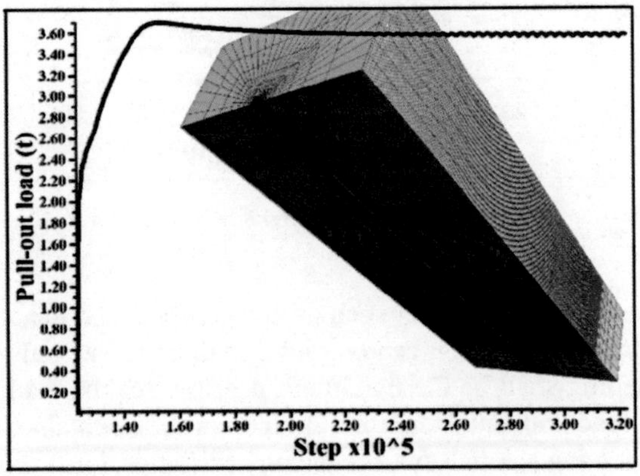

Fig. 5. A typical pull-out test result

4 Field Validation

Some field measurements were conducted to validate the numerical models. The numerical modelling assumes an ideal condition, which includes uniform and consistent bolt diameter, bolt surface, borehole diameter, bolt annular thickness, etc. However practically such ideal conditions may not exist in the field. Drill string undergoes some wobble that can result in higher diameter at the collar. This will result in non-uniform contact and frictional force developed between the borehole and the split-set. The bolt may

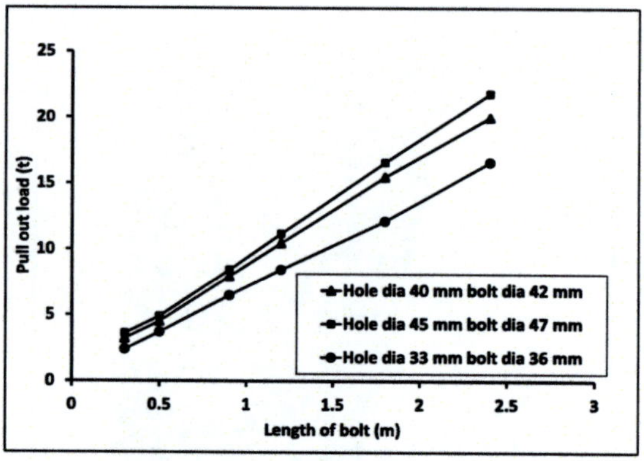

Fig. 6. Pull-out load versus bolt length

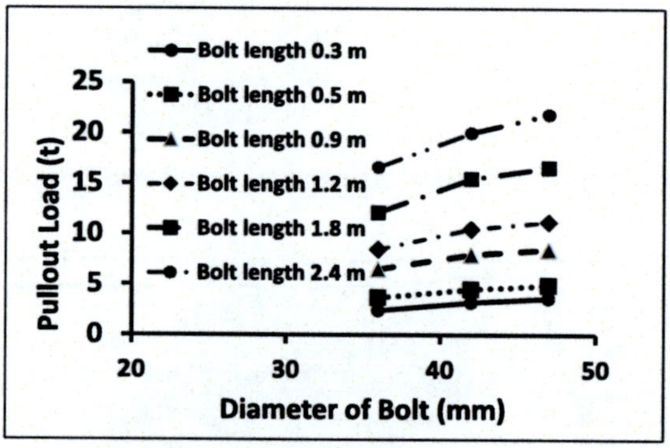

Fig. 7. Pull-out load versus bolt diameter

also have some manufacturing defect such as non-uniform surface and uneven diameter and annular thickness. Hence it can be said that the ultimate pull-out load obtained in numerical modelling shall be the maximum possible for that particular bolt. In the field, the pull-out loads obtained shall be lower than those obtained in the numerical model. A photo showing the pull-out test being conducted for a split-set is shown in Fig. 8 and a partially pulled-out split-set is shown in Fig. 9. The results of some pull-out tests conducted in a Lead-Zinc underground mine site is given in Table 2. A correlation between the pull-out load obtained for the field tests with respect to the numerical modelling results is presented in Fig. 10. The figure indicates, that on an average, about 82% efficacy is obtained in the field installation of the split-sets as compared to the idealistic values obtained from numerical modelling, which may be considered as very good split-set installation quality.

Fig. 8. Pull-out test being conducted in a mine

Fig. 9. A split-set after pull-out test

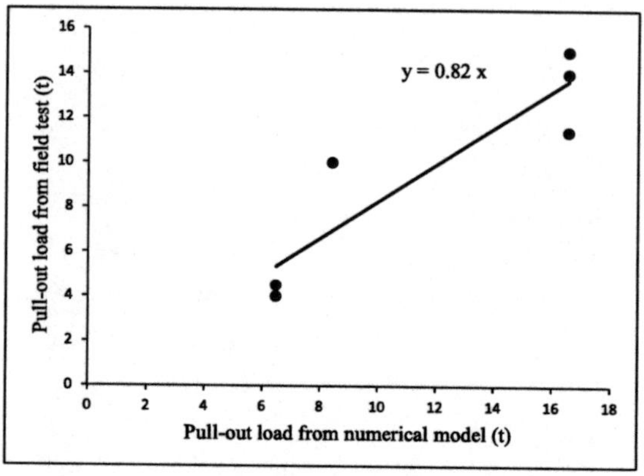

Fig. 10. Comparison between numerical modelling and field pull-out tests

5 Conclusion

In this paper the working principle of a friction type split-set bolt has been successfully simulated using elasto-plastic finite difference numerical modelling. The modelling methodology consisted of initially simulating bolt and rock mass separately, and later on coupling them together to transfer stresses to each other. This method has been found to successfully simulate the split-set installation and pull-out mechanism successfully. The developed modelling methodology has been used to find out the pull-out load for various lengths and diameter of the split-set. The results showed that the pull-out load increased linearly with increase in length and curvilinearly with increase in diameter of the bolt. The results have been compared with some pull-out tests conducted in the field, which indicated about 82% efficacy of the split-set bolting system in the field as compared to the ideal condition.

References

Brady, B.H., Brown, E.T.: Rock Mechanics: For Underground Mining. Springer Science & Business Media (2006)

Davis, R.L.: Split-set rock bolt analysis. Int. J. Rock Mecha. Mining Sci. Geomech. Abstr. **16**(1), 1–10 (1979)

Hassell, R., Villaescusa, E.: In-situ corrosion assessment of galvanised friction bolts using overcoring techniques. Morgantown, WV, s.n., pp. 349–356 (2005)

He, M., et al.: Development of a novel energy-absorbing bolt with extraordinarily large elongation and constant resistance. Int. J. Rock Mech. Min. Sci. **67**, 29–42 (2014)

Komurlu, E., Kesimal, A.: Improved performance of rock bolts using sprayed polyurea coating. Rock Mech. Rock Eng. **48**, 2179–2182 (2015)

Komurlu, E., Kesimal, A.: Experimental study on usability of friction rockbolts with plastic bodies. Int. J. Geomech. **17**(9), 04017058 (2017)

Li, C.C.: Field observations of rock bolts in high stress rock masses. Rock Mech. Rock Eng. **43**(4), 491–496 (2010)

Li, C.C., Stjern, G., Myrvang, A.: A review on the performance of conventional and energy-absorbing rockbolts. J. Rock Mech. Geotech. Eng. **6**(4), 315–327 (2014)

Scott, J.J.: Split Set Friction Rock Stabilizers I, p. 136 (1980)
Stacey, T.R.: Addressing the consequences of dynamic rock failure in underground excavations. Rock Mech. Rock Eng. **49**, 4091–4101 (2016)
Tomory, P., Grabinsky, M., Curran, J., Carvalho, J.: Factors influencing the effectiveness of Split Set friction stabilizer bolts. CIM Bull. **91**(1018), 205–214 (1998)

Mine Waste Management

Utilization of Fly-Ash in the Backfilling of Void in Underground Mines

Vivek Kumar Kashi[1(✉)], Atma Ram Sahu[2], and Roshan Kumar[3]

[1] Department of Mining Engineering, Faculty of Science and Technology, The ICFAI University Jharkhand, Ranchi 835222, India
vivekkk.rs.min15@itbhu.ac.in, vivekkumar.k@iujharkhand.edu.in

[2] Department of Mining Engineering, Aditya College of Technology & Science, Satna 485001, Madhya Pradesh, India
atmars.rs.min15@itbhu.ac.in

[3] Department of Mining Engineering, BIT Sindri, Dhanbad, Jharkhand 828123, India
roshan.min@bitsindri.ac.in

Abstract. Coal-fired thermal power plants worldwide are encountering significant challenges in handling and disposing of the ash generated during the combustion process. The problem is further exacerbated by the high ash content (30–52%) present in Indian coal. Thermal power plants face major concerns regarding the safe disposal of coal ash and the need for large storage areas. Consequently, efforts have been made to find alternative uses for the ash, rather than simply dumping it in nearby ash ponds. One promising approach involves utilizing underground mine voids to accommodate the bulk of the fly ash. Underground mine spaces have the greatest potential for storing fly ash in large quantities compared to other possible applications. The process involves using filling materials to reinforce the surrounding rock and provide support to the overlying strata, thereby reducing surface subsidence. This paper presents a study conducted in this context, focusing on the bulk utilization of fly ash in backfilling applications. The research aimed to find viable solutions for effectively utilizing fly ash, thereby addressing the challenges faced by coal-based thermal power plants in managing and disposing of this by-product. This innovative approach presents an encouraging solution for sustainable mining practices and minimizing the environmental impact associated with coal extraction.

Keywords: Fly ash · Underground Mines · Mine Void · Subsidence and Backfilling

1 Introduction

Growth of the any nation depends on the availability of energy resources. Country like India, where 65% of population are below 35 age group, automatically the necessity of energy is too high to meet the aspiration of young generation. Due to this reason, advancement and introduction of emerging technology in energy sector still not meeting the presence need of the country. Therefore, 70% of energy requirement is achieved by

A. Sinha et al. (Eds.): AMC 2023, SPEES, pp. 267–275, 2023.
https://doi.org/10.1007/978-3-031-46966-4_21

coal based thermal power plant. It is seen that all over the country, coal have 30–52% ash content (Dutta et al., 2009). This tremendous amount of ash produce from the thermal power plant are meeting severe problems of handling, management and disposal. Safe dumping of the coal ash and the large storage area required are major concerns of thermal power plants. Hence current requirement of coal based thermal power plat is to identify the alternative to disposed fly ask in safe, economical without occupied in large space.

Majority of the coal and other valuable minerals deposit present deep inside the mother earth. However, most of the coal mines extracted the coal through opencast mining method rather than underground mining method in India due to major safety concerns (Kashi et al., 2023). Currently in 2021–22, 95% of production is through opencast mining, whereas, previously 97% was done through it in 2020–21 (Kashi et al., 2020). Also, most of the coking coal are present too deep inside the earth i.e. below 500 m depth. Still, India is the 3rd largest producer of coal, but although 80% of coking coal is imported from other countries. Innovation and new technology will completely changing the scenario of current mining methods and empowering the mining industry to extract minerals through underground operation. Day-by-day huge mechanisation and application of portable safety related machines are enhancing the working condition of underground mines. In underground, Lots of the area is left empty and void due to extraction of coal and minerals from there. This will create subsidence problem and not only affect the upper surface of ground, but also destabilise the working roof strata. Those reduce production by wasting of coal left to control strata and enhance overall cost of production (Goswami and Mahanta, 2007).

To overcome from the above such situations, backfilling of the goaf made through excavation of coal from mined area is mandatory. Owing to technical and environmental reasons, backfilling is practiced in many countries such as China, Poland, USA etc. (Sivapullaiah et al., 1995). The utilization of filling materials serves the purpose of reinforcing the caving rock and providing support to the overlying strata, thereby mitigating the rate of surface subsidence. In mining operations where appropriate support measures are not implemented, the broken roof can experience failure and collapse. This poses a challenge for conducting continuous backfill mining in such conditions using conventional mining techniques. This voids are backfilled with materials such as sand, waste from coal preparation plant and waste rocks (Fig. 1).

A review of existing literature indicates that hydraulic sand stowing is a widely employed technique in underground coal mines across the globe to mitigate surface subsidence. However, the mining industry in India is currently facing a significant scarcity of river bed sand due to its increasing demand in civil engineering projects. Consequently, numerous underground coal mines remain unfilled after coal extraction, resulting in surface subsidence issues.

To address this challenge, the mining industry has been actively seeking an alternative filling material that is readily available near the mine site and cost-effective. Fly ash has emerged as a potential solution to meet these requirements. Fly ash is a by-product of coal combustion in thermal power plants and occupies considerable land space in the form of ash ponds near power plants. These ash ponds contribute to various environmental issues and ecological imbalances. By exploring the use of fly ash as a filling material, the mining industry aims to minimize surface subsidence in underground coal mines while

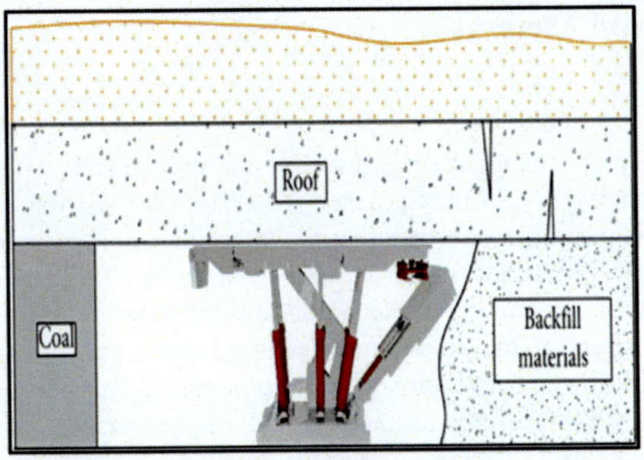

Fig. 1. Visualization of backfilling of underground longwall mine void area

also addressing the environmental concerns associated with the accumulation of fly ash in ash ponds. This approach offers the potential to repurpose fly ash, transforming it from a waste product into a valuable resource for mine filling applications.

The effective and comprehensive utilization of the vast quantities of idle fly ash has become a substantial task for coal based plant authorities. Fly ash possesses various geotechnical applications, including its use in bricks, cement, embankment construction, aggregates, fill material (Banita & Mishra, 2011). Several fruitful efforts have been made to utilize fly ash both solo and in combination with cement, lime or gypsum (Mishra & Rao, 2006). It has been recognized that the bulk utilization of fly ash in mine filling applications holds significant potential. These studies have shown that the utility of fly ash can be upgraded by adding lime.

However, the aforementioned application areas have limitations when it comes to cooperative fly ash in large quantities. Filling of mine void presents a crucial and emerging zone for heavy deployment, provided that the features of fly ash can be appropriately improved. To ensure active consumption in this context, the material should be chemically non-reactive and free from burnable substance that could pose a risk of spontaneous heating and fire in coal mines. Additionally, particle size and structure, density, porosity, water content, pH, and other factors play crucial roles in the hydraulic transportation of fly ash to mine void filling areas.

Conducting studies in this field would be immensely helpful in making informed decisions regarding the selection of materials for filling underground mine cavities and developing methodologies to address finer details. Furthermore, these investigations could provide insights and supplement similar applications in various other areas.

2 Materials and Method

2.1 Fly Ash

Fly ash is typically the resultant by-product of the combustion of pulverized coal in thermal power plants. It is commonly described as a finely-distributed amorphous aluminosilicate with varying amounts of calcium. The residue, or fly ash, originates from the burning region of the boiler, where it is carried away by the flue gases and subsequently collected using either mechanical or electrostatic separators.

Fly ash exhibits pozzolanic properties, meaning it can react with calcium hydroxide in the presence of water to form cementitious compounds. It is composed primarily of oxides such as SiO_2 (silicon dioxide), Al_2O_3 (aluminum oxide), Fe_2O_3 (iron oxide), CaO (calcium oxide), K_2O (potassium oxide), and TiO_2 (titanium dioxide) (Table 1) (Ahmaruzzaman, 2010). These components can vary in concentration depending on the specific composition of the coal and the combustion process (Table 2).

Different classifications of fly ash have been suggested based on various factors such as constituent materials, characteristics, application, and chemical properties. One commonly used classification is based on the ASTM Specification (ASTM C 618, 1995), which categorizes fly ash primarily into two types: F-Type and C-Type fly ash.

The F-Type fly ash designation refers to fly ash with pozzolanic properties, meaning it can react with calcium hydroxide in the presence of water to form cementitious compounds. This type of fly ash is typically derived from burning bituminous or sub-bituminous coals. On the other hand, C-Type fly ash encompasses both pozzolanic and cementitious fly ashes (Naik & Mishra, 2018). This type of fly ash generally originates from burning lignite or sub-bituminous coals and exhibits cementitious properties, which allows it to contribute to the strength and durability of concrete when used as a supplementary cementitious material.

Table 1. Chemical Characteristics of Fly Ash

Constituents	Percentage (%)	Constituents	Percentage (%)
Carbon	2.1	P_2O_5	0.17
Volatile Matter	0.147	SO_3	0.24
Fe_2O_3	8.83	K_2O	0.79
MgO	0.84	CaO	1.11
Al2O_3	27.73	Na_2	0.14
SiO_2	58.9	TiO_2	2.09

Table 2. Physical Characteristics of Fly Ash

Parameters	Value
Colour	Light Gary
Dry Density (kg/m^3)	1380
Specific Gravity	2.54
Optimum Moisture Content (%)	38.7
Permeability (m/sec)	(3.5–3.7) * 10^{-6}
Liquid Limit (%)	40.89
Plastic Limit (%)	Non-Plastic

2.2 Lime

Calcium oxide, also known as quicklime or burnt lime, is a highly utilized chemical compound with various applications. It appears as a white, caustic, alkaline, and crystalline solid at room temperature. Quicklime and its derived materials continue to be widely employed in construction and engineering due to their valuable properties. These materials include limestone products, cement, concrete, and mortar.

The primary source materials for quicklime production are rocks and minerals such as limestone or chalk, which primarily consist of calcium carbonate (Table 3) (Mishra & Das, 2010). Lime serves as a substance that hardens and binds other materials together, making it essential in construction processes. One of its primary applications is as a binder for floors and plaster coatings on walls. The versatile nature of quicklime makes it a valuable component in various industries, ranging from construction to chemical manufacturing.

Normal mine water is used for this purpose.

Table 3. Chemical Composition of Lime

Constituents	Percentage (%)
Minimum Array (Acidimetric)	95.0
Maximum Limits of Impurities	
Chloride (Cl)	0.1
Sulphate (SO_4)	0.5
Iron (Fe)	0.1
Lead (Pb)	0.02

2.3 Methods of Mine Backfilling

The backfilling of underground mine voids can be accomplished using a slurry made of fly ash and lime. Two methods commonly employed involve the use of fly-ash slurry. The slurry is created by mixing fly ash with mine water. The flow chart illustrating the process of the slurry-based backfilling method is detailed in Fig. 2.

In the hydraulic stowing method, which is widely practiced for void filling, the fly ash and lime mixture is mixed with water at the surface (Naik & Mishra, 2018). This mixture is then allowed to flow under gravity or with the aid of pumps into the underground void to be filled. It is important to note that the fine particles of fly ash possess pozzolanic properties, which can potentially clog the pores of the soil or aquifers, reducing their permeability. Additionally, any trace elements present in the leachate from the voids are likely to undergo chemical adsorption reactions with clay materials present in the soil, which can restrict their transport.

Any excess water present in the slurry will drain out to the surface, and this water can be reused in the production of the fly ash slurry. After the slurry is deposited, it begins to settle, and within approximately one day, it solidifies and becomes integrated with the surrounding strata. This solidified slurry provides effective roof support and enhances production within the mine.

Overall, the use of fly ash slurry as a backfilling method offers advantages such as improved structural support, reduced permeability, and the potential for reusing water resources. It provides a practical and efficient solution for filling underground mine voids.

3 Results and Discussion

3.1 Filling of Mine Void with Fly Ash Lime Slurry Through Pump

The process of backfilling underground mine voids involves three key steps: stirring, pumping, and filling. During the stirring phase, water and fly ash are mixed thoroughly in a specific proportion, typically 8 parts water to 10 parts fly ash by mass. This mixing takes place in a designated mixer, ensuring the components are well blended to form a homogeneous fly-ash slurry.

Once the slurry is prepared, the pumping stage begins. The fly-ash slurry is directed to the hopper of a concrete pump, which serves as the transportation mechanism. The slurry is then pumped from the hopper through a filling pipe, facilitating its transfer to the working face located underground. This process utilizes the concrete pump's pressure to ensure the slurry reaches the desired location efficiently.

Finally, the filling stage involves delivering the slurry to the goaf, which refers to the underground void or space to be filled. This is achieved through the use of both the filling pipe and branch filling pipes associated with the working face. The filling pipe is laid along the tendency or trend of the working face, and its length typically matches the extent of the working face itself. The branch filling pipes, on the other hand, are laid parallel to the trend of the working face, with their lengths varying between 10 to 20 m. The slurry is carried through these pipes, ultimately reaching and filling the targeted goaf area.

By following these steps, stirring the fly ash with water, pumping the resulting slurry using a concrete pump and filling pipe, and finally delivering the slurry to the goaf through filling and branch filling pipes – the backfilling process can be carried out effectively, ensuring proper utilization of the fly ash and achieving the desired filling objectives within the underground mine voids.

The stirring subsystem, as depicted in Fig. 2, consists of three main components: the fly-ash bin, screw feeder, and mixer (Kumar, 2010). The fly-ash bin serves as a storage unit for stockpiling fly ash. From the fly-ash bin, the fly ash is transported to the mixer using a screw feeder. Inside the mixer, the fly ash is thoroughly mixed with mine water, resulting in the formation of a fly-ash slurry.

In summary, this system facilitates the mixing of fly ash with mine water to create a fly-ash slurry, which is then conveyed to a distant filling operational side using a pump and a network of filling pipes, including branch filling pipes. This allows for efficient and controlled deposition of the fly-ash slurry at the designated working face.

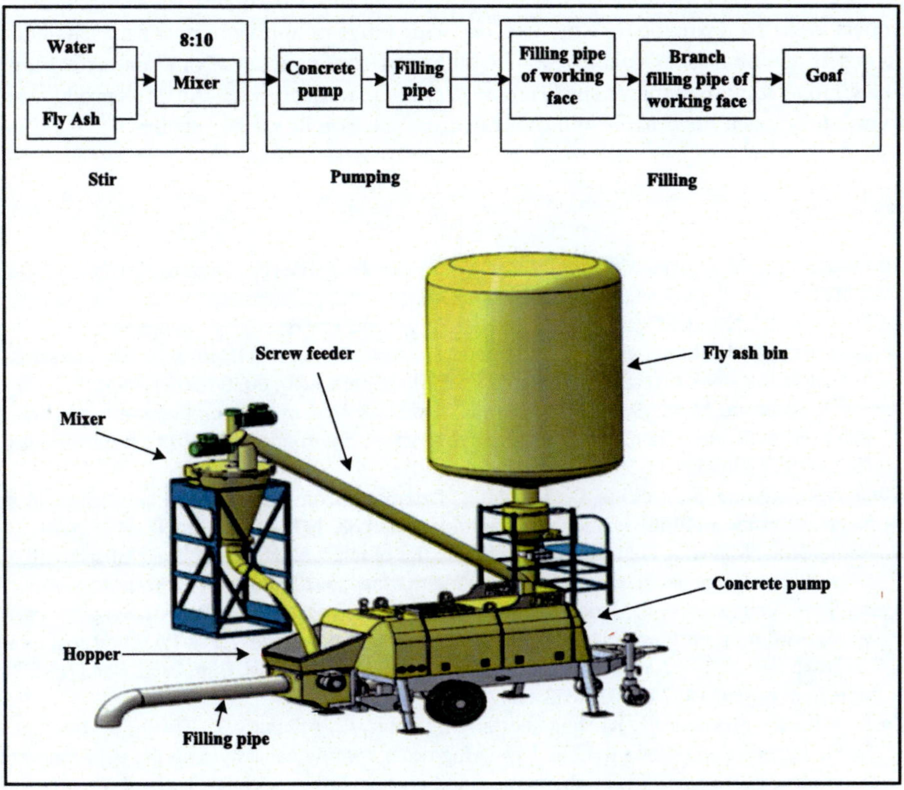

Fig. 2. Flow chart of the process of backfilling in a underground mine's void

4 Conclusions

After analyzing and researching the merits and drawbacks of existing filling materials, this paper introduces a novel filling material called fly-ash lime water slurry. The primary objective of employing this material is to exercise control over surface subsidence in mining operations while also facilitating the complete extraction of coal from barrier pillars.

The study presents the results obtained from a practical field application of the proposed technique. The approach involves the transportation of fly-ash slurry through a pipeline and subsequent grouting for backfilling purposes. The results demonstrate that this method effectively controls the movement of the overlying strata, leading to a significant reduction in surface subsidence.

The utilization of fly-ash lime water slurry as a filling material offers several advantages. It not only addresses the issue of surface subsidence but also enables the extraction of coal from previously inaccessible barrier pillars. This innovative approach presents a promising solution for sustainable mining practices and minimizing the environmental impact associated with coal extraction. The wide range of applications for fly ash underscores its potential as a valuable resource with numerous benefits. From mine void filling to civil road & embankments construction, cement manufacturing, and brick production, fly ash presents a sustainable and cost-effective solution in various sectors.

References

Ahmaruzzaman, M.: A review on the utilization of fly ash. Prog. Energy Combust. Sci. **36**, 327–363 (2010)

ASTM C 618 94. Standard specification for coal Fly ash and raw or calcined natural pozzolana for use as a mineral admixture in Portland cement concrete. In: Annual Book of ASTM Standards, American Society for Testing and Materials, Philadelphia, 4(8), 00304-306 (1995)

Banita, B., Mishra, M.K.: Strength behaviour of surface coal mine overburden–fly ash mixes stabilised with quick lime. Int. J. Mining Reclamation Environ., 1–17 (2011). https://doi.org/10.1080/17480930.2011.552285

Dutta, B.K., Khanra, S., Mallick, D.: Leaching of elements from coal fly ash – assessment of its potential for use in filling abandoned coal mines. Fuel **88**, 1314–1323 (2009)

Goswami, R.K., Mahanta, C.: Leaching characteristics of residual lateritic soils stabilised with fly ash and lime for geotechnical applications. Waste Manage. **27**, 466–481 (2007)

Kashi, V.K., Karmakar, N.C., Krishnamoorthi, S., Adhikary, P., Sahu, A.R.: A laboratory study on application of synthesised amylopectin-grafted-polyacrylamide (AP-g-PAM) on coal mine haul road dust emission at different atmospheric temperatures. J. Inst. Eng. India: Ser. D (2023). https://doi.org/10.1007/s40033-023-00473-y

Kashi, V.K., Karmakar, N.C., Krishnamoorthi, S., Sonker, E., Adhikary, P., Tiwari, R.: Reducing the dust generation of haul road by improving water holding capacity with the application of synthesised polyacrylamide at laboratory condition. Nat. Environ. Pollut. Technol. **19**, 413–419 (2020)

Kumar, V.: A comprehensive model for fly ash handling and transportation for mining sector. In: Fly Ash: An Opportunity for Mining Sector, New Delhi, India (2010)

Mishra, D., Das, S.K.: A study of physico-chemical and mineralogical properties of Talcher coal fly ash for stowing in underground coal mines. Mater. Charact. **61**, 1252–1259 (2010)

Mishra, M.K., Rao, U.M.K.: Geotechnical characterisation of fly ash composites for backfilling mine voids. Geotech. Geol. Eng. **24**(6), 1749–1765 (2006)

Naik, H., Mishra, M.K.: Characterization of Indian fly ashes for mine stowing purposes. In: Proceedings of the Eighth International Conference on Advances in Applied Science and Environmental Technology – ASET, pp. 27–36 (2018)

Sivapullaiah, P.V., Prasanth, J.P., Sreedhara, A.: A optimization of lime content for fly ash. J. Test. Eval. **23**(2), 222–227 (1995)

Underground Mine Backfilling Transition in Sukinda, Odisha: Present and Future

Santosh Kumar Behera(✉), Prashant Singh, Kanhaiya Mishra, Phanil Kumar Mandal, Sujit Kumar Mandal, and Ritesh Kumar

CSIR-Central Institute of Mining and Fuel Research, Dhanbad 826015, Jharkhand, India
skbcimfr@yahoo.com

Abstract. Mine back filling is an integral part of overall mining operation. This paper highlights the progress of underground mine back filling practices in Sukinda region, Odisha. Initially underground stope backfilling started with sand based cemented hydraulic backfilling. Slowly, the transition to other backfill material has also being done. Mine overburden, bottom ash and slag are few of the options of alternate backfill materials that can be used for underground stope filling. This study elaborates the results of strength development and water drainage characteristics of cemented hydraulic backfill prepared with sand /bottom ash. This study also highlights a future roadmap towards application of modern day paste backfill technology in underground mines of Sukinda region, Odisha.

Keywords: Underground mining · cemented backfill · sand · Overburden · bottom ash · slag

1 Introduction

Mining is one of the core sectors, which plays a crucial role in the economic development of any nation (Mishra et al. 2022). It, directly and indirectly, contributes to the growth and economic progress of the country. In India, mining also significantly contributes to the overall development of the nation. Most of the metallic mineral deposits in India are distributed in the states of Jharkhand, Odisha, Chhattisgarh, Madhya Pradesh, Andhra Pradesh, Telangana, Madhya Pradesh, Goa, and Rajasthan (Behera et al. 2019b).

Sukinda is a town in Jaipur district, Odisha, India. Odisha accounts for about 98% of the total proved chromite (chromium ore) reserves of the country, of which about 97% occur in the Sukinda Valley. In September 2007. Sukinda is flanked by the Mahagiri range and the Daitari range, Sukinda Valley spreads over an area of 50 km^2 from Kansa to Maruabil in Jaipur District. Sukinda has an abundance of chromite, around 97% of India's total deposits. The valley is abundant in the deposits of chromite and has the largest open cast chromite mines in the world (Naz et al. 2016). Around 11 mines operate in the area.

Out of these mines only one operational underground chromite mine in Sukinda is Mahagiri Mines of M/s Indian Metals and Ferro Alloys Ltd. is situated in the Sukinda

A. Sinha et al. (Eds.): AMC 2023, SPEES, pp. 276–291, 2023.
https://doi.org/10.1007/978-3-031-46966-4_22

Valley of Odisha. Opencast mining has already been exhausted and currently, underground mining is in progress. The ultimate pit limit of opencast mine is at + 185 mRL. Underground mining has been developed below the ultimate pit bottom to extract the ore after leaving a safe cap rock of about 50 m vertical thickness.

The extraction of minerals is done either by surface mining or underground mining methods. Shallow depth mineral deposits are extracted by surface mining methods, whereas the deep-seated mineral deposits are extracted by underground mining methods. At present most of the open cast mines in Sukinda have reached their ultimate pit limit. Exploration data reveals plenty of chromite deposit at higher depths below the open cast mines. It is essential to extract chromite through underground mining method. There are several methods of underground mining being practiced globally. The open stoping mining method is generally used in sublevel, long hole or blast hole stoping. Presently, most of the open stoping methods are used in conjunction with backfilling (Li 2014; Behera et al. 2019a; Singh et al. 2019).

Sublevel stoping is one of the simple mining/stoping methods, which is typically used for steep ore bodies, stable hang wall and foot wall, competent host rock and ore body, regular ore deposits (Himanshu et al. 2021; Behera et al. 2020a). The open stopes created after the extraction of ore in sublevel stoping are usually backfilled to allow maximum recovery from the stopes, provide stability to underground workings and prevent surface subsidence (Behera et al. 2023). Similarly, when the dip of a deposit is steep (greater than about 55°), ore and waste strong, ore boundaries regular, and the deposit relatively thick, Blasthole stoping is used. In this method, the miners create a vertical slot at one end of the stope and then work in the levels to drill a radial pattern of drill holes (Gonen and Kose 2011). From the drilling level, long, parallel blast holes are drilled, typically 100 to 150 mm in diameter. After charging a set of these blast holes with explosives, the blocks of ore body are blasted. After extraction of the stopes, the primary voids are backfilled with cemented backfill, and the secondary voids are backfilled with uncemented backfill.

Backfilling of the underground stopes serves the purpose of either support to the hang wall, ore recovery, as a working floor for the miner, artificial roof and as disposal of development waste rock and tailings (Villaescusa et al. 2019; Edraki et al. 2014; Grobler et al. 2019; Behera et al. 2020b; Behera et al. 2021). With progress of time backfilling in Sukinda have evolved from conventional cemented hydraulic filling using sand to present day cemented higher concentration hydraulic filling using coal ash. This article elaborates the transition of backfilling in underground mines in Sukinda, Odisha from sand based to coal ash. At the end of this article a note on future backfilling is presented.

2 Sand Based Backfill

Initially studies were conducted to investigate the feasibility of utilising Brahmani river sand in underground stope backfilling in an underground mine in Sukinda region, Odisha. Investigations were done for physical and chemical properties, water drainage characteristics and strength development of cemented sand based hydraulic backfill.

2.1 Physico-Chemical Properties

Physical Properties

The physical properties of backfill materials influence various parameters of backfill including, transportation, storage and disposal. Hence, physical properties of backfill materials were determined. The various physical properties of the sand sample are presented in Table 1.

Table 1. Physical properties of sand

Sr. no.	Parameter	Result
1.	Specific gravity	2.56
2.	Bulk density (g/cc)	1.60
3.	Porosity (%)	37.5
4.	Void ratio	0.6

The result of the particle size analysis of sand is presented in Fig. 1.

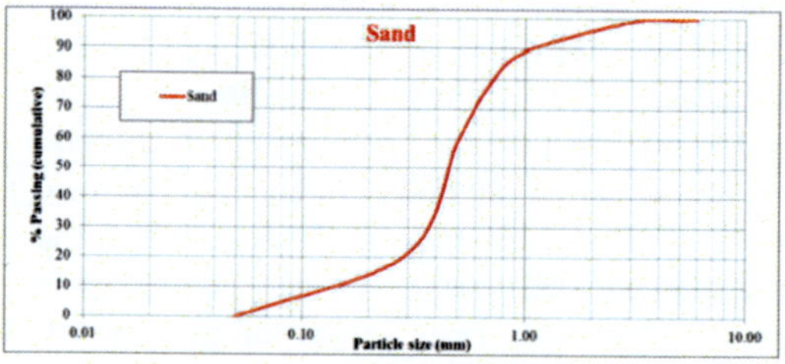

Fig. 1. Results of particle size analysis of sand.

The particle size of sand varies in the ranges of 53 μm to 5 mm. The mean particle diameter of the sand is (D_{50} = 450 μm). Further, the sand investigated in this study hardly consists of any particle below 53 μm.

Chemical Properties

The behaviour of backfill after pouring into the mine voids is greatly influenced by the chemical and elemental compositions of the backfill materials. Therefore, it is imperative to study the chemical and elemental characteristics of the sand.

The elemental composition of sand was determined by the energy dispersive X-ray spectroscopy (EDS) technique focusing at the selected points on the SEM micrograph. The mean values of the elements obtained at different points during the EDS analysis were considered as the elemental composition of sand. The results show that the sand is mostly consisting of Si, Al and Na. The EDS spectra of sand is shown in Fig. 2. The elemental composition of sand is presented in Table 2.

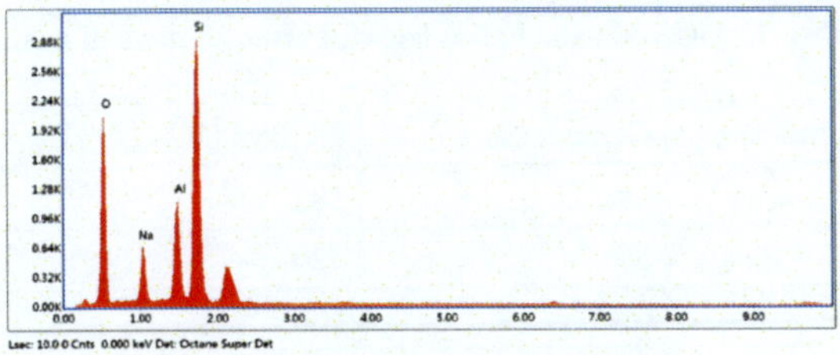

Fig. 2. EDS spectra of sand sample.

Table 2. SEM-EDS elemental composition of sand.

Element	O	Al	Si	Na
Sand	37.97	7.806	51.618	2.608

From the SEM image analysis (Fig. 3) it is observed that sand consist of irregular shaped, and sub rounded particles.

Fig. 3. SEM image of sand sample.

2.2 UCS of Backfill Specimens

In order to find the strength development of backfill using sand experiments were conducted using a hydraulic backfill mix with 5wt% (solid 55%). The back fill mix was prepared by mixing dry sand and cement for about 3 min. After dry mixing, the mixture was thoroughly mixed by adding water for about 15 min. Immediately after preparation of the backfill mix the mix was poured into cylindrical moulds of 100 mm diameter and 200 mm height maintaining a length to diameter ratio (L/D) of 2 for determination of uniaxial compressive strength (UCS) and kept for curing.

For determining the UCS, a constant loading rate of 2 mm/min [ASTM C39/C39M – 17b (2017)] was applied throughout the loading process. Majority of the samples followed hour glass type failure with few samples showing multiple fractures (Fig. 4). The results of the UCS test are presented in Table 3.

Table 3. UCS of sand based backfill after 28 days of curing.

Sr. no.	Cement (%)	28 days UCS (kPa)
1.	4%	385
2.	5%	520
3.	6%	560
4.	7%	785
5.	10%	1250

Fig. 4. Sand based backfill during UCS test.

2.3 Water Drainage Characteristics

A significant concern for any hydraulic backfill is the liquefaction potential of the backfill mass. Liquefaction occurs when the pore pressure increases dramatically, thus reducing the effect stress in the fill mass, to the point where the shear resistance of the fill mass is so low that the mass begins to flow like a liquid. If the fill mass liquefies, the backfill barricade pressure increases drastically and when the barricade pressure increases its designed strength, the backfill mass flows through the barricade. To prevent such events, hydraulic backfill is designed in such a manner that the water percolates through the barricade within a shorter period of time. Such drainage of water is achieved by percolation and decantation. Percolation of water occurs through the fill mass whereas decantation towers are used to dispose standing water over the fill mass. The presence of binder in backfill reduces the drainage behaviour. Generally a decrease in percolation rate is observed with increase in binder proportion in backfill mass.

Generally, a standard constant head permeameter is used for predicting the drainage behaviour of fill mass. A minimum of 10cm/hr percolation rate is required for hydraulic fill and is the only quality criterion ever broadly applied in determining hydraulic fill acceptability. The percolation results are well above the desired limit of 10 cm/hr (Table 4).

Sand based cemented hydraulic backfill was started for the first time in Sukinda, Odisha in the Mahagiri Mines of M/s IMFA Ltd in the year 2018. As of now more than 30 stopes have been extracted with cemented sand backfill.

Table 4. Percolation rate of sand based backfill.

Sr. no.	Recipe	Percolation rate (cm/hr)
1.	90% Sand + 10% cement	18.8
2.	93% Sand + 7% cement	20.6
3.	94% Sand + 6% cement	22.8
4.	95% Sand + 5% cement	24.2
5.	96% Sand + 4% cement	27.5

3 Coal Ash Based Backfill

Sand being a scarce material now days, there is always a need of alternate backfill material to sand. Coal ash generated by thermal power plant can be a good substitute for sand as a mine backfilling material. It is fine grained, possesses pozzolanic properties, and requires less water and energy for its transportation. Hence it is high time to utilise coal ash for underground backfilling. Bottom ash and fly ash samples were collected from IMFA power plant (PP) which is situated at Choudwar in District of Cuttack, Odisha State. The samples were collected from three different power plants with capacity 1 x 50 MW, 1 x 30 MW and 2 x 60 MW.

3.1 Physical Properties

The physical characteristics of ash depend on the quality/rank of coal used, degree of pulverization, furnace temperature, its chemical composition etc. Storage, transportation, re-handling and reclaiming of ash found to be greatly influenced by physical characteristics viz., specific gravity, bulk density, porosity, permeability, compressibility, particle size distribution etc. Hence, the physical properties of IMFA bottom ash and fly ash samples were investigated. The results are summarized in Table 5.

Table 5. Physical properties of bottom ash and fly ash

Sr. no.	Parameter	BA	FA
1.	Specific gravity	2.33	2.09
2.	Bulk density (g/cc)	1.32	0.96
3.	Porosity (%)	43.34	54.06
4.	Void ratio	0.76	1.17

*BA: Bottom ash, * FA: Fly ash

The specific gravity (SG) of fill material is determined in the laboratory using a water pycnometer as per IS: 2386 (part III) (1963). The porosity of backfill samples were determined using Mercury Intrusion Porosimeter [Quantachrome (Anton Paar), Model (PoreMaster 60)] (Panwar et al. 2022).

The particle size analysis of bottom ash and fly ash are demonstrated in Table 6. The particle size distribution curves are plotted and analysed (Fig. 5) to investigate the size distribution of these backfill materials. The particle size of bottom ash and fly ash vary in the ranges of 0.1–850 and 0.1–120 μm, respectively. From the particle size analysis, the specific surface area of the fly ash is found to be 412 m^2/kg. The mean particle diameter of the bottom ash (D_{50} = 228.45 μm) is found to be greater than the fly ash (D_{50} = 20.85μm). Hence, it is apparent that the bottom ash is coarser than the fly ash. The bottom ash used in this study consist of 3% fines (< 53 μm). This bottom ash may be a suitable material for hydraulic filling.

Table 6. Particle size analysis results of bottom ash and fly ash.

Parameters	Bottom ash	Fly ash
D(4,3), μm	–	62.26
D(3,2), μm	–	5.46
D_{90}, μm	924.84	79.74
D_{60}, μm	266.07	24.81
D_{50}, μm	228.45	20.85
D_{30}, μm	140.90	12.87
D_{10}, μm	105.61	4.90
Coefficient of uniformity, C_u	2.52	5.06
Coefficient of curvature, C_c	0.71	1.36
Specific surface area (m^2/kg)	–	412

*(C_u) = D_{60}/ D_{10}, *(C_c) = (D_{30} × D_{30}) / (D_{60} × D_{10}), *D (4,3) is volume moment mean, *D(3,2) is surface area moment mean

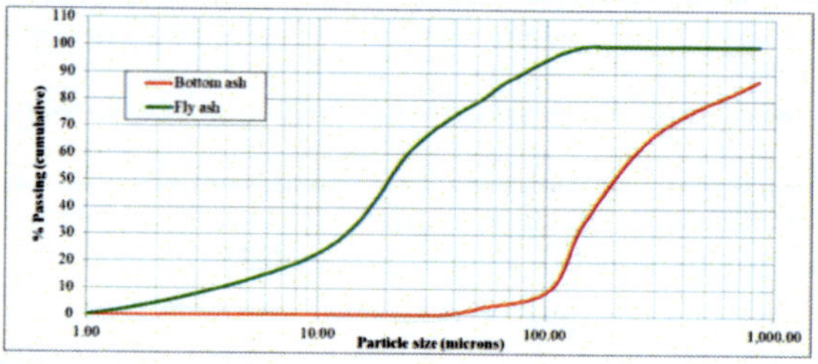

Fig. 5. Particle size distribution curves of bottom ash and fly ash samples.

3.2 Chemical Properties

The behaviour of backfill after pouring into the mine voids is greatly influenced by the chemical and elemental compositions of the backfill materials. Therefore, it is imperative to study the chemical and elemental characteristics of the bottom ash and fly ash.

The elemental compositions of the bottom ash and fly ash were determined as mentioned in Sect. 2.2. The results show that the bottom ash sample is mostly consisting of O, Si, Mg, Ca, Fe, S and Al and the fly ash is composed of O, Si, Mg, Ca, Fe, Ti and Al. The EDS spectra of the bottom ash and fly ash are shown in Fig. 6 and Fig. 7 respectively. The elemental composition of bottom ash and fly ash samples are presented in Table 7.

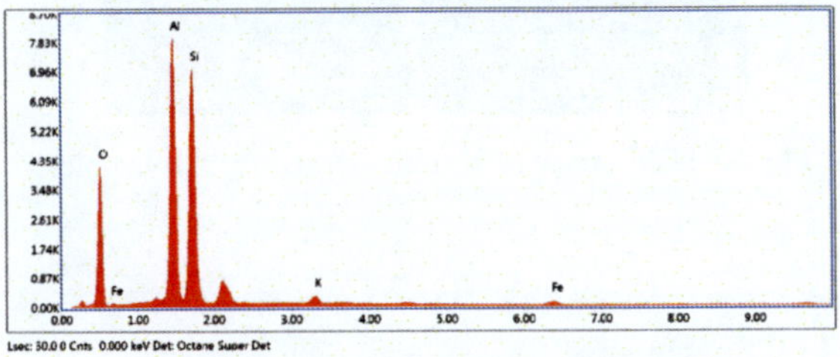

Fig. 6. EDS spectra of bottom ash sample.

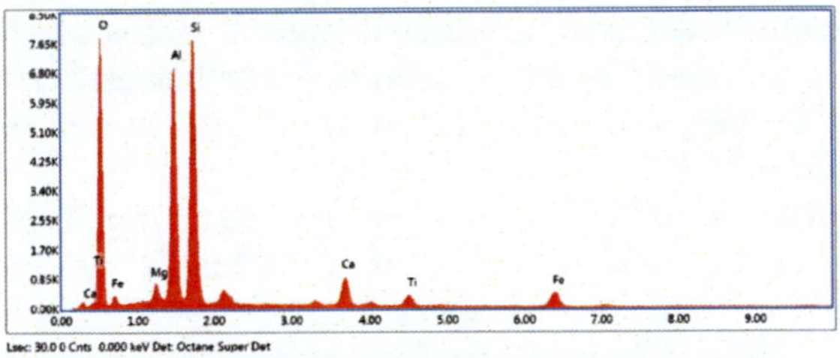

Fig. 7. EDS spectra of fly ash sample.

Table 7. SEM-EDS elemental composition of bottom ash and fly ash.

Element	Bottom ash	Fly ash
O	42.63	38.49
Al	16.43	18.11
Si	22.4	25.75
K	0.82	–
Ca	5.46	5.43
Ti	0.27	0.76
Fe	8.62	10.54
Mg	0.29	0.93
S	3.09	–

From the SEM image analysis (Fig. 8) it is observed that bottom ash consist of irregular shaped, rounded and sub rounded particles. Further, the SEM image of fly ash samples clearly show sub rounded and spherical shaped particles, which are called as cenospheres.

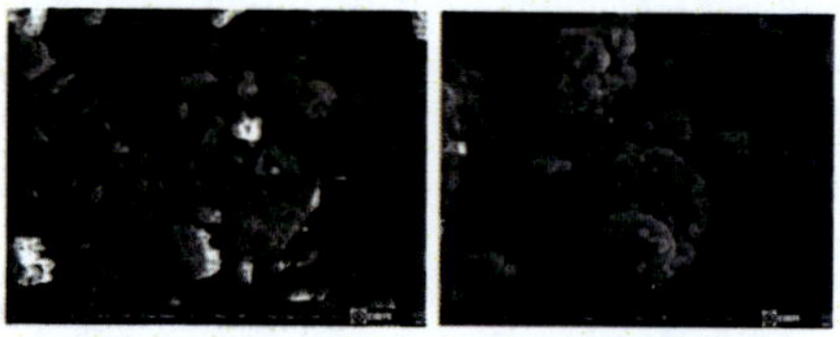

Fig. 8. SEM image of bottom ash and fly ash samples.

3.3 UCS of Backfill Specimens

The targeted UCS after 28 days of curing for bulk fill and plug are 520 kPa and 1160 kPa respectively. In order to find the strength development with bottom ash and fly ash experiments were conducted using a hydraulic backfill mix with 55wt% solid (solid 55%). The back fill mix was prepared by mixing dry bottom ash, OPC and fly ash for about 3 min. The samples were prepared and tested UCS as mentioned in Sect. 2.3. Majority of the samples followed multiple fractures during failure (Fig. 9). The results of the UCS test are presented in Table 8 and Fig. 10. The targeted UCS for plug fill was achieved by backfill mixes 89%BA + 11% cement, 90%BA + 10% cement, 90%BA + 9% cement + 1% FA, 90%BA + 8% cement + 2% FA, 90%BA + 7% cement + 3% FA for all the ash samples. The bulk filling can be done with backfill with mix recipe 93%BA + 7% cement or 94%BA + 6% cement or 95%BA + 5% cement or 94%BA + 5% cement + 1% FA or 94%BA + 4% cement + 2% FA.

Table 8. UCS test results.

Code	Mix recipe	7 days	14 days	28 days
MB1	89%BA + 11% cement	974	1087	1420
MB2	90%BA + 10% cement	828	988	1380
MB3	91%BA + 9% cement	750	872	1208
MB4	92%BA + 8% cement	612	730	1146
MB5	93%BA + 7% cement	408	580	908
MB6	94%BA + 6% cement	315	488	720
MB7	95%BA + 5% cement	220	386	562
MBF2a	90%BA + 9% cement + 1% FA	762	880	1372
MBF2b	90%BA + 8% cement + 2% FA	622	802	1356
MBF2c	90%BA + 7% cement + 3% FA	425	656	1210
MBF6a	94%BA + 5% cement + 1% FA	252	402	546
MBF6b	94%BA + 4% cement + 2% FA	186	380	522
MBF6c	94%BA + 3% cement + 3% FA	124	280	386

Fig. 9. Bottom ash based backfill under UCS test.

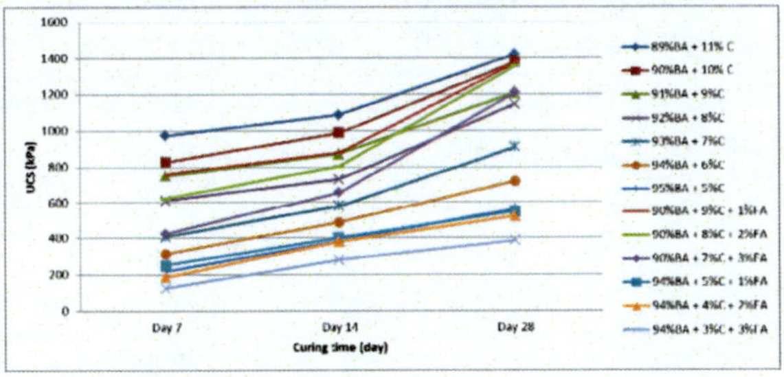

Fig. 10. UCS development of BA based backfill.

3.4 Water Drainage Characteristics

Predictions of the drainage behaviours for a given fill are made on the basis of laboratory tests using a standard constant head permeameter. A minimum of 10cm/hr percolation rate is required for hydraulic fill and is the only quality criterion ever broadly applied in determining hydraulic fill acceptability. A constant head permeameter, as shown in Fig. 11, was used for the percolation test.

Fig. 11. Constant head permeameter.

Table 9 reflects the results of percolation experiments. The percolation results are well above the desired limit of 10 cm/hr.

Table 9. Percolation rate of BA based backfill.

Code	Recipe	Percolation rate (cm/hr)
MB1	89%BA + 11% cement	17.2
MB2	90%BA + 10% cement	17.8
MB3	91%BA + 9% cement	17.8
MB4	92%BA + 8% cement	18.1
MB5	93%BA + 7% cement	18.6
MB6	94%BA + 6% cement	21.2
MB7	95%BA + 5% cement	22.8
MBF2a	90%BA + 9% cement + 1% FA	17.8
MBF2b	90%BA + 8% cement + 2% FA	17.2
MBF2c	90%BA + 7% cement + 3% FA	15.6
MBF6a	94%BA + 5% cement + 1% FA	21
MBF6b	94%BA + 4% cement + 2% FA	20.8
MBF6c	94%BA + 3% cement + 3% FA	20.2

Bottom ash based cemented hydraulic backfill was started in the year 2022. Few stopes have been extracted with Bottom ash based cemented hydraulic backfill.

4 Future Backfill Materials

Based on the exploration of potential backfill materials in Sukinda region, Odisha, slag, overburden and chromite tailings might be few such alternate backfill materials that might replace sand.

4.1 Slag

Approximately, six companies operates nearby of Sukinda region that manufacture charged chrome, Ferro chrome and steel. For instance, the charge chrome plant of M/s IMFA Ltd. Produces approximately 54000 tone/annum of Granulated slag and Jigged slag of approximately 24000 tons per annum. Granulated slag is produced during tapping from the molten material of Electric arc furnace. When molten slag is dropped in bath, the molten slag changes to nodules and granules of slag. During this process granular, glassy coarse and fine aggregate type material (Fig. 12). Such granular glassy material is termed as granulated slag. Disposal of these slags is challenging. One source for utilisation of granulated slag might be underground stope backfilling.

Fig. 12. Granulated slag produced by a charge chrome plant.

4.2 Chromite Tailings

Chromite tailings as shown in Fig. 13 is generated during the process of beneficiation of chromite ore. Though at the present scenario the generation of chromite tailings is on the lower side, in future with decreasing grade of chromite, the chromite ore has to undergo further processing. Hence, in the process of beneficiation tailings generation would increase. Surface disposal of such tailings need vast surface land and pollute air, water and soil. Hence, it's better to utilise these chromite tailings for stope backfilling.

Fig. 13. Chromite ore tailings.

4.3 Overburden

Due to existing open cast mines in Sukinda region, there are plenty of tall standing overburden (OB) dumps. The huge volume of these OB material available in the 11 mines of Sukinda region can be a major component of cemented backfill for future underground mines. These OB dumps are mainly consisting of weathered limonite, top soil and serpentinite material. Studies have resulted in poor strength development of these OB materials. Further, investigations are required to improve strength characteristics of backfill prepared using OB. Moreover, future backfill technologies need to optimised so that such huge quantity of OB can be utilised for underground backfilling.

5 Future Backfill Technology in Sukinda

Considering the changing scenario of underground hard rock mining in Sukinda region and demand of backfilling, there is always a search for better backfill material. The conventional cemented hydraulic slurry backfilling (CHSB) by adding a suitable binder is being practiced in underground mine of Sukinda region. The efficiency of this backfilling method depends on how freely numerous mining operations can take place without interfering with one another. However, the disadvantages of this method include the requirement of pump for rehandling additional water, chances of ponding, segregation of materials, accumulation of fines may impose hydrostatic pressure, increased hydrostatic pressure resulting in barricade failure, and the inrush of cemented slurry into the underground workings. Moreover, it may lead to ore dilution during the blasting and ore handling process.

High concentration filling (HCF) is an advanced version of the CHSB, which uses cemented hydraulic backfill with a solid content varying between 60 to 70 wt%. The high density of backfill minimises the water drainage. The target backfill slurry density is a function of solid wt% and particle specific gravity. It balances the requirements of transporting the slurry without plugging the pipeline and excess drainage of water. HCF is placed in underground either by gravity or using pumps depending on the site requirement. The excess ponded water inside the stopes is drained out through decantation pipes or suitable drainage system. Though the HCF is an improved version of CHSB, still the drawbacks of hydraulic fill persists up to some extent.

Hence an advanced version of backfilling technology is required to overcome all the demerits of hydraulic backfilling. One form of cemented backfilling used in underground mines is paste backfilling. The mine paste backfill is prepared by suitably mixing mill tailings or other solid waste with a specific amount of binder (OPC, OPC-fly ash, OPC-slag, OPC-lime, etc.). The paste backfill composed of an optimum water proportion so that it is flowable. As compared to other backfilling techniques, paste backfilling consists of a higher solid concentration (70–80 wt%). Potvin et al. (2005) defined "paste as a mixture of material, which akin to toothpaste, comprising 75% to 85% solid by weight and 1 wt% to 12 wt% binder, containing at least 15 wt% solid fraction less than 20 μm, when poured into voids the backfill does not bleed water, does not settle in the pipeline, and has a slump less than 250 mm". Paste backfilling is widely applied throughout the globe because of its techno-economic advantages and higher volume of industrial waste utilisation as compared to other backfilling techniques.

Paste backfilling is being practiced in different countries worldwide after its inception in 1957 by Falconbridge Nickel Mines Ltd. at the Hardy mine in Sudbury, Canada and Bad Grund Mine in Germany in the late 1970s (Tariq and Yanful, 2013). The application of mine paste backfill was later followed by the USA, South Africa, China, and Australia. India witnessed its first paste fill plant in 2015 in Sindesar Khurd (SK) Mine of Hindusthan Zinc Limited (HZL), Rajasthan. The success of paste backfill technology is witnessed in the form of increased demand over the last 40 years. In India, presently, there are four operating mines adopting paste backfill technology in Rajasthan state of Hindustan Zinc Limited (HZL). Whereas globally, more than 100 underground metalliferous mines use paste backfill technology. There is ample scope of applying paste backfilling in many underground mines in India (Behera et al. 2022).

Considering the material availability in Sukinda region for paste backfilling, few of the probable paste backfill materials are; chromite tailings, processed overburden and waste rock and fly ash. If these materials can be augmented in such manner that optimum pastes fill is prepared which have sufficient strength and at the same it is economically viable. Considering the increased demand of production from underground, high filling rate and engineered backfill material is the need of the hour.

6 Conclusions

This article highlighted about various lab based studies conducted for utilising sand and bottom ash as a backfill material for a hard rock mine. The key findings of this study are summarised below.

- Sand being the traditional backfill material and possess all favourable properties for a backfill material.

- With scarcity of sand bottom ash can be an alternate backfill material that can replace sand completely for underground stope filling. Bottom ash possesses both favourable drainage and geotechnical properties. Further, bottom possesses better strength development as compared to sand based backfill.
- Overburden use as cemented hydraulic backfill material possesses couple of challenges in terms of strength development and drainage characteristics. Hence, it is essential to search for alternate backfilling methodology so that huge volume of overburden can be used.
- The probable future backfill materials could be chromite tailings, ferro chrome slag and overburden. The field application of these materials is yet to be done.
- Considering the drawbacks of cemented hydraulic filling, paste filling is a far better option; hence, the future of underground backfilling looks forward to implement paste filling in many future underground mines in Sukinda as well as throughout India.

7 Data Availability Statement

The datasets generated during and/or analysed during the current study are available from the corresponding author on reasonable request.

Acknowledgements. The authors thankfully acknowledge Indian Metals & Ferro Alloys Limited (IMFA) management for sponsoring this study and providing necessary facilities and materials. The authors would also like to acknowledge the Director, CSIR-CIMFR, Dhanbad for his continuous guidance and support.

Conflict of Interests. The authors declare no conflict of interest.

References

1. ASTM C39/C39M – 17b. Standard Test Method for Compressive Strength of Cylindrical Concrete Specimens. ASTM 2017. https://doi.org/10.1520/C0039_C0039M-17B
2. Behera, S.K., et al.: Characterization of lead–zinc mill tailings, fly ash and their mixtures for paste backfilling in underground metalliferous mines. Environ. Earth Sci. **78**, 394 (2019a).https://doi.org/10.1007/s12665-019-8395-9
3. Behera, S.K., et al.: Strength development and microstructural investigation of lead-zinc mill tailings based paste backfill with fly ash as alternative binder. Cement Concr. Compos. **109**, 103553 (2020a). https://doi.org/10.1016/j.cemconcomp.2020.103553
4. Behera, S.K., et al.: Utilisation of lead–zinc mill tailings and slag as paste backfill materials. Environ. Earth Sci. **79**(16),389 (2020b). https://doi.org/10.1007/s12665-020-09132-x
5. Behera, S.K., et al.: Utilization of mill tailings, fly ash and slag as mine paste backfill material: Review and future perspective. Constr. Build. Mater. **309**, 125120 (2021). https://doi.org/10.1016/j.conbuildmat.2021.125120
6. Behera, S.K., et al.: Slump test: Laboratory and numerical simulation based approach for consistency of mill tailings paste. Curr. Sci. **117**(2), 235–241 (2019b)
7. Behera, S.K., et al.: Tensile strength of cemented paste backfill for lead–zinc mill tailings: Lab and in situ scenarios. Arab. J. Geosci. **16**, 451 (2023). https://doi.org/10.1007/s12517-023-11536-5

8. Gonen, A., Kose, H.: Stability analysis of open stopes and backfill in longhole stoping method for Asikoy underground copper mine. Arch. Min. Sci. **56**(3), 375–387 (2011)
9. Himanshu, V.K., Mishra, A.K., Roy, M.P., Vishwakarma, A.K., Singh, P.K.: Numerical simulation based approach for assessment of blast induced deformation pattern in slot raise excavation. Int. J. Rock Mech. Min. Sci. **144**, 104816 (2021). https://doi.org/10.1016/j.ijrmms.2021.104816
10. IS:2386–III: Indian standards: Methods of test for aggregates for concrete, Part III, Specific Gravity, New Delhi, India (1963)
11. Li, L.: Generalized solution for mining backfill design. Int. J. Geomech. **14**(3), 04014006-1:11 (2014). https://doi.org/10.1061/(ASCE)GM.1943-5622.0000329
12. Mishra, K., Paul, P.S., Ghosh, C.N., Singh, P., Behera, S.K., Mandal, P.K.: Predicting and optimising the strength of cemented paste fills through Bayesian network model. Mining Metallurgy Explorat. **39**, 2095–2120 (2022). https://doi.org/10.1007/s42461-022-00650-9
13. Naz, A., Chowdhury, A., Mishra, B.K., Gupta, S.K.: Metal pollution in water environment and the associated human health risk from drinking water: A case study of Sukinda chromite mine, India. Hum. Ecol. Risk Assess. Int. J. **22**(7), 1433–1455 (2016)
14. Panwar, N., et al.: Integration of SIP and MIP methods for pore-size distribution in carbonate rocks. In: Second International Meeting for Applied Geoscience & Energy, Society of Exploration Geophysicists and the American Association of Petroleum Geologists, pp. 2198–2202 (2022). https://doi.org/10.1190/image2022-3751698.1
15. Potvin, Y., Thomas, E., Fourie, A.B.: Handbook on Mine Fill. Australian Centre for Geomechanics, Perth, Australia (2005)
16. Behera, S.K., et al.: Paste backfilling in underground mines Present scenario in India and future perspective. In: 9th Asian Mining Congress, 4–7, Kolkata, India, pp. 301–310 (2022)
17. Singh, P., et al.: Optimisation of binder alternative for cemented paste fills in underground metal mines. Arab. J. Geosci. **12**(15), 462 (2019). https://doi.org/10.1007/s12517-019-4623-6
18. Tariq, A., Yanful, E.: A review of binders used in cemented paste tailings for underground and surface disposal practices. J. Environ. Manage. **131**, 138–149 (2013). https://doi.org/10.1016/j.jenvman.2013.09.039
19. Villaescusa, E., Thompson, A., Windsor, C.: Probabilistic estimate of rock mass static and dynamic demands for underground excavation stabilisation. J. Rock Mech. Geotechn. Eng. **11**(3), 481–493 (2019). https://doi.org/10.1016/j.jrmge.2018.08.009
20. Edraki, M., Baumgartl, T., Manlapig, E., Bradshaw, D., Franks, D.M., Moran, C.J.: Designing mine tailings for better environmental, social and economic outcomes: A review of alternative approaches. J. Clean. Prod. **84**, 411–420 (2014). https://doi.org/10.1016/j.jclepro.2014.04.079
21. Grobler, H., Chatziefstratiou, V., Mousli, O., Yumlu, M.: Design of high-strength backfill for a drift-and-fill mining method at olympias mine, Greece. In: Proceedings of the 22nd International Conference on Paste, Thickened and Filtered Tailings, Australian Centre for Geomechanics, Perth, Australia, pp. 401–410 (2019)

Coal Gasification and Coal Bed Methane

Gasification of High-Ash Indian Coal and Agro-Industry Biomass Waste in Oxy-Blown Bubbling Pressurized Fluidized Bed Gasifier

Rupesh Kumar Singh(✉), Nilesh D. Dhaigude, Arti Sahu, Vishal Chauhan, Gajanan Sahu, Pavan Kumar Gupta, Sujan Saha, and Prakash D. Chavan

Gasification and Catalysis Research Group, CSIR-Central Institute of Mining and Fuel Research, Dhanbad 826015, India
rupeshsingh@cimfr.nic.in

Abstract. India has great potential to utilize high-ash Indian coal and agro-industry-based biomass towards clean fuel technology via the gasification route which is energy efficient and environment friendly. Due to the limitation of handling low-ash coal (ash% < 30 wt.%) in all the mature gasifiers worldwide, the development of fluidized bed gasification technology is needed for the judicious utilization of high-ash Indian coal and agro-industry biomass waste to strengthen"***Methanol & Hydrogen Economy***" and "***Net Zero Emission***".

In this study, the developed BPFBG has demonstrated input fuel power of 60–80 KW_{th} and 15–25 KW_{th} for high-ash Indian coal and rice husk respectively in the presence of gasifying agents such as the mixture of pure oxygen-steam, at lower pressure and 900–1000 °C temperature. At optimal high-ash Indian coal gasification operational condition of ER of 0.32 along with steam-to-coal ratio of 0.70 and rice husk gasification operational condition of ER of 0.30 along with steam-to-rice husk ratio of 0.40, produced syngas composition was H_2:32.64, CO:22.83, CO_2:24.54, CH_4:5.87(v/v%) and H_2:9.45, CO:22.14, CO_2:40.12,CH_4:7.01(v/v%) respectively. Further, this development would give input data for the establishment of a Demo-scale facility based on Fluidized Bed Gasification Technology in India.

Keywords: Bubbling Pressurized Fluidized Bed Gasifier · High-Ash Indian Coal · Agro-Industry Biomass Waste · ER · steam-to-coal ratio · Steam-to-rice husk ratio · Oxy-blown gasification

Nomenclature

BPFBG	Bubbling Pressurized Fluidized Bed Gasifier
CCE	Carbon Conversion Efficiency
CGE	Cold Gas Efficiency
LHV	Lower Heating Value
HHV	Higher Heating Value
ER	Equivalence Ratio
BT	Billions of Tonnes

A. Sinha et al. (Eds.): AMC 2023, SPEES, pp. 295–306, 2023.
https://doi.org/10.1007/978-3-031-46966-4_23

TPD	Tonnes Per Day
MTPA	Millions of Tonnes Per Annum
GC	Gas Chromatography
KW_{th}	Kilowatt Thermal
AFT	Ash Fusion Temperature

1 Introduction

India has the 5th largest coal reserve and 2nd the biggest coal producer in the world [1]. Coal is the predominant fuel resource in India. As on April 2022, the estimated coal reserve in India up to a maximum depth of 1200 m is around 361.41 BT [2]. Commercial coal in the country is mainly consumed by the power sector (–60%). Power generation is the major source of carbon emissions, which accounts for about 40% of total green house gas emissions. Conventional coal-based power generation technologies have several drawbacks, such as lower efficiency, environmental loading and damage thereof.

Govt. of India has assured at COP-26 that the energy production based on fossil fuels will increase up to 500 GW by 2030. Therefore, clean coal technologies are of utmost importance to adopt new coal conversion technologies for improving the efficiency of coal utilization and reduction of CO_2 and other pollutant emissions in the environment. It has also decided to consume 100 MT of Indian coal by the Coal Gasification route.

In virtue of this, NITI Aayog, Govt. of India is emphasizing developing Indigenous Technology to utilize the Indian coal reserve to meet the targeted goal of "*Methanol & Hydrogen Economy*" by the Gasification route [3].

In this regard, coal gasification may be a solution to utilize coal efficiently meeting stringent environmental regulations. The clean syngas afterthe removal of various hazardous emissions can be utilized for the production of Fischer-Tropsch liquids, various chemicals such as methanol, ammonia, fertilizers, substitute natural gas, DRI application, thermal application, power generation, H_2 generation, and many more.

A developing country like India has huge renewable resources like Biomass from different sectors such as agriculture, food industries, forest residue, and MSW. The generation of biomass in India is estimated at 750 MTPA and surplus biomass availability of 230 MTPA covering agricultural residues equivalent to 28 GW of energy [4]. In India, it is mainly used for power generation. But apart from power generation, it has significant potential for the production of hydrogen, different chemicals, fertilizer, and many more by the Gasification route.

Govt. of India has assured at COP-26 to meet the country's energy demand by 50% from renewable energy as well as committed to reducing the total projected carbon emission by 1 BT by 2030. Biomass as a renewable source and carbon neutral has significant potential. Among different routes, biomass gasification technology may provide strength to different Industries by reducing carbon footprint.

In this connection, complete exploration of biomass properties in an efficient manner for its conversion to hydrogen-rich syngas by gasification route will give a break through to meet the nation's goal of sustainable energy. In virtue of this, NITI Aayog, Govt. of India is emphasizing to development of Indigenous Technology to utilize biomass resources of the nation to meet the targeted goal of "*Green Hydrogen Economy & Net Zero Emission*" by the gasification route [3].

Indian coal and agro-industry-based biomass utilization by gasification route will lead toward clean fuel technology. Conventional routes of coal and biomass utilization are less energy efficient and have adverse effects on the environment. Due to the limitation of handling low-ash coal (ash% <30 wt.%) in all the mature gasifiers worldwide, the development of fluidized bed gasification technology is needed for the judicious utilization of high-ash Indian coal.

In virtue of this, CSIR-Central Institute of Mining and Fuel Research (CSIR-CIMFR) Dhanbad has uniquely designed and developed Indigenous gasification technology known as 1.5 TPD Oxy-blown Bubbling Pressurized Fluidized Bed Gasification (BPFBG) pilot plant facility.

2 Feedstock and Methodology

2.1 Feedstock

In the Gasification process, the feedstock is considered as the least flexible choice. The particle size of coal is one of the important aspects and in case of handling it in a developed BPFBG pilot plant facility, it must be below 4 mm. In this study, one sample of high-ash Indian coal (ash% of 36.3% by wt.) and another sample of rice husk (RH) have been characterized and demonstrated in the developed 1.5 TPD Oxy-blown BPFBG pilot plant facility for gasification phenomena. Table 1 shows the feedstock characterization proximate and ultimate analysis results of considered feedstock in this study.

2.2 Experimental Setup

In this study, Indian coal and Rice husk have been used for demonstrating gasification experiments in an Indigenously developed 1.5 TPD Oxy-blown BPFBG pilot plant facility. A schematic flow diagram of the said facility is shown in Fig. 1.

The entire pilot plant facility has been divided into four different sections namely the Feeding section, Gasifying agent section, the Gasifier section, the Ash Withdrawal section, and Gas Cooling and Cleaning section. The Feeding section mainly consists of fuel feeding hopper for storing and feeding Indian coal and rice husk to the gasifier and is used for developing a feeding mechanism by controlling the flow rate of feed from fuel feeding hopper to the gasifier. The Gasifying agents section consists of an air/oxygen generator, boiler, steam accumulator, and heaters to maintain the temperature of the mixture of gasifying agents above the auto-ignition temperature of Indian coal and rice husk. The Gasifier section consists of a refractory insulated tubular reactor of the length of –4.5 m and k-type thermocouples as well as pressure gauge mounted at different locations for measuring temperature and pressure during its operation. The Ash withdrawal section consists of gasifying agent distributor for efficient heat and mass control between fuel and gasifying agent, an ash cooler, an ash extractor, and an ash collection vessel for collecting ash residue after gasification operation at room temperature. The Gas cooling and cleaning section consists of a cyclone separator to separate dust from generated syngas, a quench column for sudden cooling of generated syngas to room temperature, a ventury scrubber followed by a knock-out drum for cleaning purposes of the generated syngas, and at last it flared after monitoring the quality of the generated syngas continuously by using online gas analyzer as well as by offline GC.

The pilot plant facility is designed and developed for the conversion of Indian coal and biomass residue into suitable quality and yield of syngas for its utilization in down-stream applications like syngas to methanol, chemicals, DRI, fertilizer, power, thermal applications, etc. in India.

Table 1. Feedstock characterization

	Indian Coal	Rice Husk
General Information		
Classification	India	Agro-Industry
Bulk Density (kg/m^3)	920.00	105.00
Proximate analysis (% by weight)		
Fixed Carbon	29.30	27.50
Volatile Matter	28.30	52.10
Moisture	6.10	6.10
Ash	36.30	14.30
Ultimate analysis (% by weight)		
Total Carbon	44.40	41.80
Hydrogen	3.10	4.50
Nitrogen	1.10	1.40
Sulphur	0.90	0.08
Oxygen	8.10	31.82
Moisture	6.10	6.10
Ash	36.30	14.30
Thermal analysis (kcal/kg)		
HHV	3840	3120
LHV	3700	3200

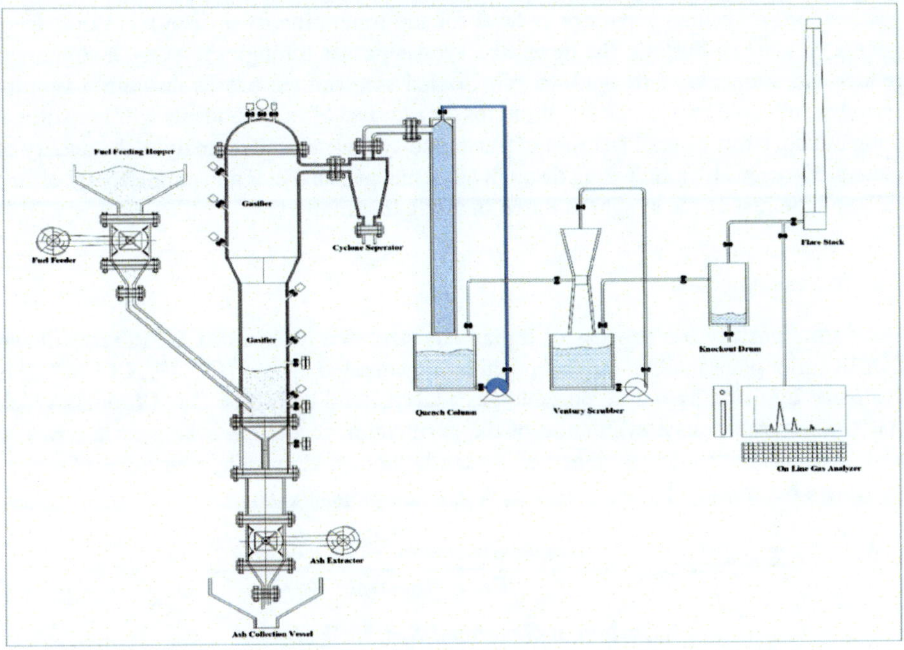

Fig. 1. Schematic Diagram of a 1.5 TPD Oxy-blown BPFBG Pilot Plant Facility.

2.3 Working Methodology

Gasification of Indian coal and rice husk has been demonstrated in an Indigenously designed and developed 1.5 TPD Oxy-blown BPFBG at a pilot scale. In this study, pre-processing of Indian coal has been performed for getting the desired size of coal of less than 3 mm. Another feedstock, rich husk has been utilized in its natural form without any pre-processing. At first, physicochemical characterization of both samples is performed to understand their gasification potential. Fluidization study has been performed for 30 cm of bed height inside the gasifier, and air as a fluidizing medium has been utilized [5, 6]. Coal ash having a density of around 1010 kg/m^3 is used as bed material for the fluidization study. Fluidization of bed material (coal ash) is established in the gasifier with continuous feeding and withdrawal synchronization mechanisms by using a fuel feeder and ash extractor respectively. This initial study of fluidization took around 1 h given the steady condition of the gasifier. Once the fluidization steady condition is achieved, electric heaters are used for the heating purpose of air as a gasifying agent to attend to the auto-ignition temperature of feedstock. A parallelly carbon-containing feedstock is charged in a controlled manner from the fuel feeder to attain the gasifier temperature –950–980 °C at atmospheric pressure in the presence of pure oxygen (92 vol.%) and steam is used as a gasifying agent for gasification. The gasification reaction takes place in the bubbling fluidized mode with a sufficient flow of gasifying agent inside the gasifier. Steam plays a vital role in controlling and maintaining the gasifier temperature and participates in water gas reactions during gasification phenomena. After gasification, a cyclone separator is used for the separation of micron-level dust from generated syngas. Further, the generated syngas passes through the quench column to reduce the temperature to −30–45 °C. Cooled syngas then passes through a ventury scrubber followed by a knock-out drum for the removal of sulfur and nitrogen constituent compounds. Now, a small fraction of generated syngas is used to analyze the quality of generated syngas by monitoring through an online gas analyzer and parallelly by offline GC, and the rest of the syngas is flared in a safe environment.

2.4 Syngas Analysis

During the gasification process of Indian coal and rice husk in a 1.5 TPD oxy-blown BPFBG pilot plant facility, the composition of generated syngas (CO, H_2, CO_2, CH_4) is evaluated by using an online portable gas analyzer and an offline Gas Chromatograph (GC). The performance evaluation of the gasification process is evaluated in terms of yield, carbon conversion efficiency, cold gas efficiency, and lower heating value of syngas which are determined by using Eqs. 1, 2, 3, and 4 respectively.

$$Yield_{Syngas} = \frac{Syngas\ Generation\ Flow\ rate\ (Nm^3/hr)}{Feed\ Flow\ rate\ (Kg/hr)} \tag{1}$$

$$CCE = \frac{Carbon\ in\ Generated\ Syngas\ (kg/hr)}{Carbon\ available\ in\ Feed\ based\ up\ on\ Ultimate\ analysis\ (kg/h)} \times 100\ \% \tag{2}$$

$$CGE = \frac{Total\ Syngas\ Gneration\ (Nm^3/Kg\ Feed) \times\ Calorific\ value\ of\ Generated\ Syngas\ (Kcal/Nm^3)}{Calorific\ value\ of\ Feed\ (Kcal/KgFeed)} \times 100\ \% \tag{3}$$

$$LHV = \frac{CV\ of\ Syngas\ (KCal/kgfuel)}{Syngas\ generated\ (Nm^3\ syngas/kg\ fuel)} \times 100\ \% \tag{4}$$

$$ER = \frac{Actual\ Flow\ rate\ of\ Oxygen}{Stiochiometric\ flow\ rate\ of\ Oxygen\ for\ complete\ combustion\ of\ C, H, N, \&S\ present\ in\ fuel} \tag{5}$$

3 Results and Discussion

In this study, Gasification of high ash Indian coal and rice husk as agro-industry residue has been demonstrated input fuel power of 60–80 KW_{th} and 15–25 KW_{th} respectively in an Indigenously developed 1.5 TPD Oxy-blown BPFBG pilot plant facility. Gasification has been performed in the presence of gasifying agents such as the mixture of pure oxygen-steam, at lower pressure and 900–1000 °C temperature and atmospheric at pilot scale. One Indian coal sample (ash% of 36.3% by wt.) and one rice husk sample having the physical properties mentioned in Table 1 have been utilized for gasification performance analysis at a pilot scale plant with fuel loading capacity up to 1.5 Tonnes per day. Gasification performance includes syngas composition, carbon conversion efficiency, cold gas efficiency, and heat value of generated syngas. Data of all parameters have been taken after 3 h of operation at the pilot plant.

3.1 Significance of Temperature on Syngas Composition in 1.5 TPD BPFBG Pilot Plant

Gasification reaction has been performed dominantly at the bottom zone of the gasifier where the fresh feed is mixed with gasifying agents like a mixture of oxygen and steam. Gasification phenomena have been driven and controlled by the gasifier temperature [7]. So, the effect of gasifier bottom zone temperature on syngas composition generated from Indian coal and rice husk has been studied and presented in Fig. 2(a) and (b) respectively. The temperature effect on syngas composition through gasification phenomena in fluidized bed gasifiers on lab scale capacity [8].

Figure 2(a) and (b) show the syngas composition distribution during Indian coal and rice husk gasification respectively. During coal gasification, the temperature at the bottom of the BPFBG varied from 834 °C to 980 °C, and the feed rate of coal, oxygen, and steam varied from 20 to 35 kg/h, 12 to 20 kg/h, and 14 to 21 kg/h respectively. Initially at the gasifier bottom zone temperature of 834 °C, generated syngas composition was found to be at lower side H_2: 11.3, CO: 7.37, CO_2: 22.82, and CH_4: 3.58 (vol.%). This might be achieved due to insufficient heat generation through combustion reaction inside the gasifier to drive the gasification reaction. Further oxygen flow rate was enhanced to increase the gasifier temperature in a controlled manner by parallelly increasing the steam flow rate. It was observed that syngas composition drastically jumped and increased exponentially i.e., H_2 from 11.3 to 32.64, CO from 7.37 to 22.83, CO_2 from 22.82 to 25.72, and CH_4 from 3.58 to 6.23 (vol.%). During rice husk gasification, the temperature at the bottom of the BPFBG varied from 840 °C to 941 °C, and the feed rate of rice husk, oxygen, and steam varied from 28 to 32 kg/h, 11 to 17 kg/h, and 8 to 12 kg/h respectively. Initially at the gasifier bottom zone temperature of 840 °C, generated syngas composition was found to be at lower side H_2: 4.91, CO: 16.86, CO_2: 38.12, and CH_4: 6.31 (vol.%). This might be achieved due to the low energy density of rice husk which results in a shortage of rice husk for combustion reaction and thus insufficient gasifier temperature is observed. Further rice husk, oxygen along with steam flow rate has increased to maintain sufficient gasification temperature. It was observed that syngas composition has increased i.e., H_2 from 4.91 to 9.45, CO from 16.86 to 22.14, CO_2 from 38.12 to 40.12, and CH_4 from 6.31 to 7.05 (vol.%).

The gasifier temperature has increased with sufficient flow of Indian coal/rice husk, oxygen along with bed height which helps to drive the endothermic reactions followed by maintaining the gasifier temperature. At steady state condition pilot plant operates for 8–12 h with constant gasifier temperature and syngas composition. The gas composition not only depends on the gasifier temperature but is significantly affected by the residence time of the gas inside the gasifier [9]. The residence time of the gas is reflected by the bed material height inside the gasifier. If the bed height is fair enough along with sufficient temperature then produced syngas quality has better yield and compositions. Residence time indicates the contact time of solid bed material with gasifying reagents along with produced gases during combustion and gasification phenomena. The gasification reaction has driven when few percentages of available carbon utilized in combustion to produce CO_2 along with heat and further left carbon in the bed has gone through gasification reactions which are endothermic in nature in the presence of gasifying reagents like steam and oxygen [10]. These endothermic reactions consumed the heat of the gasifier, so then either the flow of oxygen or the feed rate of coal increased and once this synchronism brought the gasifier temperature profile in a stable condition, the process come to an equilibrium state. At favorable gasification temperature along with sufficient bed height inside the gasifier, the best composition of syngas has been achieved by using high ash Indian coal in the developed 1.5 TPD Oxy-blown BPFBG Pilot Plant. Figure 2(a) and (b) represent the composition of generated Syngas during Indian Coal and Rice Husk Gasification respectively in 1.5 TPD Oxy-blown BPFBG Pilot Plant Facility.

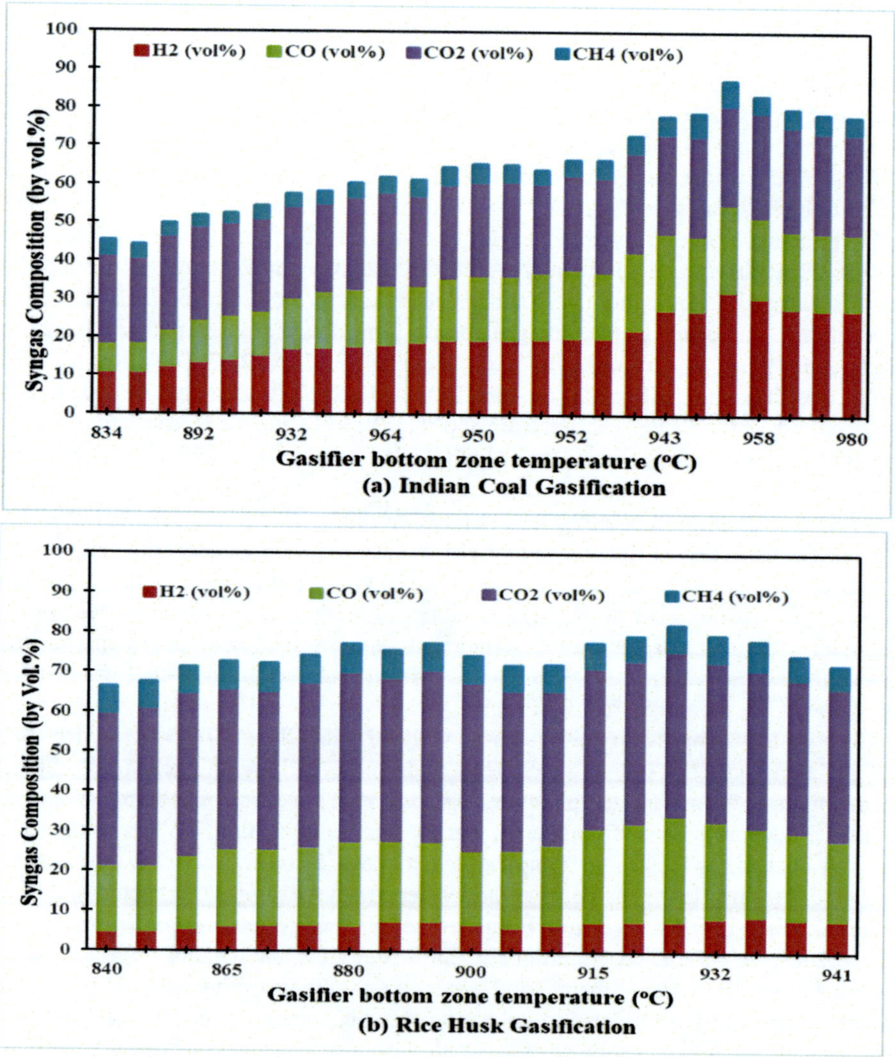

Fig. 2. (a) Indian Coal Gasification. (b) Rice Husk Gasification.

Figure 2(a) and (b) represent the composition of generated Syngas during Indian Coal and Rice Husk Gasification respectively in 1.5 TPD Oxy-blown BPFBG Pilot Plant Facility.

3.2 Significance of ER on Carbon Conversion Efficiency, Cold Gas Efficiency, and Calorific Value of Generated Syngas in 1.5 TPD BPFBG Pilot Plant

The gasification potential of any carbonaceous feedstock is determined by the parameters such as Carbon Conversion Efficiency (CCE), Cold Gas Efficiency (CGE), and LHV of syngas, and defined in Eq. 2, 3, and 4 respectively. Equivalence ratio (ER) is defined in Eq. 5 and has a significant effect on CCE, CGE, and LHV of syngas produced from the Gasification of Indian coal and rice husk in 1.5 TPD Oxy-blown BPFBG pilot plant. At a higher oxygen flow rate i.e., an increase in ER reflects more conversion of coal into syngas due to solid-gas interaction and heat transfer during the gasification of both feedstocks. This indicates that due to sufficient bed height, the solid-gas interaction and heat transfer have increased in BPFBG [8]. As the produced gas mainly, CO_2 after combustion of coal has passed through the bed in BPFBG, which helped to react with coaly material in the bed along with steam and produced CO, H_2, CH_4, and some higher hydrocarbons due to the gasification reaction, this might lead to dropping in bed temperature to some extent [11]. The temperature drop in the bed reflects that the gasification reaction has carried and produced good-quality syngas. This means if the good quality of the syngas has been produced, there has significant carbon from coal has been converted into gas either in the form of CO_2, CO, and CH_4. Figure 3(a) reflects that the Rice husk has better CCE than Indian coal at a lower ER of 0.23 to 0.30, but the conversion of carbonaceous fuel is found to be more or less equal in both the feedstock at ER of 0.32. At lower ER, unconverted carbon percentage is found more in the residue of the feedstock due to scarcity of oxygen availability. During the gasification of Indian coal, CCE varied from 73.62% to 94.23% at different ER. Whereas gasification of rice husk results in CCE varied from 77.85% to 96.42%. So, the difference in CCE has dependent on the combination of operational parameters during gasification operation like flow rate of feedstock, oxygen, residence time, and bed temperature [12]. Figure 3(b) said that the Cold Gas Efficiency has slightly more or less the same in the case of Rice husk gasification of 43.29% to 54.24% at ER of 0.23 to 0.32. Whereas, Indian coal gasification results show that the CGE has increased from 57.28% to 65.25% at ER of 0.28 to 0.32, but CGE is of inferior nature of 30.27% at ER of 0.23. It has been observed that the CGE of both feedstocks increases as the ER increases, i.e., CGE has a direct relation with ER. CGE is also enhanced when maintaining a suitable bed height of around 30–35 cm in BPFBG. Figure 3(c) has discussed the lower heating value of the produced syngas from Indian coal and Rice husk. The heat value of syngas generated from Indian coal gasification has an increasing trend from 800 kcal/Nm^3 to 2152 kcal/Nm^3 with ER of 0.23 to 0.32. Whereas the heat value of syngas generated from Rice husk gasification has more or less equal to 1700 kcal/Nm^3 at ER of 0.23 to 0.32. The reason might be the volume of the CH_4 in the produced syngas, i.e. slightly more volume of CH_4 in syngas lead to a high calorific value because of the high heat content of CH_4 [13]. At a lower ER of 0.23 i.e., lower flow rate of oxygen concentration, the produced syngas has a very low calorific value because of low CCE. Further, when the ER has increased i.e., the availability of

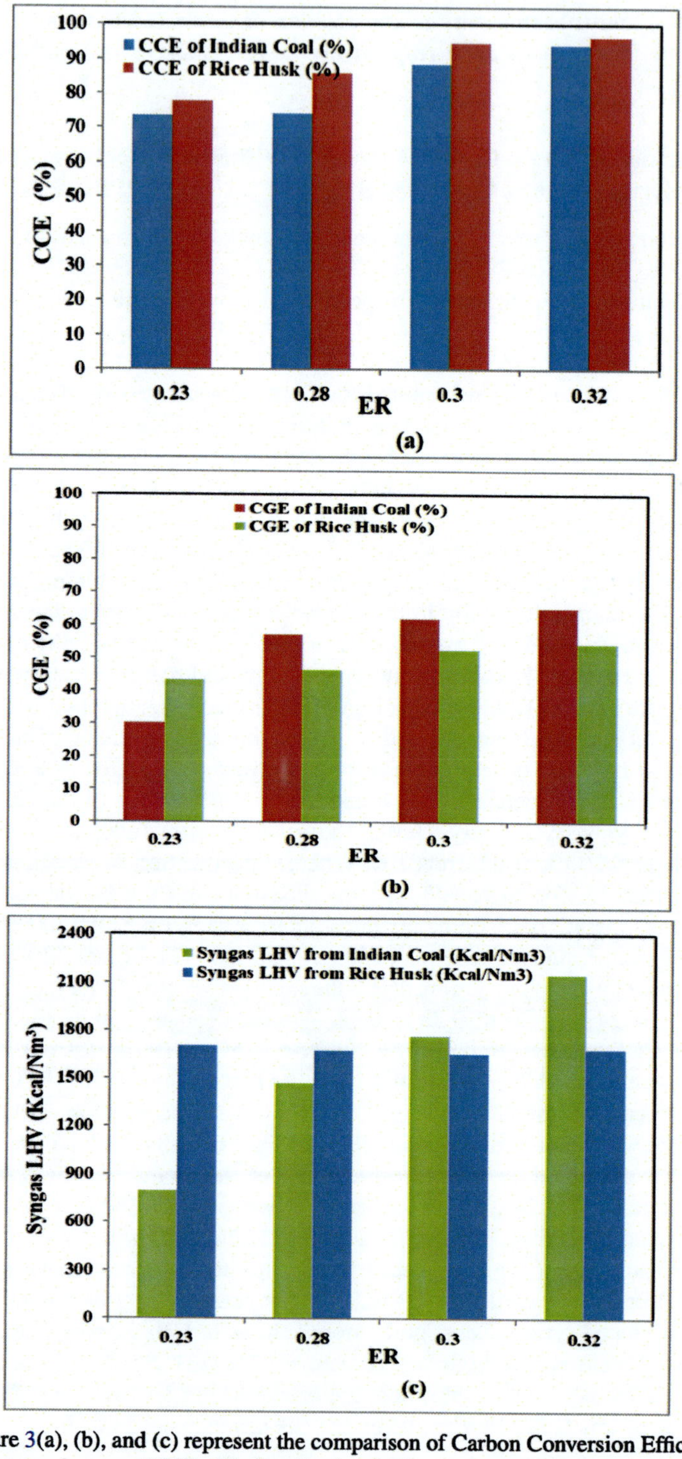

Fig. 3. Figure 3(a), (b), and (c) represent the comparison of Carbon Conversion Efficiency, Cold Gas Efficiency, and syngas LHV of Indian Coal and Rice Husk Gasification respectively in 1.5 TPD Oxy-blown BPFBG Pilot Plant Facility.

oxygen generates sufficient heat to drive all the gasification reactions and contains good heat value. Syngas generated from Indian coal and Rice husk has significant importance in Hydrogen generation, chemical production, and thermal applications respectively.

3.3 Steam Importance in Controlling Gasifier Temperature and Enhancing Syngas Quality

High-ash Indian coal is a challenging feed source for gasification in any kind of gasifier. But, a fluidized bed gasifier is an attractive option to utilize such high feedstocks with a proper understanding of the gasification phenomena and operational philosophy. Handling of high ash Indian coal in a fluidized bed sometimes leads to form agglomeration due to sudden rise in the temperature of the bottom zone of the gasifier. Agglomeration is formed when the gasifier operation temperature exceeds the Ash Fusion Temperature (AFT) of the ash particles inside the gasifier [14, 15]. This sudden rise of temperature inside the gasifier produces so much amount of heat in a fraction of seconds, so, that this situation leads to the form of agglomeration of the bed material inside the gasifier. This situation also leads to a de-fluidization state of the bed and blocks the gasifier's bottom line. Hence, further operation of the gasifier is not possible without the removal of the formed bed agglomerate. The above scenario has been controlled successfully in a 1.5 TPD Oxy-blown BPFBG pilot plant with sufficient injection of steam. First time in India, it has identified the role of steam in handling the agglomeration issues with high ash Indian coal and Rice husk in BPFBG. When the steam to feedstock flow rate increased from 0.30 to 0.70 with an oxygen purity of 92% (by wt.%) as a gasifying agent, not only agglomeration problem overcome but it also leads to getting a better quality of syngas composition with the H_2 and CO concentration, which results in better CCE, CGE, and LHV of the produced syngas.

4 Conclusions

In this study, the Gasification potential of high-ash Indian coal and Rice husk has been demonstrated in a BPFBG pilot plant facility. Gasification performance has been evaluated concerning gasifier operating temperature and ER. The results of gasification have been expressed in terms of gas composition, CCE, CGE, and LHV of generated syngas. Indian coal has great potential to be utilized in an Indigenously developed BPFBG without any operational issues such as agglomerate, and clinker formation. Gasification of Indian coal and rice husk has generated syngas composition H_2 from 11.3 to 32.64%, CO from 7.37 to 22.83%, and H_2 from 4.91 to 9.45%, CO from 16.86 to 22.14% at ER of 0.23 to 0.32 respectively. Indian coal gasification has CCE varied from 73.62% to 94.23%, whereas rice husk gasification has CCE varied from 77.85% to 96.42% at ER of 0.23 to 0.32.CGE of Rice husk and Indian coal gasification has varied from 43.29% to 54.24% and 30.27% to 65.25% respectively at ER of 0.23 to 0.32. LHV of syngas generated from Indian coal and Rice husk has varied from 800 to 2152 kcal/Nm^3 and 1655 to 1705 kcal/Nm^3 respectively at ER of 0.23 to 0.32. The steam-to-feedstock ratio of 0.30 to 0.70 plays a vital role in controlling operational challenges such as ash agglomeration and clinker formation during gasification of high-ash feedstock in an Indigenously developed 1.5 Oxy-blown BPFBG pilot plant facility. Syngas generated from Indian coal and Rice husk gasification has the potential to be utilized in different downstream applications such as Hydrogen generation, DRI, Methanol/DME, thermal application, power generation, and many more. Further, this development would give input data for the establishment of a Demo-scale facility based on Fluidized Bed Gasification Technology in India.

Acknowledgments. The authors are thankful to CSIR-Central Institute of Mining and Fuel Research (CIMFR), Dhanbad, Jharkhand, India for financial support.

References

1. Hook, M., Zittel, W., Schindler, J., Aleklett, K.: Global coal production outlooks based on a logistic model. Fuel **89**, 3546–3558 (2010)
2. Coal Availability in India: Ministry of Coal, Govt. of India (2023)
3. Economic Analysis of Methanol 020929021: NITI Aayog, Govt. of India (2021)
4. Kumar, A., Kumar, N., Baredar, P., Shukla, A.: A review on biomass energy resources, conversion, and policy in India. Renew. Sustain. Energy Rev. **45**, 530–539 (2015)
5. Singh, R.K., et al.: Significance of temperature and pressure on minimum fluidization velocity in a fluidized bed reactor: an experimental analysis. Energy Sources Part A **45**(1), 1793–1806 (2023)
6. Chauhan, V., et al.: Estimating the requirements of pressure drop during fluidization bubbling regimes in fluidized bed with different conical distributors. Part. Sci. Technol. 1–12 (2023)
7. Chavan, P., Datta, S., Saha, S., Sahu, G., Sharma, T.: Influence of high ash Indian coals in fluidized bed gasification under different operating conditions. Solid Fuel Chem. **46** 108–113 (2012)
8. Mahapatro, A., Mahanta, P.: Gasification studies of low grade Indian coal and biomass in a lab scale pressurized circulating fluidized bed. Renew. Energy **150**, 1151–1159 (2020)
9. Chauhan, V., Chavan, P.D., Gupta, N.K., Datta, S., Saha, S., Sahu, G.: Establishing fluidization parameters of different size of coal ash particles in bubbling fluidized bed. Chem. Eng. Commun. **208**, 210–219 (2021)
10. Gupta, S., De, S.: An experimental investigation of high ash coal gasification in a pilot scale bubbling fluidized bed reactor. Energy **244**, 122868 (2022)
11. Kumar, K.V., Bharath, M., Raghavan, V., Prasad, B.V., Chakravarthy, S.R., Sundararajan, T.: Gasification of high ash Indian coal in bubbling fluidized bed using air and steam: An experimental study. Appl. Therm. Eng. **116**, 372–381 (2017)
12. Datta, S., et al.: Gasification of coal and press mud blends in fluidized bed gasifier. Int. J. Oil Gas Coal Technol. **19**, 248–262 (2018)
13. Tiwary, S., et al.: Co-gasification of high ash coal biomass blends in a fluidized bed gasifier: Experimental study and computational intelligence based modeling. Waste Biomass Valor. **11**, 323–341 (2020). https://doi.org/10.1007/s12649-018-0378-7
14. Datta, S., et al.: Agglomeration behaviour of high ash Indian coals in fluidized bed gasification pilot plant. Appl. Therm. Eng. **86**, 222–228 (2015)
15. Yadav, D., et al.: A comparative review on thermal behavior of feedstocks during gasification via thermogravimetric analyzer. J. Therm. Anal. Calorim. **148**(2), 329–354 (2023)

Syngas Cleaning Technologies in Gasification Process for Downstream Applications

Pavan K. Gupta(✉), Sudipta Datta(✉), Shiva K. Saw, Shweta Kumari, Gajanan Sahu, Rupesh K. Singh, Vishal Chauhan, and Prakash D. Chavan

CSIR-Central Institute of Mining and Fuel Research, Digwadih Campus, Dhanbad 828108, India
pavan@cimfr.nic.in, sud.datta@gmail.com

Abstract. Syngas (CO and H_2) are cleaner and environmental friendly fuel as compare to fossil fuels. Generation of syngas using the availability of vast variety of feedstock (coal, biomass, natural gas, municipal solid waste etc.) make it lucrative option as an alternative to substitute for the conventional liquid fuels and other chemicals. In the industry, the word "syngas" is frequently used to refer the product gas from the gasification and reforming process from solid (coal, biomass, municipal waste, petcoke) and gaseous carbonaceous (natural gas) feedstocks respectively. During gasification process various elements like sulphur, nitrogen, halogen and alkali metals present in carbonaceous feedstock get converted to contaminants and generally referred as "impurity". The concentrations of these impurities are heterogeneous and depend strongly on the carbon source, and gasification operating parameter (e.g. pressure, temperature, residence time, types of gasifier). Those impurities can lower the activity of almost any type of catalytic reactions, so it is essential to remove them to meet the downstream process specifications. The presence of these pollutants in syngas causes a variety of technical and operational challenges, including equipment corrosion by hydrogen sulphide, and catalyst deactivation (tar, hydrogen sulphide, ammonia, Hydrogen chloride, and trace metals). Syngas cleaning methods are comprehensively reviewed in the present paper.

Keywords: Syngas · Cleaning · Carbonaceous feedstock · Gasification · Reforming · Ultra clean fuel

1 Introduction

Rapid urbanization and industrialization need huge energy, which is projected to raise by approximately 27%, worldwide from 2017 to 2040. [1] Presently, majority of transportation energy demand is met by fossil fuels, which is considered to be the primary energy source. Hence, growing energy demands and volatile oil prices, drew the attention of researcher's/academician's tofind out alternate route for liquid fuels. The conversion of carbonaceous feedstock (natural gas, coal, biomass, municipal solid waste etc.) to liquid hydrocarbons/methanol/olefinmay be considered as viable option to balance the shortage of transportation fuels.Due to abundance availability, assured supply and low

A. Sinha et al. (Eds.): AMC 2023, SPEES, pp. 307–316, 2023.
https://doi.org/10.1007/978-3-031-46966-4_24

cost of coal, gasification is presently considered as one of the most lucrative option for achieving these aims. Gasification, a clean and efficient energy utilization technologyis used for converting carbonaceous feedstocks into valuable energy products. Gasification is a thermo-chemical process with a change of the chemical structure of the carbonaceous feedstocks at 700–1100 °C in the presence of a gasifying media (e.g. O_2, steam, air, carbon dioxide, or mixtures of components).

Generally, gasification alludes to an incomplete reaction between carbonaceous feedstocks and gasifying agents (e.g. O_2, steam, air, carbon dioxide, or mixtures of components), and finally generates a mixture of gas comprising of CO, H_2, CO_2, CH_4, N_2, H_2S, COS etc. The mixture of CO and H_2 known as 'syngas' widely used in industry, and vast utilization is presented in Fig. 1.

During gasification process various elements in carbonaceous feedstocks are also converted to contaminants and generally referred as "impurity". Concentrations of impurities are heterogeneous and highly depend on the feed source, gasification operating parameter (e.g. pressure, temperature, residence time, and types of gasifier) [2, 3]. Different types of impurities and their classification has been presented in Table 1. [4].

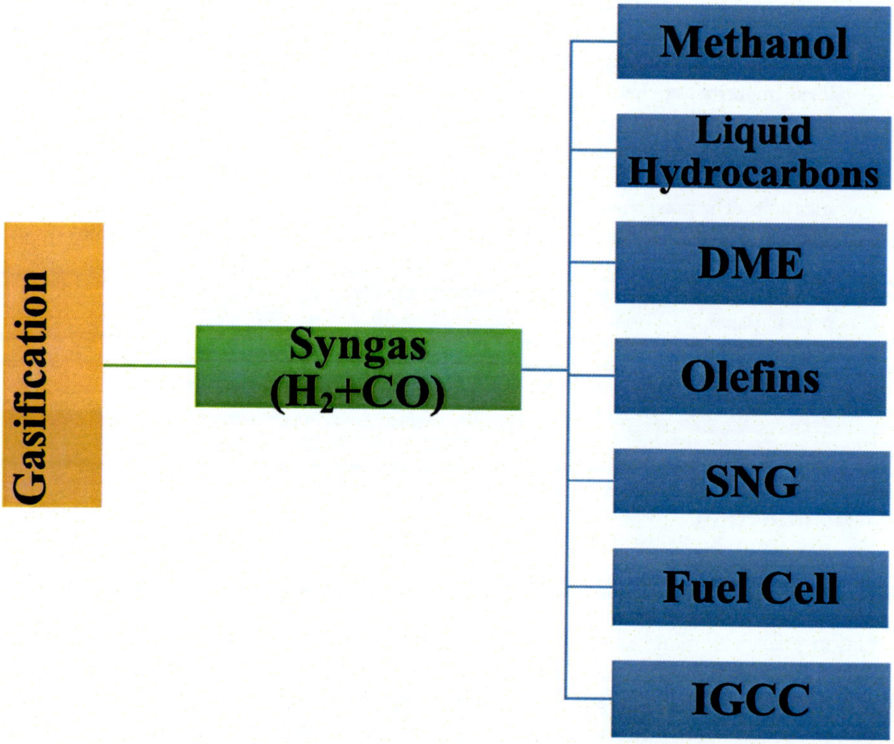

Fig. 1. Various utilization of Syngas.

The contaminated syngas may hamper the downstream process e.g. Fischer-Tropsch synthesis, Methanol synthesis, olefin synthesis etc. Even ppb amounts of these contaminants can be harmful. [5] Those impurities can lower the activity of almost any type of catalytic reactions, so it is necessary to remove them to meet the downstream process specifications as well to comply with pollution control policies and the processes requirements. Contaminants in syngas generate a wide range of technical and operational issues, including equipment corrosion by hydrogen sulphide, and catalyst deactivation by tar, hydrogen sulphide, hydrogen chloride, ammonia,trace metals), and polluted by ammonia [6, 7]. Most downstream applications have vigorous syngas cleaning requirements depending on the application [8], as indicated in Table 2. As a result, pollutant levels must be reduced through gas cleanup in order to meet the criteria of downstream applications. The contaminants such as SPM, tar, hydrogen sulphide, carbon dioxide, ammonia, carbonyl sulphide, carbon disulphide, hydrogen chloride, hydrogen cyanide in the syngas generated during power production and other uses is the reason of various operational and technical problems which are presented in Table 3 [9]. The catalysts used in the FT synthesis, methanol synthesis, and other chemical synthesis are susceptible to such impurities, and particularly sulfur and nitrogen based compounds, which contaminating the FT/Methanol catalysts. The used reactor in downstream process got fouledwith alkaline metals and tars during accumulation of these impurities on the catalysts surface also taint the end products. Thus, regressively cleaning of the syngas is essential before to entering the FT/Methanol reactor.Development of innovative cleaning processes to fulfill the demands of downstream syngas processing is a major technical hurdle.

Table 1. Classification of Impurities

Type	Example
Organic impurities	Tars, Benzene, Toluene, and Xylenes, methane, acetylene, ethylene, ethane
Inorganic impurities	Sulfur compounds (H_2S, COS, CS_2), Nitrogen compound (NH_3, HCN), Chlorine species such as HCl, KCl and NaCl, tars, ash
Other impurities	Dust, Soot, Particulate matter

Table 2. Limit of impurities for different downstream applications.

Contaminants	Gas Turbine	FT- Synthesis	Methanol Synthesis
Tar (mg/Nm^3)	NA	< 0.1–1	< 1
Sulphur Contaminants (ppmv)	< 20	0.01	< 1
Nitrogen Contaminants (ppmv)	< 50	0.02	0.1
Alkali (ppmv)	< 0.02	0.01	NA
Halides (ppmv)	< 0.1	0.01	0.1

Table 3. Contaminants and removal techniques of syngas.

Contaminants	Example	Issue	Removal Technique
Particulate	Ash, char	Erosion	Cyclone, Filter, ESP
Tar	Cyclic & polycyclic hydrocarbon	Clogging, Deposition	Physical, chemical & catalytic methods
Alkali Metal	Sodium & potassium compounds	Hot gas corrosion	Gas cooler + cyclone/ESP
Fuel Nitrogen	NH_3, HCN	NO_X	scrubbing
Sulphur	H_2S, SO_2	Corrosion	Scrubbing, activated carbon
Chlorine	HCl	Corrosion, catalyst poisoning	Scrubbing, activated carbon

2 Syngas Cleanup

Syngas cleanup is a broadarea for eliminating the impurities associated with syngas obtained from gasification. [10–12] A variety of criteria influence the selection of a gas cleaning system, including the target emission level, cost, available space, temperature, gas composition, and feedstock quality. Gas cleaning requirements for downstream applications such as FTS, Methanol, and other chemicals are indispensable, hence gasification and gas cleaning account for 60–70% of total downstream process costs. [13] Syngas cleaning methods can be categorised into types:

(i) Primary or 'in-situ' methods: This approach is used within the gasifier through specialized design solutions and careful management of the reactor operating parameters. Primary techniques are also utilized to remove tar from the gasification of biomass, whereas operating parameters (temperature, gasifying agents, equivalence ratio, and residence time) and added active bed materials are critical. The basic approaches are selection of operating condition, bed additives or catalysts in the gasifier, and gasifier design. The pictorial representation of primary method is mentioned in Fig. 2.

(ii) Secondary methods: The secondary methods do not interfere with the gasifier operation.Secondary techniques are only utilized in downstream processes, and categorized as hot (Dry) and cold (Wet) processes. Wool- cock and Brown [14] have observed that hot gas clean up generally performed at ($T > 300$ °C)and gas cleanupin cold condition ($T < 100$ ◦C).The pictorial representation of secondary method is mentioned in Fig. 3.

Secondary method is also categorized as dry gas cleaning (hot gas cleaning) and wet gas cleaning (cold gas cleaning).

(a) Dry gas cleaning: Before gas cooling at temperatures normally exceeding 500 °C, electrostatic precipitators, cyclones, hot gas filters, rotating particle separators (RPS), (mostly ceramic filters), solid sorbents, and other cleaning procedures are utilized.

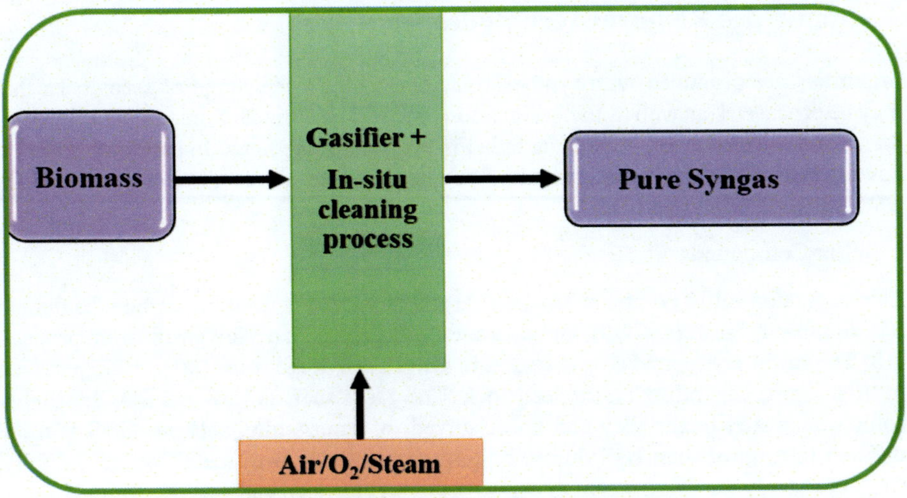

Fig. 2. Impurities removal via Primary method.

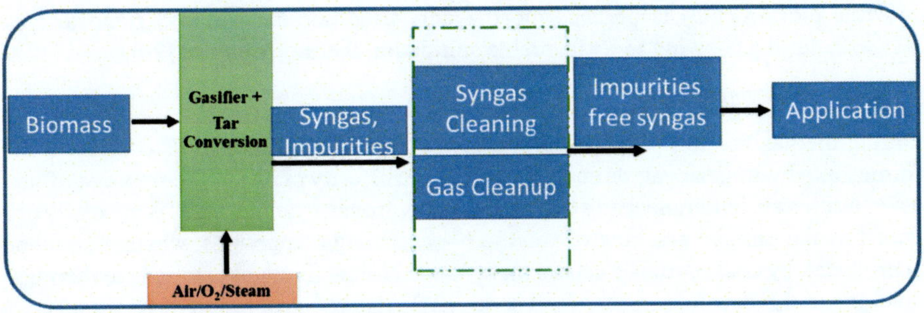

Fig. 3. Impurities removal via Secondary method.

Dry gas cleaning with bag filters, cloth filters, and sand bed filters is commonly performed after gas cooling at temperatures below 200 °C. This cleaning process advantageous as the sensible and latent heat is not wasted.

(b) Wet gas cleaning: This method of gas cleaning has been tested and used effectively in industrial applications. Traditionally, the process involves syngas cooling, low-temperature filtering, and scrubbing with several solvents at temperatures ranging from 50 to 100 °C. [15] Raw syngas from a gasifier is extremely hot and contains several contaminants such as particles, soot, undesired gas components such as acid gasses, and so on. Wet gas cleaning (in scrubbers, wet electrostatic precipitators, wet cyclones, and so on) is typically done at temperatures < 100 °C. It is inefficient in terms of thermal efficiency, consumes water, and produces waste water sludge [16]. The principal disadvantages are the energy penalty on overall plant efficiency imposed by syngas cooling, as well as the increased cost of treating or disposing of polluted streams in order to fulfil expanding environmental regulations [17].

3 Impurities Removal Techniques from Syngas

Impurities are produced with synthesis gas due to the wide range of feedstocks that may be processed, as well as the gasification procedures utilized. Some of the heteroelements and inorganic compounds included in syngas are predicted to have an impact on downstream process and performance. The cleaning process for elimination of various contaminants is detailed here.

i. Sulfur Compounds

The associated sulfur compound in the coal and biomass converted to carbon disulphide (CS_2), carbonyl sulfide (COS), and hydrogen sulfide (H_2S), during gasification process. [18] The sulfur compound concentration in syngas can range from 100 to 1000 parts per million (ppm) depending on the feedstock. The gases such as H_2S and SO_2 are highly soluble in water, hence they can be dissolved in water/amine solvent. COS is more difficult to remove than H_2S due to its limited reactivity with metal oxides. Initially, COS is hydrolyzed using steam to convert H_2S on activated carbons (Eq. 1). [19].

$$COS(g) + H_2O(g) \rightleftharpoons H_2S(g) + CO_2(g) \tag{1}$$

Zinc oxide (ZnO) is perhaps most widely used sorbent for H_2S removal from syngas/natural gas owing to its favorable sulfidation thermodynamics [20].

ii. Nitrogen Compounds

During the gasification process, the nitrogen components of the feedstocks are transformed into hydrogen cyanide and ammonia. The majority of nitrogen compounds found as impurities are ammonia, and a tiny amount is hydrogen cyanide. HCN is primarily produced by the thermal cracking of volatile N-containing compounds, whereas ammonia is produced by coal pyrolysis causes more stable N-containing compounds to thermally cracking in the developing char (and soot). [21] Also the generation of NH_3 and HCN depends on many factors [22–24], such as the nature of feedstocks, operating conditions of the gasification process (e.g. pressure, temperature), and gasifier type (fixed bed, fluidized bed, entrained flow reactor). NH_3 is highly soluble in water and may be removed by water/acidic solution. In addition, Ni-based reforming catalysts have been found to increase NH_3 breakdown to N_2 and H_2 [25–27]. HCN is soluble in water, however full elimination requires a catalytic conversion. At medium temperatures, HCN hydrolysis proceeds according to Eq. (2). [24].

$$HCN(g) + H_2O(g) \rightleftharpoons NH_3(g) + CO(g) \tag{2}$$

iii. Halogen

In syngas, gaseous halide chemicals such as hydrogen chloride, hydrogen fluoride, potassium chloride, and sodium chloride are present. In the gas phase, hydrogen chloride can react with ammonia to produce additional pollutants such as sodium chloride (NaCl) and ammonium chloride (NH_4Cl). Such contaminants generate significant deposition and fouling in downstream processes [25]. Halides and alkalis are considered the most corrosive. Chloride occurs in both organic and inorganic forms and is frequently found in coal in concentrations ranging from 50 to 2000 ppmv. [26] Thehalogen contaminants are highly soluble in water, and hence they can be readily dissolved in water through the scrubber.Two-stage basic scrubbers have a very high effectiveness in pollution absorption for gases such as hydrogen bromide, hydrogen chloride as well as for mercury. It is worth mentioning that two-stage operation of wet scrubbers may efficiently remove contaminants below regulatory standards. [27].

4 Techniques for Removal of Organic Impurities

Tar is defined as any hydrocarbon with a molecular weight larger than benzene (78 g/mol) [28]. Tar is hydrophobic in nature, so water is not used as a scrubbing medium for tar removal. Tar in syngas can clog pipelines and valves, as well as corrode metal equipment [28]. Tar condenses in the low-temperature zone of downstream applications, causing a variety of problems such as blocked narrow pipelines, filters, and other equipment, as well as catalyst deactivation. [29, 30].

One of the significant technological difficulties in the commercial development of modern gasification systems is tar cleaning [31]. The first technique pertains to coreaction within the gasifier (primary procedures) and secondary ways for cleaning the heated gas after gasification. Despite the secondary approaches' demonstrated efficacy of primary tar elimination techniques is better as compare to secondary cleaning approach. In a primary treatment, gasification process is optimized to generate a gas with the lowest tar content.Primary methods may include (i) the selection of certain operational conditions, (ii) the catalyst which is used in the bed, and (iii) gasifier modification, for minimization of the tar content. Operating parameters which includes gasifying agent, temperature, equivalency ratio, and residence duration are crucial for the minimization of the tar content [32].

In the secondary method the scrubbers, which condensate tars, are used to purify the gas. Thermal cracking, partial oxidation, catalytic removal, and plasma can all be used to convert the tar component into the incondensable gas component (carbon mono oxide, Hydrogen, Methane, carbon dioxide, and nitrogen). Their findings revealed that when the temperature reached 1300 °C [33], tar may be entirely broken. Tar can also be removed chemically using thermal (partial oxidation) or catalytic cracking (dolomite or nickel catalyst). These methods are often expensive and need an additional expenditure of energy in some way to remove the tar.

Metals Impurities in synthesis gases can include chemical species including metal atoms likeHg, Fe, or Ni. Mercury may be present in hydrocarbons in three forms: elemental mercury, mercurous oxide, and particle-bounded mercury [34]. The mercury level of petcoke and coal can range between 50–500 mass.ppb (dry basis) [34]. Numerous research has also looked at the quantity of Hg in feedstocks, studies show that mercury biomass content can range from 1 to 40 mass.ppb (dry basis). [35] Certain feedstocks have mercury concentrations as high as 40 mass.ppb, which is equivalent to the usual amount in petcoke. Several studies have been conducted on the type of mercury emitted by coal combustion and gasification. The high-temperature gasification will convert the various mercury species into Hg^0. Metal carbonyls (M-CO) are typical contaminants in CO-containing gases. The gaseous harmful species known as M-CO are produced when synthesis gases are cooled, primarily iron carbonyl $Fe(CO)_5$ and nickel carbonyl $Ni(CO)_4$ (other carbonyls, such as cobalt compounds, are solid in the working conditions of the XTL chain). They are produced by the interaction of CO with metals (or metal-containing compounds like sulfides).

The initial phase in the exhaust gas cleaning process is dust removal, which might comprise wet or dry/semi-dry fly ash separation techniques. Fly ash comprises a many inadequate combustion product (such as PAHs and soot) as well as heavy metals. Various physical forces and methods have been utilized in dry fly ash separation techniques, including centrifugal force, electrical force, gravity force, diffusion, and inertial impaction, in the removal process. Cyclones, venturi scrubbers, filters, and ESPs are the most popular types of fly ash removal systems [35].

5 Conclusion

This review article gives a syngas cleanup overview for successful utilization of syngas derived from coal, biomass, and municipal solid waste. Syngas cleaning, conditioning, and conversion account for a significant portion (30%) of the entire capital expenses for a gasification technology to downstream process. It is worth noting that the relative quantity and prevalence of various pollutants might vary greatly based on the concentration of each precursor in the feedstocks, such as the nitrogen, sulfur, halogens, and metals, as well as operational conditions.Raw syngas must be treated to eliminate contaminants that might otherwise contaminate the catalysts and hamper the downstream process. Individual pollutants can be controlled independently, but to achieve acceptable levels based on end use, an integrated, multistep strategy need to be adopted. Cooling the syngas stream and using a liquid scrubbing or absorption device is the most adaptable option. The chemical energy in the tars is lost as a result of the cooling of the syngas, and a waste stream is generated that must be treated or disposed of, the total thermal efficiency of the process is diminished. Ultraclean syngas, a process scheme with fewer separation steps has been developedand employs unique sorbents capable of removing many contaminants in a single process.

Acknowledgement. Authors acknowledge to Director, CSIR- CentralInstitute of Mining and Fuel Research, Dhanbad for allowing to publish this manuscript.

References

1. Eule, S.: A look at IEA's new global energy forecast." Global Energy Institute, U.S Chamber of Commerce (2018). https://www.globalenergyinstitute.org/look-ieas-new-global-energy-forecast
2. Gupta, P.K., Kumar, V., Maity, S.: Renewable fuels from different carbonaceous feedstocks: a sustainable route through Fischer-Tropsch synthesis. J. Chem. Technol. Biotechnol. **96**, 853–868 (2021). https://doi.org/10.1002/jctb.6644
3. Basu, P.: Combustion and Gasification in Fluidized Beds. CRC Press (2006). https://doi.org/10.1201/9781420005158

4. Hu, J., Yu, F., Lu, Y.: Application of Fischer-Tropsch synthesis in biomass to liquid conversion. Catalysts **2**, 303–326 (2012). https://doi.org/10.3390/catal2020303
5. Gupta, P.K., Kumar, V., Maity, S., Datta, S., Kishore Gupta, G.: A Review on Conversion of Biomass to Liquid Fuels and Methanol through Indirect Liquefaction Route. ChemistrySelect **7**, e202203504 (2022)
6. Cui, H., Turn, S.Q., Keffer, V., Evans, D., Tran, T., Foley, M.: Contaminant estimates and removal in product gas from biomass steam gasification. Energy Fuels **24**, 1222–1233 (2010). https://doi.org/10.1021/ef9010109
7. Ohtsuka, Y., Tsubouchi, N., Kikuchi, T., Hashimoto, H.: Recent progress in Japan on hot gas cleanup of hydrogen chloride, hydrogen sulfide and ammonia in coal-derived fuel gas. Powder Technol. **190**, 340–347 (2009)
8. Boerrigter, H., Calis, H.P., Slort, D.J.: Cleaning for Integrated Biomass Gasification (BG) and Fischer–Tropsch (FT) Systems. ECN Report ECN-C–04–056; Energy research Centre of the Netherlands (ECN): Petten, The Netherlands (2004)
9. Gray T. G., Berger M, Produced by MITRE Corporation for National Renew. Energ. Laboratory, USA, under contract number AL-4159 (1996)
10. Caballero, M.A., Corella, J., Aznar, M.P., Gil, J.: Biomass gasification in atmospheric and bubbling fluidized bed: effect of the type of gasifying agent on the product distribution. Biomass Bioenerg. **17**, 389–403 (1999). https://doi.org/10.1016/S0961-9534(99)00055-0
11. Caballero, M., Corella, J., Aznar, M., Gil, J.: A Biomass gasification with air in fluidized bed. Hot gas cleanup with selected commercial and full-size nickel-based catalysts. Ind. Eng. Chem. Res. **39**, 1143–1154 (2000). https://doi.org/10.1021/ie990738t
12. Datta, S., et al.: "Co-gasification of high ash Indian coal-biomass blends in a pilot-scale fluidized bed gasifier", Biomass Conv. Bioref. **10**, 831–838 (2020). https://doi.org/10.1007/s13399-019-00567-2
13. Martinelli, M., Gnanamani, M.K., LeViness, S., Jacobs, G., Shafer, W.D.: An overview of Fischer-Tropsch Synthesis: XtL processes, catalysts and reactors. Appl. Catal. Gen. **608**, 117740 (2020). https://doi.org/10.1016/j.apcata.2020.117740
14. Woolcock, P.J., Brown, R.C.: A review of cleaning technologies for biomass-derived syngas. Biomass Bioenergy **52**, 54–84 (2013). https://doi.org/10.1016/j.biombioe.2013.02.036
15. Hamelinck, C.N., Faaij, A.P.C.: Future prospects for production of methanol and hydrogen from biomass. J. Power. Sources **111**(1), 1–22 (2002). https://doi.org/10.1016/S0378-7753(02)00220-3
16. Tsukada, M, Abe, K, Yonemochi, Y, Ameyama, A, Kamiya, H. Kanaoka, C.: Dry gas cleaning in coal gasification system for fuel cell. Advanced gas cleaning technology, pp. 448–54 (2005)
17. Alauddin, Z.A.B.Z., Lahijani, P., Mohammadi, M., Mohamed, A.R.: Gasification of lignocellulosic biomass in fluidized beds for renewable energy development: a review. Renew. Sustain. Energy Rev. **14**, 28522862 (2010). https://doi.org/10.1016/j.rser.2010.07.026
18. Kurkela, E.: Formation, 1996. "Removal of Biomass-Derived Contaminants in Fluidized-Bed Gasification Processes," VTT Publications, p. 47−87 (1996). https://www.osti.gov/etdeweb/biblio/442486
19. Shang, G., Shen, G., Liu, L., Chen, Q., Zhiwei, Xu.: Kinetics and mechanisms of hydrogen sulfide adsorption by biochars. Biores. Technol. **133**, 495–499 (2013). https://doi.org/10.1016/j.biortech.2013.01.114
20. Hong, C., Scott, Q., Turn, V.K., Donald, E., Thai, T., Michael. F.: Contaminant Estimates and Removal in Product Gas from Biomass Steam Gasification, 24, pp. 1222–1233 (2010), https://doi.org/10.1021/ef9010109
21. Lin, J.Y., Zhang, S., ZhangL., Min Z., Tay H., Li, C.Z.: HCN and NH3 Formation during Coal/Char Gasification in the Presence of NO. Environ. Sci. Technol. **44**, 3719–3723 (2010). https://doi.org/10.1021/es1001538

22. Leppa¨lahti, J., Koljonen, T.,: Review nitrogen evolution from coal, peat and wood during gasification: literature review. Fuel Process. Technol. **43**, 1–45 (1995)
23. Orikasa, H., Tomita, A.: A study of the HCN formation mechanism during the coal char gasification by O_2. Energy Fuels **17**, 1536–1540 (2003)
24. Kröcher, O., Elsener, M.: Hydrolysis and oxidation of gaseous HCN over heterogeneous catalysts. Appl. Catalysis B: Environ. **92**(1–2), 75–89 (2009). https://doi.org/10.1016/j.apcatb.2009.07.021.
25. Kuramoto, K., Hosokai, S., Matsuoka, K., Ishiyama, T., Kishimoto, H., Yamaji, K.: Degradation behaviors of SOFC due to chemical interaction between Ni-YSZ anode and trace gaseous impurities in coal syngas. fuel process. Technol. **160**, 8–18 (2017)
26. Vassilev, S.V., Eskenazy, G.M., Vassileva, C.G.: Behaviour of elements and minerals during preparation and combustion of the Pernik coal, Bulgaria. Fuel Process. Technol. **72**, 103–129 (2001)
27. Tokumura, M., Baba, M., ZnadH, T., Kawase, Y., Yongsiri, C., Takeda, K.: Neutralization of the acidified seawater effluent from the flue gas desulfurization process: experimental investigation, dynamic modeling, and simulation. Ind. Eng. Chem. Res. **45**, 6339–6348 (2006). https://doi.org/10.1016/S0378-3820(01)00186-2
28. Maniatis, K., Beenackers, A.A.C.M.: Tar protocols. IEA bioenergy gasification task. Biomass Bioenerg. **18**, 1–4 (2000). https://doi.org/10.1016/S0961-9534(99)00072-0
29. Tao, K., Ohta, N., Liu, G., Yoneyama, Y., Wang, T., Tsubaki, N.: Plasma enhanced B. Li, C.F. Magoua Mbeugang, Y. Huang et al. Energy **244** 123172 (2022)
30. Kai, T., Naoko, O., Guiqing, L., Yoshiharu, Y., Tiejun, W., Noritatsu, T.: Catalytic reforming of biomass tar model compound to syngas. Fuel **104**, 53–57 (2013). https://doi.org/10.1016/j.fuel.2010.05.044
31. Milne, T.A., Abatzoglou, N., Evans, R.J.: Biomass Gasifier "Tars": Their Nature, Formation, and Conversion (1998). https://www.nrel.gov/docs/fy99osti/25357.pdf
32. Pereira, E.G., Da Silva, J.N., de Oliveira, J.L., Machado, C.S.: Sustainable energy: A review of gasification technologies. Renew. Sust. Energ. Rev. **16**, 4753–4762 (2012). https://doi.org/10.1016/j.rser.2012.04.023
33. Yan, N.Q., et al.: Using bromine gas to enhance mercury removal from flue gas of coal-fired power plants. Environ. Sci. Technol. **43**, 5410–5415 (2009)
34. Mark Wilhelm, S., Bloom, N.: Mercury in petroleum. Fuel Process. Technol. **63**(1), 1–27 (2000). https://doi.org/10.1016/S0378-3820(99)00068-5
35. Kamran Ghasemzadeh, S.M., Tilebon, S., Basile, A.: Conventional systems for exhaust gas cleaning and carbon capture and sequestration. In: Current Trends and Future Developments on (Bio-) Membranes, pp. 65–96. Elsevier (2020). https://doi.org/10.1016/B978-0-12-817807-2.00004-6

Check for updates

Refracturing in CBM Wells: A Case Study

Anjani Kumar[1] and G. P. Karmakar[2(✉)]

[1] Essar Oil and Gas Exploration and Production Limited, Raniganj, India
[2] Indian Institute of Technology, Kharagpur, India
gpkarmakar@gmail.com

Abstract. Essar Oil and Gas Exploration and Production Limited (EOGEPL), located in Raniganj under Paschim Bardhhaman district of West Bengal, India is spread over a 500 square kilometres area in the prolific Damodar Valley Basin with thick Permo-Carboniferous Gondwana coal seams. Nearly 350 wells have been operating since last 10 years. From the production data of more than 350 wells, it has been observed that some of the wells have been underperforming while compared with their offset wells. It has also been observed that the production from some of the wells have declined over a period of time. EOGEPL opted for refracturing as a stimulation technique to increase the production from such wells.

In inefficient initial fracturing category, 14 wells were selected for refracturing and it resulted to increase in gas production from 18,004 Standard Cubic Meters Per Day (SCMD) to 28,891 SCMD. In next category where stimulation bypass was suspected, 13 wells were selected in which gas production increased from 986 SCMD to 4,167 SCMD. In last category of wells where formation damage was suspected and due to which substantial decrease in production was observed, 11 wells were selected in which production decreased from 18,165 SCMD to 14,058 SCMD.

1 Introduction

In India, the 'demand vs. supply' gap in the energy sector has increased the focus on developing all possible energy sources. Due to low permeability seam, most of our coalfield CBM exploration rate is low. CBM and shale gas may offer the potential to contribute significantly to meet clean energy demand for the next decades. At the locations with minimum strength and minimum external resistance, hydraulic fracturing is carried out. Vertical fracturing well was mostly used to improve the exploration rate. For a long time, the hydraulic fracturing technology of CBM vertical wells has been used for developing unconventional gas resources (Hongchao, W. et al., 2011). Hydraulic fracturing plays a key role by creating artificial fractures in the reservoir rock by pumping high-pressure sand-laden fluid which has a less damaging effect on the formation and is cost-effective. Development of a CBM well involves challenges right from the start of drilling due to reservoir pressure being near to hydrostatic leading to significant mud losses while drilling activity. Since CBM reservoirs are intrinsically low in permeability, surface, subsurface and drilling conditionsare necessary to optimize the well locations for the well site optimization stage. The geological conditions including coal bed structure,

A. Sinha et al. (Eds.): AMC 2023, SPEES, pp. 317–324, 2023.
https://doi.org/10.1007/978-3-031-46966-4_25

buried depth, coal seam thickness, roof and floor lithology, gas content, permeability and hydrological conditions (Xin, Z. et al., 2016). Thus, increasing the dewatering rate and achieving critical desorption pressure. Based on the mineral identification and acid sensitivity test, acid fracturing in coal bed methane field has shown that the acid helps in increasing the permeability of coal seams (Yanga, Y. et al., 2016).The optimization of CBM production using horizontal well technology along with the key factors influencing the horizontal well productivity has been shown that the gas production rate from the horizontal well was greater than that of the fractured vertical well (Zhu, Q. et al., 2016). Combination of altering stress with shearing natural fractures, the multidirectional fracturing method could stimulate the CBM reservoirs and will increase the gas production from CBM reservoirs (Li, T. et al., 2017). Hydraulic fracturing using pressure control, flow control and sand ratio control of fracturing fluid has shown that5.67- and 12.88-times higher rate of gas production than those of the regular well, respectively been reported (Hu, G. et al., 2018).

Initial hydraulic fracture treatments can lose effectiveness over the years resulting in poor productivity because of gradual damage that occurs over the life of a well.Examples include loss of fracture conductivity from proppant crushing, proppant flow back, high skin factor, residual polymer damage from frac fluids, early screen-outs, fines or scale deposits, frequent workover operation for sand cleanout using polymer and fracture reorientation due to changes in stress regime. Hydraulic fracturing experiments for coal with triaxial core holder apparatus show that the initiation pressure is inversely proportional to the horizontal stress difference ($\Delta\sigma$) and positively related to fracturing fluid injection rate. The initiation pressure and fracture width decrease with the increasing of $\Delta\sigma$, when vertical stress (σv) is constant. In CBM reservoirs, when the fracture propagates perpendicular to the direction of minimum horizontal stress (σh), natural fractures can be connected by main fracture, resulting in secondary fractures and fracture network. The fracture morphology and propagation become complex when two stresses of crustal stress are close and far different from the third one (Zhang, F. et al., 2018). The fitting pressure using the fracturing pressure decline analysis under the condition of corrected dynamic overall leak-off coefficient have been studied. The post-fracturing permeability of coal reservoir is predicted by the injection well testing method (Zhang, X. et al., 2019).Five evaluation indexes are established to evaluate the fracturing effect of coal-bed methane wells, namely, horizontal in situ stress difference, fracturing friction, net pressure, fracture morphology, and fracturing curve shape. To predict the fracturing effect and capacity risk assessment of coal-bed methane wells, a model for evaluating the classification of fracturing effect is established by the support vector machine (Chen, Z. et al., 2020).Fracturing through coiled tubing carrying a single packer at the bottom has the advantages of simplicity, low price and good effect on development, which can be used for the L-type horizontal wells of the casing fracturing (Lian, X. et al., 2021).

Production enhancement measures like re-fracturing of existing wells have become essential in order to increase/sustain production from the existing CBM wells. Advances in completion and stimulation technology over a period of time provide opportunities to re-stimulate wells that were originally completed using an older technology. In CBM wells, re-stimulation can help to bypass near wellbore damage and increase the stimulated rock volume by connecting new fracture channels and restoring the original proppant pack permeability. An initial period of production can also alter formation stresses, allowing the new fracture to reorient along with the changed azimuth. As a result, re-fracturing often restores well productivity to near original or at even higher rates.

2 Objectives

1. To identify the methods for selecting re-fracturing candidate wells in a mature field.
2. To identify the cause of underperformance by the stimulated wells by comparing the performance of the candidate wells with offset wells.
3. To investigate the initial fracturing design and the major parameters.
4. To develop a new methodology for successful and cost-effective re-fracturing treatments.
5. To prepare operational case studies and prepare post-re-fracturing pressure diagnostic data for future use.

3 Treatment Methods

Re-fracturing is a stimulation treatment targeting the previously fractured reservoir after a significant period of production from the wells. In the Raniganj CBM field, it was found that the original fracturing treatment was inadequate or an existing proppant pack had deteriorated over time in some producing wells. This resulted in the decline of gas production from their earlier stable peak production.

The case studies carried out to fulfil the following two major objectives:

Screening Criteria of Wells for Re-fracturing: Identified candidate wells for re-fracture using production statistics, virtual intelligence, initial stimulation parameters, cumulative recovery of reservoir fluids, well spacing, coal gas content & saturation, initial production rate, drawdown, initial stimulation job size, well testing & type curves. In the Raniganj CBM field, 30 low-producing wells were selected on the basis of parameters a) Delay in putting the well on production post stimulation resulting in damage to formation due to guar residue. b) Screen-out in the middle of the job resulting in a smaller stimulated reservoir volume. c) Poor performance of the well in comparison to offset wells due to formation damage.

Designing & Execution: Developed a new technology for selective re-fracturing using Coil tubing or through work over conveyed 2 7/8" tubing with a hydraulic packer at the bottom hole with improved treatment designs using the multi-proppant system to increase the length of the fracture and prop the natural fissures with intermediate size proppants. Low viscosity fluid (15 lb linear gel) pre-pad was used as to reduce the shear on the viscous pad (30lb linear gel) resulting in achieving longer half-length. Cleaner fluids were also used to maximise the conductivity of the proppant pack. While designing the re-frac jobs, special attention was kept on maximising dimensionless conductivity values in order to increase the well deliverability. Considering the above changes while designing the jobs helped in achieving longer half-length, negative skin, and new coal cleats being connected due to stress re-orientation and thus, maximising the overall productivity of the well-being re-fractured.

The following steps were followed for the design and execution of re-fracturing treatment.

1. Rigged-up the service rig of appropriate hook load capacity and de-complete the well, ensuring the well integrity with gauge ring run via wireline unit.
2. Carried out wire-line perforations for all jobs using a mixture of deep penetration and big hole charges.
3. Rigged-up service rig and run in hole (RIH) suitable hydraulic/mechanical retrievable packer assembly on 2 7/8″ 6.5 lb per foot (ppf) N80 tubing. RIH up to target depth for Job 1. The purpose of the packer is to isolate the annulus during fracturing through the tubing.
4. Mobilized frac equipment's like frac pump (6000 hydraulic horsepower (HHP) HHP)), blender, hydration unit, frac tanks (a large, movable heavy gauge steel storage tank, typically holding up to 21,000 gallons (500 BBL), and sand king (rated at 30,000 lbs/min.), spot and rig up at the desired location.
5. Hanged-off tubing in the tubing head and hanger assembly.
6. Rigged-up the Hydrofracturing (HF) Unit, flow-back manifold: 2″ iron, and the 3″ surface line to Frac head and pressure tested at desired pressure.
7. Spotted and squeezed 18 BBL of 15% HCl acid at 3–4 Barrels per minute (BPM) on the perforations.
8. Carried out Frac job as per the treatment schedule and incorporate decisions on the fly based on the pressure response encountered from the formation during injectivity, acid job, and frac job itself.
9. Job designed as per following Table 1, Start pad stage followed by sand stages, and displace job with fresh water. Shut down the pumps and observe the pressure bleed for the fracture closure for up to 30 min. Figure 1 shows the actual job plot consisting of pumping rate, surface, and bottom hole treating pressure with time.
10. Once the job is completed, monitored the pressure decline, waited for fracture closure to arrive, then allowed to flow the well back to the surface pit till zero pressure through a fixed or adjustable choke rigged up with treating lines
11. Released the packer by shearing pins at 29,000 lbs over-pull, and waited for 15 min to deflate the element and pulled out of the hole (POOH). Took care until the packer was above perforations, and then POOH at a normal speed, laid down packer assembly, and Mast downed service rig.
12. Rigged-up Wire-line unit, RIH, and tagged sand, POOH. If necessary, designed a sand plug to isolate the previously fractured zone with sand.
13. Rigged-up the frac head and pump sand plug as designed. Repeat until the sand plug is at a desired depth. Meanwhile, service the packer, and redress the packer if required.
14. If the sand plug was at an appropriate depth, made up the tubing string with the packer and prepared for RIH for the next job.

Considerable Points During Job Execution.

1. Fluid efficiency was found to be around 50% from Minifrac analysis. So, pad percentage was optimised at 30% to create sufficient fracture geometry.
2. Used 100 mesh size sand in suitable concentration (0.25–1.0 proppant added pounds per gallon (PPA)) to minimize the effect of near well bore tortuosity based on data obtained from step-up step-down test.

Table 1. WA-D1 job 1 frac design.

	Stage	Pumping Rate	Slurry Stage Volume	Time	Cumulative Volume	Proppant Conc Stage		Stage Proppant	Stage Proppant	Proppant Type	Cumulative Proppant
			bbl	min.	bbls	ppg		lbs	lbs/min		lbs
		bpm				From	To				
			0								
	1	15	300	20	300	0	0	-	-		-
	2	15	130	8.7	430	0.5	0.5	2,669.4	307	20/40	2,669
	3	15	40	2.7	470	0.5	2	1,987.2	736	20/40	4,657
	4	15	180	12	650	2	2	13,861.6	1,155	20/40	18,518
	5	15	40	2.7	690	2	4	4,435.9	1,643	20/40	22,954
	6	15	260	17.3	950	4	4	36,968.0	2,137	20/40	59,922
RCS	7	15	45	3	995	4	4	6,398.3	2,133	RCS	66,320
Sand Plug	8	15	7	0.5	1002	6	6	1,386.4	2,773	20/40	67,707
Flush	9	15	18.9	1.3	1021	0	0	-	-		67,707

7 bbl 6 ppg underflush

Sand count	4,043	lbs/ft	Pad	30%		Normal Sand	27.2	MT
Flush Volume	71.7	bbl	Thickness	5	mtr TVD	RCS	2.9	MT
Blender tub + Lines	*9*	*bbl*	Acid	18	bbl	%RCS	10%	

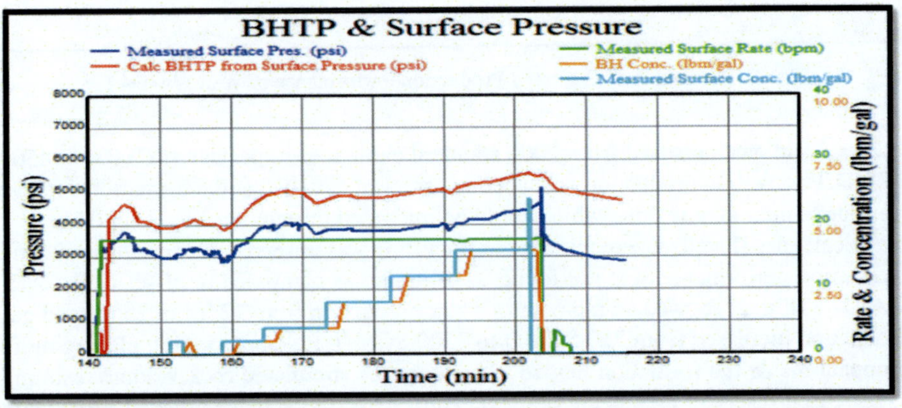

Fig. 1. Bottom hole treating pressure (BHTP) vs. Time, and Rate & Concentration (lbm/gal) vs. Time for WA-D1 Refractured Job 1

3. Planned for 4,000–6,000 lbs of proppant per net foot of coal inside the formation and executed as per plan.
4. Resin Coated Sand was used in tail slurry to minimize the sand production post re-fracturing of the wells.
5. Sand concentration was varied from 0.25 to 4.0 PPA. In some cases, extended up to 6 PPA depending upon on job pressure response.

4 Production Diagnosis

All three category of wells mentioned in screening criteria were successfully re-fractured as per above design and were put on production. Post re-frac in category I wells where delay in POP, well WA-DA1 showed an increase in water production from 10 m^3/day

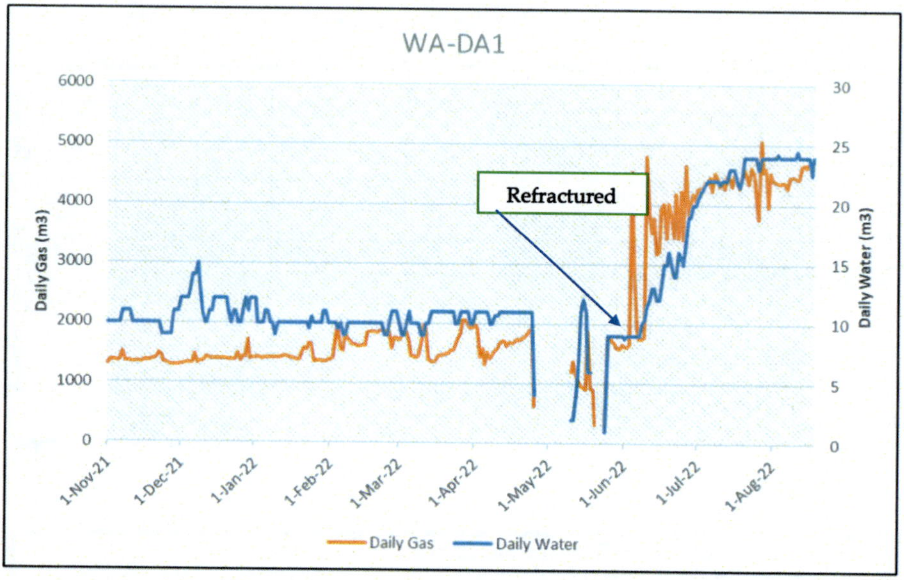

Fig. 2. Pre- and Post-Fractured Production Performance of Well WA-DA1.

to 25 m^3 and gas increased from 2000 standard cubic meters per day (SCMD) to 4500 SCMD. Post re-frac production enhancement showed clearly that the formation damage was due to guar gum residue which could be mitigated through re-fracturing (Fig. 1).

In category II wells where initial poor fracturing was observed due to screen-out and production performance was not found to be in line as compared to offset wells. Well WA-DA2 (Fig. 2) showed an increase in water production from 12 m^3 to 25 m^3 and gas production increased from 750 SCMD to 2500 SCMD. Re-fracturing by placing more proppant inside the formation helped to increase the stimulated rock volume resulting to increased reservoir area being drained. Thus, helped to achieve targeted production which was comparable to its offset wells.

In category III wells where poor performance of wells was observed due to formation damage. Well WA-DA3 was suspected to have formation damage because of XC polymer being used during well intervention. Initial fracturing parameters were good but still the well was taken for refracturing on the basis of formation damage due to polymer. Post re-frac in well WA-DA3 (Fig. 3) daily water production decreased from 40 m^3 to 30 m^3 and gas production remained constant at 900 SCMD. No gain in gas production led us to an understanding that with re-fracturing of these category wells, production could not be enhanced and thus after re-frac of 3 wells no further wells were selected for re-stimulation from this category.

5 Results and Discussion

Every re-fracturing operation is unique and economic success depends on integrating rigorous work flows for candidate selection, stimulation modelling, treatment execution and post re-fracture treatment evaluation. It was found that in wells having formation

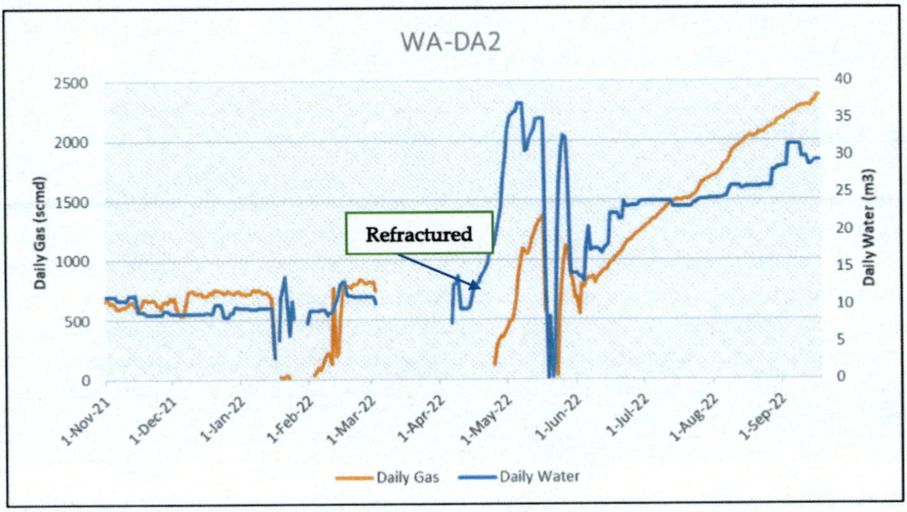

Fig. 3. Pre and Post Re-Frac Production Performance of Well WA-DA2

damage due to gel being not flowed back on time and low proppant volume pumped during initial fracturing. Productivity from these types of wells could be restored by re-fracturing of wells with optimum pumping design. In wells having polymer damage, productivity could not be restored through re-frac and another aspect beyond re-frac needs to be thought in these kinds of wells. An increase of 50% in water and gas

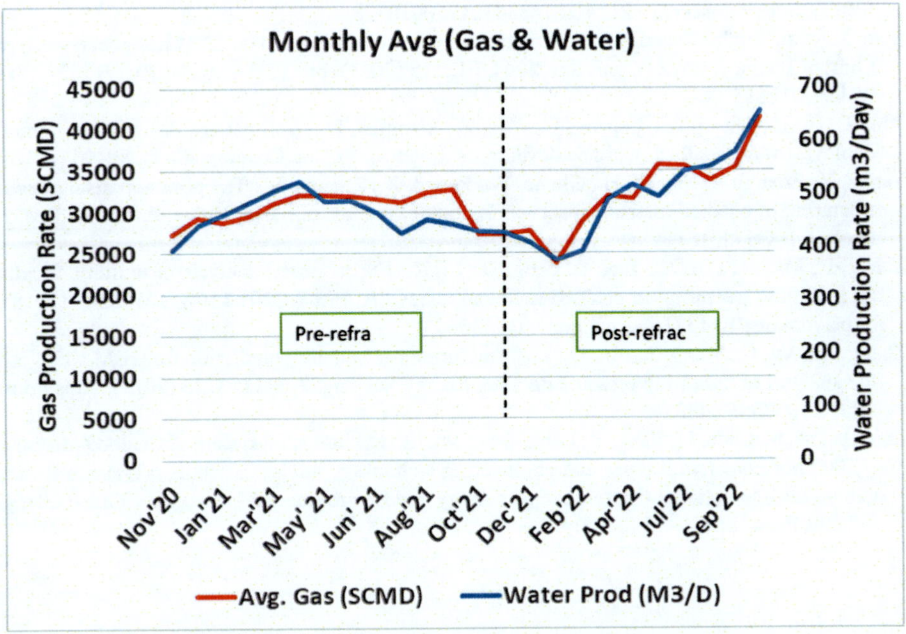

Fig. 4. Well production trend of 30 wells after re-fracing.

production was observed from the re-stimulation of 30 wells. Details of pre and post re-fracturing water and gas production are shown in Fig. 4.

6 Conclusions

This analysis will help in screening criteria of wells for re-fracturing of CBM wells in a cost-effective manner with optimum job design and thus helping to create a wider fracture and new reservoir area being stimulated resulting to an overall increase in water and gas production from the wells. Increased dewatering from re-frac wells will also help in depressurization of the seams encountered in nearby wells. Also, re-fracturing represents a viable alternative to drilling of new infill wells and will help operators to get net payback from incremental production with low capex and within specified time frame of a year.

References

Chen, Z., Sun, T.: Evaluation method of coal-bed methane fracturing in the Qinshui Basin mathematical problems. Engineering **2020**, Article ID 1545729 (2020). https://doi.org/10.1155/2020/1545729

Hongchao, W., Lianggang, L., Xiaoming, W., Yule, H.: The analysis and theory research on the factor of multiple fractures during hydraulic fracturing of CBM wells. Procedia Earth Planet. Sci. **3**, 231–237 (2011)

Hu, G., Sun, C., Sun, M., Qin, W., Linghu, J.: The case for enhanced coalbed methane using hydraulic fracturing in the geostructural belt. Energy Explor. Exploit. **36**(6), 1629–1644 (2018)

Li, T., Wan, J.: Study of cyclic fracturing in vertical CBM wells. Open Petrol. Eng. J. **10**, 108–117 (2017). https://doi.org/10.2174/1874834101710010108

Lian, X., et al.: High-efficiency fracturing technology and application on CBM horizontal wells in Qinshui Basin. J. Phys.: Conf. Ser. **2044**, 012115 (2021). https://doi.org/10.1088/1742-6596/2044/1/012115

Xin, Z., Bo, J., Qiang, X., Jiegang, L., Yue, Z., Piaopiao, D.: Well pattern design and optimal deployment for coalbed methane development. Petrol. Explor. Develop. **43**(1), 89–96 (2016)

Yanga, Y., Zhangb, C., Tianb, H., Chenc, W., Pengd, X., Zhanga, H.: The pilot appraisal of acid fracturing of coalbed methane reservoir in southeast Qinshui Basin, China. Earth Sci. Res. J. **20**(4), C1–C6 (2016)

Zhang, F., Ma, G., Liu, X., Tao, Y., Feng, D., Li, R.: Experimental analysis of multiple factors on hydraulic fracturing in coalbed methane reservoirs. PLoS ONE **13**(4), e0195363 (2018). https://doi.org/10.1371/journal.pone.0195363

Zhang, X., Wu, C., Wang, Z., Xu, D.: Post fracturing permeability prediction for CBM well with the analysis of fracturing pressure decline. Energy Sci Eng. **7**, 3111–3123 (2019). https://doi.org/10.1002/ese3.483

Zhu, Q., Tao, S., Chen, L., Yang, Y., Lang, S., Zhang, L.: Optimization system of Coalbed Methane (CBM) horizontal well types and engineering technology modes in southern Qinshui Basin, China. In: International Conference on Energy and Environmental Protection (ICEEP 2016), Atlantis Press, pp. 70–77 (2016)

Mine Health Safety and Environment

Reliable Intrinsically Safe Circuit Design Specifications for Oil Fields and Coal Mines

Amit Kumar[1](✉), Anup Kumar[2], Rakesh Kumar Mishra[1], Manoj Kumar Vishwakarma[1], Bishwajit Modak[1], Amar Kumar Singh[1], and Awanindra Pratap Singh[1]

[1] CSIR-Central Institute of Mining and Fuel Research, Dhanbad 826015, Jharkhand, India
amitkr@cimfr.nic.in

[2] National Institute of Electronics and Information Technology, Naharlagun 791110, Arunanchal Pradesh, India

Abstract. Electrical equipment installed in coal mines or oil fields must fulfil the requirements of anti-explosive or explosion-proof. The electrical sparks are frequent due to abnormal operations and these sparks may act as an ignition source for the explosive atmosphere on the field. Intrinsically safe circuits are the safest circuits that are being used in explosive hazardous environments. Since a regular advancement in the circuit is being done. Due to this, circuits are becoming complex day by day. Increasing complexity requires detailed attention while designing the printed circuit board layout. The circuitry designs of these apparatus have to meet the performance and design requirements as per the standard IS/IEC 60079-11. Most of the circuits consist of non-linear circuit components and energy storage components. The presence of these components makes the work difficult to use minimum ignition voltage current curves directly. Reliability in safety plays an important role in getting the continuous desired performance of any electrical equipment safely which is discussed in this paper. Simulation for the proposed issue was also carried out using MATLAB. The simulation results align with the output of spark test apparatus. This shows the reasonableness of the proposed reliability study.

Keywords: Explosion-proof · intrinsically safe · Output Short Circuit Energy

1 Introduction

The explosive hazardous atmosphere occurs frequently in the areas like mining and petrochemical industries. These areas generate flammable gases as by products. The chances of ignition or explosion of these flammable gases is more when the generation of spark takes place due to electrical installations. This gives rise to the need of explosion-proof equipment instead of normal electrical equipment.

This explosion-proof equipment reduces the risk of explosion by avoiding contact with ignition sources and flammable atmosphere. Two main factors on which the explosion occurrence depends are the source of ignition and the extent of hazards. Arc and

A. Sinha et al. (Eds.): AMC 2023, SPEES, pp. 327–336, 2023.
https://doi.org/10.1007/978-3-031-46966-4_26

spark in electrical installations are inherent due to various factors like insulation failure, short circuits, wrong selection of components, loose connections, switching contacts, etc. (McMillan, 1998).

Hazardous gassy areas are classified in zones (zone 0, zone 1 and zone 2) based on the duration of the occurrence of an explosive atmosphere. The international standard IEC 60079-10-1 deals with the area classification where flammable gases or vapour risks may arise. Equipment Protection Level (EPL) is an indicator that represents the safety level of any equipment. EPL for gassy hazards are classified as EPL Ga, EPL Gb and EPL Gc and for mines EPL Ma and Mb. The selection of Equipment for different zones are done as given in Table 1 (IEC 60079-11: 2023).

Table 1. Different zones and applicable equipment protection level

Zone	Duration of Occurrence	EPL	Level of Protection
Zone 0	Continuous (more than 1000 h/year)	EPL Ga	Very High
Zone 1	Occasionally (Between 10- 1000 h/year)	EPL Ga, EPL Gb	High
Zone 2	Short period (Less than 10 h/year)	EPL Ga, EPL Gb, EPL GC	Enhanced

In zone 0, only equipment with protection level EPL Ga is permitted which is highest level of protection. There are eight type of equipment protection techniques by which explosion in any hazardous may be prevented. These different type of protection techniques are Flameproof enclosures (Ex d), Intrinsic safety (Ex i), Increased Safety (Ex e), Pressurization (Ex p), Oil filled (Ex o), Sand filled (Ex q), Encapsulated (Ex m), and Non-sparking (Ex n) (Singh, 2023) (Mishra et al. 2005).

2 Overview of Effective Sources of Ignition

The source of ignition in a hazardous area can be anything that has the potential to generate a spark or high temperature. Some of the examples of Sources of ignition causes (Bottrill et al., 2005; Wang, 2011).

Sparks Mechanically created sparks (e.g., caused by friction, impact, or abrasion processes).

i. Electric sparks (Arcs Short circuit, switching operations, loose contact)
ii. Hot surfaces Heater, metal-cutting, heating up during operation.
iii. Flames and hot gases Combustion reactions, flying sparks during welding.
iv. Electromagnetic waves
v. Laser beam
vi. Lightning strike

3 Intrinsic Safety Ex 'i' and Assessment Methodology

Different methods of protection have been discussed so far. Each protection method have a different way of avoiding ignition. Increased safety method focus on the limitation of temperature and non-speaking components. Flameproof enclosures use the method of containment of explosive gases in enclosure volume. Flameproof (Ex d) enclosures are the oldest and most widely used protective method for installation in hazardous areas out of all those protection systems previously covered. In zones 1 and 2, but not zone 0, where hazardous gases are consistently present, flameproof equipment is not permitted. The safest way to guarantee that electrical equipment put in hazardous areas cannot result in a fire or explosion is through intrinsic safety. Assessment of Intrinsically safe apparatus is done on the basis of the procedure provided in IS/IEC 60079-11. Intrinsic safety is defined as the type of protection in which energy is restricted below the energy level which can cause ignition by sparking or heating effects. Two important assessment is described in this section (IEC 60079-11:2023).

Assessment of PCB Tracks thickness for permissible current:

In IS/IEC 60079-11, PCB track assessment table is provided for copper-made tracks which is highly popular in PCB making. This table is applicable to PCB having a thickness of more than 1.6mm with a single layer of copper track having a thickness 33μm. After selecting the current requirement corresponding to the given thickness the following factors of Table 2 are considered as per the standard to make the circuit safer.

Table 2. Assessment table for PCB tracks

a)	Specifications	The factor of reduction in Maximum current
b)	boards having a thickness (0.5 mm-1.6 mm)	divided by 1.2
c)	ambient temperature below 60 °C,	divided by 1.2
d)	ambient temperature below 80 °C	divided by 1.3
e)	boards having both sides conductive track	divided by 1.5
f)	copper thickness more than 18 μm	divided by 1.5
g)	copper thickness more than 70 μm	multiplied by 1.3
h)	For multilayer PCBs	divided by 2
i)	At terminations of components dissipating 0.25 W or more either normally or under fault conditions, and for 1.00mm along the conductor	divide the maximum current specified by 2 or multiply the track width by 3
j)	For tracks passing under components dissipating 0.25 W or more either normally or under fault conditions	divided by 1.5

Assessment of circuit as per annexure A of IS/IEC 60079-11

Assessment of simple inductive circuit: The assessment method is discussed in Table 3.

Given Parameters: Input voltage, current limiting resistor, load resistance.

Table 3. Assessment procedure of circuit.

i) Power supply	ii) Connection of loads
Step 1 - Calculation of Maximum Short circuit current $I_{scmax} = \frac{V_{max}}{CLR}$ Step 2 - Apply Safety Factor of 1.5 Step 3 - Compare in Table On the assessment basis, it can be said that the assessed Power supply is Intrinsically safe	Step 1 - Total Resistance after load connection ($R_{Total} = R_{input} + R_{Load}$) Step 2 - Maximum possible current $I_{max} = \frac{V_{max}}{R_{Total}}$ Step 3 - Apply Safety Factor 1.5 Step 4 - Check in the inductive load table Therefore, it can be said that the assessed circuit is Intrinsically safe

4 Factors Affecting Reliability of Intrinsic Safety

4.1 Environmental Influences

External ambient temperature of the installation area plays an important role in maintaining the explosion-proof property of an electrical equipment. Generally, there is no extra marking on equipment is required if the ambient temperature lies in the range of −20 °C to 40 °C.

Apart from the environment temperature, sometimes equipment are encountered with the heating elements like heaters, heat trace cable in pipe line etc. These external heating elements affects the cementing materials used for sealing or the gasket ('o' ring) used to block the ingress of water and dust inside the enclosure. Therefore, proper selection of Ex-proof equipment is important. Since all the explosion proof electrical equipment are designated by temperature class (T-class). T-class is specified to any apparatus according to the maximum surface temperature obtained during the testing. Temperature class is defined separately for Group-I (coal mines) and Group II (surface industries) in Table 4. (IEC 60079-0:2017).

4.2 Rating of Components

Intrinsic safety and its durability are highly dependent on the component's ratings. All the components in the circuit shall be suitably matched otherwise in long run, circuit gets over heated. This rise of temperature is not good for safety because it may reach or cross the minimum ignition temperature of the explosive gas in which it is installed. For intrinsic safety, maximum operational voltage, current and power are kept below

Table 4. Temperature classification as per IS/IEC 60079-0.

Group I		Group II	
T-class	Max. Surface temp.	T-class	Max. Surface temp.
Below 150 °C	Coal Dust layer is present	T6	85
		T5	100
		T4	135
Below 450 °C	No coal dust layer is present or Dust protected	T3	200
		T2	300
		T1	450

Table 5. Rating requirements of I.S. circuits as per IS/IEC 60079-11

Level of protection	Limits
Ex 'ia' and Ex 'ib'	$V_{operating} < 0.667 * V_{max}$ Ioperating < 0.667* Imax $P_{operating} < 0.667 * P_{max}$
Ex 'ic'	$V_{operating} < V_{max}$ $I_{operating} < I_{max}$ $V_{operating} < 0.667 * V_{max}$

the maximum rating of the components. Below Table 5, shows the limits for different protection levels.

Where, $V_{operating}$ is the voltage at which the circuit work properly.

V_{max} is the maximum voltage rating of the components.

$I_{operating}$ is the current at which the circuit work properly.

I_{max} is the maximum voltage rating of the components.

$P_{operating}$ is the required power in the circuit.

P_{max} is the maximum power rating of the components.

For the components like fuses, transformers, relays, optocouplers, and switches which only operated at its rating, there is no operation limit to work properly. Impedance matching, voltage, current, output inductance, output capacitances, inductance to capacitance ratio are important parameters to follow for external connection (IEC 60079-25:2020: Zhang et al., 2016).

4.3 Current Limiting Resistor (CLR)

Currents in the intrinsically safe circuits are limited by current-limiting resistors which are connected in series with the power source. To avoid short circuits or overcurrent in the circuit CLR are the most useful component in the circuit as shown in Fig. 1.

Current limiting resistors must be selected so that operation and safety both are maintained. According to IS.IEC 60079-11, the CLR can only be film type or wire wound or encapsulated printed resistors.

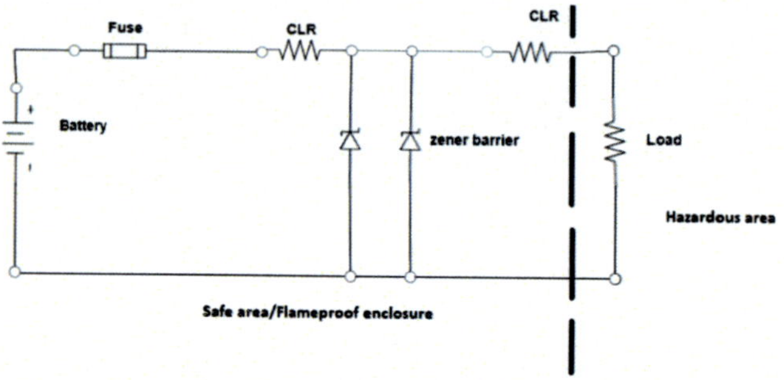

Fig. 1. A typical intrinsically safe circuit

4.4 External Connections and Separation Distances

Sometimes intrinsically safe (IS) circuits are not alone or inbuilt system, it has some association with the external system such type of systems are called associated apparatus. Associated apparatus area combination of IS and non-IS apparatus. The power supply for an IS circuit might be from a 220 V power line then a set of transformers, rectifiers, and filter regulators is required. In that condition, there must be a separation between IS and non-IS circuits according to the separation criteria provided in IS/IEC 60079-11. In intrinsic safety, it is important to have a proper separation between two conducting parts in the circuit such as terminals, PCB tracks, separation of IC legs, etc. Creepage and clearance are two parameters to represent the separation distances. Creepage denotes the distance between two conducting parts through the track on which it is mounted. Clearance is the minimum distance between conducting parts through air. Minimum requirements of creepage and clearances are varying according to the insulation materials between conducting parts as shown in Fig. 2.

4.5 More Parallel Zener Diode

Zener diodes act as a very good voltage regulator. Because of this property, it is very popular to use a Zener diode at the output terminal to get a constant voltage supply despite of fluctuations in the input side. Zener diodes also has the properties of sucking the excess current up to its Zener current rating (Singh et al., 2014).

Benefits of multiple Zener diodes.

a) Increased current observing capabilities.
b) Intrinsic safety is maintained even if one of the diodes stops working.

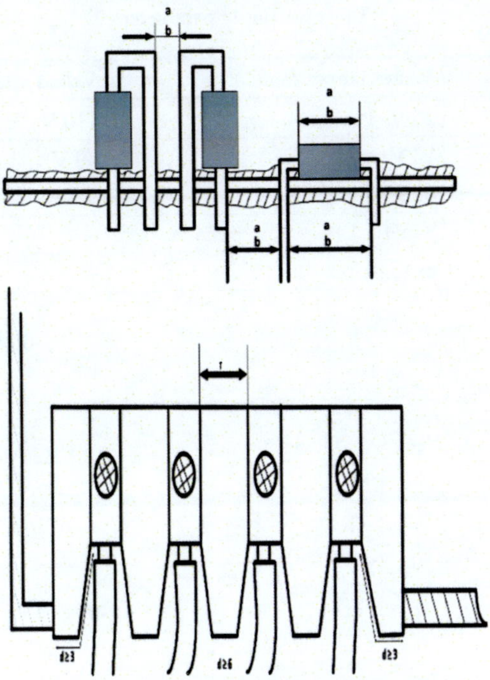

a and b=Creepage and clearance

d= creepage and T= clearance

Fig. 2. Creepage and clearance in components and terminals (from IS/IEC 60079-11)

5 Experiments and Results

An experimental model is chosen as shown in Fig. 3, for testing and studying suitability for intrinsic safety with different countable and non-countable faults in the circuit. Since an inductive circuit is common, a MATLAB Simulink model is studied with different inductive loads. DC-DC Buck converters are the most common power supply for small circuits (Liu et al., 2005). The fault was introduced at time $t = 0.125$ s in the form of inductive load in Fig. 3. The load voltage increased from 10 V to 60 V which results in the increment of current from 90 mA to 600 mA. The increased current is sufficient for high energy spark which can ignite IIC gas group. Output of the model is shown in the Fig. 4. The entity parameters of the circuit under study is given in the Table 6.

The steps followed during the experiment were as below.

Step 1: *Selection of equipment grouping*: Equipment is grouped as Group I/IIA/IIB/IIC on that basis test gas is selected.
Step 2: *Mixing of gas*: Chosen gas is mixed with air as per ratios provided in IS/IEC 60079-11.
Step 3: *Calibration*: Calibration was done to check whether the gas mixtures are prepared.

Table 6. Entity parameters

Parameter	Values before fault	Values after fault
Um	20 V (Peak)	110 V (Peak)
Uo max	10 V	60 V
Io	0.09 Amp	0.6 A
Lo	3 mH	–
Co	negligible	–

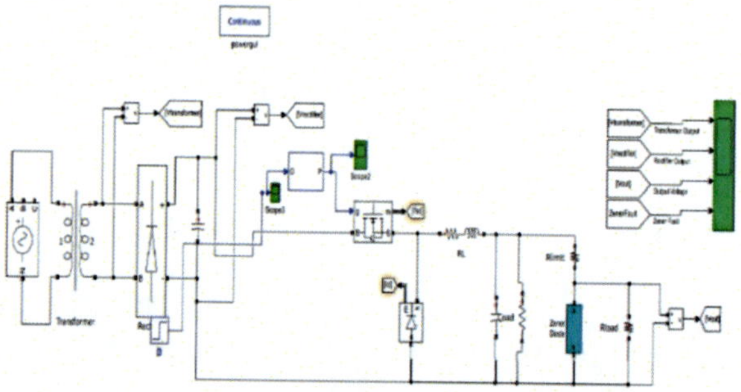

Fig. 3. MATLAB Simulink of AC-DC-DC (Rectifier + Buck Converter) power supply

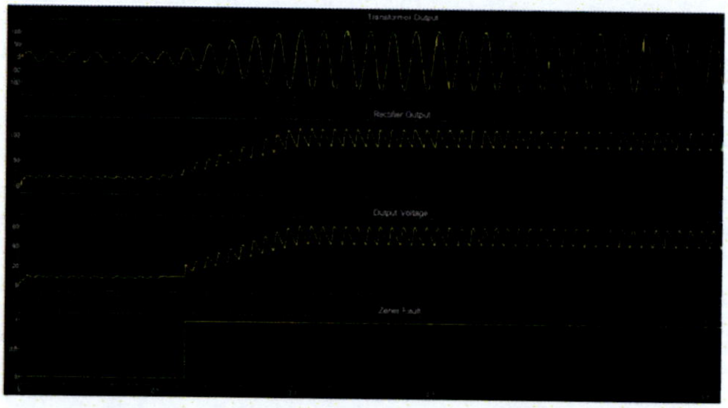

Fig. 4. Simulation result of the above model (Fault at time 0.125 s)

Step 4: *Test*: Once the calibration is found Ok and the gas in the chamber ignited within 440 revolutions then after fresh gas mixing is done for testing. During testing, the two test outlets are connected to the terminals of the spark test apparatus. After testing, again calibration was performed to check the gas mixture.

Below photographs in Fig. 5 shows the calibration and test of the above simulated circuit.

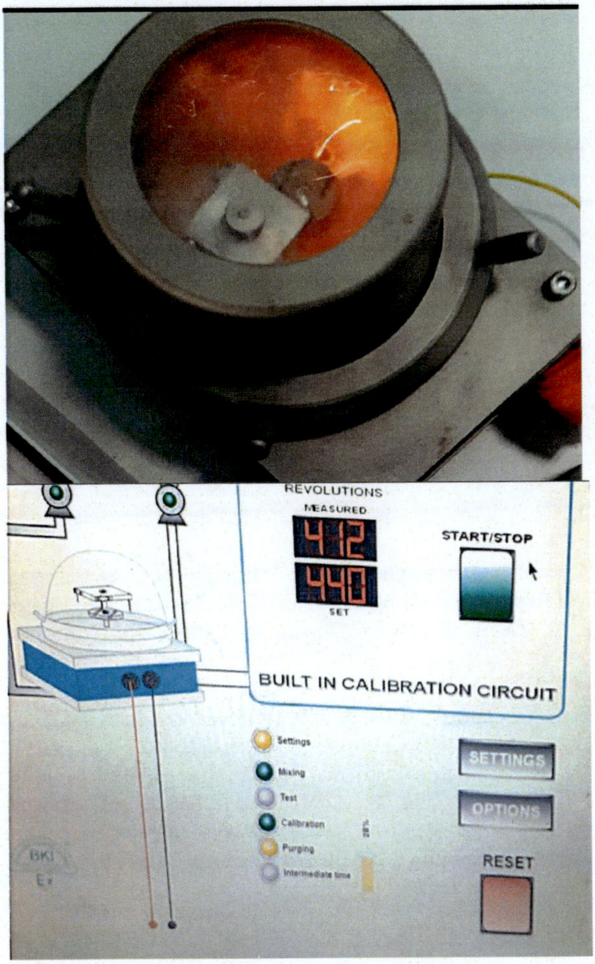

Fig. 5. Spark test apparatus results

6 Conclusion

In this paper, an overview of the electrical equipment in hazardous area and the reliability factors of intrinsic safety in a circuit have been discussed and the effect of load on the circuit is studied. The MATLAB Simulink model of an intrinsically safe circuit is exposed to a deliberate fault and the nature of curves is found inductive in nature. The study is verified with the V-I curve in IS/IEC 60079-11 and was found satisfactory. Furthermore, the study can be extended for different types of loads like capacitive, active loads with sensors, and associated apparatus with varying loads.

References

Mcmillan, A.: Electrical Installations in Hazardous Areas, 1st edn. Butterworth-Heinemann (1998). https://doi.org/10.1016/B978-0-7506-3768-8.X5000-9. ISBN978-0-7506-3768-8

IEC 60079-10-1:2020. Explosive atmospheres - Part 10-1: Classification of areas - Explosive gas atmospheres

Mishra, R.K., Kumar, A., Kumar, N., Singh, A.K., Singh, V.K.: Design construction and testing of high voltage (HV) flameproof electric motors for hydrogen explosive atmosphere (IIC) case studies. J. Sci. Industr. Res. **64**(3), 185–190 (2005). ISSN 0022-4456

Singh, A.K.: Overview of hazardous locations. In: Singh, A.K. (ed.) Explosion-Proof Equipment in Hazardous Area, pp. 1–22. Springer, Singapore (2023). https://doi.org/10.1007/978-981-99-2516-2_1-2

Bottrill, G., Cheyne, D., Vijayaraghavan, G.: Practical electrical equipment and installations in hazardous areas (2005). https://doi.org/10.1016/B978-0-7506-6398-4.X5000-9

Wang, L.: The criterion of intrinsic safe circuit characteristic based on electric arc power. In: Ma, M. (ed.) Communication Systems and Information Technology. LNEE, vol. 100, pp. 853–858. Springer, Heidelberg (2011). https://doi.org/10.1007/978-3-642-21762-3_112

IEC 60079-11:2023. Explosive atmospheres - Part 11: Equipment protection by intrinsic safety "i"

IEC 60079-0:2017. Explosive atmospheres - Part 0: Equipment - General requirements

IEC 60079-25:2020. Explosive atmospheres - Part 25: Intrinsically safe electrical systems

Zhang, H., Wang, Y., Zhang, X., Wang, D., Jin, B.: Design and performance analysis of an intrinsically safe ultrasonic ranging sensor. Sensors **16**, 867 (2016). https://doi.org/10.3390/s16060867

Liu, S., Liu, J., Yang, Y., Zhong, J.: Design of intrinsically safe buck DC/DC converters. Electr. Mach. Syst. **2**, 1327–1331 (2005). https://doi.org/10.1109/ICEMS.2005.202763

Singh, J., Singh, A., Prasad, G.M., Vishwakarma, R., Ahirwal, B.: Design of intrinsically safe circuit for use in hazardous area. In: Conference: International Seminar & Exhibition on Recent Trends in Design, Development, Testing & Certification of Ex-equipment for Explosive Atmosphere, 7–9 November 2014, DTEX 2014, Kolkata (2014)

Microbial and Biosorbent-Mediated Hexavalent Chromium Removal from the Tailing Waters Released from Chromite Mines

Puja Dokania, Sourav Maity, Tapaswini Nayak, Shingam Jayantu, and Angana Sarkar(✉)

Department of Biotechnology and Medical Engineering, National Institute of Technology, Rourkela 769008, Odisha, India
sarkar.angana@gmail.com, sarkara@nitrkl.ac.in

Abstract. Chromium (chromite) mining operations can generate specific problems related to tailing water. There is a serious threat to human health and ecosystems when chromium-contaminated tailing water is released into the environment. Reduced biodiversity and long-term ecological impact may result from these tailing water. Hexavalent chromium (Cr (VI)) being the most toxic form of chromium than Cr (III), causes serious illnesses like ulceration, irritation of the skin, lung cancer, kidney necrosis, asthma, and allergic reactions. In the present study, a total of ten Cr (VI)-reducing bacteria were isolated from these mine tailing water samples, out of which two strains exhibited a maximum tolerable concentration of 200 mg/L. The maximum reduction efficiency was found to be 40.3% for an initial Cr (VI) concentration of 50 mg/L. However, due to the lower reduction efficiency of the bacterial biomass (i.e., in-situ bioremediation), the present study also investigated the role of biosorbent using different plant-derived biomasses for the removal of Cr (VI) from the aqueous environment. The plant-derived biomasses proved to be efficient for the removal of 86.24% for an initial Cr (VI) concentration of 20 mg/L. The observations from two different isotherm models and kinetic modeling for these biosorption studies support the adsorption processes as well as indicate the endothermic and spontaneous nature of the processes. Hence, plant-derived biosorbents could be employed as an efficient, cost-effective, and eco-friendly alternative for the removal of Cr (VI) from the tailing water of chromite mines.

Keywords: Chromite mine tailing water · Hexavalent chromium · Chromium resistant bacteria · Bio-reduction · Biosorption · Eco-friendly

1 Introduction

Through improper waste-disposal practices by industries, chromium pollution has grown to be a serious environmental issue. Mining activities that take place close to chromite mines produce significant amounts of unintentional pollution, which in turn causes significant amounts of chromium to accumulate in fresh water and agricultural fields [1].

A. Sinha et al. (Eds.): AMC 2023, SPEES, pp. 337–344, 2023.
https://doi.org/10.1007/978-3-031-46966-4_27

The mining operation for chromite which is located in Orissa, India, is not an exception to this pattern. About 98% of India's chromite ore comes from the Sukinda Valley in Odisha, which contributes to all of India's chromium resources [2]. A huge area of forestland and agricultural fields have been destroyed by open-cast ore mining, which is done in the mining region for a longer length of time and results in the overburdening of soils. Studies show that these overburdened soils, which weigh 7.6 million tonnes, release 11.73 tonnes of hexavalent chromium (Cr (VI)) into the environment each year [3]. The populace who consumes the chromium-contaminated groundwater in various ways as well as the water bodies themselves are seriously threatened by the leaching of chromium from mine soil into the groundwater system through rains [4]. The heavy metal chromium is extremely poisonous and can be found in a range of oxidation states, from -2 to $+ 6$. Trivalent chromium (Cr (III)) and (Cr (VI)) are the only chromium oxidation states that are the most persistent, though [5, 6]. In contrast to Cr (III), which is much less mobile, less dangerous, and more readily soluble in water, one of the most dangerous forms of chromium is Cr (VI), which is also readily soluble in water [7]. Sedimentary deposits and geothermal activity are examples of natural sources, whereas chromite mining, leather tanning, electroplating, and metal polishing are examples of manmade sources [8]. Because hexavalent chromium is a cellular oxidizing agent, it is hazardous and can cause serious health problems in people, including cancer, organ failure, kidney disorders, and cardiovascular ailments [9].

The current work set out to determine whether it was possible to use biomass derived from both bacterial and plant sources to remove Cr (VI) from aquatic settings.

2 Materials and Methods

2.1 A: Cr (VI) Reduction using Bacterial Isolates

Sample Collection, Enrichment, Isolation, and Screening of Chromium-resistant Bacteria

Soil samples were taken from Odisha's chromium-contaminated area (Latitude: 20°40'N; Longitude: 85 52'E). The samples were enriched in 100 ml of Reasoner's 2A (R2A) medium with 100 mg/L of Cr (VI) supplied as potassium dichromate ($K_2Cr_2O_7$). For five days at 30 °C and 120 rpm, the samples were aerobically incubated. Before plating on R2A agar with the same initial Cr (VI) dosage, the samples were serially diluted. At 30 °C for 72 h, the samples were aerobically stored. After three days, the petri plates' single colonies were chosen and subcultured for microbiological analysis. Maximum Tolerance Concentration (MTC) assay identified the most superior bacteria. The bacterial isolates were inoculated in R2A broth medium with varied Cr (VI) concentrations from 100 to 250 mg/L for MTC Assay and incubated for 48–96 h. Next, the bacterial isolates' absorbance at 600 nm was measured [10].

Characterization of the Bacterial Strain: Gram Staining, IMViC Tests, and Antibiotic Profiling

Both strains underwent the Gram staining experiment to determine their gram-positive or gram-negative nature and to determine the composition of the bacterial cell wall. The assay was carried out using an IMViC test kit that included various test vials. To find out which antibiotic will be most effective against a certain bacterial infection, antibiotic susceptibility testing was carried out to ascertain the bacteria's sensitivity to several antibiotics. Here, a total of 25 different antibiotics were taken in order to test for antibiotic sensitivity of the bacterial strains [11].

Reduction Potential of the Bacterial Isolates

Two bacterial isolates were inoculated in R2A broth medium with 50 mg/L Cr (VI) and maintained at 30 °C in an incubator shaker for 24 to 48 h. After incubation, the cell culture was centrifuged for 10 min at 10,000 rpm and analyzed for residual chromium using the 1,5-diphenylcarbazide (DPC) procedure. Each bacterial strain's % Cr (VI) reduction was measured at 24-h intervals for 48 h [12].

2.2 B: Cr (VI) Removal Using Biosorbents Derived from Plant Biomasses

Preparation of Biosorbent and Adsorbate Stock Solutions

Leaf biomasses of *Artocarpus heterophyllus* and *Mangifera indica* were collected from the National Institute of Technology Rourkela, Odisha, India (Latitude: 22° 15′ 9.41807″ N; Longitude: 84°54′ 26.4455″ E). The biomass was thoroughly washed four times with regular water and twice with distilled water. After trimming, the biomass was sun-dried for two-three days and then heated in a hot air oven at 60 °C for 12 h. The dried plant material was powdered in a mixer grinder, mixed, and stored in airtight containers. The Cr (VI) stock solution of 100 mg/L was prepared using $K_2Cr_2O_7$ of 0.0282 g. This stock solution was used for the preparation of the required concentrations during the study.

Optimization Study of Different Process Variables

The different process variables like adsorbate dosage, contact, temperature, pH, and initial concentration study of the biosorption study were optimized.

$$Cr(VI)removal\% = \frac{C_i - C_f}{C_i} * 100$$

where, C_i and C_f indicate the values of Cr (VI) initial and final concentration (mg/L), respectively.

Study of Isotherms and Kinetic Models

The study of several isotherm models for the adsorption process explains the surface characteristics, adsorption mechanisms, and binding affinities of the used adsorbent. The mathematical relationship derived from several isotherm models served as the foundation for the modeling and design of biosorbents. The experimentally obtained equilibrium results for a number of linear isotherm models, including the Langmuir and Freundlich isotherm models, were explored in the current work [13].

The model equation for a Langmuir isotherm with a linear form is given by

$$\frac{C_e}{q_e} = \frac{C_e}{q_{max}} + \frac{1}{q_{\max \times K_L}}$$

Therefore, the model equation for a Freundlich isotherm with a linear form is given by

$$logq_e = logK_f + \frac{1}{n}logC_e$$

where, C_e stands for the value of Cr (VI) concentration at saturation (mg/L), q_e for the capacity of adsorption at saturation (mg/g), q_{max} is the maximum value of Cr (VI) that can be adsorbed per gram weight of adsorbent at saturation, (mg/g), and K_L for the isotherm constant of Langmuir model (L/mg), which indicates the adsorption energy, K_f denotes the Freundlich constant, which represents the efficiency of adsorption and 1/n is the another Freundlich constant, which indicates the adsorption intensity.

Studying the manner in which the process operates over time and how various factors impact the rate of adsorption requires kinetic studies of the adsorption process. The removal of Cr (VI) from the leaf biomass was assessed using two adsorption kinetic models, pseudo-first-order and pseudo-second-order.

3 Results and Discussion

3.1 A: Cr (VI) Reduction Using Bacterial Isolates

Isolation and MTC Assay of Chromium-Resistant Bacteria
Twenty separate microbial colonies emerged from the serially diluted enriched soil sample following plating on an R2A agar plate. Ten distinct colonies that did not overlap each other were chosen from this group and streaked over R2A agar plates with 50 mg/L Cr (VI). Only the two isolated isolates, S1 and S4, were shown to exhibit substantial growth in higher concentrations up to 200 mg/L using the MTC test. Consequently, it was proven to be the best of the ten isolated strains. The metal biotransformation and the metal resistance plasmid that are present in bacterial strains enable them to withstand the Cr (VI) stress [14].

Characterization of the Bacterial Strain
Gram staining gives the bacterial strains S1 and S4 a purple color, which is an obvious sign that the main stain has been retained by both strains' cell walls and that they are both gram-positive in nature [15]. According to the IMViC Test investigation, strain S4 was found to have a higher ability for utilizing glucose, arabinose, mannitol, and sucrose; however, strain S1 was unable to use the majority of carbohydrates such adonitol, lactose, sorbitol, and rhamnose. After 24 h of incubation, a distinct zone of inhibition was seen around the paper discs containing antibiotics on agar plates inoculated with the bacterial culture. According to the findings, the medicines Amoxycillin (20/10) and Rifampicin (5 mcg) were most efficient against the bacterial strain S1, while Rifampicin (5 mcg) and Roxithromycin (30 mcg) were most successful at inhibiting the bacterial strain S4.

Reduction of Cr (VI) by Isolated Bacterial Strains
According to the bio-reduction study (Fig. 1), bacterial strains S1 (40.3%) and S4 (30.9%) reduced hexavalent chromium from 50 mg/L to their maximal levels in 48 h. In 24 h, the maximum reduction was observed for bacterial strains S1 (31.7%) and S4 (26.4%) for an initial hexavalent chromium concentration of 50 mg/L.

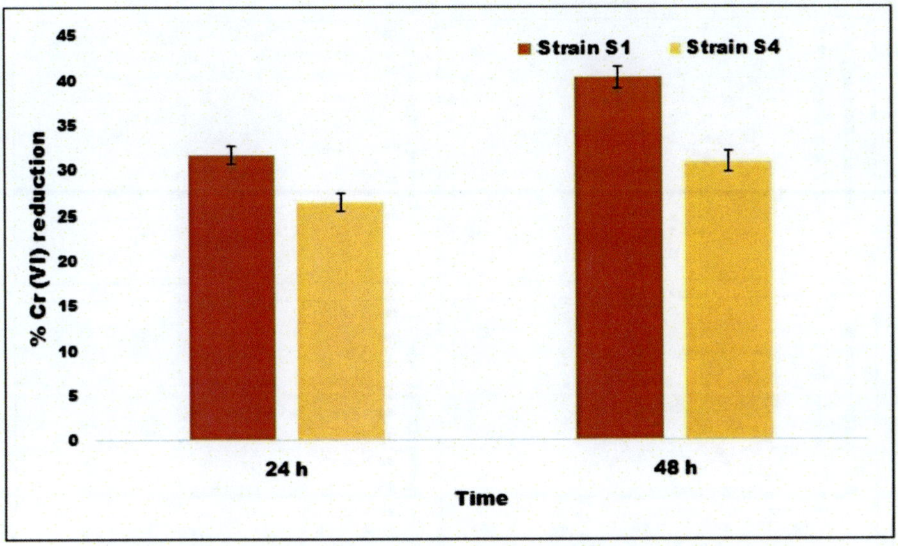

Fig. 1. Cr (VI) reduction percent by the two isolated bacterial strains

3.2 B: Cr (VI) Removal Using Biosorbents Derived from Plant Biomasses

Optimization Study of Different Process Variables

The mixed biosorbent of *Artocarpus heterophyllus* and *Mangifera indica* leaf biomasses were used to study the adsorption efficiency for Cr (VI) removal. Optimization of different process variables like adsorbent dosage, contact time, temperature, pH, and initial concentration of Cr (VI), were studied (Fig. 2). From this study, the optimum removal efficiency for Cr (VI) using the biosorbent was to be attained using 0.5 g of adsorbent dosage (83.20%), 8 h of contact time (84.30%), 30 °C of temperature (85.31%), 7.0 pH (84.25%), and 20 mg/L of initial Cr (VI) concentration (86.24%).

Study of Isotherms and Kinetic Models

The values of the isotherm constant variables from the Langmuir and Freundlich isotherm models were interpreted and shown in Table 1. The Langmuir isotherm denotes the presence of monolayer adsorption and adsorption sites that are structurally similar [16]. Since R_L is between 0 and 1 in this case, the adsorption approach was favorable [7]. The presence of multilayer adsorption and a heterogeneous adsorption surface are both indicated by the Freundlich isotherm. This model states that the adsorption process is favored and typical as long as 1/n is less than 1 [13].

The study of the kinetic models has been represented in Table 2. The lower values of R^2 show that the equilibrium data did not adequately support the pseudo-first-order kinetic model. The concept of pseudo-first-order kinetics has already been demonstrated to be only useful during the initial stages of the adsorption process by a number of studies [17]. The equilibrium results were more consistent with the pseudo-second-order kinetic model, as seen by the relative higher regression coefficient R^2 values. This

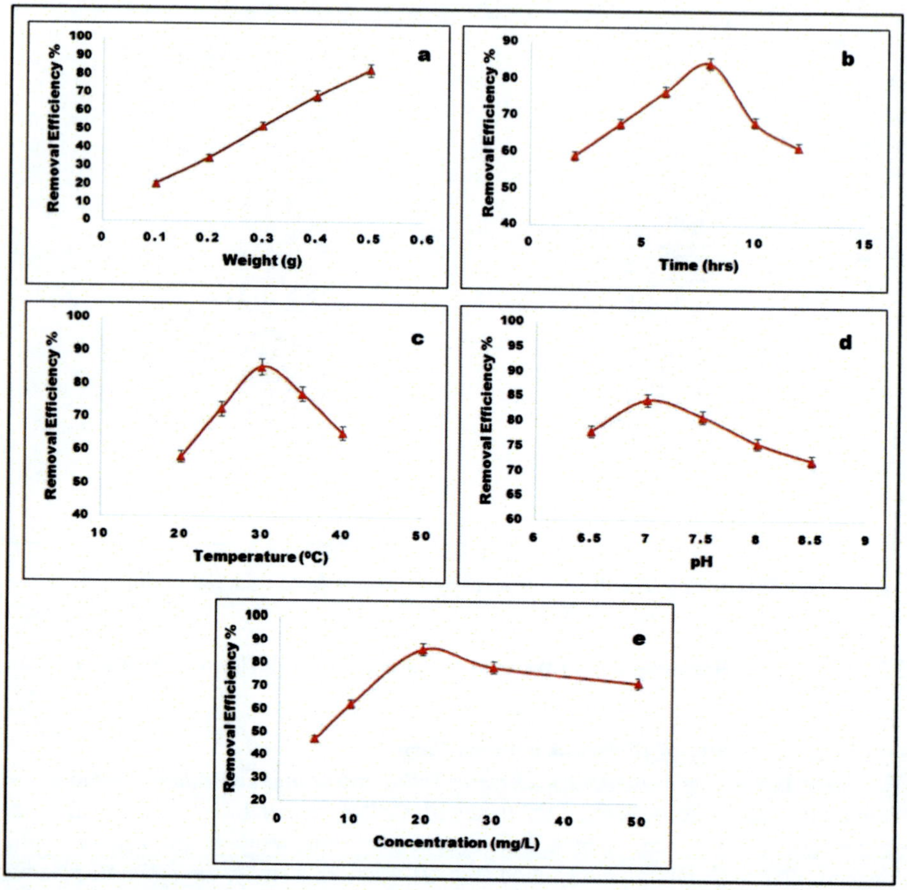

Fig. 2. Optimization results of Cr (VI) adsorption on effect of (a) adsorbent dosage, (b) contact time, (c) temperature, (d) pH, and (e) initial concentration

study supported earlier research by providing evidence that the covalent forces at the solid-liquid interface are what drive the adsorption process [18, 19].

Table 1. The values of various constants for isotherm analysis

Sl No	Isotherm model	Variables	Values	Coefficient of determination (R^2)
1	**Langmuir**	q_{max} (mg/g)	2.423	0.9899
		K_L (L/mg)	0.137	
		R_L at C_i		
		10	0.417	
		20	0.331	
		30	0.215	
		40	0.167	
		50	0.108	
2	**Freundlich**	K_f (mg/g)	0.578	0.9619
		n	1.320	

Table 2. The values of various constants for the corresponding kinetic studies

Kinetics model	Constants	Values	Coefficient of determination (R^2)
Pseudo-first-order	K_1 (min)	0.01275	0.8723
Pseudo-second-order	K_2 (g/mg/min)	0.00037	0.9912

4 Conclusion

The isolated superior bacterial biomass was used for 48 h to achieve a removal efficiency of 40.3% from a 50 mg/L Cr (VI) aqueous solution utilizing the bio-reduction mechanism. However, the acceptable Cr (VI) remediation properties of mixed leaf biomass from *Artocarpus heterophyllus* and *Mangifera indica* were investigated using biosorption. Using the current biosorbent, a removal efficiency of 86.24% from the Cr (VI) aqueous solution was achieved after optimizing all adsorption process parameters. The ideal conditions for maximum Cr (VI) remediation were determined to be a biosorbent dosage of 0.5 g, a contact period of 8 h, a temperature of 30 °C, a pH value of 7, and an initial concentration of 20 mg/L. The comparison of bacterial and plant biomass revealed that the latter was able to eliminate Cr (VI) more efficiently. In order to remove Cr (VI) from aqueous settings, raw powdered mixed leaf biomass could be used as a viable, effective, and environmentally friendly biosorbent.

References

1. Dey, S., Pandit, B., Paul, A.K.: Reduction of hexavalent chromium by viable cells of chromium resistant bacteria isolated from chromite mining environment. J. Min. **2014** (2014)
2. Dhal, B., Das, N.N., Thatoi, H.N., Pandey, B.D.: Characterizing toxic Cr (VI) contamination in chromite mine overburden dump and its bacterial remediation. J. Hazard. Mater. **260**, 141–149 (2013)

3. Dhal, B., Das, N.N., Pandey, B.D., Thatoi, H.N.: Environmental quality of the Boula-Nuasahi chromite mine area in India. Mine Water Environ. **30**(3), 191–196 (2011)
4. Das, S., et al.: Reduction of hexavalent chromium by Exiguobacterium mexicanum isolated from chromite mines soil. Chemosphere **282**, 131135 (2021)
5. Maity, S., Biswas, R., Verma, S.K., Sarkar, A.: Natural polysaccharides as potential biosorbents for heavy metal removal. In: Food, Medical, and Environmental Applications of Polysaccharides, pp. 627–665. Elsevier (2021)
6. Alam, M.Z., Ahmad, S.: Toxic chromate reduction by resistant and sensitive bacteria isolated from tannery effluent contaminated soil. Ann. Microbiol. **62**(1), 113–121 (2012)
7. Maity, S., Nanda, S., Sarkar, A.: Colocasia esculenta stem as novel biosorbent for potentially toxic metals removal from aqueous system. Environ. Sci. Pollut. Res. **28**, 58885–58901 (2021)
8. Maity, S., Biswas, R., Sarkar, A.: Comparative valuation of groundwater quality parameters in Bhojpur, Bihar for arsenic risk assessment. Chemosphere **259**, 127398 (2020)
9. Bennett, R.M., Cordero, P.R.F., Bautista, G.S., Dedeles, G.R.: Reduction of hexavalent chromium using fungi and bacteria isolated from contaminated soil and water samples. Chem. Ecol. **29**(4), 320–328 (2013)
10. Sanjay, M.S., Sudarsanam, D., Raj, G.A., Baskar, K.: Isolation and identification of chromium reducing bacteria from tannery effluent. J. King Saud Univ. Sci. **32**(1), 265–271 (2020)
11. Maity, S., Sarkar, D., Poddar, K., Patil, P., Sarkar, A.: Biofilm-mediated heavy metal removal from aqueous system by multi-metal-resistant bacterial strain Bacillus sp. GH-s29. Appl. Biochem. Biotechnol. **195**, 1–19 (2022). https://doi.org/10.1007/s12010-022-04288-7
12. Ilias, M., Rafiqullah, I.M., Debnath, B.C., Mannan, K.S.B., Mozammel Hoq, M.: Isolation and characterization of chromium (VI)-reducing bacteria from tannery effluents. Indian J. Microbiol. **51**, 76–81 (2011)
13. Maity, S., Patil, P.B., SenSharma, S., Sarkar, A.: Bioremediation of heavy metals from the aqueous environment using artocarpus heterophyllus (jackfruit) seed as a novel biosorbent. Chemosphere **307**, 136115 (2022)
14. Wani, P.A., Wahid, S., Singh, R., Kehinde, A.M.: Antioxidant and chromium reductase assisted chromium (VI) reduction and Cr (III) immobilization by the rhizospheric Bacillus helps in the remediation of Cr (VI) and growth promotion of soybean crop. Rhizosphere **6**, 23–30 (2018)
15. Wani, P.A., Omozele, A.B., Wasiu, I.A., Jamiu, K.O.: Cr (VI) reduction by indigenous Bacillus species PB5 isolated from contaminated soil of Abeokuta, Ogun State. Nigeria. Int J Soil Sci **10**(4), 203–210 (2015)
16. Kebede, T.G., Dube, S., Nindi, M.M.: Characterisation of water-soluble protein powder and optimisation of process parameters for the removal of sulphonamides from wastewater. Environ. Sci. Pollut. Res. **26**, 21450–21462 (2019)
17. Mishra, A., Dubey, A., Shinghal, S.: Biosorption of chromium (VI) from aqueous solutions using waste plant biomass. Int. J. Environ. Sci. Technol. **12**, 1415–1426 (2015)
18. Li, J., Lin, Q., Zhang, X., Yan, Y.: Kinetic parameters and mechanisms of the batch biosorption of Cr (VI) and Cr (III) onto leersia hexandra Swartz biomass. J. Colloid Interface Sci. **333**(1), 71–77 (2009)
19. Dokania, P., Maity, S., Patil, P.B., Sarkar, A.: Isothermal and kinetics modeling mpproach for the mioremediation of potentially toxic trace metal ions using a novel biosorbent acalypha wilkesiana (Copperleaf) leaves. Appl. Biochem. Biotechnol. 1–31 (2023). https://doi.org/10.1007/s12010-023-04678-5

The Case for ISO 23875:2021 W/Amendment 1: 2022 as the Path to Hemm Protection from Respirable Crystalline Silica (Rcs)

Jeffrey L. Moredock[1](✉), Biswajit Dutta[2,3], and Ankit Chatterjee[4]

[1] ISO TC 82 – Mining, Convenor WG9 – Operator Enclosures International Project Lead ISO 23875:2021, Jacksonville, USA
Jeff.Moredock@sy-klone.com

[2] MGMI(Mining, Geological and Metallurgical Institute of India), ICC, CPSI, IIFT, Bibeknagar, India

[3] Equiptech Services LLP, Eastman Crusher Co. P Ltd., Bibeknagar, India

[4] Equiptech Services and IIEST, Shibpur, India

Abstract. Respirable Crystalline Silica (RCS) is a problem prevalent in the mining industry that impacts the long-term health of the workforce across a wide range of work processes and tasks. Significant research has been conducted over time both to understand the scope of the problem in RCS-exposed environments and to engineer effective solutions to protect workers. The U.S. based National Institute for Occupational Safety and Health (NIOSH) began research into the efficacy of operator enclosures as an engineering control in the 1990's. Research done in conjunction with several industry partners, including Sy-Klone International, has resulted in multiple projects, publications, and ultimately product solutions. ISO 23875 Mining - Operator enclosures - Air quality control systems and air quality performance testing was developed by ISO Technical Committee 82 – Mining and was published in February 2021. ISO 23875 provides a comprehensive focus on controlling dust concentrations, CO_2 levels, and cab pressure over the lifecycle of the operator enclosure. Additionally, Annex B gives guidance on the operational integration of the cab into the site occupational health and safety management system. The standard's comprehensive nature and enthusiastic stakeholder adoption make it a consensus driven regulatory tool to protect machine operators working in enclosed cabs.

Respirable Crystalline Silica (RCS) is a problem prevalent in the mining industry that impacts the long-term health of the workforce across a wide range of work processes and tasks. Significant research has been conducted over time both to understand the scope of the problem in RCS-exposed environments and to engineer effective solutions to protect workers.

The U.S.-based National Institute for Occupational Safety and Health (NIOSH) began research into the efficacy of operator enclosures as an engineering control in the 1990's. Research done in conjunction with several industry partners, including Sy-Klone International, has resulted in multiple projects, publications, and ultimately product solutions, some of which are still ongoing. In 2009, Sy-Klone International produced its first generation of cab air quality products incorporating high efficiency filtration tailored to human respiration. This product was utilized in research on a sandstone cutting machine by the Queensland Australia

A. Sinha et al. (Eds.): AMC 2023, SPEES, pp. 345–354, 2023.
https://doi.org/10.1007/978-3-031-46966-4_28

Government in 2009. This study, known as RESPA™ Trial 2009 Occupational hygiene monitoring for airborne particulate matter and respirable crystalline silica (RCS) inside an excavator cabin – before and after fitting a pre-cleaner, filter and pressurization unit, was published in 2010 by the Queensland Government. The study demonstrated the value of the air quality control system in reducing RCS exposures when fitted with effective engineering controls.

NIOSH and Sy-Klone International's joint research are the foundation for operator enclosure air quality engineering controls in the NIOSH Dust Control Handbook. The International Society of Environmental Enclosure Engineers (ISEEE) advanced the science by introducing a robust field monitoring system that allowed for operator enclosures to be evaluated in real world conditions, over a work shift. With this research and field experience, ISEEE developed a course on operator enclosure air quality engineering in 2014, much of which became the basis and a catalyst for the development of ISO 23875.

ISO 23875 Mining - Operator enclosures - Air quality control systems and air quality performance testing was developed by ISO Technical Committee 82 – Mining and was published in February 2021. Lehigh Hanson, part of the Heidelberg Cement Group, sponsored a pilot installation utilizing ISO 23875 in 2021, to assess its effectiveness in granite quarry operations. The study by Lehigh Hanson was submitted to NIOSH as evidence of industry's proactivity in addressing silica exposure at their aggregate quarry. The project won the NIOSH Mine Health and Safety Technology award in 2021.

While many studies have been completed as pilots for ISO 23875 and can be seen on the ISEEE website, one of particular interest was performed by an international mining company in Canada. The yearlong study was to determine the specific work being performed when RCS overexposure was encountered. Their study found that the greatest exposure took place within operator enclosures on machines or transport vehicles which move RCS-containing materials or travel through RCS-contaminated environments. Of note were the crew buses responsible for bringing miners to their respective workstations, bull dozers moving material, light trucks working in the pit, and office workers in the lab where the material was crushed for qualitative analysis. The highest exposures, in terms of degree and as a percentage of total exposures, took place in operator enclosures. The result of this study was the decision to implement ISO 23875 to address operator enclosure air quality. This study was presented at the American Industrial Hygiene Association (AIHA) mining working group meeting in April of 2022.

The American Industrial Hygiene Association (AIHA) mining working group participated in the development of ISO 23875 Mining - Operator enclosures - Air quality control systems and air quality performance testing. The article can be viewed at https://synergist.aiha.org/20210607-standard-to-help-protect-miners.

Worker health is not a primary consideration in general business for several reasons:

(1) While there is near universal empathy for stricken workers, there is no universal agreement on where fault lies for the illness.
(2) A prevailing sentiment, regarding workers taking personal responsibility for their hygiene practices, places an unrealistic expectation on their ability to protect themselves.
(3) Investment in worker health is perceived as non-essential when workers move from job to job frequently and/or are abundantly available to the employer.

Accountability for worker health is not possible when workers do not stay in a job long enough to prove that their health condition was caused by a specific job or employment environment.

(4) The cost of implementing protective measures, if not borne by everyone in a competitive industry, creates an imbalance, which leads to price advantages for companies that choose not to invest in protective measures.

(5) Limited social/community awareness exists. While massive amounts of RCS are released in public areas during road construction, residential pool construction, benchtop cutting, in front of residences during the installation process of brick cutting, concrete removal projects etc., its risks are not widely known. There is little public outcry and no reporting mechanism for RCS releases.

Leveling the playing field through government intervention compels all businesses to invest in worker protection without a competitive disadvantage. Government can also educate workers and workplaces, bring research-based solutions to industry, strongly contribute to more positive long-term public health outcomes, as well as create social awareness and reporting mechanisms.

Government has a primary responsibility to protect its citizens. This responsibility cannot be abdicated to private enterprise, whose goals may not align with the public good.

One solution specific to operator enclosures is the inclusion/adoption of ISO 23875 with Amendment 1 in health legislation/regulation, as a readily available, practical, cost-effective, and consistent methodology to control RCS exposure in operator enclosures at all applicable types of RCS-impacted job sites.

We would recommend the inclusion of ISO 23875 with Amendment 1 as part of the regulatory regimen to control RCS exposure. The standard's comprehensive nature and enthusiastic stakeholder adoption make it a logical path forward to protect machine operators working in enclosed cabs. The bibliography to the standard is included to assist in sourcing reputable documents which support/answer many of the questions asked about the problems of and solutions to RCS overexposure in environmentally controlled operator enclosures. The mining industry in India may work to get consensus on a solution, which can be made universal within India, with the help of the Indian Government. There is widespread support for ISO 23875 among the Industrial Hygiene community globally.

Global mining companies, such as Rio Tinto, have adopted the standard as part of their self-regulation efforts. ISO 23875 represents both state-of-the-art and best practices for the control of respirable particulate in operator enclosures and provides a consensus-driven engineered solution against RCS exposure at a cost that is as low as reasonably practicable.

Worker overexposure is the result of inadequate protections. When working openly in an RCS-saturated environment, workers should use properly fitted and tested PPE in good repair on a clean-shaven face. However, workers can develop a false sense of protection when working in an environmentally controlled enclosure, like a control room or operator cabin on a piece of heavy equipment.

The international mining community has been seeking to address the universal problem of operator protection while operating a machine in an environmentally controlled enclosure. In the absence of a standard, previously, the industry was self-regulating based upon local RCS regulations. The result was a myriad of approaches to proper cab engineering. However, now that ISO 23875 with Amendment 1 is in place, it allows for a common framework for control and enforcement of air quality in an operator enclosure. Territories and countries around the world (e.g., EU, Australia) are adopting ISO 23875 with Amendment 1 and using it as a blueprint for the regulation and enforcement of RCS exposure in enclosures.

The introduction and scope of the ISO 23875 standard is included below for a more detailed overview and the full document is available from ISO at https://www.iso.org/standard/77249.html.

INTERNATIONAL STANDARD ISO 23875.
FIRST EDITION 2021–02.

Mining—Air quality control systems for operator enclosures—Performance requirements and test methods.

1 Introduction

Safety in mining operations is of concern to all involved in owning, developing, managing, and working in mining environments. Routine mining activities can generate airborne particulates which are hazardous to human health. Therefore, it is necessary to develop controls which limit the enclosure. With the rise in the number of countries regulating air quality in mining, construction, and industrial environments, machine manufacturers have become increasingly aware of the need for standard practices in the design and performance of operator enclosures. This document seeks to address the fundamental design requirements that will allow for operator enclosures to perform at a level that provides sustained air quality, reducing concentrations of respirable particulate matter and carbon dioxide that are harmful to human health. The emphasis of this document is in three areas: 1) design, 2) air quality control system performance testing, and 3) maintenance and operation instruction for the operator enclosure.

All operator enclosures, either on new machines or existing machines currently in operation, meeting the requirements of this document are expected to provide consistent air quality performance. The technical aspects of an operator enclosure are universal as are the design and performance testing methods. Therefore, every attempt has been made to make this an inclusive document which addresses the needs of fixed and mobile operator enclosures.

This document was developed to provide for the occupational health and safety of personnel who work inside operator enclosures. It primarily addresses air quality concerns by establishing parameters to determine air quality control system effectiveness. The control of these airborne contaminants is through an effective air quality control system (for both external air and recirculated air), dilution of CO2, routine testing of the air within the operator enclosure, and effective maintenance throughout the life cycle of the operator enclosure. Extensive research and subsequent publications have produced a substantial body of knowledge around the air quality control systems and are the basis of this document. See Bibliography.

As illustrated in Fig. 1, this document presents a life cycle approach to operator enclosure air quality control system design, performance testing, and maintenance.

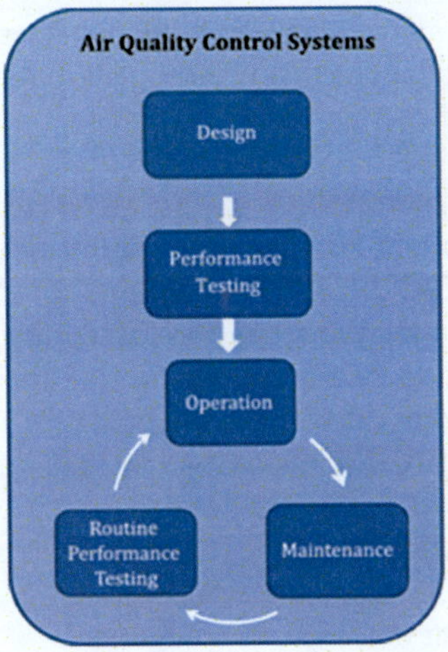

Fig. 1. Air quality control system life cycle

2 Scope

This document specifies performance and design requirements for air quality control systems for operator enclosures and their monitoring devices. The design specifications are universal in their application and do not contemplate specific mining environments. They are intended to meet identified parameters of both pressurization and respirable particulate and carbon dioxide concentrations. This document also specifies test methods to assess such parameters and provides operational and maintenance instructions. Recommendations are made for operational integration of the air quality control system.

Gases and vapors that can be a hazard in the work environment outside of the operator enclosure are excluded from this document.

The cost model for implementation of ISO 23875 with Amendment 1, assumes the following:

1) The worker wage assumption in Appendix D was used in the cost model and are assumed to be accurate.
2) Machines currently operating in RCS-saturated environments have environmentally controlled operator enclosures.
3) The cost of maintaining an ISO 23875 operator enclosure is like maintaining a non-compliant enclosure. This assumption can be challenged on the basis that a non-compliant operator enclosure does not have the higher cost, more efficient filters required to remove respirable particulates, recirculation filtration, or air quality monitoring equipment required under ISO 23875. This is a valid observation. However, for an operator enclosure to protect against RCS, it must have each of these engineering controls and continuous cab pressure. In other words, comparing an enclosure that cannot protect against overexposure to RCS and one that can, is not a fair comparison. It is assumed that the operator enclosure will be modified or manufactured to protect the operator from RCS overexposure.

4) The difference between an operator enclosure that does not protect the operator and one that does, assuming the one that is failing was intended to protect the operator, is a matter of design and maintenance.
5) The operator enclosure will be continuously maintained to ensure proper operation of the air quality system and HVAC.
6) All operator enclosures can be engineered to comply from the manufacturer or retrofit to comply with ISO 23875.
7) Resource industry operators, including Rio Tinto, have adopted ISO 23875 as part of their exposure reduction efforts and included in their Heavy Mobile Equipment (HME) tenders and new machine specifications.
8) Currently, Caterpillar, Epiroc, Komatsu, Soosan, Sandvik, Everdigm, Hyundai, JCB, and Furukawa have active ISO 23875 development programs and may deliver ISO 23875 compliant cabs when requested in the machine specification.
9) Adoption of ISO 23875 by the European Union as EN/ISO 23875 is evidence that it has value as an element in a regulatory regimen.
10) When purchased from the manufacturer, ISO 23875 compliant enclosures are delivered with the machine.
11) Local machine dealers and solutions providers are currently capable of retrofitting equipment for compliance if requested.
12) Education on ISO 23875 with Amendment 1 is available online, in many languages, at iseee.net. The education includes a course completion certificate which can be used as evidence of formal training in the application of the standard.
13) The purpose of ISO 23875 harmonizes with the desire for the "fix" to be permanent. Without a life cycle approach to maintenance and performance testing, exposure will continue to persist. Acceptable operator air quality requires vigilance and a systematic approach which weaves education, standardization, regulatory oversight, and industry commitment together to create a lasting solution to operator air quality. The absence of any of these requirements will cripple any effort to permanently address RCS over exposures in operator enclosures.

As illustrated in the Fig. 2, inhalable dust is classified into three types where the largest inhalable particles are easily trapped by the cilia hairs of the nose while the thoracic fraction and respirable fraction are small enough to enter through the sinuses and into the lung airway. The Thoracic fraction does not pose major health risk while the Respirable fraction poses to be the major health concern.

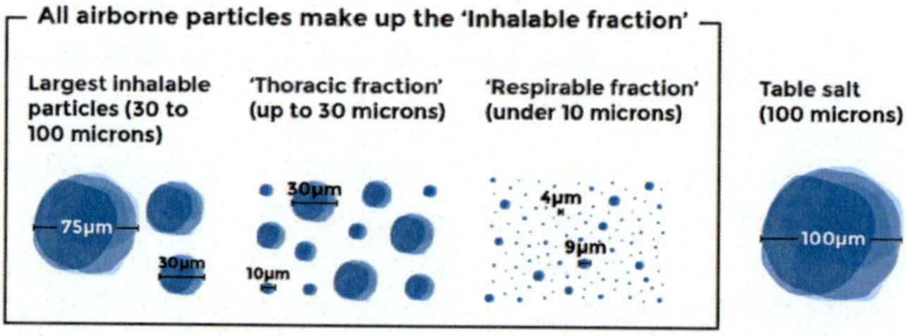

Fig. 2. Size comparison of different Respirable Dust 'fractions'.

Table 1. ISO 23875 compliant installation cost estimate.

Machine types	Variable	Input	Formula applied
All mining, construction, tunnelling, ag, and mineral processing machines with environmentally controlled operator enclosures	ISO 23875 included in new machine spec. Upgrade existing systems to comply with ISO 23875 w/Amd. 1 Annual maintenance assuming annual machine hours of 3.5K hrs Recertification to ISO 23857 done annually	**Additional cost when purchased with machine=** $2.5K to $20K **Cost to retrofit** $2.5K to $20K* **Annual Maintenance** 4 filters = $.750K** (Local filter supplier) Labour**** = $2.03 x 8hr = $0.01624K **Annual Recertification** costs: $.50K/yr. **Non-compliant maintenance cost** 4 filters = $.400K Labour**** = $2.03 x 8 hrs = $0.01624K	**Initial expense for ISO 23875 compliance** $2.5K to $20K **Annual maintenance** $0.75K Filter costs + $0.01624K Labour costs + $0.200K recertification = $0.96624K annual cost - $0.40K non-compliant filter costs - 0.01624K non-compliant labour = **$.55K additional annual maintenance costs for ISO 23875 compliance**

*Costs range for new and retrofit installations to be reconfirmed with retrofit companies and machine dealers.
** Filter costs reflect average retail prices for ISO 23875 compliant EPA filters.
***Cost figures in USD.
****Labour rate taken from F.No.1/5(2)/2023-LS-II, Government of India Ministry of Labour and Employment Office of the Chief of Labour Commissioner(C), New Delhi 03/04/2023.
< https://safesilica.eu/wp-content/uploads/2018/09/IMA-Fractions-illustration-v5-compressed.jpg >

There are administrative costs associated with managing a maintenance process. Assessing the additional time required to learn the ISO 23875 process is important, however, it should not be significantly different from standard maintenance practices. The primary cost drivers are the requirement to use higher efficiency and more costly filtration on both the fresh and recirculation airflows. Additionally, there is an administrative burden associated with maintaining an annual recertification program; however, this, too, is modest beyond the actual cost of a contractor to administer the test protocol. This cost can be mitigated by doing the recertification test in house, which the standard allows.

3 Conclusion

Higher precedence of pulmonary diseases will continue to be a tax on public health and legal resources and may result in increased medical costs and rising medical trends (insurance costs) for both the private and public sector if not addressed comprehensively. Industry-wide solutions are rarely popular or aggressively enforced. However, regulatory compliance is greatest when the regulatory framework aligns with industry best practices based upon consensus standards.

Adoption of ISO 23875 with Amendment 1 addresses overexposure in environmentally controlled operator enclosures. This is a comprehensive solution which addresses the manufacture, retrofit, validation testing, ongoing performance, maintenance, and recertification process. The standard aligns with industry best practice and creates regulatory enforcement of an industry developed consensus standard. This has the benefit (as previously mentioned), of leveling the playing field. The studies confirming the effectiveness of the standard are listed in the bibliography.

4 Normative References

The following documents are referred to in the text [of ISO 23875] in such a way that some or all of their content constitutes requirements of the document. For dated references, only the edition cited applies. For undated references, the latest edition of the referenced document (including any amendments) applies.

ISO 18158, *Workplace air—Terminology.*
ISO 29463–5:2011, *High-efficiency filters and filter media for removing particles in air—Part 5: Test method for filter elements.*
ISO/IEC 17000, *Conformity assessment—Vocabulary and general principles.*
ISO/IEC 17050–1, *Conformity assessment—Supplier's declaration of conformity—Part 1: General requirements.*

5 End Notes

[i] RESPA™ Trial 2009 Occupational hygiene monitoring for airborne particulate matter and respirable crystalline silica (RCS) inside an excavator cabin – before and after fitting a pre-cleaner, filter and pressurization unit

[ii] American Industrial Hygiene Association https://synergist.aiha.org/20210607-standard-to-help-protect-miners

[iii] Adopted as a European norm EN ISO 23875:2022

[iv] Adopted as a national standard in Colombia, SA NTC/ISO 23875 – Norma Tecnica Colombiana

[v] ISO 23875:2021 and ISO 23875 Amendment 1: 2022 can be purchase from ISO.org/standards

Bibliography

Standards

1. ISO 5353, *Earth-moving machinery, and tractors and machinery for agriculture and forestry — Seat index point*
2. ISO 6750–1, *Earth-moving machinery — Operator's manual — Part 1: Contents and format*
3. ISO 16000–26:2012, *Indoor air — Part 26: Sampling strategy for carbon dioxide (CO2)*
4. ISO 18758–2, *Mining and earth-moving machinery — Rock drill rigs and rock reinforcement rigs — Part 2: Safety requirements*
5. ISO 19296, *Mining — Mobile machines working underground — Machine safety*
6. ISO 20474 (all parts), *Earth-moving machinery — Safety*
7. ISO 24095, *Workplace air — Guidance for the measurement of respirable crystalline silica*
8. IEC/IEEE 82079–1, *Preparation of information for use (instructions for use) of products — Part 1: Principles and general requirements*
9. EN 474, *Earth-moving machinery — Safety — Part 1: General requirements*
10. ASHRAE 62.1*Annex D – Ventilation for Acceptable Indoor Air Quality*
11. ASTM D6245–18*Standard Guide for Using Indoor Carbon Dioxide Concentrations to Evaluate Indoor Air Quality and Ventilation*

Operator Enclosure Technical Works

12. Cecala, A.B., O'Brien A.D., Schall J., Colinet J., Franta R.J., Schultz M.J. et al.: Dust Control Handbook for Industrial Minerals Mining and Processing, Second Addition, Publication No. 2019–124 (RI 9701), Centers for Disease Control and Prevention, National Institute for Occupational Safety and Health, Pittsburgh, PA (2019)
13. Organiscak, J.A., Cecala, A.B., Hall, R.M.: Design, Testing, and Modelling of Environmental Enclosures for Controlling Worker Exposure to Airborne Contaminants, IC 9531 NIOSH Mining Program Information Circular, Centers for Disease Control and Prevention, National Institute for Occupational Safety and Health, Pittsburgh, PA (2018)
14. Organiscak, J.A., Cecala A.B., U.S. Department of Health and Human Services, Public Health Service, Centers for Disease Control and Prevention, National Institute for Occupational Safety and Health, Office of Mine Safety and Health Research. (2009). Key Cab Factors. Pittsburgh, PA (2009)

Research Works and Case Studies

15. Morin, K., Lachapelle, G, Hon, C.-Y.: *Pilot evaluation of novel engineering control to reduce airborne exposure to particulates inside drilling cabs*, Canadian Institute of Mining Journal Volume 8, Issue 2, 2017, Canadian Institute of Mining, Metallurgy and Petroleum. Ontario Canada
16. Noll, J., Cecala, A.B., Organiscak, J.A., The effectiveness of several enclosed cab filters and systems for reducing diesel particulate matter. Society of Mining, Metallurgy and Exploration, Inc., Transactions 2011, Vol. **328**, pp. 408–415, Littleton, Colorado (2011)
17. Cecala, A., Noll, J., Organiscak, J.A.: U.S. Department of Health and Human Services, Centers for Disease Control and Prevention, National Institute for Occupational Safety and Health, Office of Mine Safety and Health Research. (2013). *Key components for an effective filtration and pressurization system for mobile mining equipment.* Pittsburgh, PA (2013)

18. The State of Queensland, Department of Employment, Economic Development and Innovation, 2010. *RESPA® Trial 2009, Occupational hygiene monitoring for airborne particulate matter and respirable crystalline silica inside of an excavator cabin – before and after fitting a precleaner/filter/pressurizer unit.* Queensland, Australia
19. Cecala, A.B., Organiscak, J.A. Zimmer, Hillis, M.S., Moredock, D.J.: National Institute for Occupational Safety and Health. (2009). *Maximizing air quality inside enclosed cabs with a unidirectional filtration and pressurization system* (NIOSHTIC-2 No. 20035150). Pittsburgh, PA (2009)
20. Satish, U., et al.: Is CO2 an indoor pollutant? direct effects of low-to-moderate CO2 concentrations on human decision-making performance. Environ. Health Perspect. (2012). https://doi.org/10.1289/ehp.1104789
21. Jung Heejung University of California Riverside. *Modeling CO2 Concentrations in Vehicle Cabin.* SAE International published 04/08/2013 publication number 2013–01–1497
22. Michael L. Grady, Heejung Jung, Yong Chul Kim, June Kyu Park and Bock Cheol Lee, Hyundai Motor Company. *Vehicle Cabin Air Quality with Fractional Air Recirculation.* SAE International published 04/08/2013, Publication number 2013–01–1494
23. Wang, J., Tronville, P.: Toward standardized test methods to determine the effectiveness of filtration media against airborne nanoparticles. J. Nanoparticle Res. **16**(6), 1–33 (2014). https://doi.org/10.1007/s11051-014-2417-z
24. Zhu Na, Ph.D., Dingyu Zhang, M.D., Wenling Wang, Ph.D., Xingwang Li, M.D.,Bo Yang, M.S., 2020. *A Novel Coronavirus from Patients with Pneumonia in China.* s.l. : The New England Journal of Medicine, 2020
25. Kunkel, S.A., Azimi, P., Zhao, H., Stark, B.C., Stephens, B.: *Quantifying the size-resolved dynamics of indoor bioaerosol transport and control.* s.l. : John Wiley and Sons A/S, 2017 (2017)

Reference and Educational Works

26. Moredock, J.L.: International Society of Environmental Enclosure Engineers. (2020). *Best Practices Guide for Developing Highly Effective Environmental Enclosures* a/k/a *Advanced Cab Theory Workbook.* Jacksonville, Florida USA. www .iseee .net
27. Moredock, J.L. International Society of Environmental Enclosure Engineers. (2016). *ISEEE 1.002 In-Field Testing Method,* Jacksonville, Florida USA. www .iseee .net
28. Moredock, J.L. International Society of Environmental Enclosure Engineers. (2017). *ISEEE 1.003 OEM Cab Performance Level Certification Protocol,* Jacksonville, Florida USA. www .iseee .net
29. Vijayakumar R., 2014. Aerosol Mechanics for the Rest of Us. Liverpool, NY : R. Vijayakumar, 2014
30. Moredock, J.L. International Society of Environmental Enclosure Engineers. (2020). *Operator Enclosures and Covid-19,* Jacksonville, Florida USA. www .iseee .net
31. ACGIH Guide to Occupational Exposure limits and its companion volume ACGIH TLVs and BEIs Book
32. ICMM. HEALTH AND SAFETY CRITICAL CONTROL MANAGEMENT GOOD PRACTICE GUIDE, © 2015 International Council on Mining and Metals

Enabling Net Zero Industrial Targets Through Decarbonizing Lubricant Life-Cycle

Anshuman Agrawal(✉)

Minimac Systems Pvt. Ltd., Gat No 448/15, Success Industrial Park Nighoje, Khed, Pune 410501, Maharashtra, India
anshuman@minimac.in

Abstract. Lubricants play an important role in keeping the machinery running smoothly and efficiently. The demands for lubricants are increasing and eventually impacting the environment. There is a need to balance the increased demand with the impact of lubricants on the environment. As the world is facing a climate emergency, industries are urged to decarbonize. With innovative technologies, industries can lower the carbon emissions from their operations and reduce their carbon footprints. The technologies like Lubricant Reconditioning Systems which helps in increasing the oil's life and reliability ultimately leading to the lesser oil replacement costs and new oil purchase. What people are doing now is taking materials from the earth, making products from them, and giving them back to the environment in the form of waste. The 'Circular Economy' tackles global challenges like these while addressing important social needs. Through Oil Reclamation, the oil is rescued from normal degradation; it involves removing the contaminants from the oil and restoring it to good health. This offers many benefits like increased machine reliability, cost savings, saving of time on oil change-outs, decreased oil disposal costs, and reduced environmental impacts.

Keywords: Circular Economy · Sustainability · Lubricant Life Cycle · Decarbonization · Carbon Footprints · Lubricant Reconditioning System · Net Zero · Machine Reliability

1 Introduction

As the world races towards achieving Net Zero emissions, industries play a crucial role in transitioning to sustainable practices. India has already declared its goal to achieve Net Zero by 2070. Niti Aayog, the policy maker of the GoI in its recent notification/memo No. **M-11039(18)/1/2021** has formed 11 committees for implementing Circular Economy concepts in 11 focus areas. One of these is for **"Circular Economy in Used/Hazardous Oil Wastage"**. By decarbonizing the lubricant life-cycle, industries can make substantial progress towards their net-zero targets. The metal, mining, and mineral industry equipment requires a higher demand for lubrication. These industries operate non-stop, which subjects components to greater friction and wear than in other sectors. With this article, let's explore and understand the various strategies and technologies that can be implemented to achieve this goal.

A. Sinha et al. (Eds.): AMC 2023, SPEES, pp. 355–361, 2023.
https://doi.org/10.1007/978-3-031-46966-4_29

1.1 Hazardous Waste Management Hierarchy

Whenever Hazardous Waste Management is cited, the first reaction is Recycle and Recover. However, the preventive concept of HW Management is aimed at minimizing waste generation at source by all possible means:

a. Improvement in **process technology** & equipment which may completely eliminate waste streams;
b. Improvement in **plant operations**; and.
c. Promoting prolonged use of materials through **recovery/recycle/reuse of waste.**

Following are the recommended steps in decreasing order of priority for HW Management (Fig. 1).

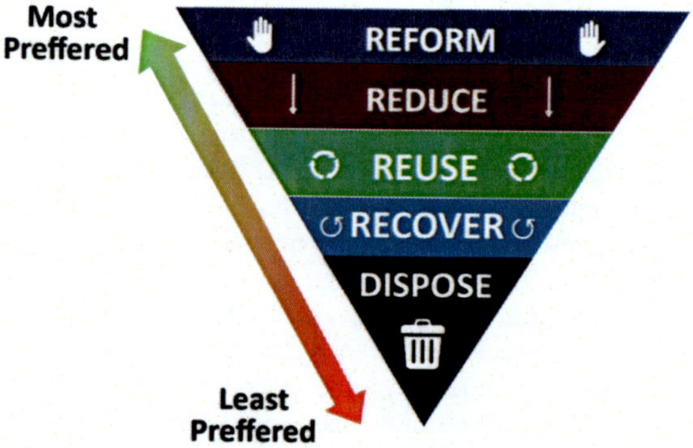

Fig. 1. Hazardous Waste Management Hierarchy (Source: hazardous waste management rules-2016 of CPCB and NPC, MoEF, GoI) [1].

2 Used Oil (Lubricant) Hazardous Waste

Lubricants to rotary and hydraulic equipment with other plant machinery are analogous to what blood is to the human body. The criticality of lubricants properties directly impacts operational efficiency, reliability, and un-halted functioning of the lubricated machinery. Generation of Used Oil Lubricant is due to following reasons:

- Periodic Replacement of Lubricant fully or partially as part of capital overhauls / turnarounds.
- Leakages from equipment
- Unplanned shutdowns / breakdowns

Lubricants of various grades collected in common non-segregated tanks lead to generation of Used Oil Waste. Generally, the waste thus generated is.

1. Disposed, or
2. Incinerated, or

3. Sold off to waste collection companies with no traceability of end-use.

All the above practices are considered to be of Linear Form and lead to high risk of uncontrolled GHG Emission into the environment and being paradox to the corporate ESG goals (Fig. 2).

Greenhouse Gas	Formula	100-year GWP (AR4)
Carbon dioxide	CO_2	1
Methane	CH_4	25
Nitrous oxide	N_2O	298
Sulphur hexaflouoride	SF_6	22,800
Hydrofluorocarbon-23	CHF_3	14,800
Hydrofluorocarbon-32	CH_2F_2	675
Perfluoromethane	CF_4	7,390
Perfluoroethane	C_2F_6	12,200
Perfluoropropane	C_3F_8	8,830
Perfluorobutane	C_4F_{10}	8,860
Perfluorocyclobutane	$c\text{-}C_4F_8$	10,300
Perfluoropentane	C_5F_{12}	13,300
Perfluorohexane	C_6F_{14}	9,300

Fig. 2. GHG comparison of Global Warming Potential [3]

2.1 Reducing Used Oil Wastage in the Mining Industry [4]

It is a critical endeavor that can significantly impact both environmental sustainability and operational efficiency. Implementing advanced filtration systems and adopting proactive maintenance practices can extend the lifespan of lubricants, curbing the need for frequent replacements. Additionally, embracing cutting-edge technologies like remote monitoring and predictive analytics enables timely identification of potential equipment failures, optimizing oil change intervals. Proper oil storage and handling protocols, including spill containment measures, further minimize oil loss and prevent contamination. By adopting these innovative approaches, mining companies can bolster their commitment to ecological preservation, while simultaneously maximizing equipment performance and reducing overall operational costs.

From Linear to Circular Economy: Understanding the Carbon Footprint of Lubricants in order to decarbonize the lubricant life-cycle, it is essential to first understand a few terms – Cradle to gate, Cradle to grave, Nano, Micro & Macro circularity (Fig. 3).

Cradle to Gate: It refers to manufacturing and distribution of the lubricant. Oil is ready to reach the customer's gate through barrels or tanks. This happens through direct or indirect channels of the lubricant manufacturers.

Cradle to Grave: It refers to the entire life-cycle of lubricants, from their manufacturing to their disposal. It surrounds the extraction of raw materials, manufacturing, distribution, usage, and end-of-life management. By adopting this approach, companies can minimize their environmental footprint and promote a circular economy rather than a linear economy.

Nano Circularity: This term refers to **Mid-life Treatment of Oil** through Oil Reconditioning Solutions. Enhancing the life of in-service lubricant right from its first fill till the

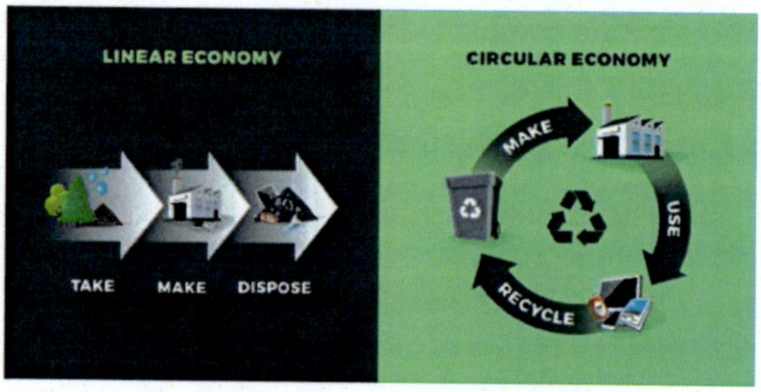

Fig. 3. Linear to Circular Economy

end of oil's life. This leads to reduction in new oil purchases, increasing the drain interval, lesser part failures like valves, better machinery reliability, and increase in productivity. With every 1% increase in energy efficiency there is a reduction of 2.5% of GHG emissions. This concept promotes "REFORM" & "REDUCE". To see how the Reconditioning solutions have helped various organisation lower their downtime, breakdown, and new lubricant consumption cost and CO_2e, refer attachments from [5–15].

Micro Circularity: This represents the **End-life Treatment** of Oil through Used Oil Reconditioning Solutions. When the in-service lubricant is nearing its life cycle, the oils must be collected & segregated in used oil storage tanks as per its grade. The used oil after reclamation is fit for reuse in the lubricated machineries maintaining its oil properties. The lubricant's new-like state can be obtained by Reconditioning which can be used again in the machinery. 100% of the used oil no longer converts into hazardous waste. This advancement is the future of lubrication. This concept promotes "REUSE". To see how the Reconditioning solutions have helped various organisation lower their downtime, breakdown, and new lubricant consumption cost and CO_2e, refer attachments from [5–15].

Macro Circularity: This represents **Waste Oil Collection and Recycling.** Out of a billion gallons of used lubricant, only 15% approximately is re-refined currently. The remaining 85% is either improperly disposed of or is burnt further. The last milestone in decarbonizing the lubricant sector is through unadulterated waste oil collection and management. The re-refiners then recycle the waste oil and produce RRBO (Re-Refined Based Oil) which acts as a feedstock to the lubricant manufacturers. This concept promotes "RECYCLE" (Fig. 4).

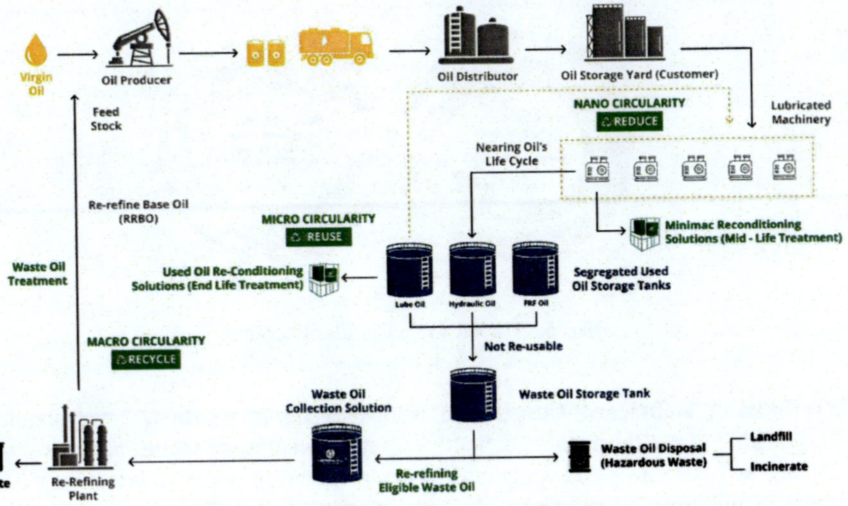

Fig. 4. The Circular Economy Model

Adopting a circular economy approach with Nano, Micro and Macro Circularity through Oil Reconditioning Solutions at every stage from Cradle to Grave is the way ahead. Let's stress upon sourcing, logistics, optimum utilization & disposal.

a. **Sustainable Sourcing & Production,** the first step is to implement a sustainable procedure to source raw materials, and do production. To support this statement, here are a few points:
- **Transitioning to Renewable Feedstock:** Traditionally, lubricants are typically derived from fossil fuels, but by shifting towards RRBOs, industries can significantly reduce their carbon footprint. Also, the current lubricant production ratio is 2:1, it should be certainly changed to 1:1.
- **Energy-Efficient Production Process:** By implementing energy-efficient and IIoT enabled technologies and processes in lube oil production – minimizing carbon emission.

The critical 4 R's to a Circular Economy (Fig. 5).

REFORM
REDUCE
REUSE
RECYCLE

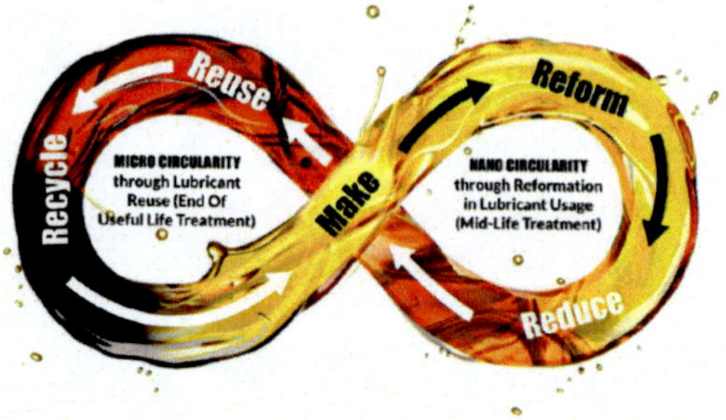

Fig. 5. The 4R's to a Circular Economy

b. **Optimizing Lubricant Usage**: The behaviour needs to change; we should be responsible to use lubricants efficiently. A few strategies are mentioned below:
- **Effective Total Lube Management:** By implementing proper lubrication management techniques and practices – we can maximize the efficiency and lifespan of lubricants as well as machineries thus achieving Asset Reliability.
- **Mid-Life Oil Treatment:** By enabling Mid-Life Oil Treatment, the oil in use is regularly monitored, tested, and reconditioned for better usage and to achieve the desired lifespan of oil. This also helps to maintain the good health of the machines and increases uptime.
- **Training and Education:** Proper training and education about correct usage and maintenance of lubricants must be given to the team handling – this will lead to better practices and reduction in wastage and GHG emission.

c. **Sustainable distribution and supply chain** also plays a crucial role in decarbonizing lubricant life-cycle. A few major focus areas are:
- **Efficient Logistics:** By optimizing and figuring out the best route for the distribution of lubricant along with sustainable packaging will help reduce carbon emissions associated with the distribution process.
- **Adoption of Renewable Energy:** With the help of companies and suppliers, the transition from petroleum vehicles to renewable energy-sourced vehicles can be done; thus, reducing a lot of carbon footprint while transportation.

d. **End-Life Management** & proper recycling management of lubricant oil is vital to achieving net-zero goals. A few important points are mentioned below:
- **Organized Used Oil Collection Centres:** Pilferage proof solutions should be made available easily and economically in front of the customers. Organize the unorganized sector.
- **Recycling & Reprocessing:** Technological advancements and re-refining plants must be set up to produce better quality of RRBO (Re-refined Base Oil) which is cost efficient and better in quality.
- **Proper disposal:** The waste lubricant which is not fit for recycling must be disposed of through re-refiners or organized channels using technology.

3 Conclusion

In order to reach the net-zero industrial aim, a thorough strategy that takes into account the entire lubricant life-cycle is needed. To contribute to a sustainable future, all stakeholders should work together to decarbonize the production, usage, distribution, and end-of-life management of lubricants. Especially, Mining industries can leverage the following benefits: extended equipment life, improved overall performance, reduction in Replacement and Maintenance costs, increase in operational hours by 3000 to 10000 hours, increase in profit, controls contamination, eliminates downtime cost and breakdown.

As the world races towards achieving net zero emissions, industries play a crucial role in transitioning to sustainable practices. India has already declared its goal to achieve net zero by 2070 [2]. By decarbonizing the lubricant life-cycle, industries can make substantial progress towards their net-zero targets and enjoy various benefits like increased machine reliability, cost savings, saving time on oil change-outs, decreased oil disposal costs, and reduced environmental impacts. The technologies and tactics discussed in this article will undoubtedly lay out a clear path for the sector to follow as it strives to adopt more eco-friendly practices and solutions and move towards a net-zero carbon future.

References

1. Hazardous waste management rules-2016 of CPCB and NPC, MoEF, GoI
2. Press release by Press Information Bureau (Govt. of India)
3. IPCC, 1996a
4. EPA.gov
5. Product Performance Certificate from Shubh International
6. Product Performance Certificate from MK Corporation
7. Product Performance Certificate from Jindal Saw Ltd.
8. Product Performance Certificate from IL&FS Tamil Nadu Power Company Limited
9. Product Performance Certificate from Hindustan Electricals
10. Product Performance Certificate from Hindalco Industries Ltd.
11. Product Performance Certificate from East Coast Railway
12. Product Performance Certificate from Lawas Lube Specialities
13. BPCL Empanelment
14. IOCL Empanelment
15. Case studies

Critical Minerals

The Impact of Recent Geopolitical Events on Supply of Critical Minerals for an Early GET

S. P. Banerjee(✉)

IIT (ISM), Dhanbad, India
sakti.banerjee@gmail.com

Abstract. The most important step in fighting the climate change menace is to have an early switch over to a decarbonized global economy. Fossil fuels currently supply about 80% of the world's energy needs and hopefully, a major part of the same can be replaced by energy from renewables within a few decades. It is a gigantic task complicated by the variety of conditions and priorities in different countries. An adequate supply of some specialized minerals is essential to usher in the Green Energy Transition (GET). The accelerated induction of electrical vehicles in many countries in the last two years has necessitated a relook at the availability of battery metals. The R&D efforts for improving battery technology are going on in full swing. However, two recent world events, namely the Covid-19 pandemic and the Russia-Ukraine war, have adversely affected the world's progress towards an early GET. The latter event especially, has caused the major economies of the world to pay greater attention to energy security and assured supply chain arrangement for critical minerals. The paper examines the issue of the availability of minerals for GET in the coming decades from the global as well as the Indian perspective.

Keywords: Green Energy Transition · battery metals · critical minerals

1 Introduction

Thanks to the series of IPCC Assessment Reports No.1 to 6, the whole world has united in reducing GHG emissions to save mankind from the looming climate change disaster. At the Paris Climate Conference in 2015, all nations agreed to take such steps as would "limit the increase in the global average temperature to well below 2 °C and pursue efforts to limit the temperature increase to 1.5 °C above pre-industrial levels in this century" (Paris Agreement 2015). However, the IPCC AR6 Policy makers' summary report of 2022 showed that the world has failed to take the necessary steps and we are likely to hit the 1.5 °C rise mark by the early 2030s itself (IPCC AR6 2022). It is recognized that the most important step in fighting the climate change menace is to have an early switch over to a decarbonized global economy. Fossil fuels currently supply about 80 per cent of the world's energy needs and hopefully, a major part of the same can be replaced by energy from renewables within a few decades. It is a gigantic task and considering the variety of conditions and priorities in different countries, many challenges have to be overcome

A. Sinha et al. (Eds.): AMC 2023, SPEES, pp. 365–375, 2023.
https://doi.org/10.1007/978-3-031-46966-4_30

before a satisfactory Green Energy Transition (GET) is achieved. An adequate supply of some specialized variety of minerals is essential to usher in the green energy transition. The subject has been covered in the 2021 IEA special report "The Role of Critical Minerals in Clean Energy Transitions" (IEA, 2021) and also in a paper by Banerjee "On the availability of Minerals needed for a Green Energy Transition" presented at the 9th Asian Mining Congress (Banerjee 2022).

IEA has reported an accelerated induction of Electrical Vehicles (EV) in many countries in the last two years needing a relook at the availability of the battery metals. The R &D efforts for improving the battery technology and finding cheaper substitutes for Li-Ni-Co batteries are going on in full swing. However, two recent world events, namely the Covid-19 pandemic and the Russia-Ukraine war, have adversely affected the world's progress towards an early GET. The latter event specially, has caused the major economies of the world to pay greater attention to energy security and assured supply chain arrangement for strategic and critical minerals. It is proposed to consider the impacts of the above developments and examine afresh the issue of availability of minerals for the GET in the coming decades from the global as well as the Indian perspective.

2 The Impact of Covid 19 and the Russia-Ukraine War on Global GET

The first case of someone suffering from Covid-19 was reported from Wuhan in December 2019 and the virus spread rapidly all over the world by early 2020. WHO declared the official end of the pandemic on 4th May 2023. At the height of its prevalence in 2020 and early 2021, it created economic havoc for most countries as huge expenses had to be incurred in dealing with health and unemployment issues. Investments relating to GET were drastically curtailed and the rate of growth of renewable energy and energy efficiency of appliances were reduced considerably. The COP 26 meeting scheduled to be held in Glasgow in 2020 was crucial for progress towards GET as many developed countries were expected to make emissions-reduction commitments at this conference. However, the Covid-19 related travel and meeting restrictions caused a full one-year postponement of this crucial meeting.

Another international event that is greatly affecting the progress towards an early GET is the Russian invasion of Ukraine that started on 24th February 2022 (and is continuing till the time of penning this paper in July 2023). Before this invasion, EU countries have been importing high volumes of natural gas and oil from Russia but after the imposition of sanctions on Russia in March 2022, EU countries had to arrange the import of oil and gas from other costly and distant sources. In some of the EU countries mothballed coal-based thermal power stations have been brought back to service. In the international market, coal, oil and gas prices, increased greatly and the price of food grains also went up. The resulting inflation and supply chain disruptions have caused a slowing down of the world economy, and that in turn has been a major setback for achieving an early GET. Many countries have been compelled to review their energy security and assured availability of critical minerals. The Net Zero Industry Act 2023 of the EU and the Inflation Reduction Act 2022 of the USA are examples of legislations with provisions for incentives for local production and assured availability of critical minerals and equipment needed for GET which is discussed later in more detail.

3 Some Positive Developments Helping GET in the Past Couple of Years

3.1 Accelerated Induction of Electrical Vehicles (EVs)

The number of EVs on the road in the world has gone up from 5 million in 2018 to 26 million in 2022. Under the IEA Stated Policies Scenario (STEPS), the global outlook for the share of EV sales based on existing policies and firm objectives have increased to 35% in 2030, up from less than 25% in the previous world EV outlook of 2021. According to the IEA's World EV Outlook 2023, global EV car stock (excluding two/three-wheelers) grows from almost 26 million in 2022 to about 240 million in 2030 in the STEPS Scenario (IEA 2023).(Fig. 1).

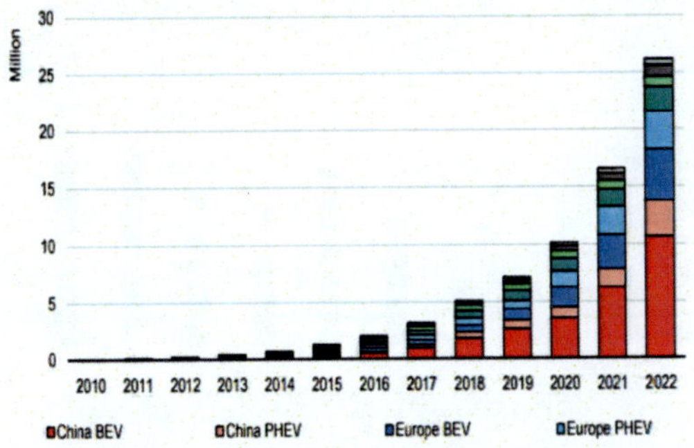

Fig. 1. Growth of EV stocks in the world 2010–2022 as per World EV Outlook (IEA 2023)

3.2 Increase in Production of Battery Metals and Improvements in Battery Technology

The rapid increase in demand for EVs is driving demand for batteries and related critical minerals. A Lithium-ion battery is the preferred choice for a majority of EV manufacturers as such batteries provide a high energy density allowing a decent range of travel before recharging. The shares of three main types of batteries in use today are - NMC (lithium nickel manganese cobalt oxide) battery (60%), followed by LFP (lithium iron phosphate) battery (30%), and NCA (nickel cobalt aluminium oxide) battery (8%) (IEA 2023). With a demand-supply imbalance, the price of Lithium carbonate has skyrocketed in the last two years (Fig. 2a) even though the world production of lithium metal has gone up from 82,500 t in 2020 to 130,000 t in 2022 (Fig. 2b).

After lithium, nickel and cobalt are the next two most expensive battery metals and global production of both have substantially increased in the last few years. Surprisingly

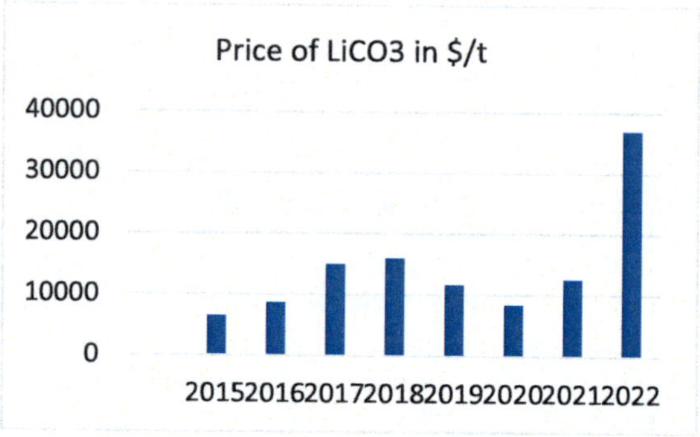

A . The price of $LiCO_3$ during 2015-2022
(based on USGS Data)

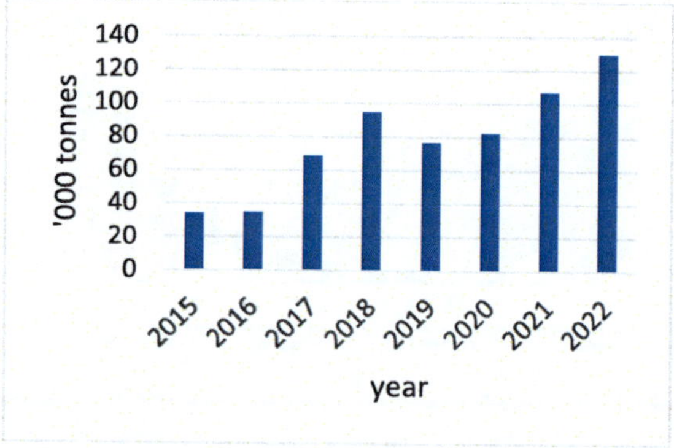

B . World Production of Li during 2015-2022
(based on USGS Data)

Fig. 2. a The price of $LiCO_3$ during 2015–2022 (based on USGS Data). b World Production of Li during 2015–2022 (based on USGS Data)

the price of all three expensive battery metals (Li, Co and Ni) showed a downward trend in 2023 presumably due to a temporary oversupply in the market (Fig. 3) (IEA 2023). Another interesting conclusion from Fig. 3 is that the continuous downward trend of battery prices since 2015 seems to have ended in 2022.

The efforts to develop alternatives to conventional lithium-ion batteries have also intensified. The LFP batteries in contrast to that of NCA and NMC batteries use cheaper iron and phosphorus rather than the costly nickel and cobalt. The downside of LFP

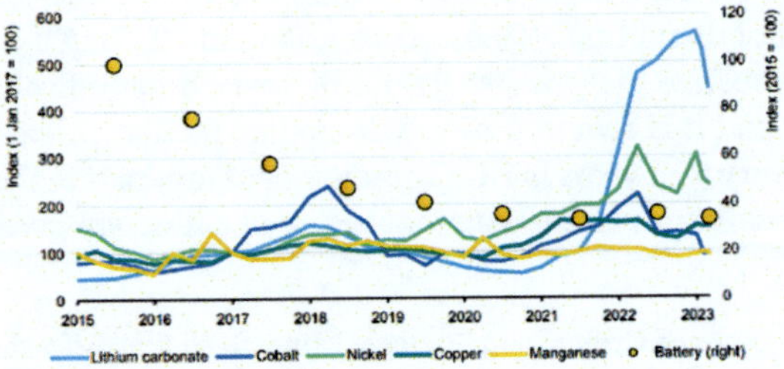

Fig. 3. Price of selected battery materials and lithium-ion batteries, 2015–2023 [Courtesy: IEA, The Global EV Outlook 2023 (IEA 2023)]

batteries is that the energy density tends to be lower than that of an NMC battery (IEA 2023). It is reported that Na-ion batteries are much cheaper than NMC batteries and China has started using such batteries in the cheaper models of EVs in 2023.

3.3 Some Recent Legislations in Developed Countries Supporting EV Induction

On 18 April 2023, EU Parliament approved a set of "Fit for 55" legislative packages paving the way to reduce net GHG emissions by at least 55% by 2030 compared to 1990 levels and to achieve climate neutrality by 2050. European Union in March 2023 passed the Net Zero Industry Act, which contains a clause paving the way for raising EU battery manufacturing capacity from 69 GWh in 2022 to at least 550 GWh in 2030. Similarly, the 2022 Inflation Reduction Act of the USA has provisions for strengthening domestic supply chains for EVs, EV batteries and battery minerals, laid out in the criteria to qualify for clean vehicle tax credits (IEA 2023). Canada's Federal budget presented on 28 March 2023 included tax credits for GET-related investments expected to cost some $55 billion through to the 2034–35 fiscal year (Canada 2023). Japan increased its EV subsidy scheme amount to $ 530 million in 2022, and South Korea has passed its Carbon Neutrality Act in 2020, under which a "Green New Deal" has been announced (IEA 2023). All the above legislations in different countries will accelerate the growth of EVs in the world and consequent increase in demand for battery metals.

4 Different Critical Minerals List in Different Countries

Critical minerals refer to mineral resources which are essential to sustain the production process of an economy, or are required for national security, and whose supply chain is vulnerable to disruption. Within this general description, the legal definition of critical minerals in different countries varies from each other depending upon the special needs, perceived vulnerabilities, or simply due to preference of some wording by the policy makers in each country. The EU Critical Raw Materials Act 2023 identifies raw materials which are important for the economy and face a high risk of supply disruption as critical materials. Within this group of critical materials, the EU further identifies a list of raw materials characterized by high strategic importance and calls them strategic materials. Australian Geoscience defines a critical mineral as "a metallic or non-metallic element that is essential for modern technologies, economies or national security, and has a supply chain at risk of disruption" (Geoscience Australia 2022), All countries with a list of critical minerals realize that the list should be revised periodically to take stock of changed situations and it is required by law to update the list at least once every three years in USA and every four years in EU. The number of minerals in the Critical Minerals list of five selected countries along with the year of updating are given in Table 1.

Table 1. Number of Critical and Strategic Minerals in five different countries

Country	Number	Year of listing	Authority	Remarks
Australia	26	2022	*Australian Geoscience*	
India	30	2023	Ministry of Mines, GoI	*List of June 2023*
Japan	12	2021	Ministry of Economy, Trade and Industry (METI)	
EU	34	2023	Critical Mineral Act of EU	
USA	50	2022	USGS, Dept. of Interior	

N.B. *Compiled from Australian Geoscience Report 2022 (Geoscience* Australia, 2022*) and for India Ministry of Mines Report on Critical Minerals June 2023* (*Ministry of Mines,* Govt. of India, 2023*)*

The updated Critical Minerals list of Australia contains names of 26 minerals and the report also contains a table showing how the minerals considered as critical in Australia fare in the critical minerals list of four other countries, namely, USA, EU, Japan and India. The Indian list in the Australian document gives names of 18 minerals (taken from an old CEEW report of 2016) and for the current Indian position, one should refer to the June 2023 list of the Ministry of Mines, GoI containing names of 30 critical minerals (Ministry of Mines, Govt. of India 2023).

5 Projection of Critical Minerals Needed for GET

Efforts to decarbonize the global economy have resulted in a plethora of reports dealing with the additional mineral needs for the task. But the reports have come out with a diversity of results because of different conditions and scenarios assumed by the researchers, in the choice of the target year, their inability to fathom the impact of rapid development in the battery and other technologies, and the extent of financial and regulatory support for GET in different countries. Soon after the Paris Climate Agreement, a World Bank Group started working on the additional mineral needs for the GET and came out in 2017 with a report projecting growth of metal needs by two, three and ten times for solar panels, wind turbines and batteries by 2050 (World Bank 2017).

The intermittent nature of solar and wind power means some form of energy storage system is essential for their integration into the broader grid system and most analysts have predicted a rapid growth of the stationary Battery Energy Storage Systems (BESS).

But the predictions by different authorities for BESS capacity in 2030 vary widely. The 2021 IEA Report on the need of critical minerals predicted the demand for BESS under its STEPS scenario as 420 GWh in 2040 (IEA 2021). The business intelligence company Rystad Energy, in a 2023 report, predicted a similar BESS capacity of 400 GWh but the need arises by 2030. Darminto & Herculano (Darminto et al. 2022) on the other hand, have projected a whopping demand of 1,857 GWh for BESS in 2030. More than BESS, it is the exploding demand for EV batteries that is causing concern as to the availability of various battery metals in future. The battery metal demand depends upon the number of EV sales per year and the average metal needed for a car battery. In both aspects, a variety of demand projections have been made by different market analysts (Table 2).

Table 2. Demand projections made by different market analysts

Reporting organization	Year of report	Sales estimate in millions	Target year
IEA Global EV outlook	2019	23	2030
Deloitte, Woodward et al	2020	25	2030
IEA Role of Critical Minerals	2021	70	2040
Research And Markets	2021	35	2030
Research And Markets	2022	39	2030
Darminto et al., iClima	2022	50	2030
IEA Global EV outlook	2023	40	2030

Based on the projected data IEA predicted the total mineral demand to double in the STEPS and quadruple in the SDS in the 2020–2040 period as shown in Fig. 4 below (IEA 2021).

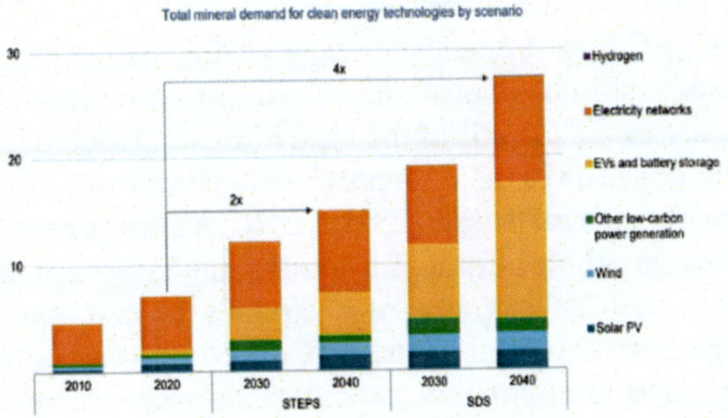

Fig. 4. Mineral demand set to double in STEPS, and quadruple in SDS by 2040 (IEA 2021)

The 2021 IEA Report singled out Lithium, graphite, Cobalt, Nickel and Rare Earth Elements as critical metals for the GET and predicted their demand growth as 42, 25, 21, 19 and 7 times respectively in the SDS scenario in the 2020–2040 period (IEA 2021).

We can now discuss the demand and availability of these five critical minerals in the near-term future of up to 2030.

The forecast for lithium given by Mckinsey (McKinsey-GBA 2022) is 3 to 4 Mt of LCE (lithium carbonate equivalent) and by Albemarle Corporation (2023) is 3.7 Mt LCE. Darminto et al. (iClima, Darminto et al. 2022) gave their estimate as 3 to 3.7 Mt of lithium metal and Cochilko (2023) of Chile gave their estimate as 3.8 Mt but by the year 2035. The demand estimate for graphite in batteries in 2030 comes to 2.45 Mt (Northern Graphite 2023). Regarding Cobalt, Xinkai Fu et al. 2020 of MIT in a 2020 publication made a demand estimate ranging from 235 to 430 kt (Xinkai Fu et al. 2020). Regarding Nickel, the world production in 2021 was 3.3 Mt and Vale S A, a major producer of nickel, has estimated a total demand of 6.2 Mt. in 2030 (Vale 2022). World production of Rare Earth Elements in 2021 amounted to 280 kt. Considering a seven times growth in 20 years the demand for REE in 2040 is expected to be approximately 2 Mt. However, a point of significance is that the demand growth of the Heavy REEs (the minerals needed in the manufacture of permanent magnets) is much faster than that of the lighter REEs.

6 The World's Efforts to Meet the Increased Critical Metal Needs

Over the last 40 years world GDP has grown at a CAGR of approximately 5 per cent and the consumption of most minerals has also steadily increased to sustain the economic growth. The normal law of demand and supply applies to the mineral sector also although the response may be a bit slower than for other commodities due to the long gestation period of mining projects. In case of the critical minerals for GET, another factor of importance is the extremely high rate of demand growth expected to continue over a long period. Hence strong policy support from the governments of different countries will be needed to bolster the normal market forces to ensure the availability of critical mineral commodities in future. Events in the last couple of years show different governments and mining companies are aware of the situation and initiated many steps to increase the production of the critical minerals for the GET.

Canada adopted a Critical Minerals Strategy in 2022 listing a number of focus areas including (a) driving research, innovation, and exploration, and (b) accelerating project development relating to the growth of the critical minerals sector (Canada 2022). The 2022 Inflation Reduction Act of the USA provides tax incentives and other financial support to develop critical mineral deposits. Since 2016, Australia is running an "Exploring for the Future" program to help inform decision-making by government, community and industry on the sustainable development of Australia's mineral, energy and groundwater resources. Many governments (e.g. Australia, Quebec etc.) and mining companies (e.g. BHP) are offering grants to exploration and critical minerals companies to conduct mineral exploration and mine development. Presently nearly half the production of lithium in the world comes from the four Latin American countries of Argentina, Peru, Bolivia and Chile. All four countries are witnessing an inflow of huge investments for further exploration and mining of lithium resources.

Because of the frenzied exploration activities, many new lithium deposits have come to light in the last couple of years. These include deposits from the USA (Nevada,

Arkansas, Oregon), Canada (Galaxy lithium, Quebec), Mexico (Sonora), Czech Republic (Cinovec), Serbia (Jadar), Australia (Mount Holland), DRC (Manono) and India (Reasi, J&K). Increased price for lithium has permitted the restart of a closed lithium mine in Australia.

The story is similar for other battery metals where the sudden increase in demand has raised prices and interest in exploration activities. In Sweden, a large rare-earth deposit has been recently discovered and LKAB is planning to apply for an exploitation concession. Brazil's Pela Ema rare earths mine is expected to start production by the end of 2023. Chinese company Harita Nickel is trying to apply improved HPAL (High-Pressure Acid Leaching) technology to extract battery-grade nickel from low-grade Indonesian ore. The success of this experiment is likely to flood the world with low-cost nickel. A similar development is taking place for lithium in the form of direct lithium extraction or DLE technology which has the potential to greatly reduce the cost of lithium.

World leaders are also aware of the need to diversify the sources of supply chains of critical metals. The 13-member Mineral Security Partnership program was announced by the USA in June 2022 and now includes India in the club. One of the decisions taken by the G20 Energy Ministers at their meeting in Goa on 22nd July 2023 was to ensure the availability of low-cost financing to ensure early green energy transition.

7 Indian Efforts to Ensure Supply of Critical Metals for the GET

India started its fascinating journey with renewable energy (RE) in 2010 with the establishment of the National Solar Mission. As a part of its climate pledge after the Paris Climate Agreement, India set a target to install 175 GW of renewable energy capacity by 2022. A CEA report shows that India had an installed capacity of 171.8 GW from renewable energy sources on 31st March 2023, thus marginally failing to achieve the target of 175 GW by 2022. In November 2021 at COP 26 meeting in Glasgow, India promised to have an installed capacity of 500 GW of non-fossil energy capacity and to meet at least 50% of its energy requirements through RE by 2030. India is currently implementing the second largest renewable energy expansion programme in the world, second only to that of China. According to Mr. B.S. Bhalla, MNRE Secretary, India plans to have 100 GW of solar module manufacturing capacity per annum by 2026.

Such rapid growth of RE in India has brought to the fore the urgent need for Battery Energy Storage Systems (BESS). India currently has only 37 MWh of BESS capacity. According to an estimate by CEA, India is likely to need 41 GW of BESS capacity by 2029–30. In early 2023 India announced it will provide incentives totalling Rs 37.6 billion to companies manufacturing battery storage projects.

The government of India has declared its intention to become a hub of electric vehicle manufacturing and has been supporting the deployment of EVs in the country through incentives paid to the purchasers and manufacturers of EVs under its 'FAME' (Faster Adoption and Manufacturing of Electric Vehicles) scheme. Although penetration of EVs in the passenger car market in India has been slow, recent data indicate it is picking up fast. In 2022 over one million EVs were sold in India. In June 2023, it was reported in the media that Tesla is considering setting up a 500,000 unit per year capacity EV manufacturing plant in India. Perhaps the forecast of 1.9 million EV sales in India in 2030 by IEA or 2.05 million forecast by CSE of New Delhi will need a revision soon. India aims to boost domestic manufacturing of solar cells, wind turbines, electric vehicles, and batteries for BESS and EV through Production Linked Incentive (PLI) schemes. To fulfil this desire, India will have to ensure an assured supply chain for large quantities of the critical minerals such as copper, lithium, nickel, cobalt and Rare Earth Elements required for the green energy sector.

KABIL (Khanij Bidesh India Limited), a JV Company of NALCO, HCL and MECL, was formed in 2019 to focus on identifying and sourcing battery minerals like Lithium and Cobalt abroad. KABIL is continuing its engagement with a few companies and government agencies in Australia, Argentina, Bolivia and Chile. The partnership between India's KABIL and the Critical Minerals office of Australia has been formalized on 10/3/2023.

In June 2023, India became a member of the Mineral Security Partnership, a strategic grouping of 13 member states including Australia, Canada, Finland, France, Germany, Japan, the Republic of Korea, Sweden, the United Kingdom, the USA, the European Union, Italy (and now India becomes the 14th member).

The Geological Survey of India has intensified its search for critical mineral deposits in India and has reported finding "inferred resources" of 5.9 million tonnes of lithium in Reasi district in Jammu and Kashmir. If ongoing further exploration of the deposit confirms the economic viability of mining the deposit, it would be a big boon for the RE and EV sectors of the country.

8 Concluding Remarks

The world is making good progress in its GET programme and the need for critical minerals will increase manyfold. With the price rise of the metals and higher investments in exploration and mining, the discovery of many new mineral deposits has taken place. It appears that the spectre of a very large metal shortage in the near future has disappeared, at least temporarily, from the horizon. However, the present widespread financial and policy support for the GET must continue in the future also to avoid critical mineral shortages in the mid-term and long-term future. The efforts of the major economies to diversify supply sources to avoid possible disruption and abnormal price rise in critical times are also bearing fruit. In the context of India, the discovery of a large inferred lithium deposit in Jammu and Kashmir may help the country in its GET pursuit. India becoming a member of the Mineral Security Partnership in June 2023 is a major development and India hopes it will have an assured supply of critical minerals to fulfil its ambitions to achieve an early Green Energy Transition.

References

Albemarle Corporation 2023. Strategic Update, 25 Jan 2023. https://www.greencarcongress.com/2023/01/20230125-albemarle.html

Banerjee, S.P.: On the availability of minerals needed for a green energy transition. In: Proceedings of 9th Asian Mining Congress, MGMI, Kolkata, April 2022

Canada: The Canadian Critical Minerals Strategy, p. 6 (2022). https://www.canada.ca/en/campaign/critical-minerals-in-canada/canadian-critical-minerals-strategy.html

Canada: Federal budget 28 March 2023. https://www.budget.canada.ca. home-accueil-en

Darminto et al.: 'Forecasting Demand for Batteries until 2030 and Considerations on Supply' (2022). https://www.iclima.earth/article/forecasting-demand-for-batteries-until-2030-and-considerations-on-supply-different-applications-technologies-minerals-and-costsforecasting-demand

Geoscience Australia: 'Critical minerals at Geoscience Australia', Australian Government, March 2022. https://www.ga.gov.au/scientific-topics/minerals/critical-minerals

IEA: World Energy Outlook Special Report: The Role of Critical Minerals in Clean Energy Transitions, Paris, p. 103 (2021)

IEA: The Global EV Outlook 2023 'Catching up with climate ambitions', Paris, p. 9, 11, 12, 57, 58, 61, 78 (2023)

IPCC AR6: Summary for Policymakers, Pörtner et al in Climate Change 2022: Impacts, Adaptation, and Vulnerability. Working Group II, Cambridge University Press, p. 10 (2022)

Mckinsey-GBA: 'Battery 2030: Resilient, sustainable, and circular.' Joint article by Mckinsey and Global Battery Alliance (2023)

Ministry of Mines, Govt. of India: Notification re Critical Metals, June 2023

Paris Agreement: UNFCC Paris Agreement, English (unfccc.int), p. 3 (2015)

Research and Markets, Dublin (2022)

Vale: 'Vale sees 44% increase in Nickel Demand by 2030 (2022). https://www.reuters.com/markets/commodities/vale-sees-44-increase-global-nickel-demand-by-2030-2022-09-07/

World Bank: 'The Growing Role of Minerals and Metals for a Low Carbon Future', Arrobas et al., Washington, D.C. (2017). http://documents.worldbank.org/curated/en/207371500386458722/The-Growing-Role-of-Minerals-and-Metals-for-a-Low-Carbon-Future

Xinkai Fu et al.: 'Perspectives on Cobalt Supply through 2030 in the Face of Changing Demand' Environmental Science and Technology, 54(5) MIT Open access article (2020). https://dspace.mit.edu/bitstream/handle/1721.1/142581/2020_cobalt_resource_review.pdf?sequence=2&isAllowed=y

Design of Operators' Cabins for Mineral Processing Plants Using Specially Fabricated Acoustic Material With Simple Expansion Chamber Structure

Nihar Ranjan Sahu[1](✉) and Bibhuti Bhusan Mandal[2]

[1] Noise Lab, Department of Mining Engineering, Indian Institute of Technology, Kharagpur, West Bengal 721302, India
niharranjansahu1995@gmail.com

[2] Department of Mining Engineering, Indian Institute of Technology, Kharagpur, West Bengal 721302, India

Abstract. Mineral processing plant operators are frequently exposed to noise levels that can reach up to 100 dB(A). This research aims to find appropriate acoustic materials for the construction of cabins that can protect the operators by reducing noise levels by at least 15 dB(A) in the specific frequency range of 0.5–5 kHz, as the human ear is more sensitive to these frequencies. 3D printed material made from Ultraviolet (UV) cured standard resin with an internal Simple Expansion Chamber (SEC) structure was selected for the design. The ratio of the length of SEC to the total thickness of samples varies from 80%–20% keeping the thickness of the sample constant to investigate the optimum length of SEC. Acoustic properties like Sound Absorption Coefficient (SAC) and Sound Transmission Loss (STL) were measured using the Impedance Tube Apparatus. A maximum SAC of 98% was observed at 3.075 kHz when the sample having a ratio of the length of SEC to the thickness of the sample is 80%. The peak SAC decreases when the length of SEC is reduced from 8–2 mm. STL increases when the length of SEC decreases from 8–2 mm. Based on our findings, it is evident that our designs hold significant promise for constructing operator's cabins, offering effective attenuation of at least 20 dB within the tested frequency range of 0.5–6.3 kHz.

Keywords: Simple Expansion Chamber · Sound Absorption Coefficient · Sound Transmission Loss · Impedance Tube Apparatus

1 Introduction

Excessive noise levels may lead to tinnitus and hearing loss in the exposed population. It is also a major cause of annoyance and irritation in communities around industrial establishments. Noise exposure is a common occupational hazard in mines and mineral processing plants. Worldwide, 16% of disabling hearing loss in adults is attributed to occupational noise, ranging from 7 to 21% in the various sub-regions (Nelson et al. 2005). In India, health surveillance studies indicated that many workers engaged in mining

A. Sinha et al. (Eds.): AMC 2023, SPEES, pp. 376–386, 2023.
https://doi.org/10.1007/978-3-031-46966-4_31

operations suffer from Noise-induced hearing loss (NIHL). The affected population was found to vary from 12.4–25.7% in various mines (Nandi and Dhatrak 2008). NIHL is bilateral and symmetrical, usually affecting the higher frequencies (3–6 kHz) and then spreading to the lower frequencies (0.5, 1, or 2 kHz) (Noise and hearing loss conference 1990). NIHL usually begins in the frequency region around 4–6 kHz, the upper levels of the speech region (Nandi and Dhatrak 2008). Even though there is a lack of consensus among the researchers, it can be concluded from the available literature that the human ear is more sensitive at 0.5–6 kHz (Gelfand 2011, p.87).

Crusher plants and machinery are major noise sources in the mining industry (Madahana et al. 2020). Several authors across the world reported that noise level inside the crusher plants ranges from 80–115 dB (A) (Manwar et al. 2020; NIOSH 1999). The present authors also conducted a noise survey in a mineral processing plant of an open cast iron ore mine located in central India. We found that the equivalent sound pressure level (L_{eq}) at the operator's location of the jaw crusher, cone crusher, vibrating screens, roll crusher, etc. ranged between 87.7 and 98.6 dB(A). The noise was continuous over the entire period of measurement. Fast Fourier Transform (FFT) also revealed that the SPL was dominant at the lower side whereas it was subservient towards the upper side of the audible frequency domain (20 Hz–20 kHz).

An eight-hour time-weighted average noise exposure of 85 dB(A) has been stipulated by national and international organizations across the world to be safe (WHO 2011, 2014; OSHA; NIOSH; JOSHRC 2017; DGMS 1975). Readings from nine out of 11 locations showed the L_{eq} values exceeding 85 dB(A) in the survey conducted by the present authors in a mineral processing plant. SPLs were higher in the frequency range of 0.5–6 kHz. Hence it is imperative to apply one or more noise control measures to protect the workforce from the ill effects of the exposure.

Various re-engineering methods may be implemented at the source to reduce noise emissions. Alternatively, modification of the path of noise transmission or the use of Personal Protective Equipment (PPE) by the recipient can provide an effective reduction of noise exposure (Madahana et al. 2020). The mineral processing plant under study is recently built with modern technology. Hence it would be difficult to opt for any substantial modification mainly due to techno-economic constraints. Similarly, the protection of the employees by providing PPEs may not be effective since it largely depends on regular use and various ergonomic issues. To address this problem, the authors wanted to design appropriate cabins for the operators to protect them from the high noise levels generated by the mineral processing equipment. Since the noise level in the plant is about 100 dB(A) at most of the measured locations, the customized acoustic material was required to provide an attenuation of at least 15 dB to bring it down to 85 dB(A). Moreover, the attenuation is desirable in the frequency range of 0.5–6 kHz.

However, traditional soundproofing materials have limitations in terms of their effectiveness and practicality in these applications. In recent years, there has been an increasing interest in the development of new acoustic materials that can effectively attenuate sound waves across a wide range of frequencies, particularly those generated by industrial equipment. The use of a Simple Expansion Chamber (SEC) structure is an effective method for sound attenuation, and the principle of SEC) has gained attention as a promising area of research for developing acoustic materials with improved performance.

The principle of the Simple Expansion Chamber (SEC) structure involves the use of an expansion chamber along the direction of sound transmission. The expansion chamber is designed to trap and dissipate sound energy, resulting in a significant reduction in noise levels within the structure. The design of the SEC can be optimized to target specific frequencies, allowing for tailored acoustic performance.

This research article aims to present a novel acoustic material specifically developed for use in control rooms of mineral processing plants to protect the operators from the adverse effects of noise exposure. The design is based on the principle of SEC structure, which aims to effectively attenuate ambient noise by a minimum of 15 dB in the frequency range of 0.5–6 kHz. Notably, this material needs to be lightweight but strong, and capable of maintaining a certain amount of ventilation through the pores across the walls.

2 Materials and Methods

2.1 Materials

Ultraviolet (UV) cured standard resin (brand: WOL3D) with density of 1.05–1.25 g/cm^3, tensile strength of 36–52 MPa, tensile modulus of 1.78–2.38 MPa, thermal expansion of 95×10^{-6} and viscosity of 100–450 MPa.s is used for 3D printing of samples (WOL 3D Technical specification of standard UV-cured resin 2023).

2.2 Methods

2.2.1 Design of Unit Cell

Acoustic material having an SEC structure exhibits wider and periodic Sound Transmission Loss (STL) properties and a larger area expansion ratio(m) exhibits higher peaks (or domes) in STL (Munjal 2013, p. 169). Perforation ratio takes an important role during the design of acoustic material and a higher perforation rate allows more sound energy to enter into the material which causes higher absorption of sound energy. These two factors influence the authors to design acoustic material having internal SEC structures. A unit cell can be designed and then used for creating a periodic structure for the design of a sample. For example, octet truss, diamond lattice, and Kelvin cell geometries are already explored for mechanical and thermal applications (Zheng et al. 2014; Heinl et al. 2008; Ekade and Krishnan 2019; Deshpande 2001). In this research Computer aided design (CAD) of the unit cell is carried out in SOLIDWORKS. A cuboid unit cell of 5 × 5 × 10 mm is designed along (x × y × z) direction. A solid cylindrical of diameter 4 mm and length of 8 mm is subtracted from the center of the unit cell along the z direction, which results in an expansion chamber structure. Then a solid cylinder of diameter of 1 mm and length of 10 mm is subtracted from the center of the unit cell along the z-direction, which results in inlet and outlet ports of the expansion chamber. The steps involved in the design of the unit cell-A are given below:

Step-1: Cuboid unit cell of size 5 × 5 × 10 mm is created by the extrude feature available in SOLIDWORKS (Fig. 1(a)).

Step-2: SEC of diameter 4 mm and length of 8 mm is created at the center of the unit cell along z direction by cut extrude feature (Fig. 1(b)).

Step-3: A cylinder of diameter 1 mm and length of 10 mm is extracted by cut extrude feature along the z direction (through the center of the expansion chamber). This extraction results in the unit cell of the cylindrical expansion chamber along with two cylindrical ducts which connect SEC along the z direction and the unit cell is named unit cell-A (Fig. 1(c)).

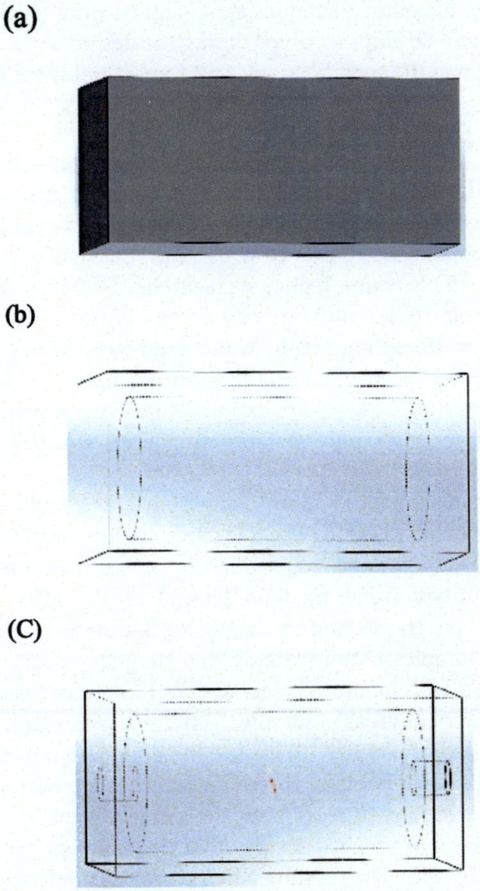

Fig. 1. Steps involved in the design of unit cell

The authors try to optimize the length of SEC concerning the thickness of sample so the length of the expansion chamber decrease to 6 mm, 4 mm, and 2 mm which results in the formation of unit cell -B, C, and D.

2.2.2 Design of Sample from the Unit Cell

CAD design of samples is carried out using SOLIDWORKS. Samples are designed by the periodic pattern of unit cells. For measurement of acoustic properties (Sound Absorption Coefficient (SAC) and Sound Transmission Loss (STL)) in Impedance tube apparatus (ITA) (brand: Holmarc) it requires a cylindrical sample of diameter 29.8 mm. For the design of the first sample named sample-A (29.8 mm diameter & 10 mm thickness), the authors imported unit cell-A to SOLIDWORKS and then made 6 periodic patterns of the unit cell along the x and y-axis. This resulted in the formation of a 30 mm × 30 mm × 10 mm rectangular cuboid as shown in step-2 (Fig. 2). Now the cuboid was cut to obtain a cylindrical sample of 29.8 mm diameter and 10 mm thickness as shown in step - 3 (Fig. 2). Similarly, the other 3 samples are designed from unit cell B, C, and D and named sample-B, C, and D. Steps involved during the design of the sample from the unit cell is given in Fig. 2 and the specification of all four samples are presented in Table 1.

2.2.3 Fabrication of Samples

Stereolithography (SLA) 3D printing is the most common resin 3D printing process that has the highest resolution, accuracy, sharpest details, and smoothest surface finishes of all 3D printing technologies (Saran et al. 2022). The SLA 3D printing process first appeared in the early 1970s, when Japanese researcher Dr. Hideo Kodama invented the modern layered approach to stereolithography, using ultraviolet light to cure photosensitive polymers. The term stereolithography was coined by Charles (Chuck) W. Hull, who patented the technology in 1986. SLA 3D printers use light-reactive thermoset materials called "resin". When SLA resins are exposed to certain wavelengths of light, short molecular chains join together, polymerizing monomers and oligomers into solidified rigid or flexible geometries.

The CAD designs of samples are done using SOLIDWORKS software, and then the designs are converted to an STL file. The STL files are imported to the slicing software CHITUBOX and printing parameters, layer height, exposure time, bottom exposure time, choosing support type (light/ medium/ heavy), infill density, and support density. After slicing the files are transferred to ELEGOO Saturn S 3D printer through a pen drive for printing of samples. After printing, the samples undergo a cleaning process where they are washed with Isopropyl alcohol (IPA) to remove residual resin from the internal structure and surface areas of samples. After washing supports are removed from the sample with the help of a cutter. To ensure the complete removal of resin particles present in the internal structure a compressed air jet is used to remove any foreign object and resin from the internal structure. Finally, the samples are UV-cured to achieve their full strength. This thorough process ensures that the samples are produced with high quality and are ready for use. The printing parameters considered during the fabrication of samples are shown in Table-2. The figure of the ELEGOO Saturn S 3D printer is shown in Fig. 3.

2.3 Experimental Details

Samples were tested for SAC and STL characteristics in impedance tube apparatus.

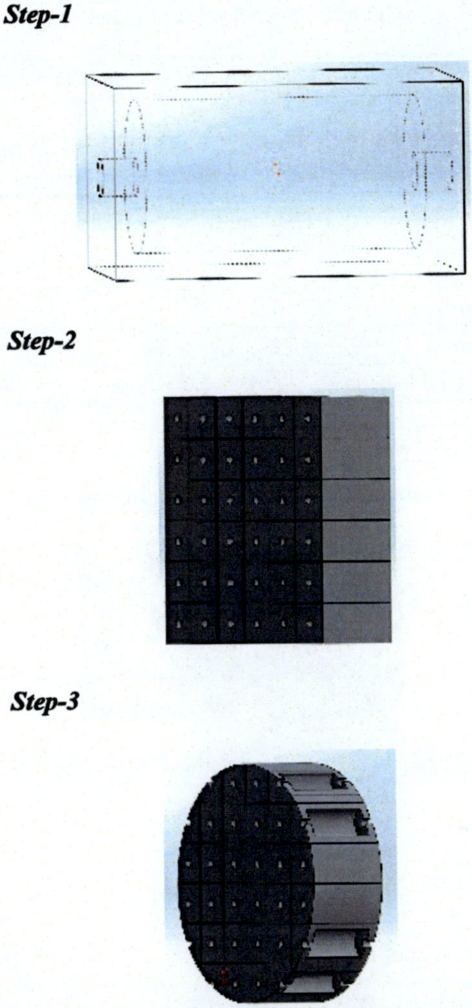

Fig. 2. Steps involved in sample design from unit cell

2.3.1 Sound Absorption Coefficient (SAC)

The SAC of all samples is measured in the frequency range of 0.5–6.3 kHz using a two-microphone impedance tube, following ASTM E1050-19 and ISO 10534-2 (Fig. 4) (ASTM E1050, 2019; ISO 10534-2, 1998). Plane waves are generated by an amplifier and fed to a 30 W speaker in the tested frequency range at an interval of 25 Hz. Incident type 1/4"microphones were used in this apparatus with a sensitivity of -3 dB at 1.5 V having a frequency response of 20–16,000 Hz. A Holmarc (India) make 8-channel data acquisition system (DAQ) was used with two source outputs which are connected to a laptop via USB 3.0. The Wave Analyzer 4C software is used to control the measurement process and also to access the data. In total, 4 cylindrical samples are tested. SAC is

Table 1. Specification of samples

	Sample-A	Sample-B	Sample-C	Sample-D
Diameter of sample (mm)	29.8	29.8	29.8	29.8
Thickness of sample (mm)	10	10	10	10
Diameter of duct (mm)	1	1	1	1
Diameter of SEC (mm)	4	4	4	4
Length of SEC (mm)	8	6	4	2
Length of inlet duct (mm)	1	2	3	4
Length of outlet duct (mm)	1	2	3	4
Ratio of SEC length to total length (%)	80%	60%	40%	20%

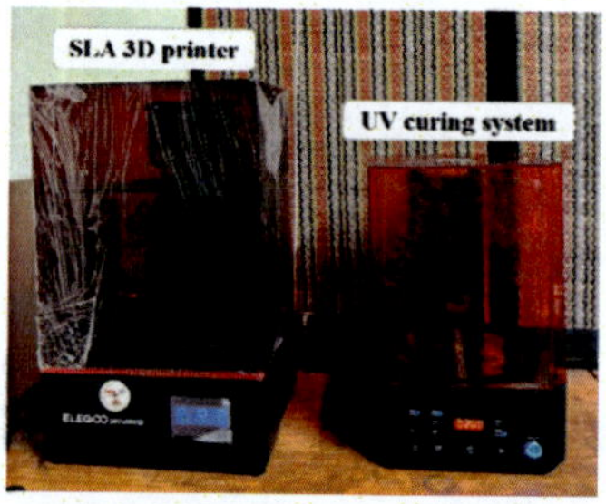

Fig. 3. Stereolithography (SLA) 3D printer and Ultraviolet (UV) curing system

given by,

$$SAC = 1 - R^2$$

where, R is the reflection coefficient

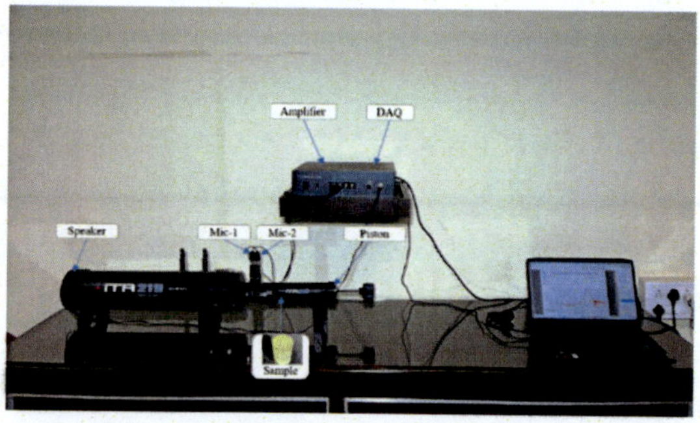

Fig. 4. Experimental setup for SAC measurement

2.3.2 Sound Transmission Loss (STL)

The STL of samples is quantified through four microphone impedance tube tests following ASTM E2611-09 (ASTM E2611, 2009). The experimental setup for STL measurement is depicted in Fig. 5, which shows the position of the four microphones relative to the sample. The sample is fixed between the upstream (reflection side) and downstream (transmission side) sections of the tube. A loudspeaker attached to the left side of the tube generates a plane wave. The other side of the tube has an anechoic material with a rigid cap. Using microphones, the complex sound pressures (a combination of waves traveling in the positive and negative direction in both upstream and downstream directions), p1-p4 are measured at different positions (mic 1- mic 4).

Normal sound incident STL is given by,

$$\boldsymbol{STL} = \mathbf{10}\boldsymbol{log}\frac{1}{|\tau|^2} \tag{1}$$

where, τ is the normal incident pressure transmission coefficient in case of an anechoic termination.

To minimize ripples in the STL curves, high-quality components, and accessories are used in the impedance tube setup. The measurement tube is made of anodized aluminum for a better surface finish and to avoid vibrations, and the test is conducted with necessary damping to eliminate oscillations in the sound absorption curves. The experiment is conducted under strictly controlled lab environmental conditions (temperature 27^0 C and humidity 50%) to minimize human errors, and repeated trials are performed to obtain consistent results. To prevent sound leakage from upstream to downstream, the edges of the test sample are tapped with Teflon, which tightly fits in the impedance tube.

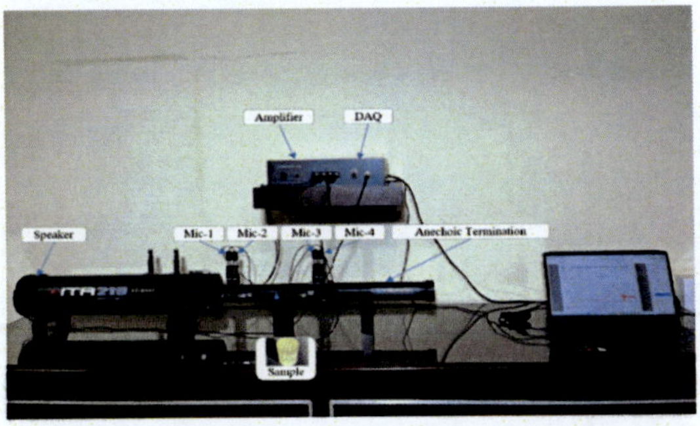

Fig. 5. Experimental setup for STL measurement

3 Results and Discussion

3.1 Effect of SEC Length on SAC

For all the samples, the length of SEC and the length of inlet and outlet ports has been shown in Table 1. It is observed that for sample A, the peak absorption coefficient reaches up to 98% at 3.025 kHz. By decreasing the length of SEC from 8–2 mm peak

SAC decreases from 98% to 79%. It can be concluded that the sample having a higher ratio of the length of the expansion chamber to the thickness of the sample exhibits higher absorption properties. The variation of SAC with frequency for 4 samples is shown in Fig. 6.

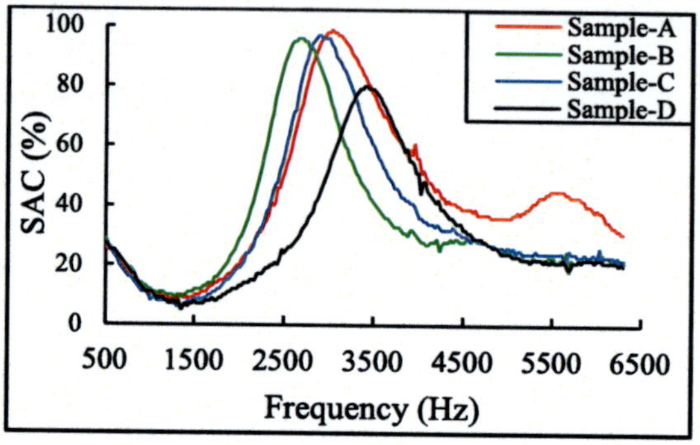

Fig. 6. Variation of SAC with the frequency for Sample-A, B, C, D

3.2 Effect of SEC Length on STL

With decreasing the length of SEC from 8 -2 mm STL increases 4–7 dB across 0.5–6.3 kHz. With decreasing the ratio of the length of the expansion chamber to the thickness of the sample from 80–20% the STL gradually increases. The variation of STL with frequency for 4 samples is shown in Fig. 7.

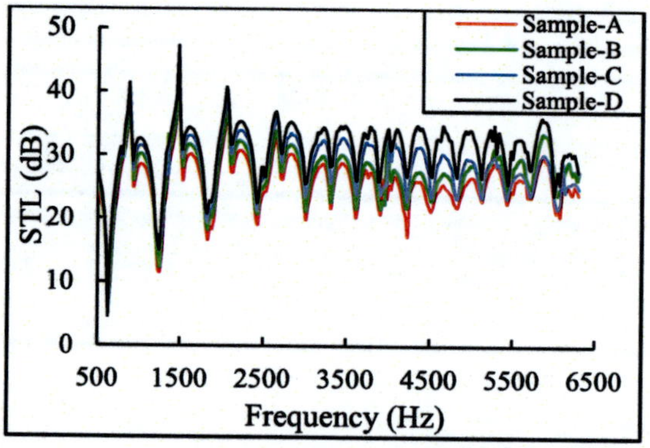

Fig. 7. Variation of STL with the frequency for Sample-A, B, C, D

4 Conclusion

The acoustic properties of periodic structures produced by additive manufacturing were investigated using the impedance tube. Peak absorption of 98% is observed at 3.025 kHz for sample-A. The overall sound transmission loss for samples A, B, C, and D ranges from 25 to 35 dB in the broad frequency range of 2.1–6.3 kHz. The outcome of this research encourages us to experiment with certain identified parameters for manufacturing customized acoustic samples for practical applications in industry.

The developed acoustic material is effective in attenuating noise levels across the 0.5–6 kHz frequency range, making it suitable for use in operator cabins in mineral processing plants. Such new acoustic materials based on the inline cavity structure principle are important in the field of applied acoustics. The lightweight, well-ventilated, easy-to-install, and durable properties of this material make it an ideal solution for many industrial noise control applications. The new acoustic material can reduce the risk of noise-induced hearing loss among workers in the mining industry, improving their health and safety in the workplace.

References

ASTM standard: Standard Test Method for Impedance and Absorption of Acoustical Materials Using a Tube, Two Microphones and a Digital Frequency Analysis System. ASTM standard E1050-19

ASTM standard: Standard test method for measurement of normal incidence sound transmission of acoustical materials based on the transfer matrix method. ASTM standard E2611-09

Directorate General of Mines Safety, Ministry of Labour (1975). Protection of workers against noise and vibration in the working environment. Technical Circular no. 18 of 1975. Dhanbad. India

Gelfand, S.: Essentials of Audiology. Thieme. p. 87. ISBN 978–1604061550. "hearing is most sensitive (i.e., the least amount of intensity is needed to reach threshold) in the 2000 to 5000 Hz range" (2011)

ISO Standard: Determination of sound absorption coefficient and impedance in impedance tubes-Part 2: Transfer-function method. ISO 10534–2 (1998)

Japan Occupational Safety and Health Resource Center. (2017). Guidelines for noise control at mines. https://www.jniosh.go.jp/icpro/josh_data/pdf/noise_control_mine.pdf. Last accessed on 2023-04-02

Madahana, M.C., Moroe, N.F., Nyandoro, O.T.: Engineering noise control for mines: Lessons from the world. South African J. Commun. Disorders **67**(2), 1–5 (2020)

Manwar, V.D., Mandal, B.B., Pal, A.K.: Environmental propagation of noise in mines and nearby villages: a study through noise mapping. Noise Health **18**(83), 185 (2016)

Munjal, M.L.: Noise and Vibration Control. World Scientific Publishing Co. Pte. Ltd, p. 169 (2013)

Nandi, S.S., Dhatrak, S.V.: Occupational noise-induced hearing loss in India. Indian J. Occupational Environ. Med. **12**(2), 53 (2008)

National Institute for Occupational Safety and Health (NIOSH). (1999). Noise Control Resource Guide - Surface Mining. Retrieved from https://www.cdc.gov/niosh/mining/userfiles/works/pdfs/ncrgsmin.pdf. Last accessed 2022-08-24

National Institute for Occupational Safety and Health. (n.d.). Noise and hearing loss prevention. https://www.cdc.gov/niosh/topics/noise/default.html. Last accessed 2023-01-12

National institute of health: Consensus conference-noise and hearing loss. JAMA **263**, 3185–3190 (1990)

Nelson, D.I., Nelson, R.Y., Concha-Barrientos, M., Fingerhut, M.: The global burden of occupational noise-induced hearing loss. Am. J. Ind. Med. **48**(6), 446–458 (2005)

Occupational Safety and Health Administration (OSHA). (n.d.). Occupational Noise Exposure. Retrieved from https://www.osha.gov/noise

Ekade, P., Krishnan, S.: Fluid flowand heat transfer characteristics of octet truss lattice geometry. Int. J. Therm. Sci. **137**, 253–261 (2019)

Heinl, P., Körner, C., Singer, R.F.: Selective electron beam melting of cellular titanium: mechanical properties. Adv. Eng. Mater. **10**(9), 882–888 (2008)

Saran, O.S., Reddy, A.P., Chaturya, L., Kumar, M.P.: 3D printing of composite materials: A short review. Materials Today: Proceedings **64**, 615–619 (2022)

Deshpande, V.S.: Effective properties of the octet-truss lattice material. J. Mech. Phys. Solids **49**(8), 1747–1769 (2001)

WOL 3D Technical specification of standard UV-cured resin. Retrieved from https://worldoflilliputs.com/product/itech-lcd-standard-resin-grey-500ml/. Last accessed 2022-08-24

World Health Organization. (2011). Burden of disease from environmental noise: Quantification of healthy life years lost in Europe

Zheng, X., et al.: Ultralight, ultrastiff mechanical metamaterials. Science **344**(6190), 1373–1377 (2014)

Mine Fire and Ventilation

Design of an Effective Ventilation System for Deep Underground Coal Mines Affected by Spontaneous Heating and Fire Using Double Booster Fans – A Practical Approach

Debashish Mishra[1,2](✉), N. K. Mohalik[1], D. P. Mishra[2], S. K. Ray[1], Asfar M. Khan[1], and J. K. Pandey[1]

[1] CSIR-Central Institute of Mining and Fuel Research, Barwa Road, Dhanbad, Jharkhand 826001, India
dmishra200rout@gmail.com

[2] Indian Institute of Technology (ISM), Dhanbad, Jharkhand, India

Abstract. Deep underground coal mines in India are particularly vulnerable to spontaneous heating/fire accidents, which put miners' safety and mines at risk. Spontaneous heating leading to fire and explosion results in substantial financial losses for the mining industry. This study examines practical methods for the prevention of spontaneous heating/fire in deep underground coal mines of the Jharia coalfield using ventilation control techniques. It identifies the major factors impacting the design of a ventilation system for the prevention of spontaneous heating/ fires in one of the deep underground coal mines of the Jharia Coalfield. An attempt has been made for the prevention of spontaneous heating/fire to improve the workplace environment by incorporating a double booster fan into the ventilation circuit. The computer simulation study reveals a viable solution involving the use of a double booster fan to produce the requisite air quantity. The finding of the study is successfully implemented at the 2-pit Jamadoba colliery of Tata Steel Limited (TSL).

Keywords: Ventilation system · multi-seam workings · double booster fan · Computer simulation

1 Introduction

The underground coal mining industry in India has experienced technical breakthroughs to meet the ever-growing energy demand, which has sparked worry about workplace safety. Deeper depth, workings in multi-seam, gobbed out sections in the upper seam, spontaneous heating/fire in upper seams are just a few of the challenges Indian underground coal miners must overcome to operate safely and productively (Roghanchi and Sunkpal 2015). The brief history of fire in India shows that there have been several incidents of underground coal mine fire that exists from 1860 till date. There are 77 registered incidents of coal mine fire as per the annual reports published by DGMS till 2014

A. Sinha et al. (Eds.): AMC 2023, SPEES, pp. 389–400, 2023.
https://doi.org/10.1007/978-3-031-46966-4_32

(Directorate General of Mines Safety 2014) and it revealed that the contribution of spontaneous heating is 96% of the total reported cases. However, different researchers have proposed different factors that kept influencing the spontaneous heating/fire in underground coal mines. In deep underground coal mines with multi-seam workings, sluggish ventilation, frictional heat, electrical sparks, or spontaneous combustion in coal piles sitting in airways that are far from the active ventilation circuit of the mine are the main causes of mine fires (Mcpherson 1993; Sahay et al. 2014). Researchers have suggested diverse solutions to improve the workplace environment in mines, including booster fan installation, supplementary ventilation, and periodic upgrades of the primary ventilation system with larger capacity fans (Banerjee 1986; Mcpherson 1993; Mishra 1986; Vutukuri and Lama 1986). However, auxiliary fan installation won't bring in any new air and will only compel the current air to move at a higher speed, which may not be ideal for a better working environment. Thus, there are only two additional ways to enhance the amount of fresh air in workplaces adjusting the MMV or installing underground booster fans (Mishra et al. 2022). The first option of changing the existing MMV with a higher capacity involves a huge capital requirement in comparison to underground booster fans. So, the use of booster fans inside underground can provide sufficient air quantity so that sluggish ventilation can be avoided to prevent spontaneous heating/fire of coal.

The use of the booster fan technique to improve airflow in certain mine workings was adopted in coal mines across the world. Major coal-producing nations have established regulatory standards to have adhered to during the installation of booster fans in underground mines i.e., a basic online monitoring system to track the toxic gases make-up accompanied by recirculation of air quantity, quarterly operations inspections for problems, at least one number of methane gas detector installed next to the fan for continuous measurement, stops the fan once the concentration breached above 2.0% (Gillies et al. 2010, Martikainen and Taylor 2010, Gangrade 2014). According to the Coal mine regulations (CMR) 2017, in India, the installation of booster fans in underground is only acceptable following an inclusive pressure-quantity survey of the mine area that could be affected. There are no explicit regulations governing the installation of booster fans in underground coal mines as well as with complex ventilation circuits having multi-seam workings. Although many underground coal mines with single-seam working implemented booster fans for the workplace environment. Few researchers successfully applied single booster fans in deep underground coal mines with multi-seam workings of Jharia coalfield (Mishra et al. 2022). There is no case study available on the application of a double booster fan in any Indian deep coal mines for improvement of the workplace environment with multi-seam workings. The authors have attempted the use of high-capacity booster fans in Indian deep underground coal mines with multi-seam working to resolve the sluggish ventilation leading to spontaneous heating/fire. It has been successfully implemented at one of the deep underground coal mines belonging to Tata steel limited, India.

2 Field Investigations

2.1 Specifics of the Study Site

The ventilation survey was done at Jamadoba Colliery, 2-Pit of Tata steel ltd. Jharia Division. The mine is operating in XI seam at a depth of almost 700 m by bord & pillar method of mining with an annual production of 0.3 million tonnes. There are huge graved-out areas, which spreads across XVIII, XVII, XVIA, XVI, XVA, XV, and XIV seams. Three dynamic sealed-off fire areas are present in XVI and XIV seam respectively, under regular supervision. The mine is operating with two downcast shafts, viz., No-2 (depth = 374.8 m and Diameter-5.5 m) sunk up to XIV seam intersecting XVIII, XVII, XVIA, XVI, XVA, XV and No-4 pit (depth = 80.49 m and Diameter-5.33 m) sunk up to XVII seam. The return circuit of the mine is connected to the No-3 pit (depth = 270.73 m and Diameter-4.80 m) sunk up to XVI seam. The ventilation system of the mine is of exhaust homotropal nature and driven by a Voltas make VF-3000 fan of axial flow nature installed on the surface. The fan is sucking return air through the No-3 pit via fan drift. The ventilation circuit of the mine is shown in Fig. 1. The diagram depicts the goaved-out areas and sealed-off fire areas in zone -1 and zone-2 as current workings of XIV and XI seam.

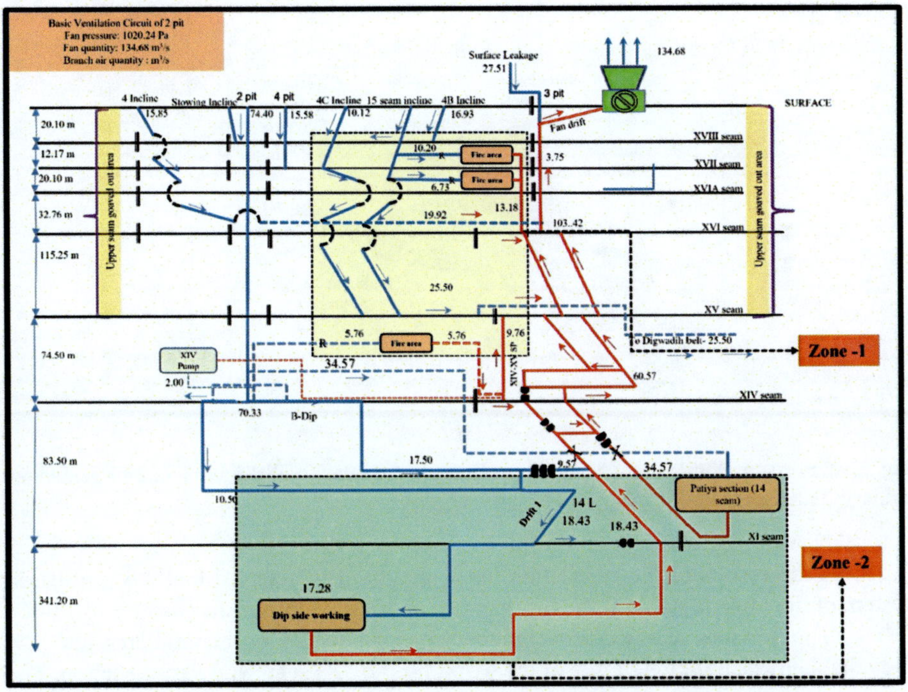

Fig. 1. Line diagram of the ventilation circuit of 2-pit Jamadoba colliery (not to scale)

2.2 Ventilation Investigation

In this case, a thorough ventilation investigation including pressure, air quantity, temperature survey, and MMV study was completed. The mine's ventilation network has been prepared using this field investigational data. The details of the ventilation survey findings, including the pressure, and air quantity measurements along the ventilation circuit, are represented in Figs. 2a and b. A detail of the MMV working performance is given in Annexure-I.

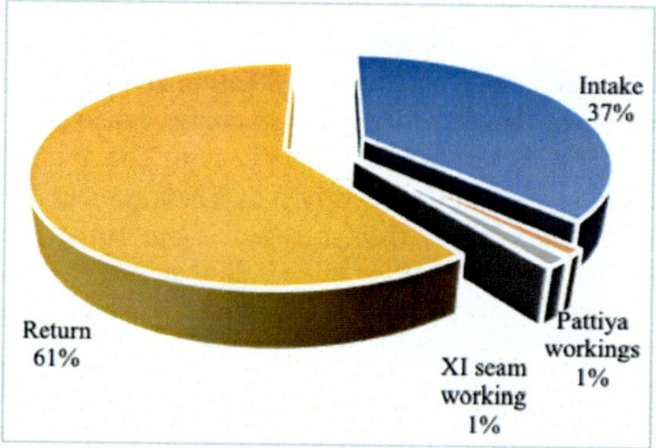

a

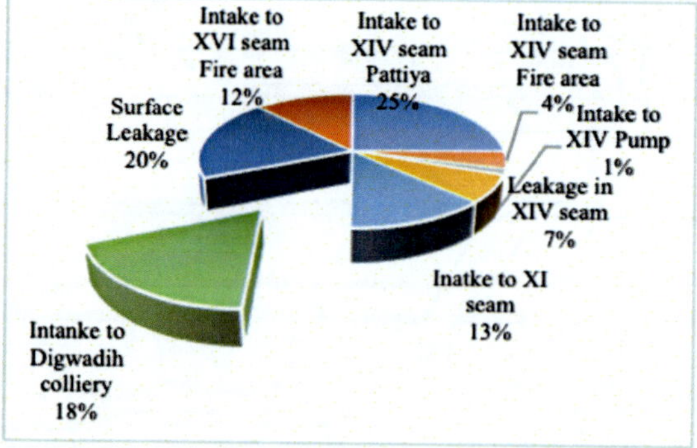

b

Fig. 2. a. Pressure Profile of the Ventilation circuit, b. Air quantity distribution in 2-pit Jamadoba colliery, Tata Steel Ltd.

From the results of the ventilation investigation, it is detected that the ventilation system of the mine is stable at the fan pressure of 1020.24 Pa and air quantity of 134.68 m^3s^{-1}. The pressure survey discovered that there is only 2% of the total pressure drop is available for ventilation of two of the workings, i.e., XI seam and Patiya section. The pressure loss in the return and intake circuit of the mine is of the order of 61% and 37% of the total fan pressure respectively as shown in Fig. 2(a). The presence of entry loss in the main intake, pressure loss due to sharp bends, and single airway in XI seam and Patiya section returns are the prime sources for pressure drop along the circuit Fig. 1.

The total intake to the mine is 132.88 m^3s^{-1} via No-2-pit, No-4-pit, 4-Incline, 4C incline, 15 seam incline and 4B incline. The total air quantity of the mine is distributed into different sections of the mine as shown in Fig. 1. Effective air available at No-2-pit bottom is 70.33 m^3s^{-1}, which is 56% of the total air intake into the mine. Further, the effective air quantity available from the No-2-pit is 51.85 m^3s^{-1}, after being provided to the XIV-seam pump, fire area, and some part in leakage. This makes the VEQ of the mine to be 39% which is very poor according to regular practice in the mining industry which falls to around 60%. So, keeping in view the poor ventilation efficiency quotient (VEQ) of the mine it has been considered to be improved by the mine management with the scientific interventions CSIR-CIMFR, Dhanbad.

3 Methodology

In almost all coal-producing countries regulatory norms form the core of planning and designing its ventilation system. In the present case, there are two imperative issues with the ventilation system design. The first involves estimating the amount of air necessary for a comfortable working environment, and the second involves the method to be adopted in this case considering all the limitations in the form of multi-seam workings, sealed-off fire areas, and multiple goaved-out areas.

Gauging the seriousness of the issues, the authors have studied the findings discussed in (Mishra et al. 2022) and tried to extend the work by installing a double booster fan in series to deal with the problem of ventilation. In the present case, we have prepared a simple and realistic approach for the installation of booster fans in underground coal mines with multi-seam working as shown in Fig. 3.

In the first step, the ventilation network of No-2-pit Jamadoba colliery is modeled using SINET software based on the pressure-quantity survey data for a true representation of the existing conditions. Thereafter, keeping in view air quantity shortage at the workings being the major bottleneck of the ventilation system in No-2-pit Jamadoba colliery, different simulation schemes/alternatives have been worked out. The schemes/alternatives are, viz. a. introduction of double intake and return in both XIV seam and XI seam workings (for reduction of resistance to airflow), b. installation of two booster fans in the returns of XIV seam and XI seam (to support the main MMV and reduction of leakage). The updated ventilation network of the mine is developed to conduct airflow simulation using SINET software designed and developed at CSIR-CIMFR, Dhanbad, based on Atkinson's equation and solved using the Hardy-cross iterative technique (Mcpherson 1993).

The computer simulation of the ventilation network of the mine is carried out incorporating the above scheme/alternatives. After computer simulation installation of a double booster is carried out following the steps described in Fig. 3. The details of the procedure are as follows:

a. Step-1: Establishment of the neutral point (the neutral pressure zone where the intake and return pressure zone nullifies) (Mcpherson 1993). If the location is safe concerning air recirculation, sealed-off fire areas, and goaved-out areas then go for installation otherwise find an alternative safe location.

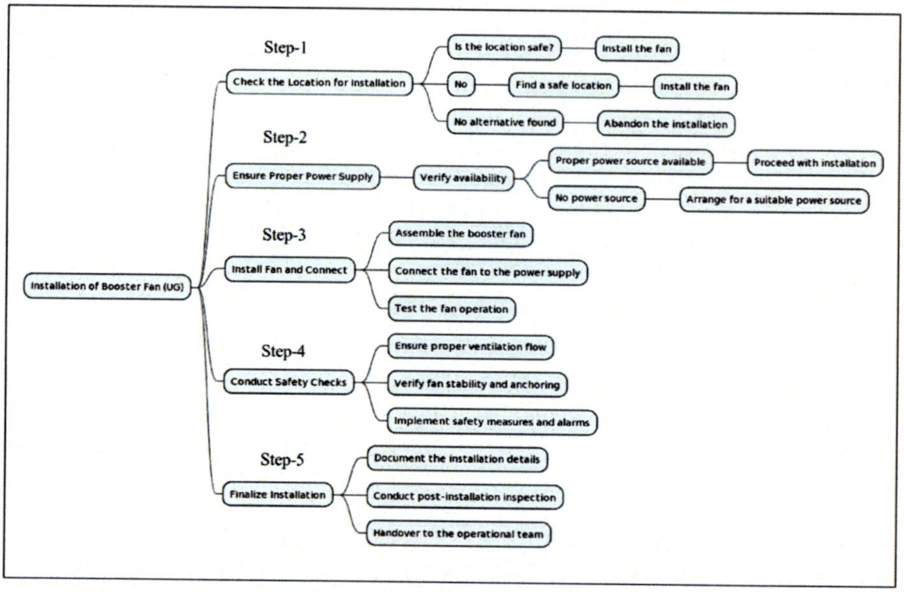

Fig. 3. Systematic approach for installation of booster fans in underground (UG) coal mines

b. Step-2: Ensure proper power supply for the installation of heavy power motors associated with the booster fans.
c. Step-3: Install the fan to the power supply and start the test run.
d. Step-4: Conduct all the safety checks associated with the booster installation, viz. Proper ventilation airflow without any recirculation, verification of fan stability and anchoring, and implementation of safety measures and alarms for any untoward situation, viz. Power failure, motor issues, noxious gases, air recirculation, etc.
e. Step-5: Finalization of the booster installation after full-scale post-installation inspection and handover to the operational team.

4 Result and Discussions

This section discusses the briefs of the computer simulation results after the incorporation of different schemes/alternatives as discussed in the above section. The details of the computer simulation of the ventilation network are depicted in Table 2. The ventilation network containing 108 numbers of basic branches and 56 junctions is simulated in 'SINET' software satisfying McPherson's criterion of simulation. The computational model is successfully simulated with fifteen iterations satisfying $error = \frac{\Delta Q}{Q} \leq 0.05$. The MMV operates at a pressure of 1020.00 Pa delivering 134.00 m^3s^{-1} (Table 1). Thus, the simulated ventilation network is a true representation of the ventilation circuit of No-2-pit Jamadoba Colliery. In this ventilation network, the alternative of a double booster of capacity along with double intake and return in XIV and XI seam workings has been worked out and results are summarized in Table 1. In the alternative, two booster fans, of Voltas make and Model-VF-2000 of rated capacity 40–105 m^3s^{-1} at 392.4–686.7 Pa have been incorporated into the ventilation circuit. The 1st booster is incorporated at 1-cross-cut of XIV seam main return and the 2nd booster at 1-cross-cut of XI seam main return as shown in Fig. 4.

Table 1. Details of the computer simulation scheme

Sl.No	Details of the parameters observed	Present condition	Alternative with double booster fan	Percentage change (%) ***(-) is for reduction**
1.	Main fan pressure (Pa)	1020.24	957.46	-6.00
2.	Main fan quantity (m^3s^{-1})	134.68	153.31	14.00
3.	XIV Seam - Booster fan pressure (Pa)	-	608.22	-
4.	XIV Seam - Booster fan quantity (m^3s^{-1})	-	88.74	-
5.	XI Seam - Booster fan pressure (Pa)		313.92	
6.	XI Seam - Booster fan quantity (m^3s^{-1})		47.74	
7.	Intake through No-2-pit (m^3s^{-1})	74.40	107.47	44.00
8.	No.4 Pit intake to XVII seam (m^3s^{-1}) - Intake to Digwadih colliery	15.58	15.02	-4.00
9.	4 Incline intake to XVI seam (m^3s^{-1})	15.85	15.25	-4.00
10.	4C Incline Intake to XV seam (m^3s^{-1}) - Intake to Digwadih colliery	10.12	7.43	-27.00
11.	4B Incline Intake to XVIII (m^3s^{-1}) mainly for upper seam fire areas shown in Zone-1, Fig. 1,2	16.93	16.37	-3.00

(continued)

Table 1. (*continued*)

Sl.No	Details of the parameters observed	Present condition	Alternative with double booster fan	Percentage change (%) ***(-) is for reduction**
12.	XIV Seam panel (Patiya section) intake (m^3s^{-1})	34.57	41.00	19.00
13.	XI seam Intake (m^3s^{-1})	18.43	46.17	151.00
14.	XI seam Panel face quantity (m^3s^{-1})	17.28	45.60	164.00
15.	Surface leakage (m^3s^{-1})	27.51	14.22	–48.00
16.	The air power of MMV (kW	137.41	146.78	7.00
17.	Airpower of booster fan in XIV seam (kW)	-	53.97	-
18.	Airpower of booster fan in XI seam (kW)	-	14.99	
19.	Total air power (kW)	137.41	215.75	57.00

From the computer simulation results of the ventilation circuit of the mine, the following points emerged:

There is a significant increase in the overall air quantity handled by the MMV after the introduction of a double booster fan in underground which is 14% more than the existing one. The effective intake of fresh air quantity through No-2-pit increased to 103.62 m^3s^{-1}, which is 47% more than the existing condition. Further, due to the introduction of double booster fans in underground, the fresh air quantity available for XIV seam and XI seam workings increased by 19% and 164% respectively. The air quantity in the XI seam working has increased almost 2.64 times the existing air quantity.

After, a successful simulation of the ventilation network of the mine authors approached the installation of double booster fans as per the procedure described in Fig. 3. A detailed risk assessment study for the installation of double booster fans is also conducted along with a proper monitoring system to ensure its safe operation.

5 Field Validation and Discussion of Results

A local ventilation survey has been carried out in the XIV seam and XI seam intake and return circuit after the installation of a double booster fan. The ventilation circuit is depicted in Fig. 4 and the detail of the validation study is given in Table 2. It is observed that the ventilation circuit of the mine is stabilized at the MMV pressure of 967.12 Pa

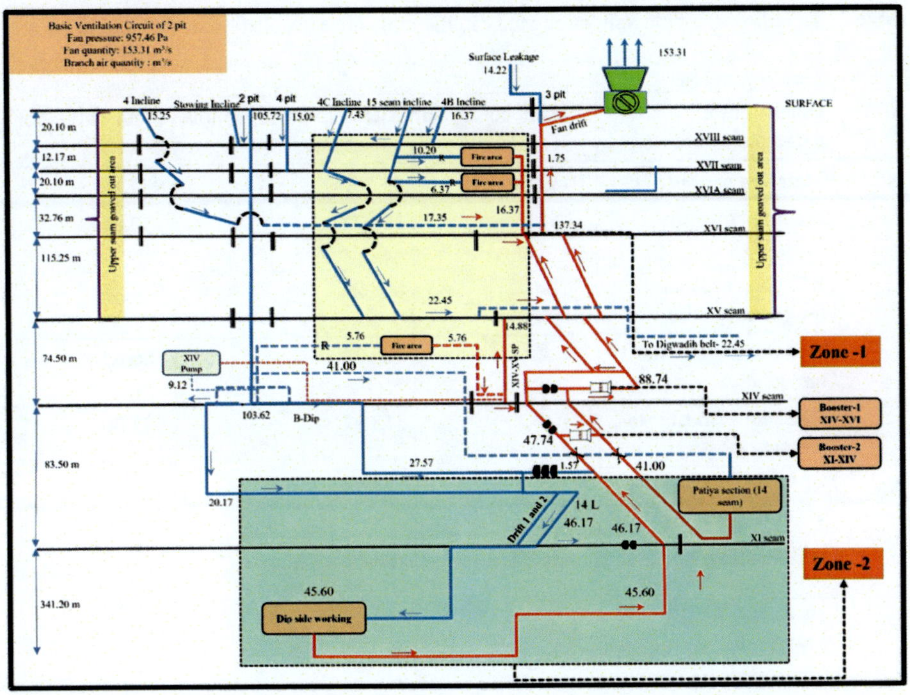

Fig. 4. Schematic diagram of the ventilation network of No-2-pit Jamadoba colliery with double booster fans

delivering 150.61 m^3s^{-1} of air. The total intake of the mine through No-2-pit to XIV is 104.63 m^3s^{-1}. The air quantity available for XIV and XI seam workings is of the order of 38.85 and 43.69 m^3s^{-1}. The booster fan in the XIV seam operates at 628.88 Pa handling 86.65 m^3s^{-1} of return air and the booster in the XI seam return operates at 333.00 Pa and 44.56 m^3s^{-1} of air. From the results of the field validation study, it can be concluded that the application of a double booster fan is quite effective in increasing air quantity at deeper sections of the mines with multi-seam workings. The experimentation of double booster fans is successfully implemented at No-2-pit Jamadoba colliery which paves the way for extensive use of booster fans in underground coal mines in India.

Table 2. Details of the field validation study in the mine after installation of the double booster fan

Sl -No	Details of the parameters observed	Present condition	Alternative with double booster fan	Field Validation
1.	Main fan pressure (Pa)	1020.24	957.46	967.12
2.	Main fan quantity (m^3s^{-1})	134.68	153.31	150.61
3.	XIV Seam - Booster fan pressure (Pa)	-	608.22	628.88
4.	XIV Seam - Booster fan quantity (m^3s^{-1})	-	88.74	86.65
5.	XI Seam - Booster fan pressure (Pa)		313.92	333.00
6.	XI Seam - Booster fan quantity (m^3s^{-1})		47.74	44.56
7.	Intake through No-2-pit (m^3s^{-1})	74.40	107.47	104.63
8.	4B Incline Intake to XVIII (m^3s^{-1}) mainly for upper seam fire areas shown in Zone-1, Fig. 1, 2	16.93	16.37	15.58
9.	XIV Seam panel (Patiya section) intake (m^3s^{-1})	34.57	41.00	38.85
10.	XI seam Intake (m^3s^{-1})	18.43	46.17	43.69
11.	XI seam Panel face quantity (m^3s^{-1})	17.28	45.60	41.26

6 Conclusion

This paper put forward a practical solution to increase the fresh air quantity requirement in the deeper section of underground coal mines, where multi-seam workings have been practiced with the liability of sealed-off fire areas and goaved-out workings. The computer simulation study revealed that the implementation of a double booster in the mine will increase the face air quantity in the deepest working of the mine in XI seam by 164% of the existing value. The total fan quantity of MMV is also increased by 14% with a net decrease in fan pressure by 6% and reduces surface leakage by 48%. The VEQ of the mine improved from 39% to 69% which is quite impressive for such a complex ventilation circuit. The installation of a double booster fan is implemented in such a way that the ventilation of upper seam goaved out areas and sealed-off fire areas remain unaffected. The application of double booster fans in deep underground coal mines will help practicing ventilation engineers.

Acknowledgment. The authors are grateful to the Director, CSIR- CIMFR, Dhanbad, for his continuous support and encouragement. Further, the contribution of the supporting staff of Mine Fire, Ventilation, and Miner's Health Research Group is duly acknowledged. The authors are also grateful to the mine officials of No-2-pit Jamadoba Colliery, M/s Tata Steel Ltd, for their continuous support and cooperation in the field investigation. This research work is part of an industry-sponsored project funded by M/s Tata Steel Ltd, Jharia Division (Grant No.: SSP- 469–2020-21), and a part of the Ph.D. work of Mr. Debashish Mishra at IIT(ISM), Dhanbad.

Appendix -I Main fan performance study

The performance study of the fan has been carried out by running the fan on full load. Details of the fan parameters are detailed in Table 3.

Table 3. Details of fan parameters

Parameters	Fan – 1 in Pit. No.3
Fan Type	Adjustable blade axial flow
Make & Model	M/s Voltas-VF-3000–8-725
Mode of Installation	Exhaust
Impeller diameter (mm)	3000
Fan Casing	Good condition
Blades	8
No. of Belt	3
Blade Angle	20^0
Fan RPM (designed)	725
P. F	0.97
Fan Motor, kW	337
Voltage, kV	3.3
Current, Amp	54
Fan pressure, Pa	1020.24
Fan quantity, m3s-1	134.68
Input Power, kW	275.04
Air Power, kW	137.41
Overall efficiency, %	49.96
Maximum Rated Power, Kw	300

References

Banerjee, S.P.: Mine Ventilation. Lovely Prakashan (1986)

Directorate General of Mines Safety: Annual Report (DGMS) 2014. Directorate General of Mines Safety (2014). http://annualreport.marksandspencer.com/M&S_AnnualReport_2016.pdf

Gangrade, V.: Monitoring systems for coal mines utilizing booster fans (2014)

Gillies, S., Slaughter, C., Calizaya, F., Wu, H.W.: Booster Fans-Some Considerations for their Usage in Underground Coal Mines-Mining Innovation (2010)

Martikainen, A.L., Taylor, C.D.: Breaking the ice on the booster fan dilemma in U.S. underground coal mines. Min. Eng. **62** (2010)

Mcpherson, M.J.: Ventilation and Environmental Engineering. University Press (1993)

Mishra, D., Mohalik, N.K., Mishra, D.P., Ray, S.K., Pandey, J.K.: Improving workplace environment of a deep underground coal mine with Multiseam workings: an alternative approach. Min. Metall. Explor. **39**(4), 1429–1443 (2022). https://doi.org/10.1007/s42461-022-00639-4

Mishra, G.B.: Mine Environment and Ventilation. Oxford University Press (1986)

Roghanchi, P., Sunkpal, M.K.K.: Understanding the human thermal balance and heat stress indices as they apply to deep and hot US mines. In: Jong, E., Sarver, E., Schafrik, S., Luxbache, K. (Eds.) 15th Northern American Mine Ventilation Symposium, pp. 1–6 (2015)

Sahay, N., Sinha, A., Haribabu, B., Roychoudhary, P.K.: Dealing with open fire in an underground coal mine by ventilation control techniques. J. South. Afr. Inst. Min. Metall. **114**(6), 445–453 (2014)

Vutukuri, V.S., Lama, R.D.: Environmental Engineering in Mines. Cambridge University Press (1986). https://doi.org/10.1007/978-3-642-41714-6_51314

Assessment and Control of Concealed Fire in Underground Coal Mine: A Case Study

Jitendra Pandey[1,2](✉), A. Khalkho[1], S. K. Ray[1,2], Aditya Kumar[1], and J. K. Pandey[1,2]

[1] Mine Fire, Ventilation, Miners Safety and Health Research Group, CSIR – Central Institute of Mining and Fuel Research, Dhanbad, India

[2] Faculty of Engineering Sciences at Academy of Scientific and Innovative Research (AcSIR), Ghaziabad, India

jitu.cimfr@gmail.com

Abstract. Coal, the most used fossil fuel, is a gift of nature to mankind. However, the Phenomenon of auto oxidation / fire in this fossils fuel has a long history on international magnitude in the destruction of valuable natural resources and dominantly contributes towards adverse impact on environment by liberating obnoxious and toxic gases. The fire in underground coal mines creates hazardous environment for the mining operation and endangers the life of miners. It is also a major problem for the Indian coal mining industry and exists since beginning of the mining. A huge amount of the precious coal reserves of different coalfields are blocked due to existence of fire and thus cannot be extracted without extinguishing it. In the case of concealed underground coal mine fire, without locating its exact location and root cause, its control effort may be futile.

In the present paper, a case study has been discussed for assessment of concealed underground coal mine fire locations and its major cause of initiations. Furthermore, the methodologies adopted for control this menace to overcome the safety threats of mines are also presented.

Keywords: Underground coal mine · auto-oxidation · ventilation pressure · goaved out workings · inertization

1 Introduction

Energy security is prime issue to fast economic growth and upholds the gross domestic product (GDP) of the nation. To meet up energy demand of country, coal is predominantly used for energy generation in India [MOC, 2023]. Coal is carbonaceous remains of vegetable matters. Depending upon the extent of carbonization in presence of heat, burial pressure and duration; coal matures in different rank from peat to anthracite [WCI, 2015]. The spontaneous combustion phenomenon occur in all ranks of coal but more frequent in low rank coal [Banerjee 2000; Roy 1996]. Presence of several others constituents like high level of pyrites, volatile matters, etc. increase its propensity towards spontaneous combustion [Graham 1923; Hodge and Hinsley 1964; Smith and Glasser 2005]. The self-heating or spontaneous combustion of coal has been explained as the interaction

A. Sinha et al. (Eds.): AMC 2023, SPEES, pp. 401–411, 2023.
https://doi.org/10.1007/978-3-031-46966-4_33

(oxidation) between coal and oxygen (oxidation) with exothermic reaction, leading to heat build-up [Banerjee 2000]. The imbalance between the heat generation by oxidation reaction and heat transfer rate to the surrounding atmosphere leads to increase in further chain reactions. The continuation of chain reaction leads to spontaneous combustion in coal resulting formation of gaseous products like CO, CO_2 and H_2, water vapours followed by generation of heat [Pandey et al. 2016].

Presently, occurrences of coal fires reported in numerous countries worldwide and spontaneous combustion remains one of the main causes for initiation of these coal mine fire [Stracher and Taylor 2004, Pandey et al. 2017]. The coal mining industries in India are facing several challenges due to spontaneous combustion/ fire in both underground and opencast mine workings resulting loss of lives and huge natural resources. It also creates several distresses on environment, economy, society and productivity in mining operation [Sinha and Singh 2005; Pandey et al. 2016]. Another major ramification of spontaneous heating in underground mine are endangering safety of the miners due to evolution of toxic gases, fumes, explosion due to product gases, making mining difficult or impossible and locking of valuable natural resources [Mishra et al. 2012]. In Indian coal mines during last century around 40% of the total mining disasters and about 50% of the total miner's fatalities were reported due to coal mine fire and explosion [DGMS, 2011]. The problem is gradually spreading towards new areas and creating more complexity to the safe mitigation and management of coal mine fire [Pandey et al. 2016].

Early detection and dealing of spontaneous heating/ fire particularly in underground coal mines is foremost important to manage the fire in early stage before getting it more complex [Singh et al. 2004]. In underground coal mines, if the fire became advance stage and complex; then all preventive measures becomes futile and remains only option to isolate/seal the entire mine from surface [Singh et al. 2016]. There are several coal mines fire prevention and control methods being used in different coal mines of India. The selection of method or combination of methods depends upon several site specific situations and prevailing condition of fire [Mohalik et al. 2006]. The success of dealing with spontaneous heating/fire also depends upon identification of exact location /seat of fire, magnitude/extent of the fire, approachability to the seat of fire, gassiness of the mine, availability of resources and ventilation control system [Tripathi 2008, Pandey et al. 2015]. The present case study discussed the detection of concealed underground coal mine fire locations along with its major cause of initiations and the modus-operendi adopted to deal this menace in safe and effective manner.

2 Background of the Problem

Hindusthan Lalpeth Colliery, No.1 (HLC No.1), Chandrapur Area of Western Coal Fields Limited is situated at 10 km in south-west direction of Chandrapur town adjacent to Babupeth railway station on broad gauge line in western part of India. The HLC No-1 colliery started in the year 1924. The mine is having only one working seam named as Lalpeth seam categorised as first degree of gassiness. The thickness of Lalpeth seam is varying from 8.5 to 16.45 m steeping with a gradient of 1 in 3.75 and the surface cover ranging from 24 to 315 m. The leasehold area has eight numbers of major and minor faults (F-1 to F-8) having maximum throw of 108 m. There are numbers of dykes exist within the boundary of the mine ranging between 5 and 15 cm thick.

The mine is permitted to extract the coal seam by Bord and Pillar system of working and depillaring with hydraulic sand stowing with an average production of 300 TPD. The rise side working of the property were depillared in different panels with hydraulic sand stowing leaving very limited numbers of approach gallery (4–5 Nos) towards dip side for ventilation and transportation purposes. These panels are lying at average depth of 45–140 m below the surface. The main intake and return air circuit of the present working are passing through nearby old depillared panels.

HLC No. 1 mine has history of fire within these old sealed off panels. Several incident of spontaneous heating were previously reported in different panels (Panel G-1, D-1, D-3, D-5, and B-1) in both main intake and return circuit. These fires were successfully dealt with rescue cover by the mine management. However, in Panel B-1, several incidents of frequent reoccurrences of spontaneous heating have been reported by management. In most of the occasions, fire in the sealed off panel becomes dormant with application of suitable preventive steps and inertization with liquid nitrogen. However, again after lapse of few months' indication of heating has been reoccurred. Panel B-1 is located adjacent to main intake used as travelling roadways for the entire mine. Therefore, in case of any outbreak of fire in this panel endangered the life of miners. Keeping safety threats into account, mine management requested for comprehensive study to identify the root causes of reoccurrences of spontaneous heating in the sealed off Panel B-1. The location details of HLC No. 1 Mine along with panel B1 and B-2 are illustrated in Fig. 1.

3 Spontaneous Heating Assessment and Control Methodology

The Panel D-1, D-2, D-3, D-6, L-1, B-1 with its extension and Panel B-2 were located along main intake ventilation circuit. The height of these intake galleries are 5–6 m. At some locations isolation stopping is very near to intake airway and have no space for further erection of stopping. In case of any jumping/crossing of fire, entire mine become unapproachable and force to seal off. Symptoms of spontaneous heating were observed previously in panel B-1 at 16L/40D and controlled by flushing of sand & water after erecting additional stopping. Again after lapse of 4–5 months spontaneous heating has been reoccurred in Panel B-1. Several incidents of frequent reoccurrences of spontaneous heating have been reported by management and dealt with flushing of Liquid Nitrogen inside the sealed off panel from surface borehole.

For the assessment of concealed fire exact locations in goved out panel and root causes of its frequent re-initiation, all the concern parameters were comprehensibly studied. Substantially, on the basis of these analysis results, the root cause of heating/fire was identified and measures for its control have been implemented. The different methodology conducted during the study to make it successful has been discussed in successive sections.

A. Study of Coal Characteristics

Coal samples of Lalpeth seam, HLC, Mine No.1 of Chandrapur area, WCL has been collected and its different coal characteristics have been studied in laboratory. Coal samples from coal seam were collected from running panel by channel and chip sampling

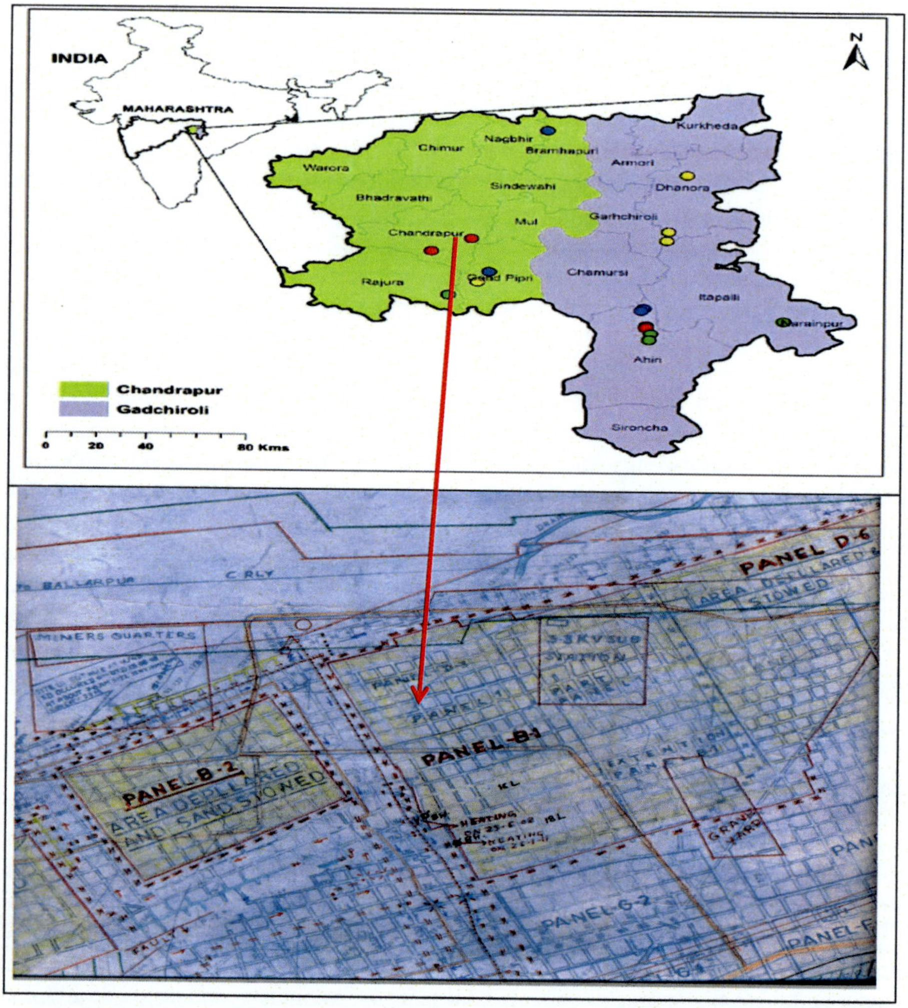

Fig. 1. Location map of Panel-B1 of HLC No-1 mine of Chandrapur Area, WCL.

method. Before the experiments, representative samples were prepared from the coals obtained by applying closed circuit crushing to -72 Mesh (BSS). The different analysis of coal sample are conducted as specified in I. S. (Indian Standard).The results of Proximate Analysis, Ultimate Analysis, Crossing Point Temperature (CPT), Ignition Point Temperature (IPT) and Gross Calorific Value (GCV) are furnished in Table 1.

The study of coal characteristics of HLC No.1 colliery indicates that percentage of V.M. (36.46%) and Moisture (9.95%) is in higher side, which indicates more proneness of coal towards spontaneous heating. This was corroborated with ultimate analysis result data wherein oxygen percent found to be more than 10% [Banerjee 2000]. However,

Table 1. Result of coal characteristics analysis of HLC No.1 coal mine, (a) proximate analysis, (b) Ultimate analysis, (c) CPT, IPT and (d) GCV.

Coal sample location	Proximate analysis					Ultimate analysis				Thermal analysis		Gross calorific value
	C	H	N	S	O	A	M	VM	FC	CPT	IPT	GCV
	(%)	(%)	(%)	(%)	(%)	(%)	(%)	(%)	(%)	(°C)	(°C)	(Cal/gm)
62 1/2 D/ 35L	**57.47**	**4.59**	**1.12**	**0.25**	**12.38**	**14.24**	**9.95**	**39.35**	**36.46**	**144**	**158**	**5223.3**

the crossing point and Ignition point temperature result yield the coal is moderately susceptible to spontaneous heating.

B. Study of Ventilation Network of Mine

For the assessment of root cause of spontaneous heating /fire, the comprehensive study of ventilation network of the mine was carried out. The purpose of ventilation network study was to determine the pressure losses at different locations particularly in old sealed off areas and distribution of air quantity flowing through the different networks, to identify leakage paths and pressure behaviour of the stoppings in both intake and return circuits etc.in the mine. The observations of ventilation network study were described in successive sections.

(i) Pressure and Quantity Survey

Ventilation survey of HLC Mine No.1 was carried out to understand distribution of air pressure and quantity in the entire ventilation network and determination of segment wise pressure and air quantity losses. The longest ventilation circuit is selected for ventilation network analysis and accordingly stations are marked. The pressure difference between the stations were measured starting from the surface through intake circuit and completed at the fan drift through return circuit covering winding shaft, Pit Bottom, 6L, 42D, 27L, 57D, 31L, 62D, 36L, 63R, 34L, 65R, Panel- K1 and return back to the Fan Drift in return air network. Accordingly the quantity and humidity is also measured at every station. Finally cumulative pressure drop is calculated and the result is plotted on the graph taking cumulative distance along X-axis and corresponding cumulative pressure drop along Y-axis as shown in successive Fig. 2. The analysis of pressure-quantity survey result reveals that distributions of pressure in the mine were almost uniform except few bottle-necking locations. The loss of air-quantity in the entire network was found almost normal. The wet bulb and dry bulb temperature, humidity condition in the entire ventilation circuit were found within the prescribed limit.

(ii) Study of Pressure Behaviour of Stoppings

To understand the leakage condition and environmental condition within the sealed off Panel B-1, 24 hours monitoring of pressure behaviour of stoppings were carried out

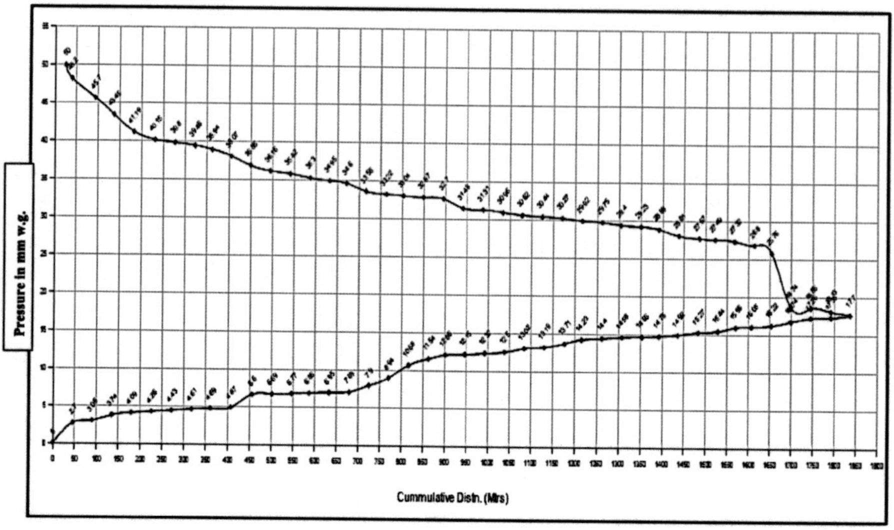

Fig. 2. Plot illustrating distribution of Air-Pressure in entire mine handled by mine fan starting from winding shaft in intake side to fan drift in return side.

at different locations. It has been observed that the pressure behaviour at stopping no. 7LN/40D, 16LN/40D, 17LN/40D and 22LN/39D of the panels is responding towards diurnal variation in the atmosphere pressure. However, pressure at stopping 18L/40D has showed high positive pressure throughout varying from 10 to 24 mmwg even after stoppage of main mechanical ventilator (main mine fan). This reveals that there is source of leakages near to the stopping 18L/40D. The plots of 24 hours pressure behaviours recorded at different stoppings of Panel B1 are shown in Fig. 3.

C. Thermo-Compositional Analysis

Thermo-compositional monitoring using multi-gas analyzer, toxi-meter and thermocouple on the spot as well as chemical analysis of air samples in the laboratory had been carried out to detect the presence of spontaneous heating/ fire. The spot monitoring of air sample during positive pressure using multi gas analyser has been carried out of stoppings no. 7LN/40D, 16LN/40D, 17LN/40D, 18LN/40D and 22LN/39D. The spot analysis of the air sample during underground visit corroborate the preliminary data of the air sample laboratory analysis results of sealed off panels. The oxygen percentage varied from 15 to 17% in sealed off panels which confirm the leakage through cracks, fissure and cleats of coal. It has also been observed that the pressure behaviour of the panels is responding towards diurnal variation in the atmosphere pressure. However, pressure at stopping 18L/40D were recorded high positive pressure throughout varying from 10 to 24 mmwg. The CO, H_2 and H_2S were observed from air sample analysis of 18L/40D of panel-B1 indicates the presence of spontaneous heating/ fire. The result of temporal thermo-compositional study reveals the following observations:

I. No traces of CO were found from sample points of 7- level north of 40 dip and 22-level north 39 dip of adjoining depillared panels.

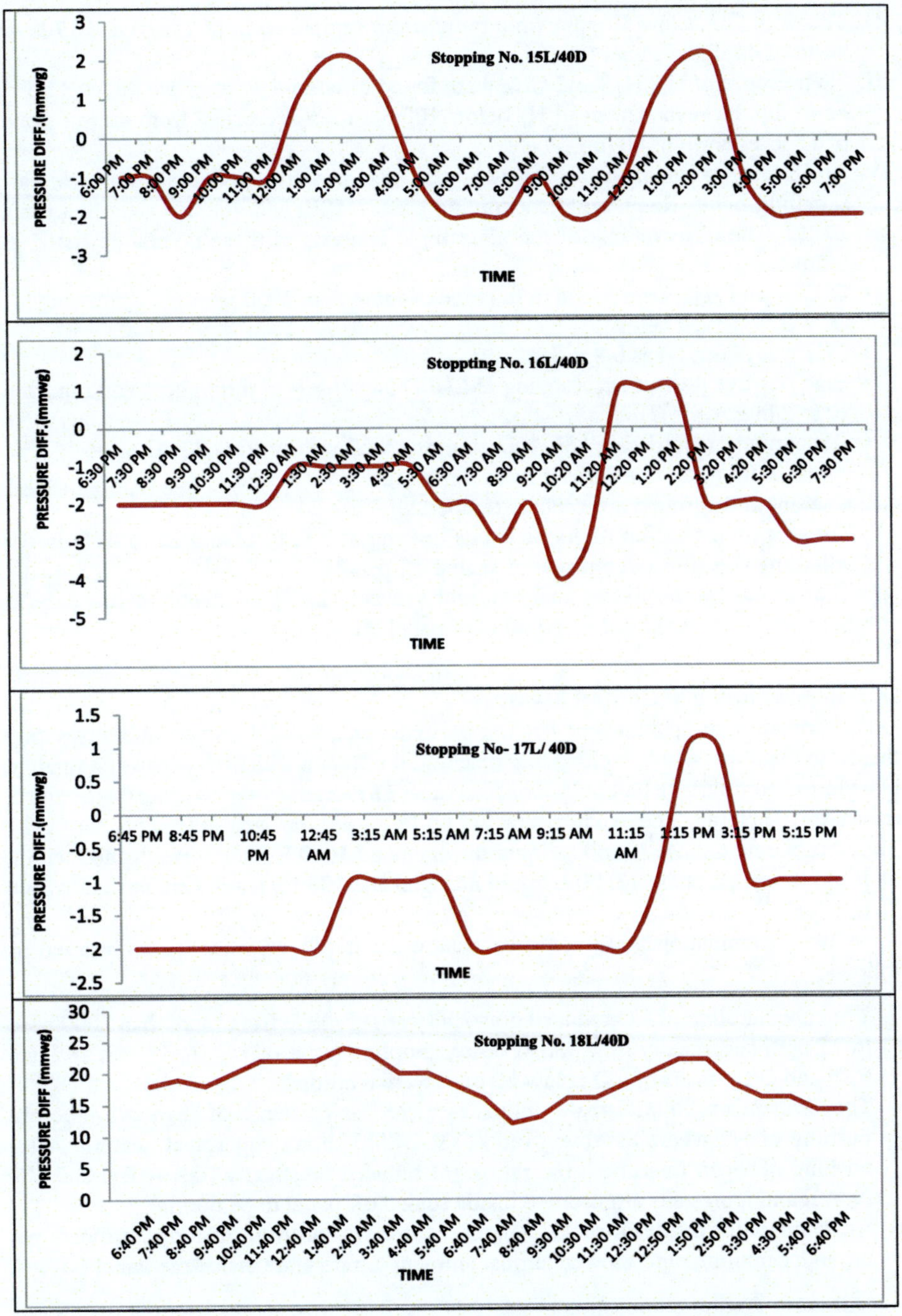

Fig. 3. Plot showing 24 h Pressure behaviours recorded at different Stoppings of Panel- B1.

II. Traces of CO below 30 ppm were found from sample point of 15, 16 and 17 level north of 40 dip of Panel B-1.
III. No presence of H_2, H_2S and CH_4 were found at sample point of 15, 16 level north of 40 dip, however, traces of H_2 below 100 ppm were recorded from sample point of 17 level north of 40 dip.
IV. At 18-level north of 40 dip, there were three sampling points such as rise side sampling point (RSP), middle sampling point (MSP) and dip side sampling point (DSP). The observations of the air sample analysis of these points are listed as follows.
  - CO, H_2 and H_2S were found in the tune of more than 2000 ppm
  - O_2 was observed below 5%
  - CH_4 was observed as 0.8 to 1%.
  - High positive pressure at stopping 18L/40D had observed throughout varying from 10 to 24mmwg.
  - Temperature recorded at 18L/40D stopping wall was varying 32 °C to 35 °C and temperature of air coming out from sampling pipe was near to 40 °C, whereas the ambient underground atmosphere temperature was 24 ^{0}C.
V. There is no indication of spontaneous heating at 7 L/40 D and 22 L/40 D in the adjoining rise side and dip side of sealed off panels.
VI. The thermo-compositional analysis results reveals that spontaneous heating exist in between 15LN/40 D to 18 LN/ 40 D of panel B1.

D. Implementation of Control Measures

On confirmation of spontaneous heating existence in panel-B1 and considering safety threats to the mine, it was decided to infuse liquid nitrogen with high expansion foam in two phases from surface borehole connected near 18 level of 39 dip as immediate control measures to heating. The experimental set up for liquid nitrogen flushing is shown by Fig. 4. In the first phase 12000 L liquid nitrogen and 1000 L foam were flushed within 38 h and in second phase 13000 L liquid nitrogen and 1000 L foam were flushed within 70 h.

After implementing these control measure the trends of gases were observed as follows;

- The concentration of CO reduced drastically at 18L/40 D from more than 2000 ppm to 75 ppm specifically in MSP. At other sampling point of 15, 16, 17 and 18 level RSP and DSP traces of CO below 50 ppm were recorded.
- The Graham's ratio (GR -1.00) indicates there was existence of heating/fire before flushing of N2 where as Young ratio (YR-24.88) shows superficial heating. After flushing nitrogen foam both the ratios got reduced drastically [GR (0.03) and YR (9.69)] and represents atmosphere inside sealed off panel to be normal.
- Air-temperature inside the sealed off area recorded by inserting 5m long thermocouple through sampling pipe showed almost ambient temperature inside the Panel B1.

All these thermo-compositional analysis results represents that present conditions inside the sealed off panel are within dormant or under controlled condition.

After flushing of nitrogen foam in panel-B1, shift wise monitoring of air sample, pressure and temperature has been carried out continuously. After three months,

Fig. 4. Experimental set up for flushing of liquid nitrogen from surface borehole.

it was observed that percentage of oxygen again showing increasing trend and pressure behaviour of 18LN/40D stopping has almost similar (High Positive Pressure) as observed before Nitrogen Flushing. However, no symptom of spontaneous heating/ fire has been observed. Therefore, using smoke tube detection of leakages in stoppings and side of barrier pillar has been carried out. But no major leakage sources have been observed.

Meanwhile, at the surface borehole connected to panel B1 near to 18LN/ 40D stopping some hissing sound has been observed. High Suction of air from surface to Panel B1 through bore hole has been observed during pressure monitoring of borehole using Inclined Manometer. Therefore, after details discussion with mine management, it has been decided to plug the bore hole by mud slurry from surface to avoid such ingress of air inside the panel. After lapse of time, it was observed that the pressure at 18LN/40D stopping pressure decreased (up to ± 3 mm) and it behaves similar to other stoppings of panel-B1. It is also observed that oxygen percentage of the panels shows decreasing trends and no traces of spontaneous heating / fire has been observed thereafter. Hence, on identification of root cause of the problem, the frequent occurrences of spontaneous heating were completely resolved and production of mine became resumed.

4 Conclusions

Problem of spontaneous heating/fire in coal mines has become a socio-technological problem of mining fraternity of the world at large and India in particular. Spontaneous combustion problem particularly in underground coal mine creates several mining complications on productivity and safety of mine, which leads to a huge economic burden to the nation. To overcome this menace, early detection and dealing of spontaneous heating/ fire particularly in underground coal mines is foremost important to manage the spontaneous combustion in early stage.

In the present case study, concealed fire in goved out panel-B1 of HLC no mine WCL continuously endangering the safety of miners because of it was located very close to main intake circuit. On the basis of comprehensive study prime cause of spontaneous heating occurrences was identified and after application of suitable preventive steps the carbon monoxide reduced from 2000 ppm to below 50 ppm within the few days and finally became zero ppm. Similarly other explosive gases like H2 and CH4 percentage became comes down within safe limit. Overall on these efforts the panel-B1 of HLC no-1mine became safe in all respect thereafter. Therefore, it is concluded that the success of prevention and control methods particularly in concealed coal mining condition, management of spontaneous heating primarily depend upon the identification of its exact locations. Once the seat of spontaneous combustion located and cause of its occurrences identified, the safe dealing became easy and effective. In this case study, once the root cause and exact location of concealed smouldering combustion in depillared sealed off panel of HLC No-1 mine was identified, the small effort of control approaches were completely solve the frequent safety threats of mines and its frequent re-occurrences was permanently prevented. Hence present study concludes that in case of dealing of concealed underground coal mine spontaneous heating/ fire, major focus should concentrated on identification of its cause of initiation and its exact location for complete management of this menace.

Acknowledgments. The authors are obliged to the Director, CSIR-CIMFR, Barwa Road Dhanbad for his kind permission to publish the paper. Authors also acknowledge thanks to management of Hindustan Lalpeth Colliery, Chandrapur Area of WCL for providing their help during field investigation. The views expressed in this paper are those of authors.

References

Ministry of Coal: Report on Major statistics of Coal – Indian Energy Choice (2023). https://coal.nic.in/en/major-statistics/coal-indian-energy-choice

World Coal Institute: The coal resource: a comprehensive overview of coal, London, UK (2015). http://www.worldcoal.org

Banerjee, S.C.: Textbook on Prevention and Combating Coal Mine Fires. Oxford & IBH, New Delhi **2000**, 33–34 (2000)

Roy, T.R.: Evaluation of spontaneous fire risk potential of underground coal mine panels. CIM Bull. **96**(1073), 63–67 (1996)

Graham, J.I.: Pyrites as a cause of spontaneous combustion in coal mines. Trans. Inst. Min. Eng. **1923–24**(67), 100 (1923)

Hodge, D.J., Hinsley, F.B.: The influence of moisture in spontaneous heating of coal. Min. Eng. **1964**, 40–45 (1964)

Smith, M.A., Glasser, D.: Spontaneous combustion of carbonaceous stockpiles. Part II, Factors affecting the rate of the low temperature oxidation reaction. Fuel, 2005, 84, pp. 1161–1170 (2005)

Pandey, J., Kumar, D., Singh, V.K., Mohalik, N.K.: Environmental and Socio-economic impacts of fire in Jharia coalfield: an appraisal. Curr. Sci. J. **110**(9), 1639–1650 (2016)

Pandey, J., Kumar, D., Panigrahi, D.C., Singh, V.K.: Temporal transition analysis of coal mine fire of Jharia coalfield, India, using Landsat satellite imageries. Environ. Earth Sci. J. (Springer Sciences), **76**(12), (76:439) (2017). https://doi.org/10.1007/s12665-017-6765-8

Stracher, G.B., Taylor, P.T.: Coal fire burning out of control around the world: thermodynamic recipe for environmental catastrophe. Int. J. Coal Geol. **59**, 7–17 (2004)

Sinha, A., Singh, V.K.: Spontaneous coal seam fires: a global phenomenon. In: International Conference on International Research for Sustainable Control and Management on Spontaneous Coal Seam Fires: Mitigation a Global Disaster, Beijing, P.R. China, 29 November–1 December 2005, pp. 42–66 (2005)

DGMS Report on Statistics of Mines in India, Volume 1 (Coal), published by Directorate- General of Mines Safety, Ministry of Labour & Employment, Government of India (2011)

Mishra, R.K., Pandey, J., Chaoudhry, S.K., Khalkho, A., Singh, V.K.: Estimation of air pollution concentration over Jharia coalfield based on satellite imagery of atmospheric aerosol. Int. J. Geomatics Geosci. **2**(3), 723–729 (2012)

Pandey, J., Mohalik, N.K., Mishra, R.K., Khalkho, A., Kumar, D., Singh, V.K.: (2015), An Investigation into the role of different fire retardant materials for preventing spontaneous heating as well as controlling existing fires in Indian coal mines. Fire Technol. **51**(2), 227–245 (2015)

Singh, R.V.K., Pandey, J., Mohalik, N.K., Sural, G., Barnwal, R.P., Singh, V.K.: Early detection of occurrences of spontaneous heating in blasting gallery method – a case study. J. Mines Metals Fuels **52**(5 & 6), 95–101, May–June 2004

Singh, R.V.K., Tripathi, D.D., Mohalik, N.K., Khalkho, A., Pandey, J., Mishra, R.K.: Environmental issue due to fire in coal mines: its impact and suggestion for implementing precaution and control measure. In: Environmental Pollution, Water Science and Technology Library, (Springer Sciences, Singapore), 2018, vol. 77, Chapter 3, pp. 27–35 (2016). https://doi.org/10.1007/978-981-10-5792-2-3

Mohalik, N.K., Singh, V.K., Pandey, J., Singh, R.V.K.: Proper Sampling of mine gases, analysis and interpretation – a prerequisite for assessment of sealed off fire area. J. Mines Metals Fuels **54**(10, 11), 210–217, October–November 2006

Tripathi, D.D.: New approaches for increasing the incubation period of spontaneous combustion of coal in an underground mine panel. Fire Technol. **44**, 423–426 (2008)

Explosive and Blasting

Strategic Planning and Guidelines to Control Dust and Post-Detonation Fumes Arising Due to Blasting in Opencast Coal Mines

Pijush Pal Roy(✉)

Gammon Engineers and Contractors Pvt. Ltd. and Former Outstanding Scientist, CSIR-Central Institute of Mining and Fuel Research, Dhanbad 826001, Jharkhand, India
ppalroy.cimfr@gmail.com

Abstract. To exploit coal from opencast mines, commercial explosives, in large quantities, are used for blasting of overburden rock. The increased use of commercial explosives creates a lot of problems due to the sprawl of urbanization and man's increased sensitivity to blasting that ultimately necessitates cost-effective and safe rock excavation techniques. It is aimed at exploiting coal safely and productively from opencast mines without any health risk to the mine workers and people residing in the vicinity of the mine.

The objective of the paper is to develop a strategic planning, categorization of different subsystem parameters and framing of guidelines to control dust and post-detonation fumes arising due to blasting which would directly help the mine management to control such unwarranted hazards by putting them below the regulatory safe levels. It would also help in generating a huge amount of multilateral data for formulating new sets of guidelines for the mining industry. Non-compliance of regulatory guidelines would otherwise lead to silicosis and other respiratory diseases to mine workers and neighbouring residents.

Keywords: Strategic planning · fugitive dust · toxic fumes · risk assessment · oxygen balancing

1 Introduction

In India, there are about 211 operating opencast coal mines, which require small, medium, large and epic size of blasts in overburden rock and coal seams under constrained and unconstrained conditions. When explosives are detonated to break the rock, its energy is partitioned in different components. Out of the energy released some is useful energy and the rest is waste energy. While useful energy is capable of doing work, the waste energy cannot be used in a constructive way and it creates some environmental problems and annoyance.

Research revealed that as far as the sensitivity is concerned, human beings are much more sensitive to vibrations, noise, fugitive dust and fumes than the structure they reside in. While considering human sensitivity, along with ground vibration and air-blast; dust and fumes must be considered concurrently. In recent years, in India, human response

A. Sinha et al. (Eds.): AMC 2023, SPEES, pp. 415–424, 2023.
https://doi.org/10.1007/978-3-031-46966-4_34

to blast induced vibrations, flyrock, air blast, dust and fumes have been identified as important features for any successful mining operation. Numerous complaints arising due to unoptimized blasts ultimately lead to many unsavoury situations, as a result of which majority of the mines are either running in loss or approaching to closure. Continual noise, dust, flyrock and overall vibration from blasting operations may be upsetting and annoying to the residents. This may go to any stage of violence if such operations involve any physical or property damage. The overall aggravation may result first to mass complaints, then violent resistance and when they mix-up with political interference it may ultimately lead to the closure of the blasting operation.

For a long period of time, normal blast-impacts like ground vibration, air-overpressure/noise, ejection of flying fragments beyond the safety zone are dealt by many researchers, quite elaborately, and as a result of that many countries have developed their own standards approved by the respective regulatory agencies. It is a matter of fact that no such work has so far been conducted involving the fugitive dust and fumes, which escape beyond the mine boundary to reach the inhabitants residing in the vicinity of the mining property. Many times, concerns were raised to the authors on the impacts of fugitive dust and fumes arising due to surface blasting at different sites of Singrauli Coalfields particularly upon the quality of life in the surrounding area. Lot of emotions have surrounded this problem but astonishingly very little attention has been given to tackle this issue and therefore, till date, no sufficient data available to establish a unified standard for implementation. Complaints persist, but there is no tangible way to correlate these complaints to any specific levels of dust and fumes with that of the size of blast vis-à-vis explosive quantity, their composition, distance of the object, wind direction, and direction of material throw. Currently, there is no way to regulate which complaints are justified and which are not. This very specific problem is entirely different from that of blast vibration complaints, which can easily be resolved by vibration monitoring followed by regulatory guidelines. But the concerns are evolved principally on the point how far can fugitive emissions be expected to travel and at what concentrations? The answer, till date, is not scientifically validated.

Blasting is a transient phenomenon therefore its impact does not last long; it is always short-lived. But whatever be the duration, if it is a high-level exposure, it will definitely have cascading effects on eyes, nose and throat irritation and coughing resulting in dizziness and headache. This actually hints at the requirement of an elaborate study covering the probable hazards postured by the toxic fumes. Both toxic gaseous products primarily carbon monoxide (CO) and nontoxic being oxides of nitrogen (NOx) shall be of prime concern. While detonation, the formulation, confinement, ageing and degree of contamination of explosive etc. determine the quantity of toxic gasses to be produced by an explosive. It is therefore proposed to formulate the techniques which would protect the workers and surrounding community from the potential hazards of post-detonation toxic fumes.

As a matter of fact, the study should address: • How the quantity of toxic fumes can be minimised. • How a planning can be chalked out to move workers and neighbours so that they are not exposed. • How to prevent the fumes from moving towards neighbouring villages. • Monitoring the air at surroundings to activate safety measures. • Ventilating confined spaces or exposed structures until the concentration of CO level falls below the statutory threshold level • Generation of data-base under varying size of the blast vis-a-vis total explosives fired in the round. • Risk assessment • Development of new maximum permissible limits of SPM during blasting and non-blasting operative conditions of the mine. • Colour identification of post-blast fumes and its relation with the oxygen balancing of the explosive products.

Presently, 500 m radius is considered as a statutory safety zone for blasts from which all personnel are required to be evacuated. There is a need to modify this safety zone through rigorous scientific study supported by a multivariate data-base to substantiate it logically.

2 Sovereignty of the Problem

Continual noise, dust, fly-rock and overall vibration from blasting operations may be upsetting and annoying to the residents. This may go to any stage of violence if such operation involves any physical or property damage.

It is a well-known fact that toxic fumes create probable hazards in human health. There are several large opencast mines wherein explosives up to one million kg (1000 ton) are detonated in one shot. Many of such shots yield either red- or orange-coloured clouds by the presence of NO_2 (Barnhart 2003; Lawrence 1995). Due to unavailability of published reports of direct measurements, even today, it is still unknown whether the orange cloud contains toxic levels of NO_2 or not? Irrespective of the colour of the fumes, concerning safety and health hazards, it should be assumed primarily that such a post-detonation cloud is unsafe to breathe and has an immense impact on health. It is true that for surface blasting operations, the carbon monoxide (CO), a major pollutant amongst all gaseous products released immediately after a blast is not of great concern since CO is much less toxic than NO_2; the IDLH (Immediate Dangerous to Life or Health) values are established to ensure that the worker can escape from a given contaminated environment in the event of failure of the respiratory protection. The IDLH for CO is 1,200 ppm compared to 20 ppm for NO_2. CO lies with the gas that remains in the fragmented mass and starts releasing during loading operations. It can also migrate hundreds of meters through the ground and accumulate in confined spaces (Santis 2001).

Proper and improper explosive energy utilization may be visualized in the field by observing the nature of gases released just after detonation. Whenever coloured fumes are seen just after detonation, it indicates that the explosive reaction was inefficient and some useful energy was lost. Proper mix of fuel and oxidizer minimizes the production of CO and NO_x. The composition and the physical structure of the chemical mix determines the performance of explosives. When there is excess fuel oil, increased quantities of CO would be generated and if there is not enough fuel, increased quantities of NO_x will be generated on detonation of the explosive products. Figures 1 and 2 illustrate the effect of the percentage of fuel oil content on the quantity of carbon monoxide (CO) and nitrogen oxides (NO_x) produced due to detonation of ANFO. Figure 3 shows the relative size of dust particles carried suspended in air vs. wind velocity and particle density. It is apparent from the graph that the very small particles do not require much air speed to remain in suspension (Hesketh and Cross 1983). The question at this point comes how to correlate this information in an actual blasting event. In the case of dust particles, the condition is little different as these particles get an enormous quantity of momentum by the blast making the air and gases very turbulent. The graph also implies, when the dust cloud is heavy, it will show some gas-like properties whereas in motionless air, the particles will diffuse rather than straight drop down. This type of phenomenon is seen usually in the cast blasting in absence of turbulence, the cloud continues to expand and become thinner. Different meteorological conditions which may dominate the propagation of post-detonation dust, fumes and air-blast are shown in Fig. 4 (Bhandari 1997). As a matter of fact, when explosive composition is improper besides creating noxious gasses, the blasthole may not have sufficient energy to break the rock properly in front of the hole. Blastholes are therefore under-loaded for the amount of rock present. Such a situation may lead to air-blast and fly-rock and the blasthole may not break up to the desired depth.

Presently, there has been a common safety zone of 500 m radius for blasts from which all personnel are evacuated. This common safety zone needs to be re-looked under prevailing conditions with proper scientific understanding and thorough scrutiny of data.

For surface blasting, no specific standards exist for fumes and dust under the environment protection laws. As for example, the blasting in mines of Singrauli Coalfields under Northern Coalfields Ltd. Produce a very visible dust and fume cloud with the following apparent characteristics (Fig. 5)-

1. The dust cloud dispensed from the overburden blast is substantially high.
2. The fume cloud dispensed from the blasting shot may not necessarily move with the dust cloud.
3. The cloud thins out and may diffuse, with the fumes intermingling and gradually moves out of the pit.

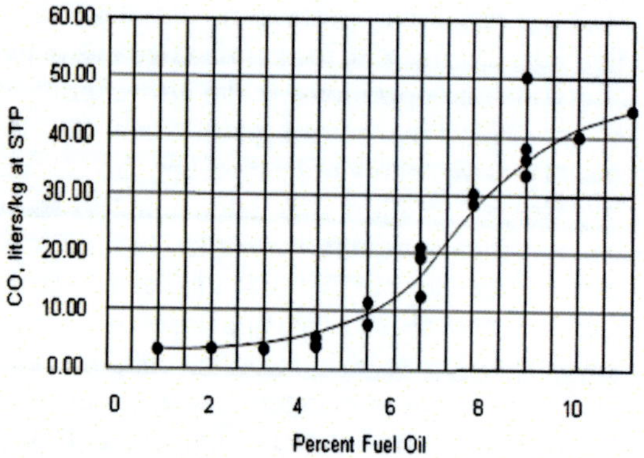

Fig. 1. The effect of fuel oil content in percent on the quantity of carbon monoxide produced by detonating ANFO (Rowland III &Mainiero 2000)

Prediction of wind velocity and identifying the adequate sites for instrumentation become a real challenge. Though it is presumed that a visible dust cloud would contain

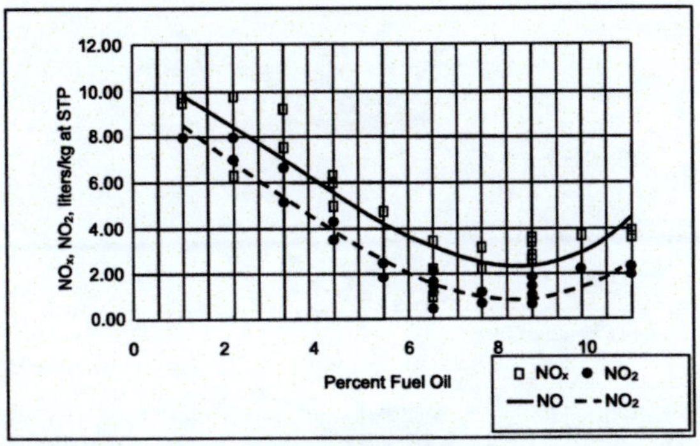

Fig. 2. The effect of fuel oil content in percent on the quantity of nitrogen oxides produced by detonating ANFO (Rowland III &Mainiero 2000)

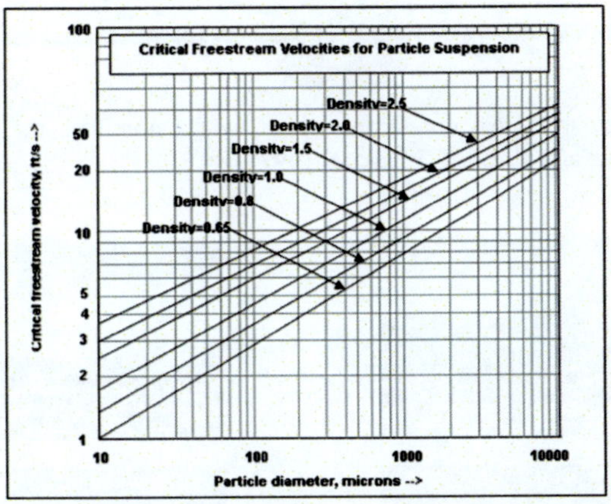

Fig. 3. Particle suspension velocities against particle diameter (adapted from Hesketh and Cross 1983)

more than 200 ppm (0.02%) of nitrous oxide, apparently should have been a very serious threat but no cases of death or serious health impairment have so far been surfaced from any of the mine-adjoining villages rather mining personnel pose a very optimistic view to say that they had driven through and worked in such clouds without any adverse health impacts.

In nutshell, this is an undefined problem with no known concentration data, no factual evidence of health deterioration, casualties, or even temporary impairment; only a few stray circumstantial incidents which can't be taken for granted without some sort of official and scientific assessment. It has remained a debatable issue without being truly defined and therefore requires substantial research activities to strengthen the logical output.

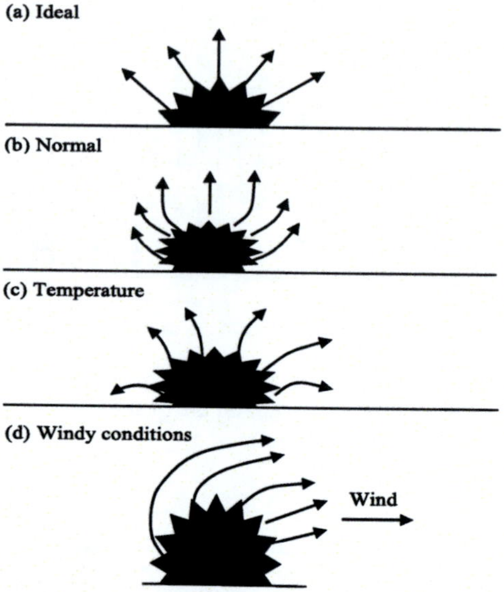

Fig. 4. Different meteorological conditions which would dominate the propagation of post-detonation dust, fumes and air-blast (Bhandari 1997)

Fig. 5. Pictorial view of the generation and spreading of blast-induced dust in opencast coal mines

3 Present Scenario

Unlike many scientific studies on the propagation, prediction and control of ground vibration and air overpressure/noise; very few have so far been undertaken in India and abroad correlating blast-design parameters, initiation patterns and explosive properties with that of blast performance, generation of fugitive dust and fumes, blast-impacts and human sensitiveness. There is no such known dust concentration data; no real evidence of damage in eye, nose and throat irritation and coughing, dizziness and headache due to high level exposure. Strong debate and circumstantial activities are revolving around on that issue without any deterministic approach and established guidelines.

As per the directive of the Hon'ble Supreme Court of India dated 23.09.2013 on SLP 21211 of 2012, a comprehensive study to assess the impact of mining using explosives on structural integrity of Chittorgarh Fort Complex was initiated by CSIR-CBRI (CSIR-Central Building Research Institute), Roorkee, in collaboration with CSIR-CIMFR (CSIR-Central Institute of Mining & Fuel Research), Dhanbad. Hon'ble Supreme Court of India directed CSIR-CBRI to undertake the following issues: (a) Whether blasting including the cumulative effect of blasting beyond a specified distance has any impact whatsoever upon the structure of the Fort? (b) What appeared to be the causes that have led to cracks and other damage caused to the Fort, other than ageing simplicitor? (c) Whether the uncontrolled access to tourist has any adverse impact upon the structure and if so, any suggested steps to regulate this activity (d) Whether the activities within the colony situated in the Fort as well as the flow of traffic including heavy traffic in the vicinity of the Fort have any adverse effects upon the structural integrity of the Fort, and if so the suggested measures to deal with the problems (e) General recommendations on the steps to be taken to restore the structural integrity to repair the cracks and generally ensure that no damage in future is caused to the structural integrity (f) A comprehensive Environmental Impact Assessment (EIA) covering all kind of pollution-air, ground water, noise etc. by the complete cycle of mining activities including its transportation.

In another collaborative study conducted by CSIR-CIMFR, CMPDIL (Central Mine Planning and Design Institute Ltd.), WCL (Western Coalfields Ltd.), DGMS (Directorate General of Mines Safety) and DRDO (Defence Research Development Organisation), observations were made regarding dust clouds generated during blasting in dragline and shovel-dumper faces of Ghugus Opencast Project of WCL. The measurements of suspended particulate matter (SPM) were carried out using 47 mm open-ended air samplers (controlled valve) downstream of the sample filter in the suction line.

The sampling rate was 25, 30 and 50 LPM (litres per minute) respectively at three locations and sampling time was 5 to 50 min to cover the drift of the dust cloud. From the background data collected, it was seen that even during non-working period, the SPM level of 749 $\mu g/m^3$ in the mine area exceeded the 500 $\mu g/m^3$ specified limit set by CPCB (Central Pollution Control Board), Govt. of India for industrial area. Table 1 shows the maximum permissible SPM limits established by CPCB for different areas.

Table 1. Maximum permissible SPM limits established by CPCB

Type of Area	Concentration in ($\mu g/m^3$)
Industrial Area	500
Residential Area	200
Sensitive Area	100

The recommendations of the said study included- (1) time-bound water spraying before and after blast, especially in blasting area and nearby haulage roads to control dust,

(2) The use of DTH (down-the-hole) delay and NTD/TLD (trunk-line-delay) systems (or in other words shock-tube system) for effective reduction of dust and noise hazards by nearly 30 to 40% and (3) Proper planning, time-bound water spraying and plantations to reduce the environmental hazards arising due to blasting.

4 Strategic Planning

It is thus far the questions arising in every sensible person's mind what could be expected travel distance of fugitive emissions and at what concentrations? There is, till date, no scientific answer which can be relied upon for strategic planning. As such, there has been a strong requirement of categorization of different parameters and their influence on health-hazards so as to determine the precise guidelines for control of dust, post-detonation fumes and fragmentation arising due to opencast blasting in coal mines that would directly help the mine management to optimize the blasting parameters controlling the adverse impacts amicably within the threshold levels and parallelly generate huge multilateral data for formulating new set of guidelines applicable to the mining industry.

Such a gigantic study would require a few subsystem studies to achieve the desired objective, likewise (Fig. 6)-

a) Study of different blast design parameters with different blasthole depths preferably between 10 and 30 m to evolve safe and optimum blasting patterns to achieve higher production and productivity.
b) Study of blast-induced ground vibrations, air overpressure, noise, dust sizes and suppression distance, post-blast fumes and demarcation of safety zones.
c) Study of fugitive dust and fumes at minimum 10 mines with 10 blasts in each mine under varying rock-geologic conditions.
d) Study on explosive properties viz. Composition, density, VOD (velocity of detonation), oxygen saturation, colour of post-detonation fumes etc. and its consequent effects.
e) Evaluation of a new set of guidelines for dust and fumes and blasting strategies during different climatic conditions in conformity with the wind directions.
f) Determination of safe, safer and safest zones of disturbance.

5 Action Plan Flowchart

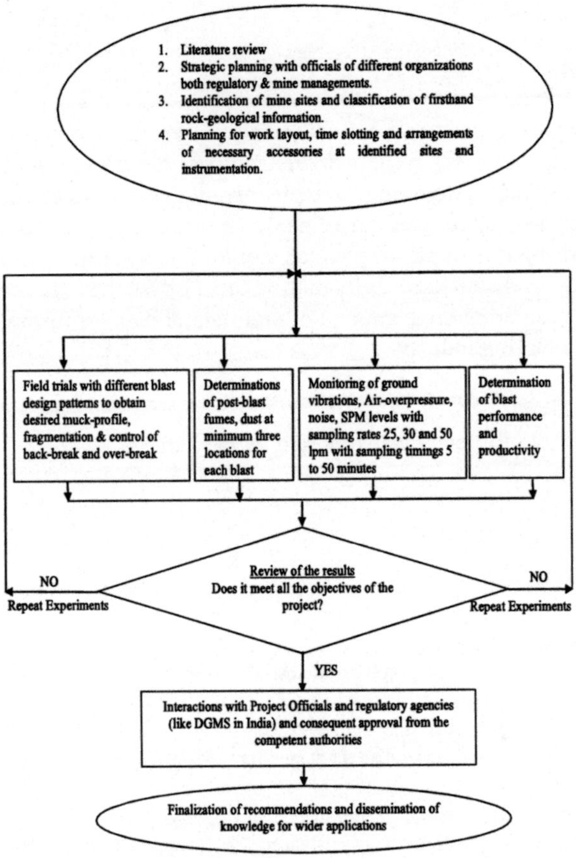

Fig. 6. Flowchart of the work-plan

6 Suggestive Work Elements

The primary aim should be to create authentic data-base to tangibly assess the extent of dust and fumes emanating from surface coal mine blasting under varying rock-geologic conditions, and identify if these levels create either a health risk (as defined by existing regulations) and/or a nuisance. Concurrently it should focus on the blasting performance and fragment size distribution involving the parameters (a) Environmental (water, geology, confinement, rock competency, wind speed/direction; (b) Application (powder factor, hole diameter, hole depth, burden & spacing, initiation type, fragment size distribution); (c) Product (sensitivity, loading contamination, prill quality, explosive energy, density and reactivity, additives, primer size, timing, sleep time).

Hence, there would be requirements of measurements for the parameters like nitrogen dioxide; nitrous oxide; carbon monoxide; ammonia; total dust; respirable dust and fragment size.

7 Discussions

The proposed idea is focused to exploit coal safely and productively from opencast mines without any health risk to the mine workers and people residing in the vicinity of the mine. It would also help developing strategic planning, categorization of different sub-system parameters and framing of guidelines to control dust, post-detonation fumes and fragmentation arising due to blasting which would directly help the mine management to control such unwarranted hazards by putting them below the regulatory safe levels. It would also help generating huge amount of multilateral data for formulating new set of guidelines for the mining industry.

Disclosure statement
No potential conflict of interest was reported by the author.

References

Bhandari, S.: Engineering rock blasting operations. In: Balkema, A.A., (ed.) Rotterdam, Netherlands, p. 375 (1997)

Barnhart, C.R.: Understanding the 'orange smoke' problem in cast blasting. In: Proceedings of the 29th Annual Conference on Explosives and Blasting Technique. February 2–5, Cleveland, OH: International Society of Explosives Engineers, pp. 1–11 (2003)

Hesketh frank, L.J., Cross Howard, E.: Fugitive emissions and controls. Ann Arbor Science. ISBN-13: 978–0250404483, June, p. 145 Internet surfing (1983). www.osmre; www.cdc.gov; www.healthnsw.gov.au

Lawrence, D.L.: A study of post-blast generation of nitrogen dioxide. In: Proceedings of the 22nd Annual Conference on Explosives and Blasting Technique. February 2–5, Cleveland, OH: International Society of Explosives Engineers, pp. 1–12 (1995)

Rowland iii, J.H., Mainiero, R.J.: Factors affecting ANFO fumes production. In: Proceedings of the 26th Annual Conference on Explosives and Blasting Technique. Cleveland, OH: International Society of Explosives Engineers (ISEE), vol. 1, February 13–16, pp. 163–174 (2000)

Santis, L.D.: A summary of subsurface carbon monoxide migration incidents. In: Proceedings of the 27th Annual Conference on Explosives and Blasting Technique. International Society of Explosives Engineers. January 28–31, Cleveland, OH (2001)

The Influential Role of Confinement, Charge Factor and Segmentation of Explosive Charge in a Blasting Round: A Perspective for the Blasting-Induced Ground Vibration Compliance

Narayan K. Bhagat[1,2(✉)], Rakesh K. Singh[1], P. Hembram[1], C. Sawmliana[1,2], and A. K. Mishra[1,3]

[1] Rock Excavation Engineering Division, CSIR-Central Institute of Mining and Fuel Research, Dhanbad, India
nkbhagat@cimfr.nic.in, narayan_bhagat@yahoo.co.in
[2] Faculty of Engineering Sciences, Academy of Scientific and Innovative Research (AcSIR), Ghaziabad, India
[3] CSIR-Central Institute of Mining and Fuel Research, Dhanbad, India

Abstract. This paper focuses on evaluating the effects of confinement, charge factor and segmentation of holes in a blast round on blast-induced ground vibrations in a mine. The study involved conducting a series of experiments with different drillhole diameters and modifying design parameters. The results indicated that the initial blast designs were not optimised, resulting in to generation of higher magnitudes of ground vibrations. Based on the vibration data recorded, regression analysis was carried out to establish a best-fit relationship between the scaled distance and vibration. On the basis of the threshold values of peak particle velocity prescribed by the DGMS, the permissible maximum charge per delay at specific distances were determined. The permissible charge values were not workable and matching with the mine productivity. The subsequent experimental blast with a smaller drill hole diameter also did not significantly reduce the magnitudes of ground vibrations. However, the experiments conducted with reduced confinement conditions with less number of holes and optimized charge factors resulted in a notable reduction in the ground vibration magnitudes. An innovative technique of segmentation of total holes in a blasting round using electronic detonators demonstrated a significant decrease in vibration magnitudes with the potential of increased productivity. The findings of the study emphasize the importance of re-assessing the blast design parameters and charging patterns to mitigate ground vibrations and optimize the blast outcomes in mining operations.

Keywords: Blast-induced ground vibration · design parameters · drill diameter · Mitigation strategies

A. Sinha et al. (Eds.): AMC 2023, SPEES, pp. 425–440, 2023.
https://doi.org/10.1007/978-3-031-46966-4_35

1 Introduction

The safety of residential houses/dwellings ensuring the well-being of nearby communities located within the danger zone of blasting operations in open-pit mines is of utmost importance for continuing the mining operations smoothly.

Numerous factors have impact on the magnitude of blast-induced vibrations including, uncontrollable such as site geology and blast-to-structure distance and controllable parameters like maximum charge per delay, hole diameter, confinement, spatial distribution of energy source, delay timings and blast orientation (Bender 2007; Monjezi et al. 2013; Hasanipanah et al. 2017; Hosseini et al. 2023; Roy et al. 2023). To minimize vibration intensity, various measures can be employed, such as reduction in maximum charge per delay, decrease in confinement, optimizing the spatial distribution of explosive energy, refining the delay timings and implementing the explosive decoupling techniques within blast holes (Yue et al. 2022).

This paper presents a comprehensive case study conducted in an opencast coal mine, wherein extensive trial blasts were conducted to determine the optimal unit charge factor (specific charge), maximum charge per delay and optimum blast design parameters for the safety of residential dwellings present within the danger zone (100–500 m from the mine workings). The objective was to keep the ground vibration magnitudes within safe limits while maintaining the production and productivity of the mines.

This case study underscores the significance of developing tailored blast schemes to effectively manage vibration impacts on nearby structures. By implementing specific blast design parameters derived from extensive trials, this study offers a practical and feasible approach to ensure safe mining operations while safeguarding the residential structures/dwellings within the danger zone of mine workings.

1.1 About the Study Area

The trial blasts for the present research work were conducted at Pachwara (Central) Coal Block, a captive coal mine of M/s Punjab State Power Corporation Limited (PSPCL) in Pakur District of Jharkhand State in India. The mine benefits from a relatively flat topography and sandstone-dominated overburden, necessitating careful blasting techniques for excavation. With a coal seam thickness of around 20 m, the mine has substantial energy production potential. Geological features such as the Barakar, Dubrajpur, and Rajmahal formations along with various fault structures contribute to the region's diverse geological landscape. Within the mine, ten correlatable coal seams are present including the prominent Seam Zone II.

Drilling and blasting system has been employed in the mine for excavation of overburden and expose the coal surface. Thereafter, a surface miner is utilised for extraction of various coal seams present. Initially, the drilling operations as laid in the mine planning were done by using 260 and 160 mm diameter drilling machines. Later on, it has been replaced by 160 mm diameter drilling machines. The typical depth of the hole ranges from 5–7 m, with a burden of 4–5 m and a spacing of 5–6 m.

Site-mixed emulsion (SME) explosives in combination with non-electric (Nonel) system of initiation known as Down-the-Hole-Delay (DTH) and Trunkline-Delay (TLD) are used at the mine for charging of holes with explosives. Occasionally, electronic detonators are also used as an alternative initiation method.

This mine is surrounded by many residential houses/ dwellings of nearby village. The houses/hutments are present within 300–500 m distance from the mine workings and in near future, they may come up to 100 m distances from the mine workings. These structures/houses require precautions from the side-effects of the blasting operations viz. Ground vibrations, noise/air-overpressure, flyrock.

2 Blast-Induced Ground Vibrations

Blast-induced ground vibrations are generated by the detonation of explosives during blasting operations. Excessive vibrations can result in cracks, shifting foundations, and other forms of structural damage (Seferoglu et al. 2023), compromising the stability and longevity of buildings.

The propagation mechanisms involve compression waves, shear waves and surface waves, each contributing to ground motion. The characteristics and intensity of these vibrations are influenced by various blast design parameters, such as the type and quantity of explosives, drill hole diameter and patterns, blast hole spacing and initiation sequence. The prevailing geological conditions and rock properties including density, stiffness and presence of fractures, also affect the propagation of vibrations. The distance from the blasting site to the point of measurement or structures, as well as soil type and moisture content, impact the intensity and transmission of vibrations (Siskind et al. 1985). The dominant frequency content of the vibration waves determine their potential of causing damage or discomfort (Siskind et al. 1980). Therefore, a clear-cut understanding of the causes and mechanisms of blast-induced ground vibrations is crucial in implementing the mitigation measures and designing the blasts to minimize their impact on nearby structures and communities.

3 Role of Confinement of Explosive Charge in Blasting

The confinement of explosive charge in blasting operation is very critical for an effective and safe blasting operation. A proper confinement contains the blast within the designated area and reduces hazards from flyrock and uncontrolled fragmentation. It allows for efficient rock fragmentation and helps in controlling blast-induced ground vibrations. However, excessive confinement can lead to higher energy transmission to the surrounding rock mass thereby resulting in elevated vibration levels and stronger reflections at rock interfaces. It can also affect vibration dispersion, causing broader frequency content and longer propagation distances.

The stiffness ratio (Hole depth/Burden ratio)) is a crucial factor in generating the higher magnitude vibrations. A stiffness ratio between 1 to 2 is considered as critical in view of the severe vibration, fragmentation, flyrock and air-overpressure (Konya and Walter 1991; Bhagat et al. 2020a) and it should be more than 4 for better blast results.

To mitigate the impacts of over confinement, careful blast design and optimization, including selection of explosives, reduction in hole diameter and blast hole configuration, are necessary (Bhagat et al. 2020b).

3.1 Role of Maximum Charge Per Delay and Total Charge on Vibration Generation

The maximum charge per delay and total charge in a blasting round are very important factors that influence the intensity and extent of ground vibrations during blasting operations (Duvall 1963; Mandal et al. 2008; Himanshu et al. 2018; Roy et al. 2020). Increase in the charge per delay and total charge generally leads to higher vibration amplitudes to greater propagation distances. A balance of these two explosive charges is very crucial to control and minimize ground vibrations considering site-specific factors and regulatory standards.

3.2 Role of Charge Factor (Specific Charge)

The charge factor is also a critical parameter in blasting operations, as it determines the amount of explosive required to break per unit volume of rock. It plays a significant role in the generation of vibrations, affecting energy release, stress waves, fragmentation, frequency content, and duration of vibrations. It is important to note that the specific charge is influenced by various factors, including the type and properties of the explosive material, rock characteristics, and the desired blasting objectives. The charge factor is often determined through extensive trial blasts and empirical studies to find the optimal ratio of explosive charge to break the rock.

Studies have shown that higher specific charge generally result in increased energy release and larger ground vibrations (Bender 2007; Dindarloo 2015). However, exceeding the optimal specific charge factor can lead to diminishing returns, where the increase in vibrations becomes disproportionate to the additional energy input. Therefore, careful consideration and evaluation of site-specific conditions are necessary to determine the appropriate charge factor.

3.3 Regulatory Framework and Vibration Standards

The regulatory framework and vibration standards related to blasting in mining operations vary across different countries and jurisdictions. Permissible limits for blast-induced vibrations are often specified in regulatory frameworks. These limits usually involve criteria for peak particle velocity (PPV) and frequency content. PPV measures the maximum velocity of ground vibrations at a specific location, while frequency content criteria may restrict certain frequency ranges. In case of India, the DGMS Technical Circular 7 of 1997 (Table 1) governs the threshold values of vibrations for different types of structures near mines. The circular sets safe values of PPV for residential dwellings/houses, with limits of 5 mm/s and 10 mm/s for dominant excitation frequencies below 8 Hz and between 8–25 Hz, respectively. Overall, these regulatory frameworks and standards aim to ensure the safety of nearby structures and communities, while providing guidelines for mining companies to conduct blasting operations responsibly and in compliance with legal requirements.

Table 1. DGMS standard (*Technical Circular Number 7 of 1997*)

Type of structure	Safe PPV in mm/s considering Dominant excitation frequency, Hz		
	<8 Hz	8–25 Hz	>25 Hz
(A) Buildings/structures not belonging to the owner			
1. Domestic houses/structures (Kutcha, brick & cement)	5	10	15
2. Industrial buildings	10	20	25
3. Objects of historical importance and sensitive structures	2	5	10
(B) Buildings with limited span of life and belonging to owner			
1. Domestic houses/structures	10	15	25
2. Industrial buildings	15	25	50

4 Experimental Works

The initial set of experiments (1st set) involved conducting four rounds of trial blasts in the mine with 260 mm diameter holes. The trial blasts were conducted with varying design parameters and charging patterns. The number of holes varied between 30–53 with hole depths ranging between 5.0–7.0 m. The burden and spacing values varied between 4.5–5 m and 5.5–6 m respectively. The maximum charge per delay varied between 55–130 kg whereas total explosive charge between 1593–2895 kg. A summary of the blast design parameters and explosive charging patterns are given in Table 2. During the first set of trial blasts, the charge factor varied from 0.29 to 0.4 kg/m^3. Ground vibrations resulting from the blasts were monitored on ground surfaces towards the back side of the free faces of blasts at distances varying between 90 to 480 m. The blast vibration results are also shown in Table 2. The magnitudes of recorded vibration data ranged between 46.3–3.32 mm/s at distances varying between 90–480 m.

After getting higher magnitudes of ground vibrations, particularly at nearby distances, it was planned to conduct few trial blasts with smaller hole diameter. The diameter of the blast holes were reduced to 160 mm from the 260 mm. In the 2nd set of experiments with 160 mm drill hole diameter, four numbers of experimental blasts. The number of holes in the trial blasts varied between 45–58 with depths ranging from 5 to 6 m. The burden and spacing values varied between 3.0–4.0 m and 3.5–5.0 m respectively. The average charge per hole ranged from 30 to 50 kg, with a maximum charge per delay of 70 to 150 kg and a total charge of 1475 to 2790 kg. The explosive charge factor in this set of trial blasts was increased a little, varying between 0.41 to 0.48 kg/m^3. The summary of the blast design parameters and explosive charging patterns experimented during this second set of trial blasts are given in Table 3.

Ground vibrations resulting from the blasts of this 2nd set were monitored on ground surfaces towards the back side of the free faces of blasts. The blast vibration results are also shown in Table 3. The magnitudes of recorded vibration data ranged between 20.8–1.09 mm/s at distances varying between 100–500 m. The magnitudes of vibration

Table 2. Details of blast design parameters used in the 1st set of experimental blasts

Blast No.	NH	HD [m]	B × S [m × m]	CPH [kg]	TC [kg]	Qmax [kg]	SC [kg/m³]	Ground Vibration Results	
								D [m]	PPV [mm/s]
B1	53	6.0–7.0	5.0 × 6.0	50–65	2895	130	0.29–0.31	130	31.4
								170	18.0
								210	16.8
								260	10.5
								310	10.1
B2	42	5.5–7.0	4.5 × 5.5	45–60	1934	125	0.29–0.33	100	30.7
								150	10.0
								200	8.02
								250	5.1
								290	4.38
B3	30	5.0–7.0	4.5 × 5.5	50–65	1593	65	0.38–0.40	140	18.6
								200	11.0
								245	6.91
								280	4.98
								330	4.29
B4	50	5.5–7.0	4.5 × 5.5	45–55	2445	55	0.31–0.35	100	46.3
								150	21.84
								300	5.25
								400	3.46
								480	3.32

recorded were comparatively lesser than that recorded in the first set of trial blasts. However, at nearby distances, the recorded vibration data were still in higher side. Therefore, some more experimental blasts were planned to be conducted with reduced confinement having reduced burden and spacing values with a lesser number of holes. The blast diameter was kept similar as 160 mm considering only available choice.

In the 3rd set of experimental blasts, four more blasts were conducted with reduced burden and spacing values and lesser number of holes. The number of holes varied between 10– 20 having depths ranging between 4.0—5.8 m. The burden values varied between 2.25 - 3.0 m whereas spacing varied between 2.5–3.5 m respectively. The maximum charge per delay ranged between 25—50 kg with total explosive charge in the blasting round ranging between 250—873 kg. After seeing the results of the trial blasts conducted in 2nd set of blasts, the charge factor used in this 3rd set of blasts were further increased, ranging between 0.72—0.89 kg/m^3. The details of blasts design parameters experimented during the 3rd set of trial blasts and the recorded ground vibration data at different distances are presented in Table 4. The magnitudes of recorded PPV values at distances ranging between 90–510 m varied between 11.97—0.959 mm/s. Further

Table 3. Details of blast design parameters used in the 2nd set of experimental blasts

Blast No.	NH	HD [m]	B x S [m × m]	CPH [kg]	TC [kg]	Qmax [kg]	SC [kg/m^3]	Ground Vibration Results	
								D [m]	PPV [mm/s]
B5	58	5 - 6	4.0 × 5.0	40–50	2790	150	0.44	130 200 375 480	16.6 10.6 6.17 1.67
B6	49	5.8	4.0 × 5.0	45–50	2307	95	0.41	100 200 300 450	20.4 9.75 5.06 1.63
B7	45	6	3.0 × 4.0	30–35	1475	70	0.46	125 200 400 500	20.8 16.6 8.63 2.17
B8	55	6	3.0 × 3.5	30	1650	90	0.48	100 150 300 400 500	15.3 10.1 4.94 4.21 1.09

Note: NH-No. of holes, HD-Hole depth, B-Burden, S-Spacing, CPH-Charge per hole, Qmax-Maximum charge/delay, TC-Total charge, SC-Specific charge or charge factor, D-Distance, PPV-Peak particle velocity (vibration)

decrease in the vibration magnitudes at nearby distances was observed in this 3rd set of trial blasts.

Encouraging with the results of trial blasts with reduced confinement and lesser number of holes or total explosive charge in the blasting round, a set of two experimental blasts were planned for increasing the productivity of mine and to observe the impact of total charge on vibration magnitudes. In this 4th set of experiment, electronic detonators were used to fire the explosive charged holes in place of non-electric (NONEL) initiation system and the total number of holes in a blasting round was splitted into three smaller segment of holes by providing a delay time gap of more than 1500 ms between those segments.

In this 4th set of experimental blasts, two blasts were conducted for comparison. One was conducted with 91 holes having 5005 kg explosive as total explosive charge in the blasting round whereas second blast was conducted with 86 holes splitted in three sub-groups of 30, 29 and 27 holes, with 4730 kg as total explosive charge in the blasting round keeping all other parameters almost same. The details of these two experimental blasts are given in Table 5. The vibration magnitudes in these two blasts, one with segmented total charge and one without segmenting, were recorded at different distances varying

Table 4. Details of blast design parameters used in the 3rd set of experimental blasts

Blast No.	NH	HD [m]	B x S [m × m]	CPH [kg]	TC [kg]	Qmax [kg]	SC [kg/m³]	Ground Vibration Results	
								D [m]	PPV [mm/s]
B9	20	4.0	2.5 × 3.0	20–35	448	35.0	0.75	150 235 350 450	4.018 3.737 1.727 0.907
B10	10	5.5	3 × 3.5	45–50	486	50	0.84	90 175 300 400	11.97 7.948 1.497 0.907
B11	10	5.0	2.25 × 2.5	25	250	25	0.89	100 250 400 490	8.962 1.611 0.813 < 0.5
B12	20	5.8	3.0 × 3.5	45–50	873	50	0.72	100 250 400 510	10.94 4.687 1.454 0.959

between 100–400 m. The recorded vibration magnitudes showed a clear-cut decrease in the vibration magnitudes with the blast conducted with splitted total charge.

5 Results and Analysis

The 1st set of experimental blasts was conducted with an aim to evaluate the effects of various parameters and charging patterns on blast outcomes in the mine. The analysis of blast design parameters revealed that the blast design parameters are not optimised. The stiffness ratio (hole depth/burden) has an important role in generation of vibration apart from the hole diameter. For a stiffness ratio ≤ 1, the rock breaks with causing over-break and generating higher vibration magnitudes. The stiffness ratio of the blasts conducted in this 1st set of experiment varied between 1.2–1.4 with the charge factor values ranging between 0.29–0.4 kg/m^3. The blast results clearly showed that the used charge factor was not sufficient to break the rock efficiently. The ground vibration magnitudes were relatively higher, ranging from 46.3 mm/s to 10.1 mm/s at 90 m and 300 m respectively. For further analyses, the vibration data generated were grouped together correlating the maximum charge per delay and distance of vibration monitoring locations from the blasting faces for obtaining a best-fit regression equation using the Blastware software of M/s Instantel Inc., Canada. The regression plot is shown in Fig. 1.

Table 5. Details of blast design parameters used in the 4th set of experimental blasts

Blast No.	NH	HD [m]	B x S [m × m]	CPH [kg]	TC [kg]	Qmax [kg]	SC [kg/m^3]	Ground Vibration Results	
								D [m]	PPV [mm/s]
B13	91	5.7–6.3	4 × 5	55	5005	55	0.46	100 200 300 400 500	25.13 14.47 6.001 4.658 2.28
B14	86	6	4 × 5	55	4730	55	0.46	–	–
Split 1	30				1650	55	0.46	100 200 300 400	16.51 7.01 2.95 1.93
Split 2	29				1595	55	0.46	100 200 300 400	14.99 7.03 2.59 1.65
Split 3	27				1485	55	0.46	100 200 300 400	14.32 8.07 1.99 1.65

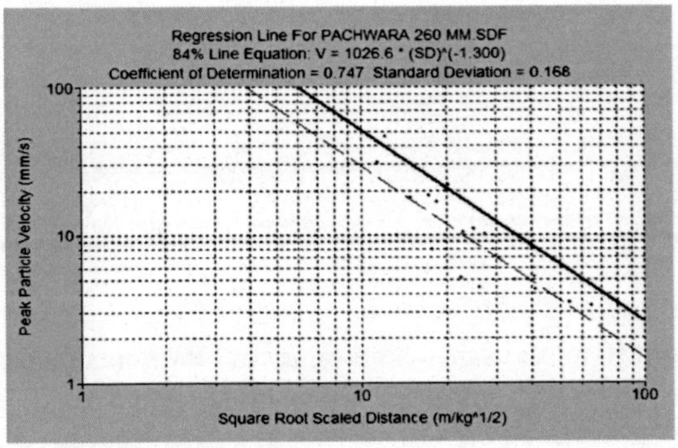

Fig. 1. Regression plot of data generated in 1st set of experimental blasts with 260 mm hole diameter

On the basis of the established predictor equation for this 1st set of experiment, the safe values of maximum charge per delay can be determined for any structures at a concerned distance considering the safe value of PPV based on the DGMS guidelines.

Considering 5 mm/s and 10 mm/s as threshold values of PPV depending upon the dominant frequencies of blast waves, the permissible maximum charge per delay comes to 2.8 kg and 8 kg at 100 m whereas 11 kg and 32 kg at 200 m respectively. These findings suggest that the blasting operations in the 1st set of experiment conducted with the 260 mm diameter and existing blast design parameters of the mine are generating increased vibration magnitudes. For keeping the vibration magnitudes within safe limits, the suggested permissible charge per delay are not workable considering the productivity of mine. Therefore, the design parameters and charging patterns were re-assessed to mitigate the adverse effects of high ground vibrations on the surrounding environment and structures.

In the 2nd set of experiments utilizing a smaller 160 mm drill diameter, four additional blasts were conducted. Similar to the initial set of experimental blasts, variations were made in the design parameters and charging patterns having reduced drill diameter of 160 mm. The stiffness ratio in this set of experimental blasts varied between 1.25–2.0 with the charge factor varying between 0.41–0.48 kg/m^3. The blast results showed that despite the reduction in the size of drill hole diameter, the magnitudes of ground vibrations remained high. However, the vibration magnitudes were decreased to some extent, the recorded PPV values, ranging from 20.8–8.63 mm/s at 125 m and 400 m respectively, were higher than the DGMS prescribed safe PPV levels. The data generated with these experiments were grouped together once again for regression analysis. The regression plot is shown in Fig. 2.

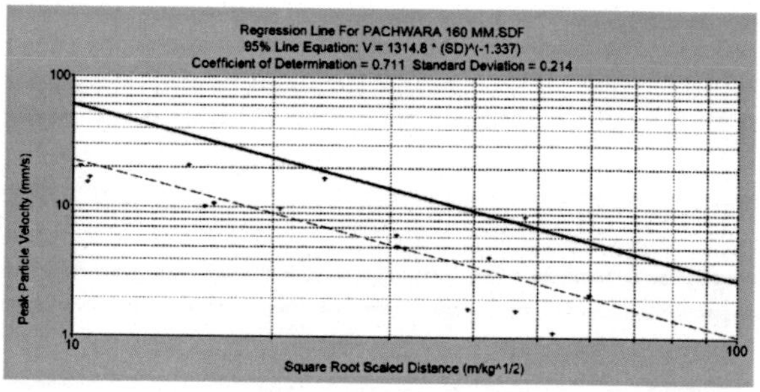

Fig. 2. Regression plot of data generated in 2nd set of experimental blasts with 160 mm hole diameter

Considering 5 and 10 mm/s as threshold values of PPV depending upon the dominant frequencies of blast waves, the permissible maximum charge per delay come to 3.6 kg and 10 kg at 100 m whereas 14 kg and 40.5 kg at 200 m respectively. These values of maximum charge per delay were more than those obtained from the regression equation developed for 1st set of experimental blasts. These findings suggest that although, the use of smaller drill hole diameter reduced the magnitude of PPV and increase the maximum charge per delay which can be fired safely at nearby distances, but, it did not completely mitigate the adverse effects of ground vibrations. Therefore, further adjustments in the design parameters and charging patterns were done to effectively reduce the magnitude of ground vibrations and minimize their potential impacts on the surrounding structures.

In the 3rd set of experimental blasts, four more blasts were conducted with reduced confinement and a smaller number of holes using 160 mm drill hole diameter. The stiffness ratios were slightly higher and varied between 1.6–2.2 with increased charge factor varying between 0.72–0.89 kg/m^3. The results of all the blasts indicated a significant reduction in the ground vibration magnitudes compared to the previous experiments. The recorded PPV values ranged from 11.97 mm/s to 0.959 mm/s at 90 m and 510 m, respectively. Once again data generated modified experiments were grouped together for regression analysis. The regression plot between is shown in Fig. 3.

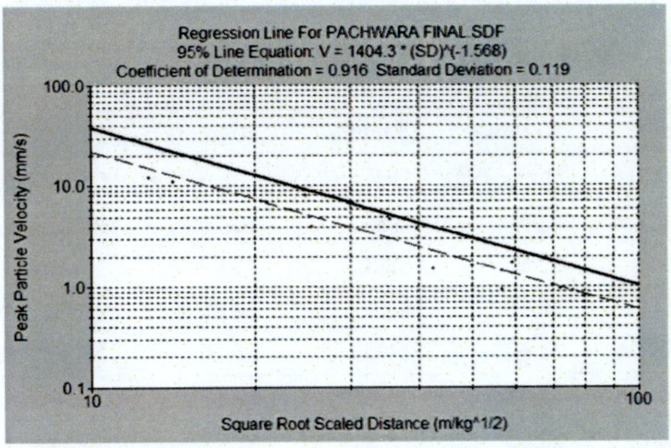

Fig.3. Regression plot of data generated in 3rd set of experimental blasts with less confinement and optimum charge factor

The permissible maximum charge per delay according to 5 and 10 mm/s as threshold values of PPV, now increased to 9.2 and 22.1 kg at 100 m whereas 36.5 and 88.3 kg at 200 m respectively.

The safe values of maximum charge per delay at different distances from the blasting point keeping 5 and 10 mm/s as threshold value of PPV have been calculated from the regression equations developed for all the three sets of experimental blasts and are given in Table 6. The detailed analyses of Table 6 revealed that the when the hole diameter is reduced only, from 260 mm to 160 mm, and same confinement and charge factor are kept, there is no much increase in the safe maximum charge per delay between 100–300 m distances. The suggested maximum charge per delay cannot be worked out for safe blasting operations at nearby distances within 100–300 m from the blasting faces. But, when the hole diameter is reduced to 160 mm and the blast design parameters are optimised, the maximum charge per delays increased significantly, i.e. 229% and 175% in case of 5 and 10 mm/s as safe threshold. It also revealed that when blast design parameters were polished by reducing the burden and spacing with optimal charge factor, the advantage was impressive and workable maximum charge per delay obtained beyond 150 m distance in case of 5 mm/s and 100 m distance in case of 10 mm/s as threshold of vibration.

A detailed analysis was further done with the help of established predictor equations for different sets of experimental blasts. The predicted levels of ground vibrations at different distances from the blasting sites for various maximum charge per delay have been plotted for each set of three experimental blasts conducted. The plot of predicted PPV levels at different distances are given in Figs. 4, 5, 6 and 7. It is clearly evident from the plots that the predicted PPV in case of blasting with 160 mm diameter along with modified blast designs (3rd set of experimental blasts) has produced the least magnitude of ground vibrations at distances varying between 100–300 m.

Table 6. Predicted maximum charge delay using regression analysis for all three sets of experiments

	1st Set of Experimental blasts		2nd Set of Experimental blasts		3rd Set of Experimental blasts	
PPV –>	5 mm/s	10 mm/s	5 mm/s	10 mm/s	5 mm/s	10 mm/s
Dist. (m)	Qmax (kg)	Qmax (kg)	Qmax (kg)	Qmax (kg)	Qmax (kg)	Qmax (kg)
100	2.77	8.05	3.60	10.16	9.12	22.1
125	4.33	12.57	5.63	15.87	14.25	34.51
150	6.23	18.10	8.10	22.85	20.53	49.69
175	8.48	24.64	11.03	31.10	27.94	67.63
200	11.08	32.18	14.40	40.63	36.49	88.34
225	14.02	40.73	18.23	51.42	46.18	111.80
250	17.31	50.28	22.51	63.48	57.01	138.02
275	20.94	60.84	27.23	76.81	68.99	167.01
300	24.93	72.41	32.41	91.41	82.10	198.75
325	29.25	84.98	38.04	107.28	96.35	233.26
350	33.93	98.55	44.11	124.41	111.75	270.53
375	38.95	113.14	50.64	142.82	128.28	310.55
400	44.31	128.72	57.62	162.50	145.96	353.34
425	50.03	145.32	65.04	183.45	164.77	398.89
450	56.08	162.92	72.92	205.66	184.73	447.20
475	62.49	181.52	81.25	229.15	205.82	498.26
500	69.24	201.13	90.03	253.91	228.06	552.09

The results obtained from the 4th set of experimental blasts clearly demonstrated a significant decrease in the vibration magnitudes by using splitting of total charge in a round of blast. A plot of vibration data recorded at different distances from blast conducted with splitted total charge and without splitting is shown in Fig. 8. From the plot, it can be seen that a significant decrease (34–58%) in the magnitude of vibrations has been achieved by splitted total charge blast using electronic detonators. The decrease in recorded PPV was 34% at 100 m which, then gradually increased up to 58% at 400 m.

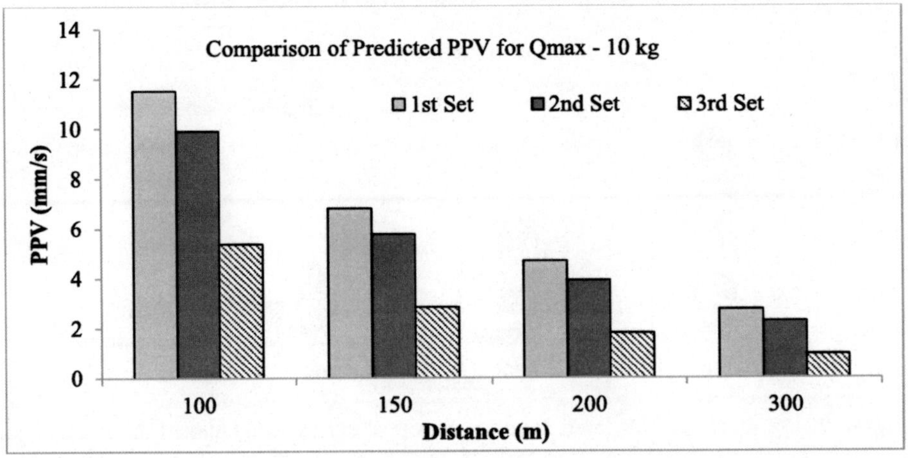

Fig. 4. Plot of predicted PPV levels for different sets of experimental blasts [Max. Charge per delay - 10 kg]

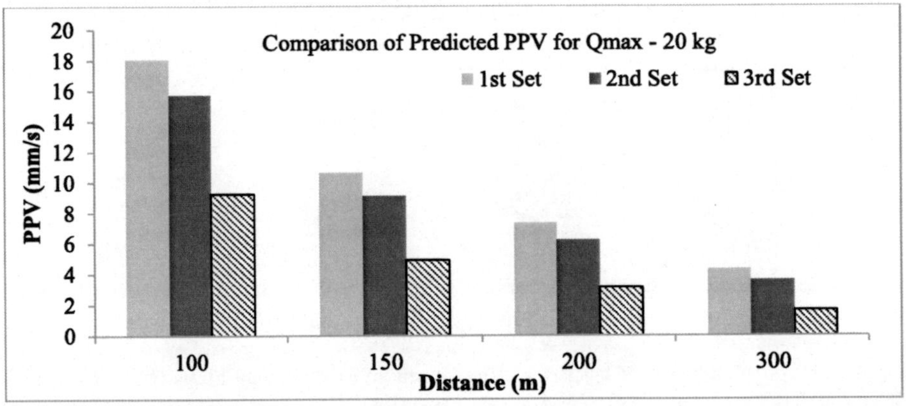

Fig. 5. Plot of predicted PPV levels for different sets of experimental blasts [Max. Charge per delay - 20 kg]

More research in this area is required to be carried out with splitting of total charge in a round of blast to establish a standard operation practice.

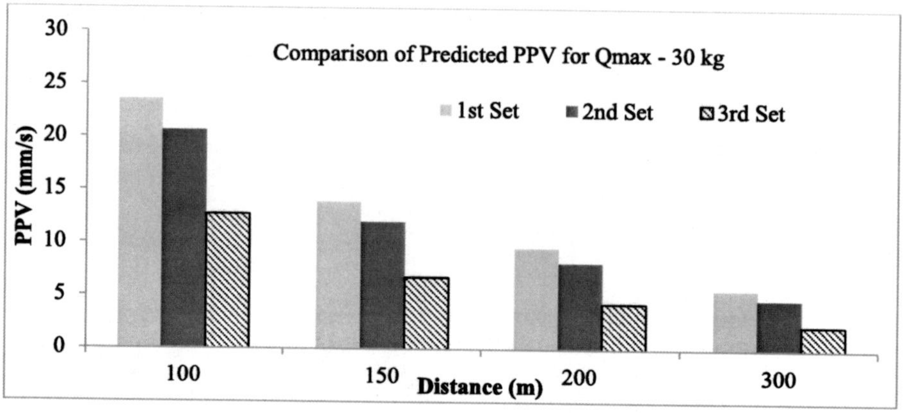

Fig. 6. Plot of predicted PPV levels for different sets of experimental blasts [Max. Charge per delay - 30 kg]

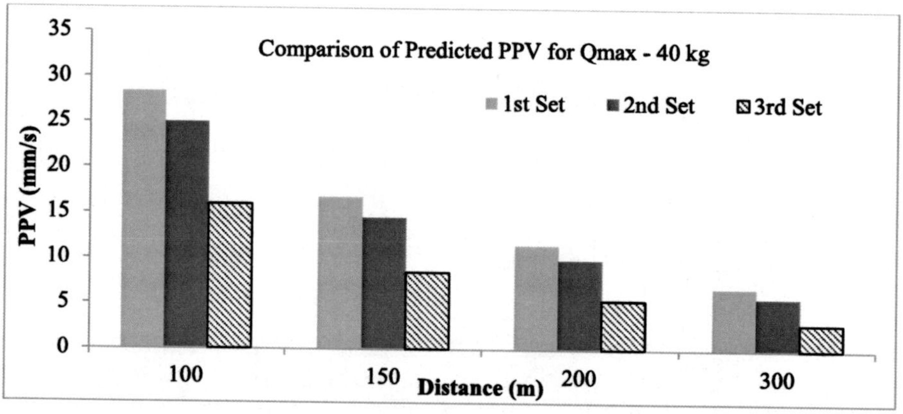

Fig. 7. Plot of predicted PPV levels for different sets of experimental blasts [Max. Charge per delay - 40 kg]

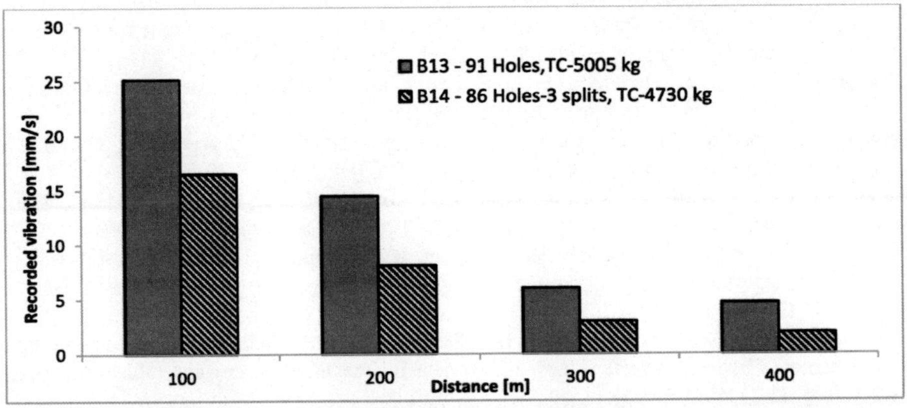

Fig. 8. Plot of recorded vibration data with respect to distance from the blast conducted with splitted total charge and blast without splitting

6 Conclusion

The results obtained in the experimental blasts and analysis of the recorded vibration data provide valuable insights into the effects of various parameters and charging patterns on blast-induced ground vibrations in the mine. The findings indicate that the initial blast designs using 260 mm hole diameters were not optimized thereby generating higher ground vibration magnitudes. The subsequent experiments with smaller hole diameter (160 mm) also did not reduce the magnitude of ground vibrations significantly. However, when combined with the reduced confinement values with a lesser number of holes, conversely, total charge in the blasting round and optimized charge factor, a significant decrease in the ground vibration magnitudes was observed. The innovative technique of segmenting the number of holes in a blasting round using electronic detonators reduced the vibration magnitudes significantly compared to other blasts conducted without segmented total charge. The potential of this technique in reducing vibration magnitudes can be explored further.

Overall, these findings emphasize the importance of optimized blast design parameters, charging patterns, and confinement to minimize ground vibrations. Further research and optimization are needed to achieve further reduction in vibrations and establish standard practices for splitting the total charge in the blasts into a number of small segments maintaining production and productivity of the mine ensuring complete safety of the structures in the vicinity.

References

Bender, W.L.: Understanding blast vibration and airblast, their causes, and their damage potential. In: Workshops of the Golden West Chapter of the International Society of Explosives Engineers (2007)

Bhagat, N.K., Mishra, A.K., Singh, M.M., Rana, A., Singh, P.K.: Innovative directional controlled blasting technique for excavation of unstable slopes along a busy transportation route: a case study of Konkan Railway in India. Min. Metall. Explor. **37**, 833–850 (2020a)

Bhagat, N.K., Singh, M.M., Rana, A., Tewari, S., Singh, P.K.: Blasting technique for stabilizing accidentprone slope for sustainable railway route. Curr. Sci. **118**(6), 901–909 (2020b)

Dindarloo, S.R.: Peak particle velocity prediction using support vector machines: a surface blasting case study. J. South. Afr. Inst. Min. Metall. **115**(7), 637–643 (2015)

Duvall, W.I.: Vibrations from Instantaneous and Millisecond-Delayed Quarry Blasts (No. 6151). US, Department of the Interior, Bureau of Mines (1963)

Hasanipanah, M., Faradonbeh, R.S., Amnieh, H.B., Armaghani, D.J., Monjezi, M.: Forecasting blast-induced ground vibration developing a CART model. Eng. Comput. **33**, 307–316 (2017)

Himanshu, V.K., Roy, M.P., Mishra, A.K., Paswan, R.K., Panda, D., Singh, P.K.: Multivariate statistical analysis approach for prediction of blast-induced ground vibration. Arab. J. Geosci. **11**, 1–11 (2018)

Hosseini, S., Pourmirzaee, R., Armaghani, D.J., Sabri Sabri, M.M.: Prediction of ground vibration due to mine blasting in a surface lead–zinc mine using machine learning ensemble techniques. Sci. Rep. **13**(1), 6591 (2023)

Konya, C.J., Walter, E.J.: Rock blasting and overbreak control (No. FHWA-HI-92–001; NHI-13211). United States. Federal Highway Administration (1991)

Mandal, S.K., Singh, M.M., Bhagat, N.K.: Impact of single-hole and multi-hole blasting on vibration parameters. J. Mines Metals Fuels **56**(7&8), 122–128 (2008)

Monjezi, M., Hasanipanah, M., Khandelwal, M.: Evaluation and prediction of blast-induced ground vibration at Shur River Dam, Iran, by artificial neural network. Neural Comput. Appl. **22**, 1637–1643 (2013)

Roy, M.P., Mishra, A.K., Agrawal, H., Singh, P.K.: Blast vibration dependence on total explosives weight in open-pit blasting. Arab. J. Geosci. **13**(13), 531 (2020)

Roy, P.P., Sawmliana, C., Singh, R.K.: Strategic planning to reduce ground vibration, air overpressure and flyrock in a mine at a sensitive area. Curr. Sci. **123**(8), 995 (2022)

Seferoğlu, M.T., Seferoğlu, A.G., Akaryali, E.: Investigation of the structural damage caused by blasting activities in tunnels to surrounding buildings: the case of Gümüşhane belt highway. Gümüşhane Üniversitesi Fen Bilimleri Dergisi **13**(1), 221–234 (2023)

Siskind, D.E., Stachura, V., Nutting, M.: The Generation of Low-Frequency Long-Duration Vibrations from Surface Mine Blasting at Blanford. US Bureau of Mines, Washington, DC (1985)

Siskind, D.E., Stagg, M.S., Kopp, J.W., Dowding, C.H.: Structure Response and Damage Produced by Ground Vibrations from Surface Blasting, RI 8507. US Bureau of Mines, Washington, DC (1980)

Yue, H.Z., Yu, C., Li, H.B., Zhou, C.B., Chen, S.H., Shao, Z.S.: The effect of blast-hole arrangement, delay time, and decoupling charge on rock damage and vibration attenuation in multihole blasting. Shock. Vib. 1–18 (2022)

Innovative Blast Design for Reduction of Boulder Generation in Opencast Excavations

Vivek K. Himanshu(✉), Ashish K. Vishwakarma, C. Sawmliana, R. S. Yadav, M. P. Roy, and A. K. Mishra

CSIR-Central Institute of Mining and Fuel Research, Dhanbad 826001, India
vivekbit07@gmail.com

Abstract. The overall economy in mining operation is being affected by rock fragmentation. The size of fragments affects the productivity of excavators, crushers and mills. The generation of oversize boulders additionally need secondary breakage mechanism. So, the reduction of boulder generation during the bench blasting is one of the prime objectives of a blast designer. The scientific methods may be utilised to reduce the boulder generation. The geological and geotechnical parameters are assessed in this process. The charging pattern, delay timing and free face condition primarily affects the resulting blast induced fragmentation. In this paper, an innovative blast design methods used at Kanchan opencast coal mine to reduce the generation of oversize boulders, has been discussed. The sandstone rock at this mine exhibited plastic nature. This method has been designed by assessment of the reason for the boulder generation. The assessment suggested that the blasting in such rocks needed explosive energy throughout the column. The boulder generation was seen mostly from the stemming portion. Accordingly, the methods with pocket charging and satellite hole drilling were devised. The delay patterns were optimised to improve the movement of the muckpile. The devised blasting methods have addressed the issues of the boulder generation.

Keywords: Rock Fragmentation · Boulder · Blast Design · Satellite Hole · Pocket Charging

1 Introduction

Productivity of excavation by blasting depends on cycle timing of loaders and indirectly on fragment size of blasted rock. During the recent years mining industries have changed the view to assess the overall cost of production. Initially, the aim of the industries was to reduce the cost of each unit operations viz., drilling, and blasting, hauling, transportation and crushing. However, the mine to mill concept has evolved, which encompasses the total cost of production starting from mining to milling (Rafeeian & Taji 2019; Adel et al. 2006). The optimum fragment size of the blasted muckpile plays the main contribution in the overall cost during mining to milling process. The Quantity and quality of explosive is considered as the main parameter for obtaining optimum fragmented rock. However, improper confinement/coupling/impedance results in to un-necessary

A. Sinha et al. (Eds.): AMC 2023, SPEES, pp. 441–451, 2023.
https://doi.org/10.1007/978-3-031-46966-4_36

wastage of explosive energy (Himanshu et al. 2021a). Judicious design with utilization of proper explosive plays important role over here. Over the years, researchers have developed different techniques for prediction and thereby improvement of blast induced rock fragmentation. The basic and well known fragmentation prediction model is based on Kuz-Ram model, which is a combination of three different equations as Kuznetsov's equation, Rosin Rammler equation and uniformity index equation (Kansake et al. 2016). This prediction model consists of different parameters viz., rock parameters, explosive parameters and blast design parameters. The subsequent modifications in the fragmentation prediction models have been suggested by the researchers during the years (Oucheterlony & Sanchidrian 2018). Further, Aler et al. (1996) also classified the roles of controllable and uncontrollable parameters on rock fragmentation. Otterness et al. (1991) performed an extensive study to correlate shot design parameters to fragmentation. Researchers have used different approaches to predict blast fragmentation. Koullis & Rustan (1993) developed the empirical relation for predicting the median fragment size from the blast. Morin & Ficarazzo (2006) used Monte-Carlo simulation approach for prediction of rock fragmentation. Chakraborty et al. (2004), Hudaverdi et al. (2011), Shi et al. (2012) etc. used multivariate statistical analysis approach for the prediction of rock fragmentation. Saavedra et al. (2006) used neural network based algorithms for predicting the rock fragmentation. Monjezi et al. (2009) developed a fuzzy logic based model for predicting the fragmentation from blasting. Different machine learning and hybrid algorithms have also been used by the researchers for predicting the blast induced rock fragmentation (Xie et al. 2021; Muralidhar et al. 2018). Most of the machine learning based algorithms classifies the data based on training and testing from a set of data consisting different parameters. However, the actual impact of individual parameters can be studied using numerical simulation or experimentations with the isolation of parameters. Researchers have used numerical simulation based approaches for predicting the impacts of controllable and uncontrollable parameters on blast induced fragmentation or damages (Baranwski et al. 2020; Himanshu et al. 2021b; Himanshu et al. 2022; Himanshu et al. 2023).

The analysis based on experimentation however, is still the most acceptable way to provide scientific insights of fragmentation. Singh et al. (2016) found the optimum ratio of spacing to burden in the range of 1.1 to 1.3 for excellent resultant rock fragmentation. Authors have also found the contributions of joints and fractures on uniformity index. Cho & Kaneko (2004) found using the experimentation the role of optimum delay timing induced rock fragmentation by blasting. Himanshu et al. (2021c) assessed the optimum burden during blasting with different type of explosives. The following paper is based on the analysis of experimentation done to evaluate the productivity of shovel under different density of explosive. The productivity has been correlated with the cycle timing of shovel.

In this study, the innovative blast design pattern has been formulated to improve the rock fragmentation and reduce the boulder generation from blasting. The devised techniques and experimentation has been discussed in different sections of this manuscript.

2 Site Details and Initial Observations During the Site Visit

The experimentation to reduce the boulder generation from blasting operation was conducted at one of the coal mine of South Eastern Coalfields Limited (SECL). The study site is situated in Umaria district of Madhya Pradesh. A set of 6 potential coal seams with different splits are found at this site.

The blasting at this site is practiced with blasthole diameter of 160 mm. The burden and spacing of blastholes are kept in the range of 4.0–5.0 m and 4.2–5.5 m respectively. The blastholes are charged with site mixed emulsion (SME) explosives. The charged blastholes are initiated using non-electric NONEL initiation system.

The issues of boulder generation were reported by the mine management of this site. It was observed during the visit that there is no proper movement of blast faces as well. The improper movement is one of the major cause of the boulder generation and backbreaks. The generation of backbreaks further adds to the compromise with the blasting geometry for next round of firing, and finally results into the boulder generation and thereby reduces the productivity of the excavators. The rock type at this mine site is also harder than the usual rock formation of a coal mine. The rock is mostly brittle in nature. A view of rock deposits found at this mine is shown in Fig. 1. There would be the requirement of low density explosive with less detonation pressure and higher gaseous energy for the breakage of the rock mass in such rock deposits. The requirement of explosive energy is uniform, and larger column length of the blastholes cannot be kept uncharged for the excavation of such rock deposits.

Fig. 1. View of rock formation present at the experimental site.

3 Possible Reasons for Boulder Generation

Based on the site visit and initial observations, the following may be the main reasons behind the boulder generation at this mine site:

I. **Unavailability of free faces:** In some of the blasts, there were the unavailability of the proper free faces. It is necessary to ensure the availability of free faces up to the bench height to get the proper breakage of the rock mass.

II. **Inappropriate blasting geometry:** Burden and spacing of the blastholes were not appropriate in some of the blasts.
III. **Use of higher powder factor:** The maximum powder factor for blasting at this mine site is fixed as 1.8 m^3/kg. This powder factor is not suitable for the harder rock formation of this site. Hence, the powder factor may further be reduced to get the optimum output. The column length of the explosive charge is also lesser while using this powder factor.
IV. **Higher length of uncharged column of the blasthole:** Column charging pattern is used at the mine for blasting. In the blastholes of 6 m, more than 3.5 m are uncharged column. The boulder formation takes place from this uncharged portion.
V. **Inappropriate burden movement:** The sufficient explosive charge is required in the front row of a blast face, to achieve the appropriate burden movement. Sometimes the backbreak from previous blasts increases the effective burden, which results into the inappropriate face movement. The judicious charge distribution in front and back rows of the blastholes are required to get the optimum breakage.
VI. **Restricted Maximum Charge Per Delay (MCPD) and total charge for the blasts:** The MCPD and total charge for the blasts at this mine has been fixed by the regulatory body as 40 kg and 2000 kg respectively. However, some of the blasts may require higher MCPD and total charge of explosive. Since, the area for mining at this site is small, the preparation of multiple number of faces to achieve the production target is also difficult.
VII. **Numbers of rows in a blasting face:** The number of rows in a blasting faces should not be much higher, as the higher numbers of rows will result into the restricted movement of the blast faces. However, this may be unavoidable at this mine site due to the unavailability of the larger mining area.
VIII. **Incorrect firing pattern:** The existing firing pattern at this site consisted of diagonal pattern with hole to hole delay of 42 ms and row to row delay of 17 ms. This firing pattern is incorrect, and was the main reason behind the restricted movement of the blast faces. The smaller delay between the holes and higher delays between rows with incremental delay timings will give better movement of the blast faces.

4 Experimentation with the Possible Remedies to Reduce Boulder Generation

At this mine site, the experimental trials were conducted with the possible measures to improve rock fragmentation and blast face movement. Initially the modifications in delay patterns were done. The improvements in face movement were observed with the revised firing pattern. Further trials were conducted with the pocket charging and satellite hole drilling and charging pattern.

4.1 Experimentation with the Modification in the Firing Pattern

When the explosives inside the blast holes are initiated, chemical reaction starts inside it. During the chemical reaction, huge amount of gases and shock waves gets generated. These shock waves have a direct influence over the wall of the blasthole. These shock waves are responsible for the formation of new fractures. After the passing of shock wave, the explosives gases come into play. These gases enter into the fractures and help in the formation or elongation of new or existing fractures. This energy helps to elongate the existing fractures formed due to shock wave. The energy of this blasthole has to be utilised for the breakage of next blasthole. Hence, the delay timing among the holes of a row plays an immense role in the rock breakage. The higher delay timing among the blastholes of a single row may lead to the escaping of gaseous and pressure energy. Moreover, it cannot be utilised for the breakage of adjacent holes. The lesser delay timing among blastholes of a row will increase the blast induced ground vibration. Hence the delay timing between the holes of a row should be optimised. Further, the movement of burden requires time. Different researchers have found that the movement of burden (rock from free face to upcoming row) takes around 4–12 ms per meter of burden. For example, a burden of 3 m will require around 36 ms time for its movement. Hence the delay timing between the rows should also be optimised accordingly.

To improve the burden/ face movement and rock breakage, the modifications in the delay pattern were carried out. Total 6 blasts were conducted with the modified delay pattern. In the modified pattern, the delay timing among the holes of a row was changed to 17 ms instead of 42 ms. The delay timings between the rows were modified to 42 ms, 59 ms and 84 ms instead of 17 ms. The incremental delay in the subsequent rows were given to ensure the movement of the sufficient number of blastholes of the front rows before the firing of the holes of the back row. The improvements in the movement of the blast faces were observed with the revised firing pattern. A view of the blast face movement recorded using the revised firing pattern is shown in Fig. 2.

Fig. 2. View of the face movement results while taking blasts using the modified firing pattern.

4.2 Experimentation with Pocket Charging and Satellite Hole Drilling and Blasting Pattern

The diameter of the blasthole significantly influence the resulting fragmentation size from rock blasting. It is the basic parameter based on which other associated blast design parameters viz. Burden, spacing, stemming length etc., are selected. Due to explosive detonation in a blasthole, the energy or shock waves propagate radially for rock breakage. The zone of influence (ZOI) of rock breakage basically depends upon amount of explosive detonated per meter and quality of explosive. Researchers and blasting practitioners have correlated this ZOI with the blasthole diameter. The larger blasthole diameter will have greater ZOI, if the sufficient explosive column is provided to it. However, in absence of the sufficient explosive column, there would be a chance of formation of oversize boulder from the uncharged portion. The distribution of explosive charge is necessary in such cases to improve the rock fragmentation. Two techniques viz. Pocket charging and Satellite hole drilling & charging, were used to achieve the distribution of the explosive charge during the experimentation.

4.2.1 Pocket Charging Technique

The pocket charging technique is similar to the deck charging. This technique is used to reduce the boulder formation from the stemming zone or the zones with the presence of discontinuities. In this technique, a small quantity of explosive charge is placed in the stemming zone. The amount of explosive in the pocket is usually 1/7th to 1/5th of the total explosive charge in a hole. The pocket charging column are connected with the main explosive column using detonating fuse. Extra booster charge in the pocket portion would be required if the explosive charge is non cap-sensitive in nature. This technique may lead to cause flyrock ejections, if the care is not taken in stemming. The technique is advantageous in improving the rock fragmentation even with the lesser explosive charge, and thereby enhances the utility of the explosive energy. A sample charging pattern for the pocket charging technique is shown in Fig. 3.

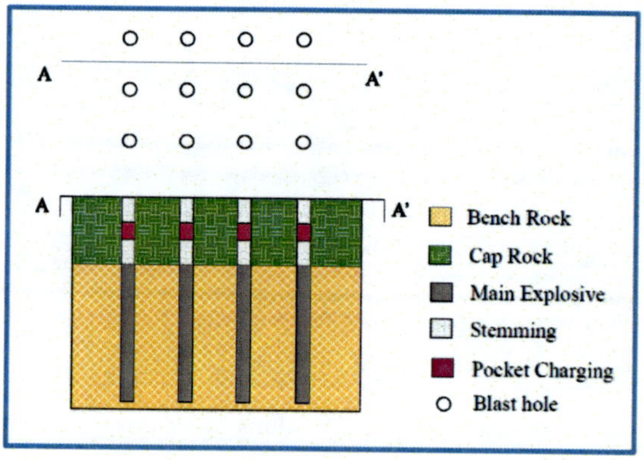

Fig. 3. Sample charging pattern for pocket charging technique.

4.2.2 Satellite/Stab Blasthole Drilling and Blasting Technique

The satellite blasthole drilling & blasting technique is another important method to reduce the boulder generation during the blasting using larger diameter blastholes. In this technique, the additional blastholes of the length less than the length of the stemming column, are drilled. Smaller explosive charge quantity (usually 1/7th to 1/5th of the total explosive charge of a blasthole) is provided in these satellite blastholes. The satellite holes should be connected to the main blasthole without any delay, for the better rock fragmentation. However, the short delay timing beyond the window of 8 ms may also be given between the blastholes and satellite holes, in order to reduce the MCPD. This technique has advantage over the pocket charging technique, as this technique gives better rock breakage due to the scattered explosive firing. However, this technique increases the drilling cost. A sample drilling and charging pattern for this technique is shown in Fig. 4.

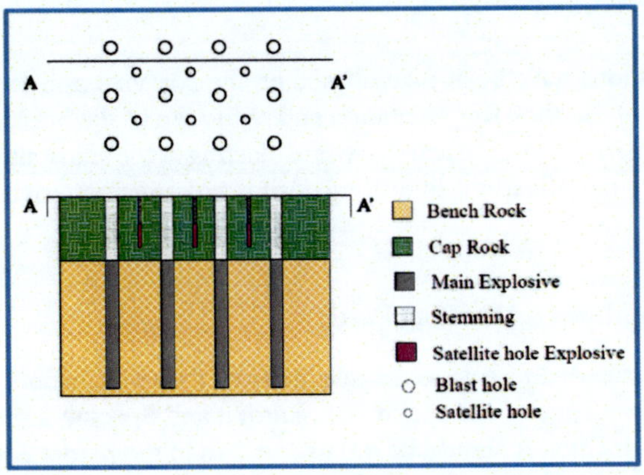

Fig. 4. Sample drilling and charging pattern for satellite hole drilling & blasting technique.

4.2.3 Results of Experimentation

During the experimentation, two trial blasts were conducted using these techniques. One trial blast was conducted using pocket charging technique. Altogether 30 numbers of blastholes were charged using this technique. The pocket was charged with the explosive quantity of 10–15 kg. The explosive charging in the front row was kept maximum with the powder factor of 1.6 m^3/kg. In the subsequent rows, it was kept 1.8 m^3/kg. In back row, the explosive was charged keeping powder factor of 2.0 m^3/kg. In this way the explosive charge was distributed to improve face movement and reduce the backbreak while keeping the overall PF of 1.8 m^3/kg. The resulting fragmentation from this blast was excellent.

In another blast, 30 satellite holes were drilled and charged. The charging of the satellite holes was done in the same fashion, as that of the pocket charging pattern. The resulting fragmentation and movement of the blast face was excellent in this blast. A view of rock fragmentation obtained using the pocket charging and satellite hole drilling & charging pattern is shown in Fig. 5.

Fig. 5. Fragmentation results from experimentation done with pocket charging and satellite hole drilling & blasting techniques.

5 Recommendations on Modified Blasting Patterns for Reduction of Boulder Generation

The recommendations have been made based on the site visit and experimentation for the improvement of the rock fragmentation and reduction of the boulder generation and thereby to improve machine productivity while blasting at the study site. The recommendations include the modifications in drilling design, charging pattern and connections of the blastholes.

5.1 Recommendations on Drilling Pattern

It has been recommended to ensure the availability of free faces upto the bench height to get the proper breakage of the rock mass. Burden and Spacing are also needed to be maintained while drilling. It should be 4.0 m × 4.2 m in the harder formation, 4.0 m × 4.5 m in medium hard rock formation and 4.5 m × 5.0 m in softer rock formation. For the blastholes of depth 4 m or less, the burden-spacing should not exceed 4.0 m × 4.0 m. It has been recommended to use the satellite hole drilling and blasting technique for effective distribution of explosive charge, and thereby improvement of the rock fragmentation. The blast design pattern shown in Fig. 6 may be used for practice.

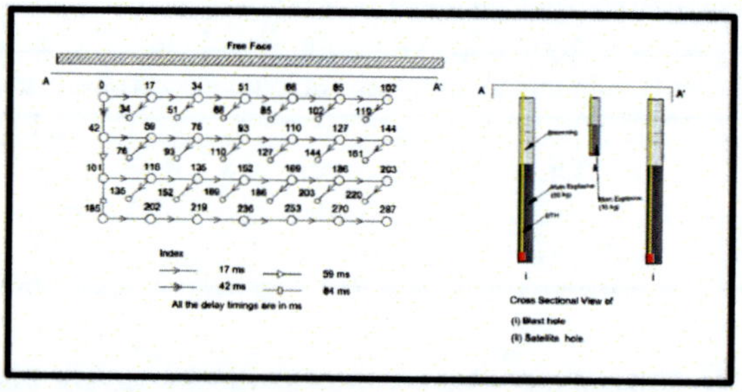

Fig. 6. Suggested charging and firing pattern with satellite hole drilling for hard rock excavation at the study site.

5.2 Recommendations on Charging Pattern

The powder factor (PF) of 1.55–1.6 m^3/kg has been recommended for blasting in the hard rock formations. The explosive charge should be distributed such that the front row holes are charged with lesser PF and subsequent rows with incremental higher PF. The explosive charge need to be distributed in a hole of 6 m depth to reduce the boulder formation from the stemming zone. The pocket charging technique may be used to distribute the explosive charge. The charging pattern shown in Fig. 3 may be used with judicious modification for practice at the mine.

The explosive charging for 4 m benches may be carried out using direct charging pattern. However, pocket charging pattern may also be used for the reduction of oversize boulder generation. The charging pattern shown in Fig. 7 may be used with judicious modification for practice at the mine.

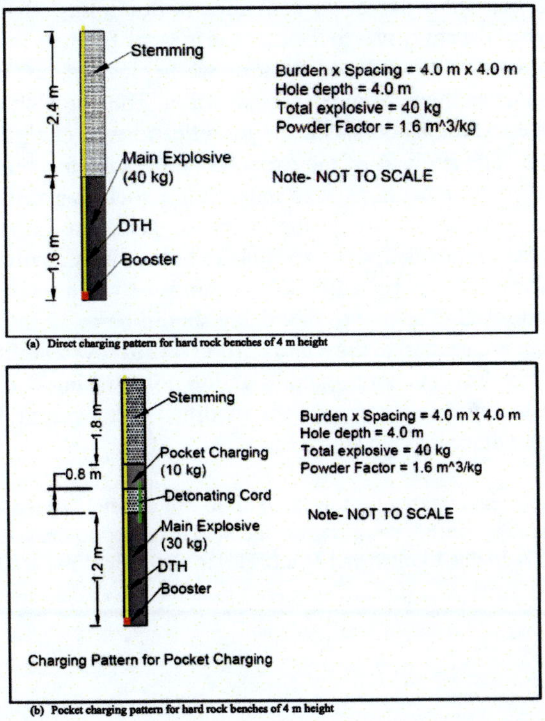

Fig. 7. Suggested charging pattern for hard rock benches of 4 m height.

5.3 Recommendations on Initiation Sequence and Delay Timings

The firing pattern with the hole to hole delay in a row of 17 ms and row to row delay of 42 ms has been recommended to be practiced for the better face movement. The incremental delay timings between second and consecutive rows have been recommended.

The row to row delay should be as 42 ms, 59 ms and 84 ms between first to second, second to third and third to fourth row respectively. The smaller delay between the holes and higher delays between rows with incremental delay timings will give better movement of the blast faces.

It has been recommended to initiate the blast from the hole having proper burden, instead of always initiating the blast from the side. The additional delay of 25 ms is recommended to be used, to reduce the MCPD while middle initiation of a blast face.

It has been recommended to use Down the Hole (DTH) delay detonator with timings of 400–500 ms instead of 200 ms. It is necessary to avoid any chances of misfire during multi-hole firing.

6 Conclusions

This study was carried out at one of the coal mine of South Eastern Coalfields Limited. The main objective of the study was to identify the reasons for the boulder generation and thereby reduced productivity of the excavators, while blasting, and thereby suggesting the suitable measures to improve rock fragmentation. The site visit was conducted for this purpose. The experimentation with the modifications in charging and connection patterns were done. The analysis of the experimental results has been carried out, and the recommendations have been made to improve the rock fragmentation. Initially the change in firing pattern and delay timings were suggested to improve the blast face movement. It was recommended to use the hole to hole delay in a row of 17 ms and row to row delay of 42 ms. The incremental delay timings between second and consecutive rows were recommended. The row to row delay should be as 42 ms, 59 ms and 84 ms between first to second, second to third and third to fourth row respectively. Further two blasting methods viz. Pocket charging and satellite hole drilling & blasting techniques were used. These techniques along with the modified delay pattern have resulted into the improved fragmentation at the study site.

Acknowledgements. The research team is thankful to M/s Baghel Infrastructures Pvt. Ltd. For sponsoring the study. The sincere co-operation and help extended to the team by the officials of M/s South Eastern Coalfields Limited and M/s BIPL are also thankfully acknowledged.

References

Adel, G., Kojovic, T., Thornton, D.: Mine-to-mill Optimization of Aggregate Production. Virginia Polytechnic Inst. and State Univ.(Virginia Tech), Blacksburg, VA (United States) (2006)

Aler, J., Du Mouza, J., Arnould, M.: Evaluation of blast fragmentation efficiency and its prediction by multivariate analysis procedures. Int. J. Rock Mech. Min. Sci. Geomech. Abs. **7**(33), 315 (1996)

Baranowski, P., Mazurkiewicz, Ł., Małachowski, J., Pytlik, M.: Experimental testing and numerical simulations of blast-induced fracture of dolomite rock. Meccanica **55**(12), 2337–2352 (2020)

Chakraborty, A.K., et al.: Parametric study to develop guidelines for blast fragmentation improvement in jointed and massive formations. Eng. Geol. **73**(1–2), 105–116 (2004)

Cho, S.H., Kaneko, K.: Rock fragmentation control in blasting. Mater. Trans. **45**(5), 1722–1730 (2004)

Himanshu, V.K., Mishra, A.K., Priyadarshi, V., Shankar, R., Yadav, R.S., Singh, P.K.: Estimation of optimum burden for blasting of different rock strata in an Indian iron ore mine. J. Geol. Soc. India **97**(7), 760–766 (2021)

Himanshu, V.K., Mishra, A.K., Roy, M.P., Vishwakarma, A.K., Singh, P.K.: Numerical simulation based approach for assessment of blast induced deformation pattern in slot raise excavation. Int. J. Rock Mech. Min. Sci. **144**, 104816 (2021)

Himanshu, V.K., Roy, M.P., Shankar, R., Mishra, A.K., Singh, P.K.: Empirical approach based estimation of charge factor and dimensional parameters in underground blasting. Mining Metall. Explor. **38**(2), 1059–1069 (2021)

Himanshu, V., Mishra, A.K., Vishwakarma, A.K., Roy, M.P., Singh, P.K.: Explicit dynamics based numerical simulation approach for assessment of impact of relief hole on blast induced deformation pattern in an underground face blast. Geomech. Geophys. Geo-energ. Geo-resour. **8**, 19 (2022)

Himanshu, V.K., Mishra, A.K., Roy, M.P., Singh, P.K.: Rock–explosive interaction during underground blasting. In: Blasting Technology for Underground Hard Rock Mining. Springer, Singapore (2023). https://doi.org/10.1007/978-981-99-2645-9_3

Hudaverdi, T., Kulatilake, P.H.S.W., Kuzu, C.: Prediction of blast fragmentation using multivariate analysis procedures. Int. J. Numer. Anal. Meth. Geomech. **35**(12), 1318–1333 (2011)

Johansson, D., Ouchterlony, F.: Shock wave interactions in rock blasting: the use of short delays to improve fragmentation in model-scale. Rock Mech. Rock Eng. **46**(1), 1–18 (2013)

Kansake, B.A., Temeng, V.A., Afum, B.O.: Comparative analysis of rock fragmentation models – a case study. In: 4th UMT Biennial International Mining and Mineral Conference, pp. 1–11 (2016)

Koulli, S., Rustan, P.A.: Computerized design and result prediction of bench blasting. In International Symposium on Rock Fragmentation by Blasting, pp. 263–271 (1993)

Monjezi, M., Rezaei, M., Varjani, A.Y.: Prediction of rock fragmentation due to blasting in Gol-E-Gohar iron mine using fuzzy logic. Int. J. Rock Mech. Min. Sci. **46**(8), 1273–1280 (2009)

Morin, M.A., Ficarazzo, F.: Monte Carlo simulation as a tool to predict blasting fragmentation based on the Kuz-Ram model. Comput. Geosci. **32**(3), 352–359 (2006)

Murlidhar, B.R., Armaghani, D.J., Mohamad, E.T., Changthan, S.: Rock fragmentation prediction through a new hybrid model based on imperial competitive algorithm and neural network. Smart Constr. Res. **2**(3), 1–12 (2018)

Otterness, R.E., Stagg, M.S., Rholl, S.A., Smith, N.S.: Correlation of shot design parameters to fragmentation. In: Proceedings of 7th Annual Research Symposium on Explosives and Blasting Technique. ISEE, Solon, pp. 179–190 (1991)

Rafeeian, N., Taji, M., Nikkhah, A.: Mine to Mill Optimisation in Sarcheshmeh Copper Mine, Kerman, Iran (2019)

Saavedra, J.C., Katsabanis, P.D., Pelley, C.W., Kelebek, S.: A neural network model for fragmentation by blasting. In Proceedings of the 8th International Symposium on rock fragmentation by blasting. Santiago, Chile, 200–206 (2006)

Shi, X.Z., Jian, Z.H.O.U., Wu, B.B., Huang, D., Wei, W.E.I.: Support vector machines approach to mean particle size of rock fragmentation due to bench blasting prediction. Trans. Nonferrous Metals Soc. China **22**(2), 432–441 (2012)

Singh, P.K., Roy, M.P., Paswan, R.K., Sarim, M.D., Kumar, S., Jha, R.R.: Rock fragmentation control in opencast blasting. J. Rock Mech. Geotech. Eng. **8**(2), 225–237 (2016)

Xie, C., Nguyen, H., Bui, X.N., Choi, Y., Zhou, J., Nguyen-Trang, T.: Predicting rock size distribution in mine blasting using various novel soft computing models based on meta-heuristics and machine learning algorithm. Geosci. Front. **12**(3), 101108 (2021)

Evaluation and Quantification of Textural Properties of a Rock and its Impact on Blast Induced Ground Vibrations

Vivek Priyadarshi[1], Ranjit K. Paswan[2], Vishal Sagar Rana[1], Sourabh Kushwaha[1], and M. P. Roy[2](✉)

[1] AcSIR, CSIR – CIMFR, Dhanbad 826001, India
mail2vivekpriyadarshi@gmail.com, {vishal.cimfr21a, sourabh.cimfr21a}@acsir.res.in
[2] CSIR-CIMFR, Dhanbad 826001, India
ranjitkp.cimfr@yahoo.co.in, mproy14@yahoo.com

Abstract. Main factors which influence the microcracking process are mineralogy, fabric and microstructures present in a rock. Crack initiation and propagation started during the dynamic loading. The objective of this paper is to investigate, how the mineralogy of rock affects the crack initiation and propagation process and transmission of induced ground vibration during the blasting of rock mass. The study was conducted in rocks containing biotite, muscovite, amphibolite, quartz, schist and gneiss. Physico-mechanical property of rock depends on the Textural coefficient which was determined to evaluate the physico-mechanical properties of the rock under study. The micro textural properties of rocks are important to understand the strength and failure behavior of the rock. The samples of metasedimentary rock were collected and their petrophysical properties and textural coefficient were determined using thin section slides and Scanning Electron Microscope (SEM). The pattern of cracks and their initiations were studied to identify the behavior of rock under induced stress conditions as well as its impact on ground vibration propagation. Crack initiation and propagation pattern studied by the digital image analysis method and measured the length and aperture of the crack. Thin section slides used to know the mineralogical composition of the rock. The study includes the petrographical and mechanical characteristics of metasedimentary rocks, various laboratory tests for mechanical characteristics and results, different types and features of explosives, variables that affect blast design, and the effects of blasting on rock mass. An empirical relation between all these parameters to know the actual behavior of rocks during the dynamic loading condition was developed to quantify the impact of microtextural properties of a rock.

Keywords: Microcrack · Dynamic Loading · Physicomechanical · Textural Coefficient · SEM · PPV · Petrographical

A. Sinha et al. (Eds.): AMC 2023, SPEES, pp. 452–466, 2023.
https://doi.org/10.1007/978-3-031-46966-4_37

1 Introduction

The petrographic characteristics of rocks play a major role in determining their technical properties. (Kowallis & Wang 1983; Åkesson et al. 2001; Åkesson et al. 2004; Rigopoulos 2009; Rigopoulos et al. 2010). Planning and designing mining and civil construction projects heavily relies on the geo-mechanically strength of the rock mass. When undertaking earth engineering studies, it is essential that rock behavior under in situ stresses must be taken into consideration (Zhang 2006). However, a rock mass cannot be viewed as a homogeneous medium because typically it is significantly heterogeneous and contains different types of rocks. Additionally, different textural characteristics (such as mineral species, grain size, form, and orientation etc.) can exist within a same rock type. Therefore, it is essential to comprehend how rock's texture affects its geo-mechanical behavior. Rock behavior is related to mechanical properties like uniaxial compressive strength (UCS), tensile and shear strength, and elastic properties, such as Young's modulus. Petrophysical properties include density, ultrasonic P-wave velocity, grain shape, grain size, binding material etc. (Tapponnier & Brace 1976; Hugman & Friedman 1979; Gupta & Sharma 2012). For essential data on rock or slope instability, failure mechanisms, strength-deformation characteristic assessment, and other engineering reasons, the mechanical characteristics and composition of the rocks are often utilized (Miskovsky et al. 2004). Strength properties of rocks are not only affected by the constituent minerals and their preferred orientation but also by the micro-cracks (Shen et al. 1995). Particularly, the failure of a rock is the dependence between the various petrographic components and micro-cracking. Many investigators attempted to determine how micro-cracks impact the Physico-mechanical properties of rocks (Onodera & Kumara 1980; Howarth & Rowlands 1986; Heap et al. 2010; Ozturk et al. 2014). Planer discontinuities in rocks that are smaller than the grain size and often exhibit some dilation but minimum displacement is called micro-cracks. (Kranz 1983). The various lithotypes' strength and durability are defined by the length, orientation, density, and distribution of their different crack types. Studies of micro-cracks may be used to predict the micromechanics of fracture and fault formation, determine the stress history of rocks, or establish comparisons between laboratory and natural environment (Nara et al. 2006; Heap et al. 2010). Ground vibration is the also one parameter which produces by the effect of the detonation of explosive in in situ rockmass which depends on textural and Physico-mechanical property of intact rock mass (Paswan et al. 2022). Blasting is the most economical viable large scale breakage process to convert intact rock to fragments (Paswan et al. 2021). In the processes of blasting explosive is used for detonation in drilled holes to breakage intact rock to fragments. Due to this energy some hazardous energy is also released which impact the environment as well as surroundings nearby structures. These hazardous energy are fly rock, dust, and ground induced vibration. A detonation of explosive in hole produce shock and vibration which may cause damage to the surrounding structures. The propagation of ground induced vibration basically depends on the heterogeneity of soil/rock, local geology of the area and the mining and excavation method. This paper deals with the evaluation and quantification of textural properties of a rock and its impact on blast induced ground vibrations.

2 Geology of Studied area

The Kayad Lead-Zinc mine is located on the Eastern fringe of Kayad village. The deposit lies between latitude N26°31′30″ and longitude E74°41′ & 74°42′. The Kayad village is 9 km NNE of Ajmer city. The location map of Kayad Mine is presented in Fig. 1.

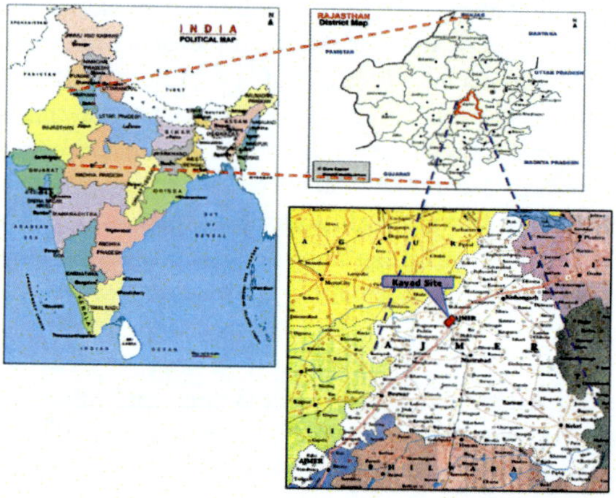

Fig. 1. Location Map, Kayad, Ajmer, Rajasthan.

Kayad is a mostly flat soil covered area except a mound of calc silicate rock (with amphibolites and pegmatite intrusions) situated to the west of the deposit. Few isolated out crops of hard and resistant migmatites are also seen further west of leasehold in the ground. In general, Meta-sedimentary formations from East to West are the host Quartz Mica Schist (QMS), Calc silicate rock (CSR) and Quartzite. A local stratigraphic succession as deciphered from surface mapping and core logging is as follows:

Intrusives Quartz/Pegmatite veins
Basic rocks (Amphibolites)
Ajmer Formation Quartz-Mica schist +/- Garnet
---------- Local shearing---------
Calc-silicate
Quartzite
----------- Shearing --------------
Anasagar Migmatites Gneisses and Migmatites
-------------------------- Base not seen-------------------------

Lead zinc mineralisation occurs mainly at the interface of Quartz Mica Schist and Calc Silicate with the earlier rock seeming to be the main host rock.

2.1 Experimental Site Details

Kayad mines of Hindustan Zinc Limited is an underground project. The project is using stope blasting with cut and fill method to excavate hard metasedimentary rock. For the experimental study we have collected 16 blast data with blasting parameter details (Table 1).

Table 1. Summary of experimental blasts details.

Blast No	No. of Holes	Explosive (Kg)	Redial Distance (m)	PVS (mm/s)
1	8	150	86.3	1.67
2	8	150	73.8	4.64
3	9	175	73.8	5.6
4	14	225	73.8	6.29
5	5	90	73.8	7.68
6	13	170	73.8	10.1
7	8	164.32	73.8	12.2
8	9	181	73.8	13.2
9	14	225	73.8	4.48
10	6	100	68.9	5.5
11	8	150	68.9	6.62
12	7	145	62	8.82
13	13	170	73.8	10.7
14	9	175	62	11.4
15	8	150	62	12.06
16	9	181	73.8	13.07

3 Texture Coefficient

An objective, quantitative way to assess grain elongation, orientation, circularity, and degree of packing is the texture coefficient. This method is introduced by Howarth and Rowlands (1987) to assess the rock fabric. The characteristics of rock texture have frequently been quantified using the texture coefficient. According to this method, the geometric relationship between the mineral grains and the matrix defines the rock's texture. Rock samples are collected from underground mines from blasting locations. These collected samples are used for making thin petrographic sections for microscopic study to determine the textural coefficient of rock samples. For the calculation of textural coefficient of rock in this paper, an open-source software Image J analysis was used. In this software petrographic image are digitally analysed with the help of the tools inside the software. For analysis, the boundary of the mineral grains was manually drawn for each sample and textural parameters like shape, size, angle of the grains was determined. Results of the calculated perimeter and area are displayed in Table 3. A representative thin section image of a rock sample of pegmatite is shown in Fig. 2a. It's digital image with marked boundary for the determination of textural parameters is shown in Fig. 2b.

Measurement and analysis of grain elongation, grain circularity, measurement and quantification of grain orientation, and weighting of results based on degree of grain packing make up the method for quantitatively assessing the texture of rocks. The formula below is used to calculate each one of these.

$$TC = AW\left[\left\{\frac{N_0}{N_0+N_1} \times \frac{1}{FF_0}\right\} + \left\{\frac{N_1}{N_0+N_1} \times AR_1 \times AF_1\right\}\right] \quad (1)$$

where, TC = texture coefficient

AW = Grain Packing Weighting

N_o = Number of grains with aspect ratio (maximum Feret's diameter or length to minimum Feret's diameter breadth) less than 2.0.

N_1 = Number of grains with aspect ratio greater than 2.0.

FF_o = Arithmetic mean of form factor of all No grains.

AR_1 = Arithmetic mean of aspect ratio of N1 grains.

AF_1 = Angle factor orientation which were computed for all N1 grains.

Area Weighting (AW) grain packing density is defiend as:

$$AW = \frac{\textit{Total grain areas within the reference area boundary}}{\textit{Total area enclosed by the reference area boundary (including matrix area)}} \quad (2)$$

By dividing the large diameter by the small diameter of the grain, the aspect ratio of the grains (AR) is calculated. The distance between two parallel lines and the tangents drawn to each grain in a variety of directions were used to calculate the maximum and minimum diameters, with the largest and smallest values being measured between these intervals. These two values are now thought of as the maximum and minimum diameter (Fig. 3).

$$AR = \frac{D_{max}}{D_{min}} \quad (3)$$

Form factor directly assesses grain roughness. If the form factor value is close to 1.0, the grain is circular. If the value is below or above 1.0, the grain is elongated as the roughness increases.

$$FF = 4\pi\frac{Area}{(perimeter)^2} \quad (4)$$

An angular parameter of grains is quantified by the Angle Factor. Only stretched and elongated grains with aspect ratios greater than 2.0 are used to calculate this. In Table 2, the angle factor (AF_1) is categorised. by a system of weights. Weighting system is defiend as angular difference between all elongated grains with high accuracy. It is divided into nine different classes, each of which is given a weight, and the absolute, acute angular difference (ß), or the range of 0° to 90°, is used to calculate the classification. An illustration of deriving the angle factor is given in Fig. 4. The calculated textural coefficient and physico-mechanical properties of studied rocks are given in Table 3.

As a result, the following equation can be used to determine the number of distinct angular differences for the group of N grains:

$$(N-1) + (N-2) + \ldots + 2 + 1 = \frac{N(N-1)}{2} \quad (5)$$

[a]

[b]

Fig. 2. [a] Thin section image of pegmatite rock sample; [b] Digital photograph with marked boundary for textural parameter determination.

By adding the products of the class weightings and the fractions of the total number of angular differences for each class, the angle factor is calculated:

$$AF = \sum_{i=1}^{n} -[\frac{X_i}{\frac{N(N-1)}{2}}]i \tag{6}$$

where:

N: Total number of elongated grains

X_i: Number of angular differences in each class;

i: Weighting factor and class number

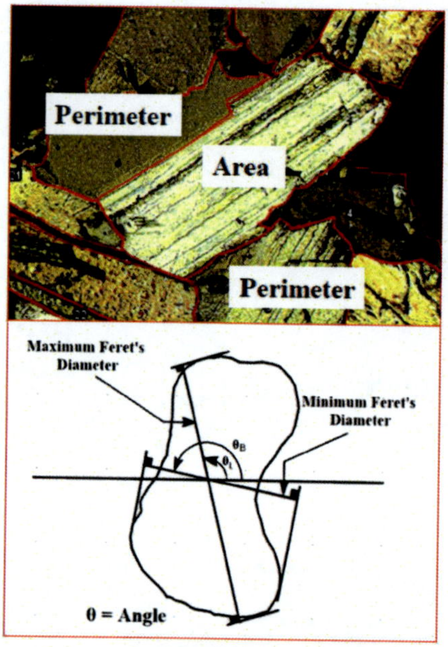

Fig. 3. The maximum and minimum diameter and grain shape of Feret.

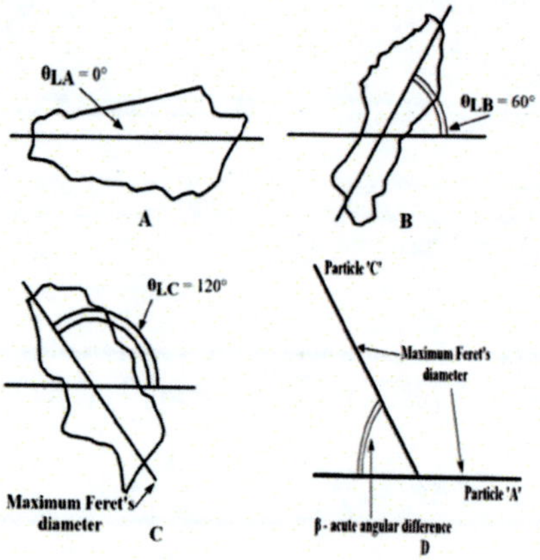

Fig. 4. Illustration of deriving the angle factor.

Table 2. Absolute, acute angular difference class range and weightings (ß).

Weight (i)	Class Range (ß)	Number (Xi)
1	0 < Qdmax < 10	6
2	10 < Qdmax < 20	6
3	20 < Qdmax < 30	3
4	30 < Qdmax < 40	5
5	40 < Qdmax < 50	2
6	50 < Qdmax < 60	1
7	60 < Qdmax < 70	2
8	70 < Qdmax < 80	2
9	80 < Qdmax < 90	4

Table 3. Texture coefficient with Physico-mechanical properties of rock collected from Kayad U/G Mine.

S. No	Sample Name	Total Area (mm^2)	AW	Form Factor	AR	AF	N0	N1	TC	UCS (MPa)	Density (g/cc)
1	Amphibole 1	2403352	0.66	0.51	2.09	0.092	10	4	0.95	66.93	2.715
2	Amphibole 2	2403363	0.98	0.47	2.68	0.075	8	12	0.96	70.13	2.961
3	Amphibole 3	2403376	0.96	0.73	2.71	0.081	17	11	0.97	71.23	3.012
4	Amphibole 4	2403394	0.8	0.68	2.83	0.089	15	13	0.99	73.41	3.107
5	CSR 1	2413722	0.71	0.49	2.70	0.038	18	26	0.64	69.87	2.657
6	CSR 2	2400282	0.98	0.54	2.01	0.027	37	20	1.19	72.94	3.225
7	CSR 3	2400452	0.81	0.72	2.11	0.033	23	11	1.23	73.45	3.314
8	CSR 4	2410463	0.89	0.76	2.19	0.45	32	19	1.26	76.37	3.375
9	Pegmatite 1	2372650	0.67	0.45	4.07	0.039	7.0	27	0.39	46.47	2.583
10	Pegmatite 2	2408279	1.02	0.5	2.00	0.039	26	15	1.33	55.71	2.629
11	Pegmatite 3	2408971	1.14	0.53	2.75	0.041	23	18	1.41	56.14	2.744
12	Pegmatite 4	2409045	1.17	0.64	2.41	0.059	29	17	1.82	59.63	2.813
13	QMS H1	2388413	0.70	0.35	2.58	0.178	4.0	7.0	0.75	20.23	2.470
14	QMS H2	2424171	0.66	0.44	2.96	0.090	13	17	0.82	21.44	2.751
15	QMS L1	2403352	0.84	0.45	2.78	0.469	13	18	0.93	29.79	2.756
16	QMS L2	2413722	0.77	0.31	3.19	0.045	12	28	1.42	32.3	2.768

4 SEM Image Analysis

The micro-crack study was perform by microscopic observation to know propagation direction of micro-cracks and SEM analysis was done to know the length and aperture of the micro-cracks. All images were processed with the help of Image J Analysis (Open-Source Software). With the help of the Image J Analysis software, the crack length and the aperture of the grain were measured. Each micro-cracks were traced manually and marked with colour for measurement.

The results of the study on micro-cracks show that under dynamic loading circumstances, deformation or collapse of rock is caused by the growth, interaction, and coalescence of a pre-exist fracture and freshly created Inter-granular and Trans-granular cracks. While Trans-granular cracks have a significant impact on crack development, inter-granular cracks have little effect on crack propagation (Rigopoulos et al. 2010). These cracks are all significant because they occurred in a direction parallel to the dominant force direction. The petrographic and SEM study and analyses revealed that the hardness of the rock is a key component in the development of fracture with trans-granular cracks developed mostly in rocks brittle by nature, which makes it break easily under loading or stress. All micro-cracks coalesce and spread through the grains, whereas inter-granular micro-cracks produced in soft, flexible, and less mineral-rich rock under loading conditions in the direction of maximum stress. These minerals are more ductile and undergo greater stress during loading.

On one side, the micro-cracks and fractures created by blasting in rocks appeared to follow the natural characteristics of the rocks without being influenced by textural or structural factors. The fractured amphibolites showed both curved-smooth and irregular erratic forms of patterns; this could be because the stress situation changed from tensile to partially shear under the control of either the rock's texture or the design and explosive parameters. Similar to this, the quartz veins associated with pegmatite had zigzag patterns of clearly defined fractures with linear and smooth profiles that may have developed under only tensile stress conditions with small changes during the rock's initiation and failure stages (Figs. 5 and 6).

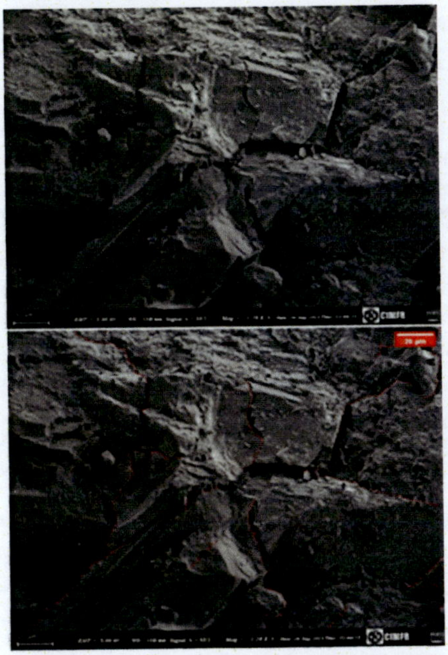

Fig. 5. Measurement of micro-crack length in SEM Images.

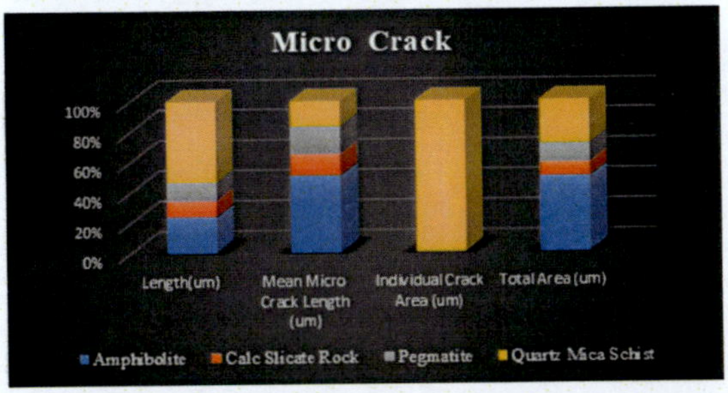

Fig. 6. Result of measurement microcrack length in SEM Images.

However, the principal axes of the propagation of cracking and fracturing along the foliation planes were visible in fragments of schist and gneissic rocks. The smooth steps-like features in the geometry and properties of the cracked surfaces may be the result of foliation planes' partial support during cracking and fracture. The majority and most clean fracture surfaces were securely supported and only fractured under tensile stress conditions.

5 Physico-Mechanical Property of Rock and Blast Induced Ground Vibration (BIGV)

Several researchers have tried to find out the influence of different rock characteristics on BIGV, generally measured in Peak Particle Velocity (PPV). Various authors have discussed the effects of rock discontinuities on blast wave propagation (Paswan et al. 2017, Roy et al. 2017, Paswan et al. 2018, Paswan et al. 2021 and Priyadarshi et al. 2022). Site factors such as geological structure, topography, and the physical and mechanical properties of rocks, in addition to the quantity of explosives used and the distance between the blasting location and the settling area, have a big impact on the degree to which ground vibrations affect a settlement area located a certain distance to the blast source. The physico-mechanical properties and recorded PPV are depicted in Table 5.

Table 5. Physico-mechanical Property of Rock and Peak Particle Velocity (PPV).

Sample Name	UCS (MPa)	Density (g/cc)	RMR	PPV (mm/s)
Amphibole 1	66.93	2.715	49	1.67
Amphibole 2	70.13	2.961	51	4.64
Amphibole 3	71.23	3.012	53	5.6
Amphibole 4	73.41	3.107	59	6.29
CSR 1	69.87	2.657	57	7.68
CSR 2	72.94	3.225	59	10.1
CSR 3	73.45	3.314	61	12.2
CSR 4	76.37	3.375	63	13.2
Pegmatite 1	46.47	2.583	68	4.48
Pegmatite 2	55.71	2.629	73	5.5
Pegmatite 3	56.14	2.744	76	6.62
Pegmatite 4	59.63	2.813	78	8.82
QMS H1	20.23	2.470	41	10.7
QMS H2	21.44	2.751	43	11.4
QMS L1	29.79	2.756	45	12.06
QMS L2	32.3	2.768	47	13.07

6 Result and Discussion

The Textural coefficient (TC) determined for all four types of rock types showed considerable variations from 0.39 to 1.82 (Table 3). These variations may be mainly due to the differences in grain shape, size and elongation of the constituent grains.

The UCS of the rocks varied from 20.23 MPa to 73.45 MPa and RMR of the rocks varied from 41 to 78. The rock densities varied in between 2.47 g/cc and 3.375 g/cc. The recorded PPVs were in the range of 1.67 mm/s to 13.2 mm/s.

The blasting-induced damage to the rock mass demonstrated the distinctive effects of textural coefficient, strengths, and foliation type on one side and the design parameters on the other, as well as the pronounced effect of geotechnical parameters of geological strata at mega-scale. Lower textural coefficient (TC) rock values were found to cause more damage. Such as Quartz-Mica-Schist may be due to lower in strengths of these rocks. Due to the movement of explosive gases in the planes created by the elongation of grains, the rocks with lower TC showed more damage. It seems that the stronger and higher TC rocks prevented the movement of explosive gases and damaging stress. Rock strata with better geotechnical qualities are seen as being resistant to damage brought on by blasting. Based on image analysis this is evident that the specific mineralogical and textural properties of rocks directly impact the initiation and growth of micro-cracks. The recorded PPV data were plotted with RMR, UCS and TC (Fig. 7, 8 & 9). This was observed that with increasing RMR, UCS and TC, PPV also increases gradually. The plot of TC vs UCS suggest that UCS also increased proportionally with TC in all the rock types (Fig. 10).

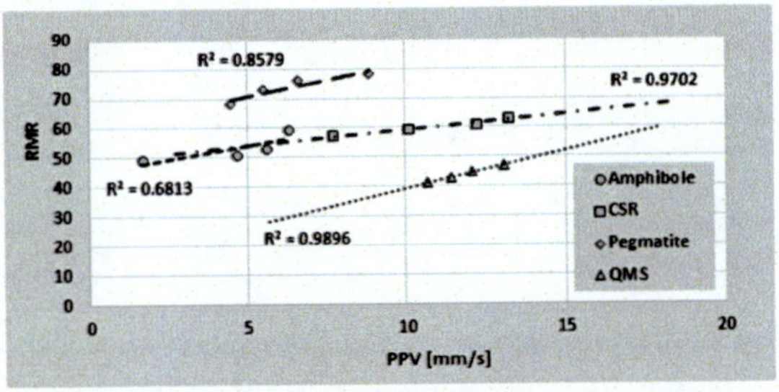

Fig. 7. PPV vs RMR.

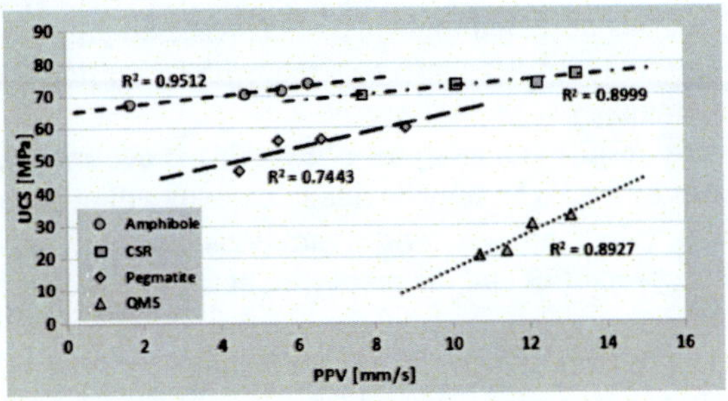

Fig. 8. Plot of PPV vs UCS.

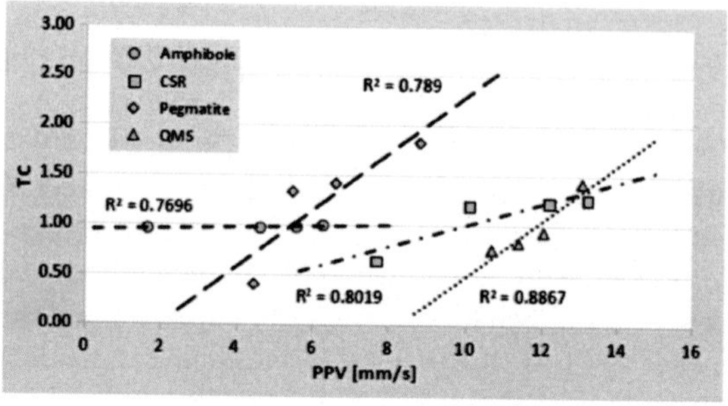

Fig. 9. Plot of PPV vs TC.

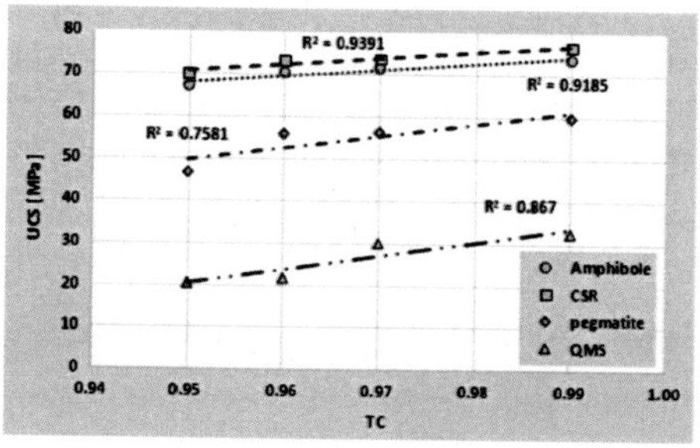

Fig. 10. Plot of TC vs UCS.

7 Conclusion

The Textural coefficient (TC) determined for all four types of rock showed considerable variations. These variations may be mainly due to the differences in grain shape, size, elongation and orientation of the constituent grains. This is evident that the mechanical properties of a rock mainly dependent of textural parameters of the rock. Under dynamic stress conditions the fractures developed in a rockmass significantly guided by the textural parameters. This was observed that UCS also increased proportionally with TC in all the rock types. The study results also reveal that with increasing RMR, UCS and TC, PPV also increases gradually.

Based on the study results and analyses of data this is evident that the variations in mechanical properties of the rocks may be due to corresponding changes in TC.

The geotechnical values showed wide variations laterally as well as vertically depending on the rock properties and the properties of foliations/ discontinuities present in the rockmass.

The scanning electron microscopy (SEM) of the major rocks associations show both smooth arched and banded types of surfacial features which indicate that the rocks have been subjected to metamorphism and deformations under directed stresses.

The microcracks and fractures developed in rocks due to blasting showed varieties of patterns appeared to follow the inherent characteristic of rocks on one side and independent of textural and structural controls. The fragmented amphibolites exhibited both curved- smooth and irregular erratic types of patterns may be due to change in stress condition from tensile to partially shear under control of either texture of rocks or design and explosive parameters.

Similarly, the pegmatite showed zig-zag pattern of well-defined cracks with linear and smooth profile may be formed under purely tensile stress conditions with minor fluctuations at initiation and failure stage of the rock.

However, fragments of QMS and CSR exhibited the major directions of propagations of cracking and fracturing along the foliation planes. The geometry and characteristics of the fractured surfaces showed smooth steps like features may be due to the partial support of foliation planes in cracking and fracturing. The clean fractures surface of strongly supported the maximum and majorities of fracturing only under tensile stress condition.

Acknowledgment. The authors express their gratitude to the mine officials for providing necessary facilities and help during field investigations. The permission of Director, Central Institute of Mining & Fuel Research, Dhanbad, India to publish the paper is also thankfully acknowledged.

References

Kowallis, B.J., Wang, H.F.: Microcrack study of granitic cores from Illinois deep borehole UPH 3. J. Geophys. Res. Solid Earth **88**, 7373–7380 (1983). https://doi.org/10.1029/JB088iB09p07373

Åkesson, U., Lindqvist, J.E., Göransson, M., Stigh, J.: Relationship between texture and mechanical properties of granites, central Sweden, by use of image-analysing technique. Bull. Eng. Geol. Env. **60**, 277–284 (2001)

Åkesson, U., Hansson, J., Stigh, J.: Characterisation of microcracks in the Bohus granite, western Sweden, caused by uniaxial cyclic loading. Eng. Geol. **72**, 131–142 (2004)

Rigopoulos, I.: Correlation between petrographic and physico-mechanical properties of the Pindos and Vourinos ophiolitic rocks and assessment of their suitability as aggregates in construction-industrial uses. PhD Thesis, University of Patras, p. 549 (2009)

Rigopoulos, I., Tsikouras, B., Pomonis, P., Hatzipanagiotou, K.: The influence of alteration on the engineering properties of dolerites: the examples from the Pindos and Vourinos ophiolites (Northern Greece). Int. J. Rock Mech. Min. Sci. **47**, 69–80 (2010)

Zhang, L.: Engineering Properties of Rocks; Geo-Engineering Book Series. In: Hudson, J., (ed.). Elsevier: Amsterdam, The Netherlands, pp. 226–230 (2006)

Tapponnier, P., Brace, W.F.: Development of stress-induced microcracks in Westerly granite. Int. J. Rock Mech. Min. Sci. Geomech. **13**, 103–112 (1976)

Miskovsky, K., Duarte, M.T., Kou, S.Q., Lindqvist, P.A.: Influence of the mineralogical composition and textural properties on the quality of coarse aggregates. J. Mater. Eng. Perform. (13), 144–150 (2004)

Shen, B., Stephansson, O., Einstien, H., Ghahreman, B.: Coalescence of fractures under shear stresses in experiments. J. Geophys. Res. **100**(B4), 5975–5990 (1995)

Onodera, T., Asoka Kumara, H.: Relation between texture and mechanical properties of crystalline rocks. Bull. Int. Assoc. Eng. Geol. **22**, 173–177 (1980)

Howarth, D.F., Rowlands, J.C.: Development of an index to quantify rock texture for qualitative assessment of intact rock properties. Geotech. Testing J. **9**(4), 169–179 (1986)

Heap, M.J., Faulkner, D.R., Meredith, P.G., Vinciguerra, S.: Elastic moduli evolution and accompanying stress changes with increasing crack damage: implications for stress changes around fault zones and volcanoes during deformation. Geophys. J. Int. **183**, 225–236 (2010)

Kranz, R.L.: Microcracks in rocks: a review. Tectonophysics **100**, 449–480 (1983)

Nara, Y., Koike, K., Yoneda, T., Kaneko, K.: Relation between subcritical crack growth behavior and crack paths in granite. Int. J. Rock Mech. Min. Sci. **43**, 1256–1261 (2006)

Howarth, D.F., Rowlands, J.C.: Quantitative assessment of rock texture and correlation with drillability and strength properties. Rock Mech. Rock Engng. **20**, 57–85 (1987)

Hugman, R.H., Friedman, M.: Effects of texture and composition on mechanical behaviour of experimentally deformed carbonate rocks. Am. Asso. Petrol. Geol. Bull. **63**(9), 1478–1489 (1979)

Gupta, V., Sharma, R.: Relationship between textural, petrophysical and mechanical properties of quartzites: a case study from North Western Himalaya. Eng. Geol. **135**, 1–9 (2012)

Erarslan, N.: Micro-structural investigation of subcritical crack propagation and fracture process zone (FPZ) by the reduction of rock fracture toughness under cyclic loading. Eng. Geol. **208**, 181–190 (2016). https://doi.org/10.1016/j.enggeo.2016.04.035

Ozturk, C.A., Nasuf, E., Kahraman, S.A.: Estimation of rock strength from quantitative assessment of rock texture. J. South. Afr. Inst. Min. Andm. **114**, 471–480 (2014)

Onodera, T.F., Asoka, K.H.M.: Relation between petrographic characteristics, engineering index properties and mechanics properties of selected sandstone. Bull. Int. Assoc. Eng. Geol. **28**, 55–71 (1980)

Paswan, R.K., Roy, M.P., Shankar, R., Singh, P.K.: Blast vibration and fragmentation control at heavily jointed limestone mine. Geotech. Geol. Eng. **39**, 3469–3485 (2021). https://doi.org/10.1007/s10706-021-01705-2

Paswan, R.K., Roy, M.P., Sarim, M.D., Kumar, S.: Blast induced damage and role of discontinuities on pre-split blasting at Rampura-Agucha Pb-Zn Open pit mine. In: International Conference on NexGen Technologies for Mining & Fuel Industries (NxGnMiFu-2017). February 15–17: Vigyan Bhawan, pp. 281–290. New Delhi, India (2017)

Roy, M.P., Paswan, R.K., Sarim, D., Kumar, S.: geological discontinuities, blast vibration and fragmentation control – a case study. In: Proceeding of 7th Asian Mining Congress. November 08–11, pp. 315–323 (2017)

Paswan, R.K., Roy, M.P., Singh, P.K.: Geotechnical characterisation of Strata for Pre-split blasting to control Pit-wall damage. Trans. Tech. Publ. Min. Geolog. Metall. Inst. India (MGMI), pp. 24–39 (2018)

Paswan, R.K., Kushwaha, S., Priyadarshi, V., Tudu, S., Kumar, S., Roy, M.P.: Characterization of blast waves based on rock type in Indian geo-mining scenario. In: International Symposium on Recent Trends in Mineral Industry – 2022. September 23–24, 2022. Department of mining Engineering, UCE, OU, Hyderabad, The Alumni of Mining Engineers' Association (OU & KU); Alumni Association of University College of Engineering (2022)

Priyadarshi, V., Paswan, R.K., Rana, V.S., Kushwaha, S., Yadav, R.S., Roy, M.P.: Impact of rock mass properties on blast induced fragmentation and ground vibration. In: International Symposium on Recent Trends in Mineral Industry – 2022. September 23–24, 2022. Department of mining Engineering, UCE, OU, Hyderabad, The Alumni of Mining Engineers' Association (OU & KU); Alumni Association of University College of Engineering (2022)

Paswan, R.K., Dubey, R.K., Roy, M.P., Singh, P.K.: Geotechnical characterization of strata to control pit-wall damage at a large Pb-Zn Opencast Mine. In: 5th Asia-Pacific Symposium on Engineering Blasting & 9th International Conference on Physical Problems of Rock Destruction. September 26–29, 2017, Zhoushan, Zhejiang, China (2017)

Controlled Blasting Within Danger Zone of 500 M From Dwelling Vis-À-Vis Production Enhancement-Problem and Challenges

Aditya Rana(✉), Saikat Banerjee, Arvind Kumar, Atul Singh, and Chhangte Sawmliana

Rock Excavation Engineering Research Group, CSIR-Central Institute of Mining and Fuel Research (CSIR-CIMFR), Barwa Road, Dhanbad 826001, Jharkhand, India
{adityarana,saikatbanerjee,akumar,csawmliana}@cimfr.nic.in

Abstract. Controlled blasting has become imperative in many Indian open cast mines as the workings benches are within/approaching the danger zone of 500 m. Accordingly the blasting impacts in the form of ground vibration, flyrock and air overpressure/noise should be controlled within the permissible limits for the safety of nearby dwellings. Flyrock can be controlled by proper blast design and application of effective muffling arrangements. Ground vibration can also be limited within the permissible limits by applying various techniques for various structures as per DGMS Circular No. 07 of 1997. However, an undue reduction in progress is often observed due to inevitable reduction in blasting geometry during controlled blasting operations. Hence, it is need of hour to pace up the production, particularly for the removal of overburden in coal mines to attain the targeted production. This paper discusses the different techniques used for reduction of blasting impacts at different opencast mines. Nevertheless, it also highlights the problems and challenges faced for maintaining the production rate within 500 m from the dwellings.

Keywords: Danger Zone · Controlled Blasting · Vibration · Flyrock · Air Overpressure · Production

1 Introduction

Drilling and blasting method is the most common method used for extraction of minerals and removal of overburden in opencast coal mines. Now a days, despite the availability of latest mechanical means to cut harder rock strata, drilling and blasting method is still preferred due to its cost effectiveness and productivity. Blasting operations are synonymous to nuisances such as ground vibration, flyrock, noise and dust generations. In India, many opencast mines are operating in the vicinity of the residential houses where the blasting impacts have to be controlled for the safety of the nearby denizens and their houses/structures. The regulation No. 196 (2b) of Coal Mines Regulation (CMR) 2017 declared the radius of 500 m from the place of blast firing as 'Danger Zone' for opencast coal mines. Further, the reg. No. 196(3) states that any building/structure of permanent

A. Sinha et al. (Eds.): AMC 2023, SPEES, pp. 467–479, 2023.
https://doi.org/10.1007/978-3-031-46966-4_38

nature (not belonging to the owner lies within the danger zone) should be protected by controlling the aggregate maximum charge fired per delay and per round. These values shall not exceed the amount fixed by the Chief Inspector. The blasting permission should be granted by Chief Inspector on the basis of a scientific study. Therefore, design of controlled blasting patterns is necessary whenever blasting operations are to be carried out within the Danger Zone.

The controlled blasts within the danger zone are designed to control ground vibration, flyrock and noise (in the form of air overpressure) within the safe level. Practically dust control in blasting operations is difficult, however, it is important considering human health safety.

In India, three ground vibration standards are stated respectively by Bureau of Indian Standard (BIS), Directorate general of Mines Safety (DGMS) and CSIR-Central Institute of Mining and Fuel Research (CSIR-CIMFR) are pertinent as indicated in Tables 1, 2 and 3, respectively. DGMS vibration standard is statutorily followed for designing controlled blasting patterns and acquiring blasting permission from the authority in opencast mines/quarries. The circular categorises the structure into two types depending upon the ownership. Each category is further classified into three types of structure. The permissible peak particle velocity is governed in the circular in accordance with the ownership, construction type and dominating frequency. Flyrock can be effectively controlled using proper blast design patterns and muffling. Several researchers also tried to predict flyrock range using empirical equations as well as artificial intelligent (Lundborg et. al. 1975; Richard & Moore 2002; Little, 2007; McKenzie 2009). The noises generated from blasting also called as air overpressure are discontinuous impulse sound, consisting of both audible and sub-audible pressure waves. In India, the standard for safe level of air overpressure for surface mines are not well established. The generally used air overpressure limits recommended by Siskind et.al. (1980) and Oriard (2002) for surface mine blasting are postulated in Tables 4 and 5, respectively. Table 6 depicts Central Pollution Control Board (CPCB), India's permissible levels for noise exposure in a work zone area as prescribed under Model Rules of Factories Act, 1948. The authors have carried out multiple scientific studies at several Indian opencast mines for developing safe controlled blasting patterns within the Danger Zone. It has been observed that production has been affected due to inevitable reduction in blasting geometry to control ground vibration within the safe level, particularly when blasting is to be carried out in the close proximity of the dwellings. This paper discusses the case study of three opencast coal mines where different controlled blasting techniques were applied for restricting ground vibrations and the associated challenges within the Danger Zone.

Note:

1. *No exposure in excess of 140 dB peak sound pressure level is permitted.*
2. *For any peak sound pressure level falling in between any figure and the next higher or lower figure as indicated in column 1, the permitted number impulses or impacts per day is to be determined extrapolation on a proportionate basis*

Table 1. DGMS-prescribed permissible limit of ground vibrations (Technical Circular No. 7, 1997)

Type of structure	Dominant excitation frequency, Hz		
	<8 Hz	8–25 Hz	> 25 Hz
(A) Buildings/structures not belong to the owner			
Domestic houses/structures (Kuchcha, brick & cement)	5	10	15
2. Industrial buildings	10	20	25
3. Objects of historical importance and sensitive structures	2	5	10
(B) Buildings belonging to owner with limited span of life			
1. Domestic houses/structures	10	15	25
2. Industrial buildings	15	25	50

Table 2. Peak Particle velocity as damage criteria for different types of rocks (after Indian Standard Institution, IS: 6922, Sect. 4.1.1.2, 1973)

Soil, weathered or soft rock conditions	70 mm/s
Hard rock conditions	100 mm/s

Table 3. CMRI prescribed limit of ground vibrations (after Dhar et. al. 1993)

Type of Structures	PPV (mm/s)	
	<24 Hz	>24 Hz
Domestic houses, dry well interior, construction Structures with plasters, bridge	5.0	10.0
Industrial buildings, steel or reinforced concrete structures	12.5	25.5
Object of Historical importance, very sensitive Structures, more than 50 years old construction and Structures in poor state condition	2.0	5.0

2 Case Study

Ground vibrations can be controlled using various mechanism. The most common method is reduction of drilling geometry i.e. use of smaller drill hole diameter, shorter hole depth and reduced burden and spacing. Reddy et. al (2001) studied the effect of trench cutting on ground vibration and found significant reduction in ground vibration by trench cutting. Nased on the numerical simulation study carried out by Venkatesh and Rao (2008), the magnitude of ground vibration could be reduced by 50 to 60% in surface mine if hole depth is equal to depth of the trench cutting. Some researchers also stated

Table 4. Air overpressure limits recommended by USBM for surface mining (RI 8485 – Siskind et. al.)

134 dB	0.1 Hz high pass measuring system
133 dB	2.0 Hz high pass measuring system
129 dB	6.0 Hz high pass measuring system
105 dB	C-slow weighting scale on a sound level meter
(Events less than or equal to 2 – sec duration)	

Table 5. Typical air overpressure criteria (After Oriard 2002)

171 dB	General window breakage
151 dB	Occasional window breakage
140 dB	Long-term history of application as a safe project specifications
134 dB	Bureau of Mines recommendation following a study of large-scale surface mine blasting

Table 6. Central Pollution Control Board (CPCB) permissible levels for noise exposure for work zone area

Peak sound pressure in dB	Permitted number of impulse or impact/day
140	100
135	315
130	1000
125	3160
120	10000

reduction in ground vibration by using air decking technique (Melnikov and Marchenko 1971, Melnikov et. al 1979 and Masoud Monjezi et. al 2022). Fei Lin et al. (2022) used a high-precision digital electronic detonators and simulated the delay sequence to reduce ground vibration by 69.70%.

In the present study, the effect of reduced drilling geometry, pre-fracture line (modified pre-splitting), a modified decouple charge using plastic conduit and segregation of blasting round using electronic detonators were investigated to control blasting vibration. The studies were conducted in Pachhwara (North) Coal Mine, Jharkhand; Talabira II & III OCP, Odisha and Chatti Bariatu Coal Mine Project, Jharkhand.

2.1 Pachhwara (North) Coal Mine

Pachhwara (North) Coal Mine is owned by M/s West Bengal Power Development Corporation Limited (WBPDCL). The mine is located in Pakur District, Jharkhand State. The residential houses and other important structures were present within 100 m from the studied blasting sites. The study was conducted to develop controlled blasting patterns for safe blasting operations within the blasting zone of 50–100 m from the nearby dwellings. In order to minimize ground vibrations, three methods (1) Reduced burden and spacing with shorter hole depth, (2) Modified decoupled charged using plastic pipes and (3) Pre-fracture line using modified pre-splitting mere studied. The blast hole diameter used in all the blast was 160 mm.

In reduced burden and spacing, burden varied from 1.5 to 2.0 m and spacing varied from 2.0 to 2.5 m. Hole depth varied from 2.5 to 3.0 m. The total number of holes in the blasting round was also restricted to 5–12 only. The explosive charge per hole and maximum charge per delay varied from 5.00 to 7.00 kg whereas the total explosive charge in the blasting round varied from 25.00 to 84.00 kg.

In modified decoupled charge, the bore hole pressure in the back side of free face side was attempted to reduce by using plastic conduit as shown in Figs. 1, 2 and 3. SME explosive charge was made in the form of cartridge with 90 mm diameter. Three plastic pipes of 40 mm diameter were tied around the explosive charge in such a manner that the maximum explosive energy is transferred towards free face with a vertical air decking in the back side and around the explosive charge. In this manner, an air gap was formed throughout the explosive charge length forming decoupled charge. Nevertheless, the explosive charge was also touching the blast hole wall towards free face side for maintaining bore hole pressure for the front burden.

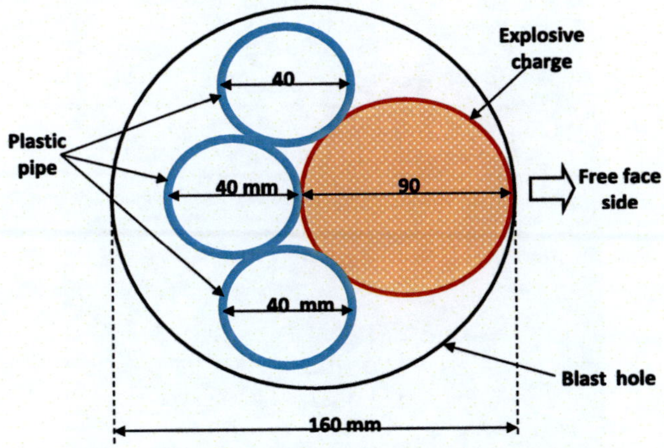

Fig. 1. Modified decoupled charge using plastic pipes

The main objective was to reduce ground vibration towards back side of the free face while achieving proper fragmentation. Three experimental blasts were carried out using this mechanism.

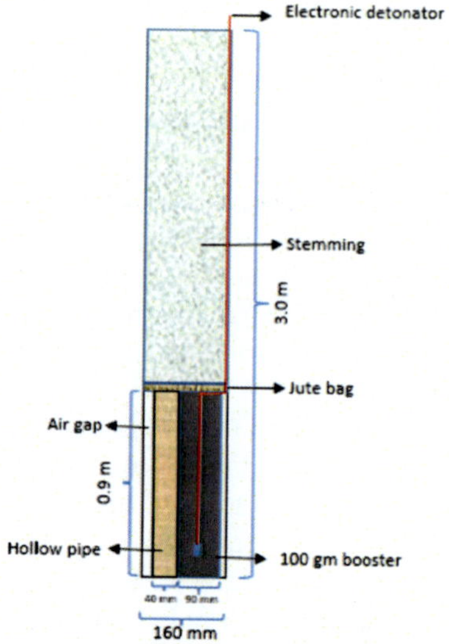

Fig. 2. Charging pattern of hole for modified decoupled charge using plastic conduit

Fig. 3. Preparation of explosive charge for modified decoupled charge

The experiments were also conducted to study the effect of fracture line on ground vibration using modified pre-splitting technique (creating a fracture line prior to the blast). The main objective was to create a fracture line between the blasting bench and residential houses to minimize the propagation of blast waves. The fracture line was blasted before the production blast. In the modified pre-split blasting, holes were drilled in a single line behind the production/main blast. Hole depth was reduced to 3.0 m (same as the main blast) in order to minimize ground vibration. Holes were charged alternately leaving uncharged hole between the charged holes. A light explosive charge was used to create a fracture line.

The ground vibrations from the different experimental blasts were monitored at the distances of 50 to 175 m from the experimental sites. The plot of ground vibration data against distance of ground vibration monitoring points for the different experimental blasts are given in Figs. 4, 5 and 6. The ground vibration recorded at the distance of 50 m from the blasting site for different experimental blast varied from 6.75 to 36.95 mm/s. The overall reduction in ground vibration due to pre-fracturing was difficult to assess. However, there was significant reduction in ground vibration due to decoupled charged using plastic conduit. The magnitude of ground vibration recorded at 50 m in two blasts were less than 10 mm/s (i.e. 6.75 mm/s and 9.81 mm/s respectively). Significant reduction in blasting vibration was also observed due to the reduced burden and spacing approach.

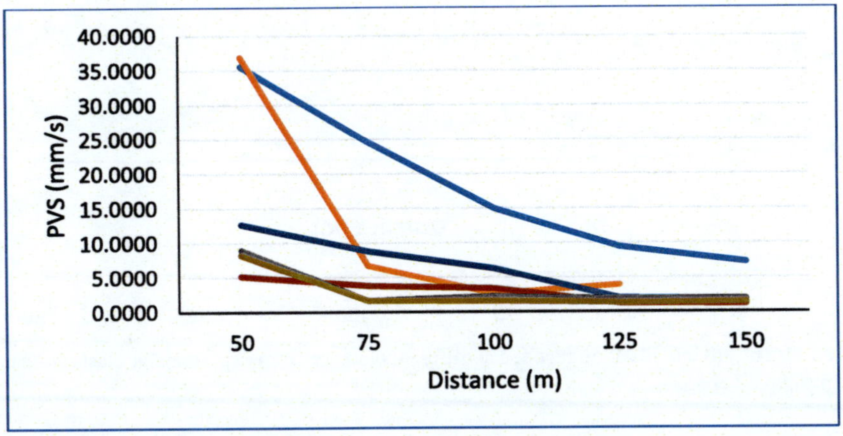

Fig. 4. Ground vibration v/s distance of measurement using reduced burden and spacing for different blasts

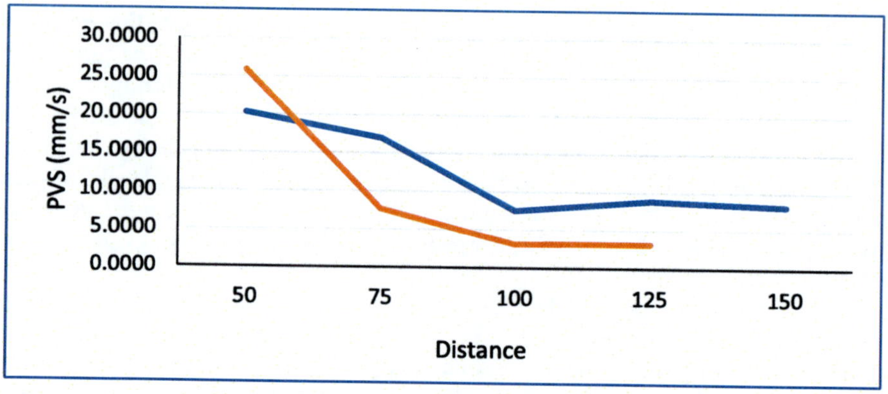

Fig. 5. Plot of ground vibration data v/s distance of monitoring point after the pre-fracture/modified pre-split blasting

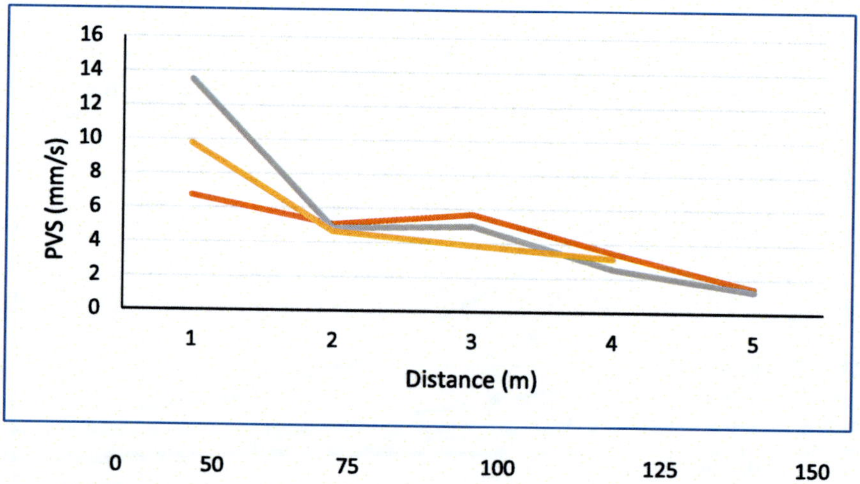

Fig. 6. Plot of ground vibration recoded at different distance from experimental blasts conducted with decoupled charge

2.2 Talabira II and III OCP

Talabira II & III OCP of M/s NLC India Limited is located in Jharsuguda District, Odisha State. It was required to developed controlled blasting patterns within 500 m from the nearby villages. The experimental blasts were conducted using reduced burden and spacing with lesser hole depth. The hole diameter used was 160 mm and hole depth varied from 3.0 to 5.0 m. Burden varied from 2.5 to 3.0 m and spacing varied from 3.0 to 3.5 m. Depend on the hole depth, the explosive charge per hole varied from 12.00 to 25.00 kg. Electronic detonators were used to divide a blasting round into several sub rounds by providing a time gap of 2 to 5 s. Hence, the total number of holes fired in a sub-round varied from 20 to 50 whereas the total number of holes in the whole rounds were up to 160. Ground vibrations were recorded at the distance of 100 to 500 m from the blasting sites.

The magnitude of ground vibration recorded at the distance of 100 m, 150 m and 200 m from the blasting site for different experimental blasts are given in Table 7. It was observed that using reduced burden and spacing and electronic detonators, the magnitude of ground vibration recorded at the distance of 100 m varied from 5.55 to 12.89 mm/s. The magnitude of ground vibration using normal burden and spacing value are higher than reduced burden and spacing. The ground vibration wave recorded by dividing the blasting round into four sub-rounds using electronic detonator is given in Fig. 7. It clearly represents that the recorded waves are separated in each sub-round and the effect of total charge on ground vibration can be controlled using electronic detonators.

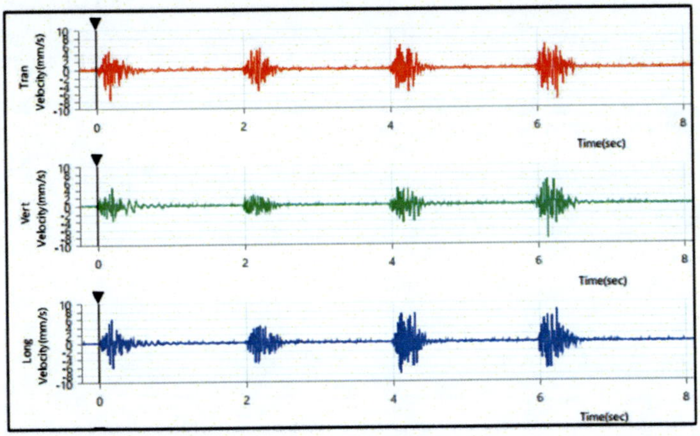

Fig. 7. Wave form recorded from the blast where the blasting round was divided into four sub-round with delay time gap of 2 s using electronic detonators

2.3 Chatti Bariatu Coal Mine Project

Chatti-Bariatu Coal Mine Projects of M/s NTPC Limited is located in Hazaribagh district, Jharkhand. The dwellings are located within 150 m from the mine boundary. It was necessary to carry out scientific study to develop controlled blasting patterns. The experimental blasts were conducted using 110 mm hole diameter as well as 160 mm diameter. Reduced burden and spacing were used in all the experimental blasts. For 110 mm hole diameter, burden and spacing used were 2.0 to 2.5 m respectively and hole depth varied from 3.0 to 4.0 m. For 160 mm hole diameter hole depth varied from 3.0 to 4.5 m, burden varied from 2.0 to 2.5 m and spacing varied from 2.5 to 3.0 m. Non-electric (Nonel) initiation system was used in all the blasts.

The ground vibration monitored at the different distances starting from 100 m in the direction of the nearby villages. Using 110 mm hole diameter with reduced burden and spacing, the magnitude of ground vibration at the distance of 100 to 150 m was

Table 7. Comparison of ground vibration data recorded at 100 m, 150 m and 200 m from different experimental blasts at Talabira II & III OCP

Sl. No	Description of Experimental blasts	Magnitudes of ground vibration recorded (mm/s)		
		100 m	**150 m**	**200 m**
1	Electronic detonators with normal burden and spacing values	14.43 17.35 – –	4.890 5.000 3.119 9.894	3.053 4.128 2.880 2.650
2	Electronic detonators with reduced burden and spacing values	5.556 12.89 9.100 3.772 8.957	3.077 6.870 4.068 2.372 5.541	2.016 2.191 2.048 2.110 3.153
3	Nonel with normal burden and spacing value	– – – – –	17.08 16.31 17.08 9.333 7.649	8.150 5.865 8.150 7.994 7.146

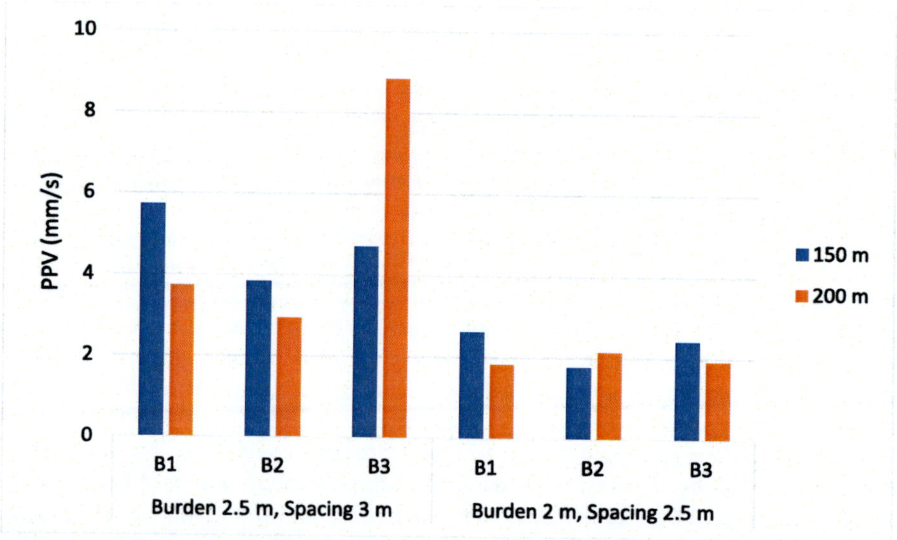

Fig. 8. Comparison of vibration data recorded at 150 m and 200 m with different burdeand spacings (B1, B2, B3 – Blast numbers)

restricted below 10 mm/s. The plot of ground vibrations recorded at 150 m and 200 m using different burden and spacing values are given in Fig. 8.

3 Problems and Challenges on Production Enhancement

The controlled blast design patterns have been developed for the three mines viz. Pachhwara (North) Coal Mines, Talabira II & III OCP and Chatti Bariatu Coal Mine Project. The safe values of maximum charge per delay for different distances have been determined based on the ground vibration predictor equations established for each mine. The safe values of maximum charge per delay for the distance of 50 to 300 m from the blasting sites to maintain peak particle velocity (PPV) level of 5 mm/s and 10 mm/s for the three mines are given in Table 8. The table clearly suggest that the safe values of maximum charge per delay in order to maintain ground vibration level of 5 mm/s is very less, particularly within the distance of 50 to 150 m zone. Similarly, PPV level of 10 mm/s can be maintained using very less values of maximum charge per delay. Therefore, it is practically difficult to blast using large hole diameter in the close proximity of structures which may also create hindrance for production enhancement.

In general, high frequencies of ground vibration are obtained at a closer distance from the blasting source. The high frequency waves are filtered and attenuated at less distance while low frequency waves can travel further distances. Therefore, higher dominant frequency waves are generally obtained up to a distance of 200 to 300 m or even more from the blasting source depending on the nature of transmitting medium (thickness of soil cover, rock geology etc.). Hence, the threshold level of PPV as per the DGMS Standards can be fixed as 10 mm/s. However, keeping 10 mm/s as threshold value of ground vibrations, the maximum charge per delay is still less for larger blast hole diameter with higher depth.

In general, 160 mm hole diameter are used in most of the opencast coal mines in India for removal of overburden as well as blasting in coal strata. However, it is often difficult to control ground vibration within safe level at closer distances. Larger blast hole depth cannot be used with limited charge per hole. Therefore, 100–115 mm diameter have been recommended for the blasting zone of 100 to 200 m from the residential houses/structures in most of the mines where the scientific study had been carried out. In order to achieve higher rate of production, large hole diameter and greater hole depth are required. But, as experienced in many opencast mines, both coal and metal mines, there is a challenge to increase production rate when the villages are located nearby the mines.

In some of the mines, more number of holes are blasted using electronic detonators by dividing the blasting round into sub-rounds with delay time gap of 2 to 3 s. In this way some improvement in production can be achieved. However, it is still a challenge to use larger hole diameter with larger depth in order to maintain safe value of ground vibration and production enhancement at closer distances from the residential houses and dwellings.

Table 8. Calculated safe value of maximum charge per delay for various distances in different mines

Distance	Safe value of maximum charge per delay for (kg)					
	Chatti Bariatu Coal Mine Project *PPV* = *6704.1*(SD)*$^{-1.732}$		Talabira II & III OCP *PPV* = *483.3*(SD)*$^{-1.228}$		Pachwara (North) Coal Mine PPV = 83351*(SD)$^{-1.914}$	
	PPV = 5 mm/s	PPV = 10 mm/s	PPV = 5 mm/s	PPV = 10 mm/s	PPV = 5 mm/s	PPV = 10 mm/s
50	0.611	1.360	1.445	4.477	1.076	2.200
75	1.374	3.060	3.252	10.07	2.422	4.996
100	2.443	5.440	5.782	17.90	4.304	8.881
125	3.817	8.490	9.034	27.98	6.727	13.88
150	5.496	12.24	13.01	40.29	9.686	19.98
175	7.481	16.66	17.71	54.84	13.18	27.20
200	9.771	21.76	23.13	71.63	17.22	35.52
225	12.37	27.54	29.27	90.66	21.79	44.96
250	15.27	34.00	36.14	111.9	26.91	55.51
275	18.47	41.14	43.73	135.4	32.56	67.16
300	21.98	48.96	52.04	161.2	38.75	79.93

4 Conclusion

Controlled blasting within the danger zone of 500 m from the residential houses/structures demands use of smaller blast geometry. Based on the studies conducted at various opencast mines, smaller blast hole diameter, shorter hole depth, less quantity of explosive charge and lesser number of holes in a blasting round are required in order to contain ground vibration within the safe level. Although safe blasting operations can be carried out nearby dwellings using different blasting techniques, the production enhancement is difficult due to limitations of large drill geometry. The introduction of high precision digitally programmable electronic detonators paves some way for production enhancement with greater safety. However, the use of smaller drill hole diameter and lesser explosive quantity are imperative in the vicinity of dwellings to address the challenge faced by the mining engineers.

Acknowledgement. The authors are thankful to the Director, CSIR-Central Institute of Mining and Fuel Research (CSIR-CIMFR) for permitting publication of the research works. They also thankfully acknowledge the cooperation and supports received from the mine management such as M/s WBPDCL and M/s BGR Mining & Infra Limited of Pachwara (North) Coal Mine Project; M/s NLC India Limited and M/s Adani Enterprises Limited of Talabira II & III OCP and M/s NTPC Limited and M/s Rithwik Project Private Limited of Chatti Bariatu Coal Mine Project.

References

Coal Mines Regulation (CMR) 2017: Regulation **196**(2)(3) (2017)

IS-6922 1973, I.S. 'Indian Standard IS-6922 1973. Criteria for safety and design of structures subjected to underground blast.', Criteria for safety and design of structures subjected to underground blast. [Preprint] (1973)

Singh, R.B., Dhar, B.B., Pal Roy, P., Singh, R.B.: Optimum blasting for Indian geomining conditions - Suggestive Standards and Guidelines. Pub. Central Mining Research Station, p. 40. ', Optimum blasting for Indian geomining conditions - Suggestive Standards and Guidelines [Preprint] (1993)

Mines Safety, D.G. (ed.) DGMS (Tech) S&T Circular No. 7. 1997. Damage to the structures due to blast induced vibration in the mining areas. Directorate General of Mines Safety, Dhanbad, 9–12., Damage to the structures due to blast induced vibration in the mining areas [Preprint] (1997)

Martos, F.: Some rock mechanical and blasting technical problems associated with large underground blasting. Int. J. Rock Mech. Min. Sci. Geomech. Abs. **11**(12), 247 (1974). https://doi.org/10.1016/0148-9062(74)92111-1

Explosives and initiation systems. Rock Fragmentation Blasting, 119–214 (2009). https://doi.org/10.1201/9781482288698-11

Richards, A.B., Moore, A.J.: ISEE conference. In: Flyrock Control-By Chance or Design. New Orleans (2002)

L.T., N., Little, T.N.: Flyrock risk. In: Proceedings of 30th ISEE Conference on Explosives and Blasting Technique, New Orleans, Louisiana [Preprint], pp 35–43 (2007)

Symposium of the international society for rock mechanics on rock fracture, Nancy, 4–6 October 1971. Int. J. Rock Mech. Min. Sci. amp Geomech. Abs. **8**(2), 179 (1971). https://doi.org/10.1016/0148-9062(71)90008-8

Mel'nikov, N.V., et al.: A method of enhanced rock breaking by blasting. Soviet Mining Sci. **15**(6), 565–572 (1979).https://doi.org/10.1007/bf02499599

Rao, R.V., Venkatesh, H.S., Venugopala Rao, R.: Reduction of blast induced ground vibrations with open trenchesin surface mines. In: The 12th International Conference of International Association for Computer Methods and Advances in Geomechanics (IACMAG), 1–6 October, 2008, Goa', Reduction of Blast Induced Ground Vibrations with Open Trenchesin Surface Mines (2008)

Lin, F., et al.: Reduction of blasting induced ground vibrations using high-precision digital electronic detonators. Front. Earth Sci. **9** (2022). https://doi.org/10.3389/feart.2021.804504

Little, O.L.: Explosive engineering, construction vibrations and geotechnology. In: Explosive Engineering, Construction Vibrations and Geotechnology. Cleveland, OH (2002)

Siskind, D.E., et al.: Explosive engineering, construction vibrations and geotechnology, USBM RI-8485 [Preprint] (1980)

Board, C.P.C. (ed.) Noise Pollution Regulation in India, Regulation 6.6 [Preprint] (2001)

Mir, H.A.: Comparison and application of top and bottom air decks to improve blasting operations (2022)

A Comprehensive Analysis of the In-Plane and Out-of-Plane Response of a Masonry Wall Under Blast Loading Conditions in the Vicinity of Coal Mines

Monika Tewari(✉), Bibhuti Bhusan Mandal, and Amit Sharma

Department of Mining Engineering, IIT, Kharagpur, India
monikatewari15@gmail.com

Abstract. Masonry structures are widely used in construction practices worldwide, but they can still suffer severe damage when subjected to blast loadings. This study aims to investigate the impact of blasting activities, a critical component of the mining industry, on nearby masonry structures. Specifically, it focuses on two commonly employed types of brick masonry walls: English bond and Stretcher bond. The research utilizes ABAQUS CAE to develop models and conduct explicit analyses to assess the response of these walls. To accurately capture the behaviour of the brick materials, a simplified micro modeling technique is employed, which incorporates the Drucker-Prager hardening criteria to account for the plastic behaviour of the bricks and mortar. Geophone sensors were used to record blast data at the foundation level of residential structures in close proximity to a coal mine (recorded ppv ranging from 7 mm/sec–30 mm/sec). A comparative analysis of the in-plane (IP) and out-of-plane (OOP) behaviour was performed, aiming to evaluate the impact of the structures' orientation relative to the blast direction. The study observed that in addition to the generally considered in-plane (IP) conditions, the out-of-plane (OOP) loadings generated due to blasting can also endanger the safety of structures and occupants. Therefore, this work provides a helpful reference for the design of blast-resistant brick masonry walls and suggests measures to safeguard them against damage caused by both in-plane and out-of-plane loading conditions.

Keywords: Blast loading · masonry structures · explicit analysis · simplified micro modeling

1 Introduction

Blasting constitutes a widely employed technique for extracting overburden and minerals in open-cast coal mines. This method relies on the controlled detonation of explosives, which serves as the principal generator of ground vibrations. About 30% of this explosive energy is productively used for rock fragmentation and the rest propagates in form of ground waves (Ouchterlony 2004). Following each blast event, shock waves propagate through the subsurface medium in the form of compression (P), shear (S) and surface

A. Sinha et al. (Eds.): AMC 2023, SPEES, pp. 480–487, 2023.
https://doi.org/10.1007/978-3-031-46966-4_39

Rayleigh (R) waves (Dowding III 1971). In the evaluation of the dynamic effects caused by Blast-Induced Ground Vibrations (BIGV), the measurement process frequently simplifies by capturing a single value known as the Peak Particle Velocity (PPV). This PPV represents the maximum particle velocity recorded throughout the entire time duration of the event (Edwards 1960) (Gad et al. 2005). These vibrations possess the potential to induce substantial structural damage to nearby constructions (Gad et al. 2005; Singh et al. 2010). In a comprehensive field survey conducted by the authors, it was observed that a significant number of residential structures situated around mine peripheries are built using under-reinforced brick masonry. The assessment of building response to blast loading has been conducted on a global scale to establish acceptable Peak Particle Velocity (PPV) limits for ensuring structural safety (BSI 1993), (ISO 1992), (Gad 2005). In India, the regulatory authority responsible for mining safety, known as the Directorate General of Mines Safety (DGMS), has established prescribed permissible values of the Peak Particle Velocity (PPV) within mining regions to ensure structural safety. Apart from the Peak Particle Velocity (PPV), several other factors can significantly impact the structural safety of masonry structures, among which an essential parameter is the condition of loading, specifically In-plane loading (IP) and Out-of-plane (OOP) loading conditions. Given the dynamic nature of blasting, the loading conditions experienced by structures around the mine periphery may vary with time. A structure that undergoes In-plane loading during one blast event may be subjected to Out-of-plane loading conditions in a different blast, highlighting the variability and complexity of the loading patterns during blasting operations. This study focuses on the response of unreinforced brick masonry to Blast-Induced Ground Vibrations (BIGV) under both IP and OOP loading conditions. Due to the combination of low tensile strength, heavy mass and improper connections between structural components, masonry is brittle by nature (Tomaževič 1994). The aim is to provide a more appropriate construction practice for areas affected by blasting activities, ensuring enhanced structural safety in such environments.

2 Methodology

Addressing the heterogeneous properties of brick masonry, accurately modeling and maintaining consistent material properties pose significant challenges. To address this issue, a simplified micro modeling approach was employed. Subsequently, the model was analysed using ABAQUS/CAE, a finite element analysis software, to study the behaviour of the structure under Blast-Induced Ground Vibrations (BIGV) and assess its response to varying loading conditions. In this particular model, the brick units were extended to account for the thickness of mortar, and homogenization techniques were applied using mathematical formulations (Fig. 1). This process allowed for the creation of a homogenized unit cell. By repeating this unit cell, the walls were assembled. To simulate the behaviour of the masonry accurately, a zero-thickness element with cohesive behaviour, enabling small sliding, was introduced at the mortar interfaces. This approach ensures a more realistic representation of the heterogeneous masonry response to Blast-Induced Ground Vibrations (BIGV) and varying loading conditions. Figure 1 shows the assembly and meshing of the wall (Fig. 2).

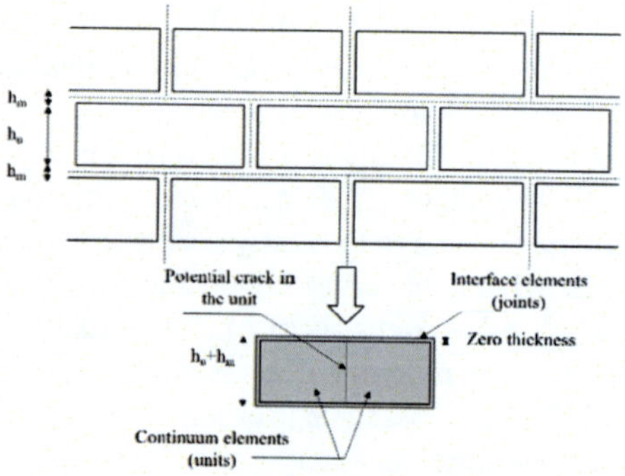

Fig. 1. Modeling Technique adopted

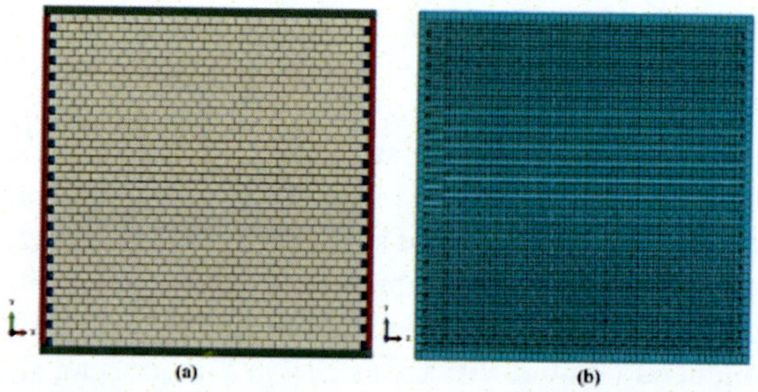

Fig. 2. (a) Assembly of brick masonry wall; (b) Meshing of the assembly

For homogenization, compression cap model was used with tension cut-off and non-associated plastic flow given by Lourenco, 1996. After homogenization, material properties of the unit are given in Table 1 and joint interface properties are given in Table 2. These cohesive interactions are basically a function of displacement separation between the edges of potential cracks (Bolhassani 2015). Drucker-Prager criteria is implemented to include the hardening and softening behaviour of bricks. This model also allows the material to have isotropic hardening and softening (Agüera 2016). Dilatancy of bricks is near to zero for brittle material like bricks as is taken as 7 degrees for this analysis.

Table 1. Material properties of a single unit

S.No.	Compressive strength	Tensile strength	Shear strength
1	3 MPa	0.3 MPa	0.020

Table 2. Interface properties of the joint

S.No.	K_{nn}	K_{ss}	K_{tt}
1	12 N/mm^3	28 N/mm^3	28 N/mm^3

3 Results and Discussions

The investigation into the response of structures to blast-induced ground vibrations was carried out employing a Nonlinear Dynamic Stress Analysis method, utilizing the C3D8R element type for bricks. This explicit analysis approach ensured a precise simulation of the dynamic behaviour of the structures subjected to blast loading. Figures 3(a) and 3(b) illustrate the difference between the models utilized here for English bond and a very common model of stretcher bond, respectively. In the English bond configuration, the arrangement comprises alternating courses of stretchers (bricks laid with their length along the wall) and headers (bricks laid with their width along the wall). Conversely, the stretcher bond involves the use of only stretcher bricks in each course.

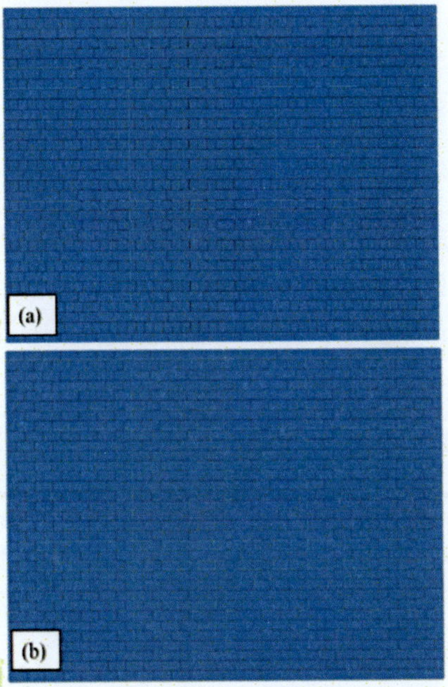

Fig. 3. (a) English bond arrangement; (b) Stretcher bond arrangement

The acceleration data was used as an input at the base of the structure and implicit analysis was done for in-plane and out-of-plane loading conditions. Figure 4 shows the in-plane and out-of-plane conditions for a brick masonry wall. If the loading is applied in the plane of the wall, it is known as in-plane loading conditions. If the loading is applied perpendicular to the plane of wall, it is termed as out-of-plane loading (Fig. 5).

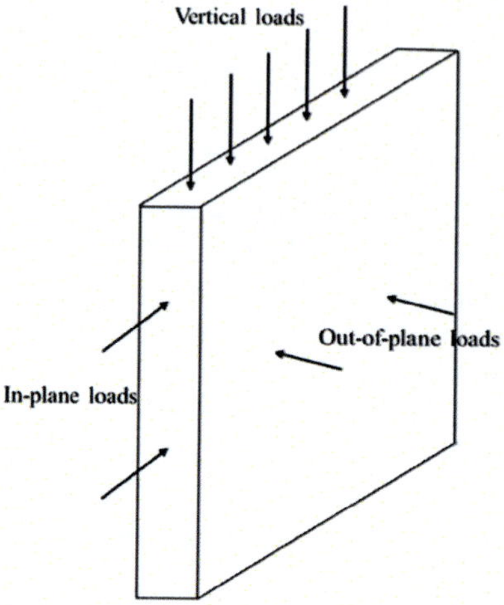

Fig. 4. In-plane and out-of-plane loading conditions.

Max Principle stresses (11.65 MPa) observed at in-plane loading conditions are less than max principle stresses (26 MPa) observed at out-of-plane loading conditions. Max strain (0.0015) at in-plane loading conditions is less than max strain (0.0045) at out-of-plane loading conditions. Displacement rate at in-plane loading conditions are also comparatively lower than the displacement rate observed in out-of-plane loading conditions. Max Displacement value observed (3.6 mm) in in-plane loading conditions are lower than the max displacement (12 mm) observed in out-of-plane loading conditions.

Figure 6 (a) and (b) illustrates the displacement contours of wall in in-plane and out-of-plane loading conditions. In case of out-of-plane loading, the combination of diagonal step cracking and flexure cracking is commonly observed. The horizontal crack on the top of the wall is due to the presence of compressive stresses. As soon as the shear stresses at the joints reach their yield strength (0.020 MPa), shear crack propagation (diagonal step cracks) initiates. As the loading continues, these cracks propagate through the joints and results into the failure of the wall.

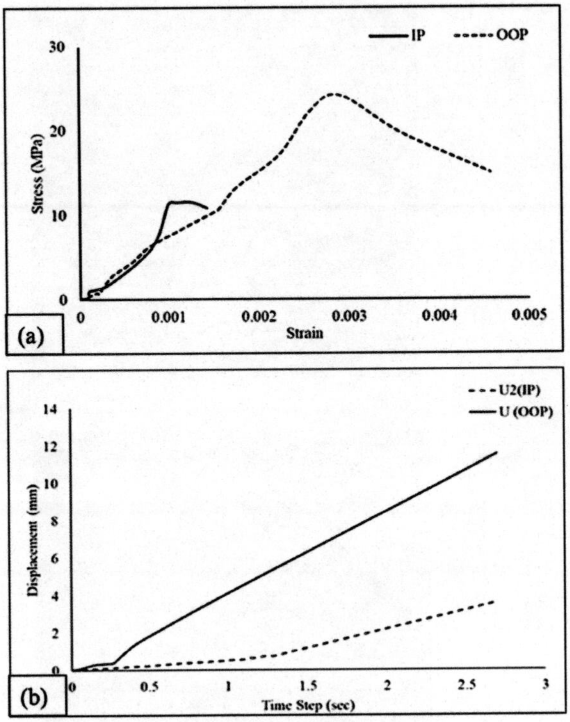

Fig. 5. (a) Stress-strain diagram for IP and OOP; (b) Displacement Vs time step plot

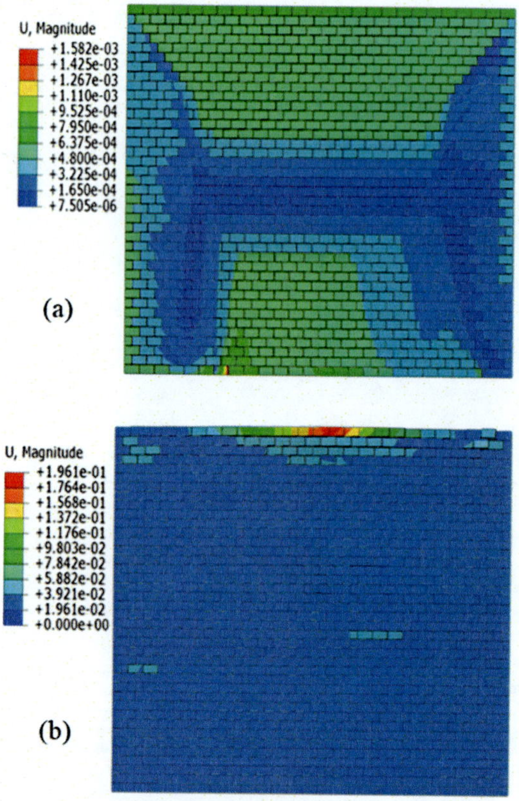

Fig. 6. Displacement contours in (a) IP loadings; and (b) OOP loading

4 Conclusion

In the context of brick masonry structures, it has been observed that in-plane loading conditions pose a relatively lower level of threat compared to out-of-plane loading conditions. When subjected to the same blast event, walls showed no significant displacements under in-plane loading. However, in the case of out-of-plane loading, the displacements are significantly higher. The stresses developed under out-of-plane loading conditions also fall within a higher spectrum. This is because the load distribution in IP loading conditions is better as compared to the out-of-plane loading conditions due to the fact that loading plane is same as the plane of wall. As the shear strength of joints is low, targeted reinforcements need to be given in the joints to stop the diagonal crack propagation.

Throughout the entire lifespan of a mine, any permanent structure situated outside the mine periphery is subjected to both in-plane and out-of-plane loading conditions. This arises from the fact that the location of blasts constantly changes, leading to alterations in the wave propagation path. As a result, structures experience varying degrees of in-plane and out-of-plane loading, making it essential to consider both loading conditions in their design and assessment for durability and safety.

References

Lourenço, P.B.: Multisurface interface model for analysis of masonry structures. J. Eng. Mech. **123**(7), 660–668 (1997)

Standard, I.: IS 1077: 1992 Common Burnt Clay Building Bricks–Specification. New Delhi, BIS (1992)

Standards, B.O.: IS 2212 - 1991: Code of practice for Brickworks. IS 2212 (2005)

BIS.: IS 2212 - 1991: Code of Practice for Brickworks. Bureau of Indian Standards (2005)

Dowding III, C.H.: Response of Buildings to Ground Vibrations Resulting from Construction Blasting. University of Illinois at Urbana-Champaign (1971)

Gad, E.F.: Effects of mine blasting on residential structures. J. Perform. Constructed Facil. **19**(3), 222–228 (2005)

Edwards, A.T., Northwood, T.D.: Experimental Studies of Effects of Blasting on Structures, The Engineer, p. 211 (1960)

Singh, P.K.: Damage to surface structures due to blast vibration. Int. J. Rock Mech. Min. Sci. **47**(6), 949–961 (2010)

Tomaževič, M.: Seismic behavior of plain-and reinforced-masonry buildings. J. Struct. Eng. **120**(2), 323–338 (1994)

Agüera, N.D.: Structural response of unreinforced masonry walls. J. Civ. Eng. Arch. **10**(2), 219–231 (2016)

Bolhassani, M.H.: Simplified micro modeling of partially grouted masonry assemblages. Constr. Build. Mater. **83**, 159–173 (2015)

Ouchterlony, F.N.: Where does the explosive energy in rock blasting rounds go? Sci. Technol. Energ. Mater. **65**(2), 54–63 (2004)

Author Index

A. Sinha et al. (Eds.): AMC 2023, SPEES, pp. 489–490, 2023.
https://doi.org/10.1007/978-3-031-46966-4